World Sustainability Series

Series Editor

Walter Leal Filho, Research and Transfer Center "Sustainability and Climate Change Management", Hamburg University of Applied Sciences, Hamburg, Hamburg, Germany

Due to its scope and nature, sustainable development is a matter which is very interdisciplinary, and draws from knowledge and inputs from the social sciences and environmental sciences on the one hand, but also from physical sciences and arts on the other. As such, there is a perceived need to foster integrative approaches, whereby the combination of inputs from various fields may contribute to a better understanding of what sustainability is, and means to people. But despite the need for and the relevance of integrative approaches towards sustainable development, there is a paucity of literature which address matters related to sustainability in an integrated way.

Notes on the quality assurance and peer review of this publication

Prior to publication, the works published in this book are initially assessed and reviewed by an in-house editor. If suitable for publication, manuscripts are sent for further review, which includes a combined effort by the editorial board and appointed subject experts, who provide independent peer-review. The feedback obtained in this way was communicated to authors, and with manuscripts checked upon return before finally accepted. The peer-reviewed nature of the books in the "World Sustainability Series" means that contributions to them have, over many years, been officially accepted for tenure and promotion purposes.

James Kanyepe · Innocent Chirisa
Editors

Advancing the Green Agenda in the Transport and Logistics Sector in Global South Countries

Pathways, Barriers, and Policy Implications

Springer

Editors
James Kanyepe
Department of Logistics and Supply Chain Management
University of Botswana
Gaborone, Botswana

Innocent Chirisa
Department of Urban and Regional Planning
University of the Free State
Bloemfontein, South Africa

ISSN 2199-7373 ISSN 2199-7381 (electronic)
World Sustainability Series
ISBN 978-3-032-31373-7 ISBN 978-3-032-31374-4 (eBook)
https://doi.org/10.1007/978-3-032-31374-4

This Springer imprint is published by the registered company Springer Nature Switzerland AG
The registered company address is: Gewerbestrasse 11, 6330 Cham, Switzerland

This book is dedicated to the scholars, policymakers, practitioners, and communities across the Global South who are actively engaged in the arduous yet indispensable task of reconfiguring transport and logistics systems toward sustainability, resilience, and equity. Their sustained efforts often undertaken within contexts of institutional constraint, fiscal limitation, and infrastructural fragility constitute a profound commitment to reimagining mobility systems that are environmentally responsible, economically viable, and socially inclusive. This volume stands as a scholarly tribute to their transformative work and enduring vision.

Preface I

Transport and logistics systems in countries of the Global South lie at the centre of an increasingly urgent tension between development imperatives and climate responsibility. While these systems remain indispensable to economic integration, trade competitiveness and spatial connectivity, they are also embedded in carbon-intensive structures marked by inefficiency, infrastructure deficits and institutional fragmentation. This volume provides a rigorous and systematic examination of the barriers constraining sustainable transformation, including weak governance coordination, inadequate infrastructure investment, limited fiscal capacity and slow technological diffusion. Moving beyond descriptive analysis, it identifies credible transition pathways encompassing modal reconfiguration, infrastructure modernisation, decarbonisation of mobility systems and the deployment of digital and intelligent transport technologies.

The policy discourse advanced in this volume is both pragmatic and strategically oriented, emphasising the need for integrated planning, robust regulatory frameworks, innovative financing mechanisms and strengthened cross-sectoral collaboration. It also highlights the importance of context-responsive approaches that reflect the structural realities of developing economies. By combining analytical rigour with policy relevance, the volume makes a substantive contribution to contemporary debates on sustainable transport transformation and offers a valuable reference for scholars, practitioners and policymakers engaged in systemic reform.

Gaborne, Botswana — James Kanyepe
Bindura, Zimbabwe — Innocent Chirisa

Preface II

This edited volume addresses a central challenge in contemporary development scholarship: advancing a meaningful low-carbon transition in transport and logistics systems across countries in the Global South. Situated at the intersection of rapid urbanisation, industrial expansion, and intensified global trade, these systems are essential to economic development while also contributing significantly to greenhouse gas emissions and environmental degradation. Across many developing economies, transport and logistics infrastructure is locked into carbon-intensive and inefficient pathways. This is evident in the continued reliance on fossil fuels, ageing and poorly maintained vehicle fleets, weak intermodal connectivity and inadequate infrastructure provision. These structural constraints are further compounded by institutional fragmentation, regulatory weaknesses, limited fiscal capacity and restricted access to climate-aligned finance.

Against this backdrop, the volume offers a critical and policy-oriented examination of the pathways, barriers, and implications of transport sector transformation. Rather than proposing simplistic or technocratic solutions, the contributions examine the structural, institutional, technological and financial conditions that shape transition outcomes. Particular emphasis is placed on system-wide interventions, including modal shifts towards rail and inland water transport, modernisation of public transport, adoption of low-carbon and electric mobility solutions and integration of digital technologies such as Intelligent Transport Systems and data-driven logistics optimisation. These transitions are understood as socio-technical transformations that require coordinated governance, strategic investment and institutional alignment. The volume is anchored in the triple bottom line approach and aligns with global sustainability frameworks, including the Paris Agreement and the Sustainable Development Goals (SDGs), particularly Goals 7, 9, 11 and 13. This volume will be of interest to scholars, policymakers, planners, practitioners and researchers engaged in transport systems, logistics and sustainable development across emerging and developing economies.

Gaborne, Botswana
Bindura, Zimbabwe

James Kanyepe
Innocent Chirisa

Acknowledgments

The editors express their profound gratitude to all contributors whose intellectual rigour, empirical depth and scholarly commitment have made this volume possible. Their collective contributions significantly enhanced the analytical sophistication and thematic coherence of the work. We further acknowledge the invaluable insights provided by peer reviewers, academic colleagues, institutional partners and research collaborators. Their critical engagement was instrumental in refining the conceptual architecture and strengthening the scholarly quality of the manuscript. Finally, we extend heartfelt appreciation to our families and personal support networks for their patience, encouragement and unwavering support throughout the demanding editorial process. Their steadfast presence provided the essential foundation for the completion of this scholarly endeavour.

Contents

Abbreviations

AI	Artificial Intelligence
CO_2	Carbon Dioxide
EV	Electric Vehicle
GHG	Greenhouse Gas
ICT	Information and Communication Technology
ITS	Intelligent Transport Systems
PPP	Public-Private Partnership
SADC	Southern African Development Community
SDGs	Sustainable Development Goals
UN	United Nations
UNFCCC	United Nations Framework Convention on Climate Change

The Green Transition in Transport and Logistics in the Global South: Conceptual Foundations and Developmental Implications

James Kanyepe, Innocent Chirisa, Tinashe Magande, and Africa Muneti

Abstract The transport systems of Global South countries stimulate the economic development and trade, yet they also contribute to the greenhouse gas emission, the environmental degradation, and the urban pollution. These countries are already under the pressure to change to low-carbon infrastructure, energy-efficient freight, and sustainable supply chains in response to international climate obligations such as the Paris Agreement and UN Sustainable Development Goals. However, institutional constraints, including lack of infrastructure, lack of funds, and institutional capacity remain in place, hindering development in the face of accelerating urbanisation and trade pressures. Nevertheless, available literature does not pay much attention to context-specific approaches to the Global South, addressing high-income models that are inappropriate in resource-restrained contexts. This creates a severe gap in knowledge regarding localised green transport and logistics transit pathways. To counter this, the chapter embarks on a qualitative investigation of drivers of sustainability, dilemmas and models that are responsive to Global South realities. The data was created as a result of a thorough qualitative review of policy documents and scholarly research and international reports. In the analysis of data, textual analysis was used in order to establish common themes, patterns, and contradictions. It was

J. Kanyepe (✉)
Department of Logistics and Supply Chain Management, University of Botswana, Gaborone, Botswana
e-mail: jameskanyepe@gmail.com

I. Chirisa
Office of the Vice Chancellor, Zimbabwe Ezekiel Guti University, Bindura, Zimbabwe

I. Chirisa · T. Magande
Department of Urban and Regional Planning, University of Johannesburg, Johannesburg, South Africa

A. Muneti
Firm Central Property Group, Harare, Zimbabwe

IC University of Technology, Phnom Pehn, Cambodia

J. Kanyepe and I. Chirisa (eds.), *Advancing the Green Agenda in the Transport and Logistics Sector in Global South Countries*, World Sustainability Series,
https://doi.org/10.1007/978-3-032-31374-4_1

found that three insights might be critical in this regard: with integrated public–private investments in electric mobility and rail, transport emissions could decline 40–60 percent in the Global South by 2050; sustainable logistics requires the use of digital tools to optimise freight routes in areas with informal economies; and to be successful in policies, capacity-building is needed to create the necessary institutional space. Such trends indicate the avenues to act: governments need to focus on blended funding and regional partnerships to scale green infrastructure, but encourage innovation in affordable, resilient supply chains. The chapter ends by showing the purpose, coverage and outline of the book.

Keywords Green transport · Sustainable logistics · Global south · Low-carbon infrastructure · Sustainable supply chains · Climate-resilient mobility

1 Introduction

The transport systems of Global South countries support the development of the economy and international trade, but they lead to the production of a significant amount of greenhouse gases and environmental damage. In countries such as South Africa, road transport contributes more than 90 percent of the industry-related CO_2, which contributes to urban pollution during rapid urbanisation (Ntuli et al. 2024). The climate demands observed globally, such as the Paris agreement force these economies to seek low-carbon infrastructure and sustainable supply chains (Kern and Jess 2021). Nevertheless, the energy efficient freight systems are challenged by the constant problems of poor rail networks and reliance on fossil fuels. This dependency increases health hazards due to air contaminants and overloads the scarce resources in third world situations. Therefore, green transitions are identified as necessary to align transport with greater goals of sustainable development.

The major thesis is that Global South transportation needs to be context-specific, and it should not be based on high-income models, playing off against stagnant regimes with niche innovations. The Multi-Level Perspective exposes the means in which the specific landscape forces such as SDGs facilitate the disruption of the fossil-based systems by electric mobility and digital logistics (Kanyepe et al., 2023). The Ecological Modernisation Theory goes further to endorse the idea of technological decoupling of growth and emissions by implementing policy changes and green financing (Bugden 2022). Scalable electrification and rail freight are proven as options in case studies of China and India. Such frameworks bring out structural challenges such as financial availability gaps that require concerted public-corporate measures. The view highlights the imperative of personalised action to attain 40–60 percent emission reductions by 2050.

The proposed research will map sustainability drivers, issues, and models of green transport in the Global South through qualitative desk review of policies and reports. Actionable niches, such as low-cost EVs, and book structure are among the key goals that were outlined and examined in greater detail. The study addresses gaps in the literature on resource-constrained strategies, which guides policymakers with regard to resilient supply chains (Khan et al. 2019). The results place importance on blended finance and capacity-building to address institutional lapses. Scope is centred on the African-based knowledge on a global basis and emphasises on empirical patterns which are peer-reviewed. By doing this, the chapter prepares a foundation of fair decarbonisation in line with the UN ambitions (Muchenje 2024).

2 Conceptual Framework

Multi-Level Perspective (MLP) of sustainability transitions can be a useful theoretical framework to use to examine the green transition in transport and logistics in the Global South. This framework theorises socio-technical transitions in terms of interactions at three analysis levels: niche innovations, socio-technical regimes, and the wider landscape. Radical innovations, including electric buses, digital freight platforms, and alternative fuels, are introduced into the niche where experimentation and development can proceed, and they may eventually become widespread (Muchenje 2024). The regimes level consists of the prevailing transport systems, infrastructure, institutional structures, and cultural norms that are presently largely dependent on fossil fuels and are hard to change because of the existence of vested interests and path dependencies (Khan et al. 2019). In the meantime, landscape pressures, including climate change necessities, global sustainability policies, international treaties, including the Paris Agreement, and accelerated urbanisation, are some of the external factors that create a sense of urgency and momentum towards transition (Muchenje 2024). The application of MLP to the Global South shows that barriers to sustainable uptake at the regime level tend to be bottlenecks, and that niche innovations may be able to use landscape pressures to slowly transform the socio-technical paradigm. Such a multi-scale view allows seeing the dynamics of transition in detail, with the opportunity to stimulate policy-level work, the introduction of technologies, and international collaboration as the key factors in the process of transforming green mobility and logistics in developing settings.

Ecological Modernisation Theory (EMT) is used as a complementary theory that highlights the possibility of balancing economic growth and environmental sustainability by using technological innovation, institutional change, and policy intervention. EMT argues against the belief that protecting the environment and economic development are opposing entities, instead suggesting that the technological advancements could decouple economic activities and environmental deterioration through the use of electric vehicles, the development of low-carbon fuels and intelligent transport systems (Bugden 2022). Environmental governance is the regulation-based approach to sustainability, incentives, and market-based solutions such as green investment, and partnerships between the government and the private sector, which is central to the promotion and scaling of sustainable transport technologies (Muchenje 2024). Among Global South countries, EMT emphasizes the vital role of the transfer of clean technologies, green infrastructure, and the creation of financing mechanisms to speed up transition, and at the same time, stimulate economic progress (Khan et al. 2019). Nonetheless, EMT states that there are certain drawbacks associated with unequal capacities, regulatory issues, and the possibility of superficial green growth. The incorporation of EMT into the framework of the multi-level transition context sheds some light on the avenues in which the institutional modernisation and innovation can convert the transport and logistics systems on a sustainable basis. Collectively, MLP and EMT offer an integrated theoretical approach to explore, plan and enact sustainable shifts in the Global South transport and logistics industry.

3 Literature Review

Thise section presents literature Review on transport sector emissions and the imperative for green transition, and structural barriers to sustainable transport in the global south. This is followed by technological innovations and sustainable logistics solutions and policy frameworks. Finally, it discusses literature on Financing Mechanisms, and Governance for Green Transport.

4 Transport Sector Emissions and the Imperative for Green Transition

The transport industry contributes considerably to the global greenhouse gases (GHG) emissions, which is approximately 15 percent of anthropogenic greenhouse gases, and about 23 percent of the global energy-induced carbon dioxide. The road transport is identified to be among the worse-performing sectors in terms of carbon footprint and energy consumption, so it is a crucial sector to be involved in the emissions reduction process (Ferrer and Thome 2023; Timilsina et al., 2025). The road transport is the primary contributor of the transport-related CO_2 emissions in the country, which is approximately 90% of all CO_2 emissions, and it is an example of the effect of the sector in the Global South (Ntuli et al. 2024). In the absence of significant policy and technological changes, the energy demand and emissions of the transport sector are expected to grow significantly, which is why sustainable interventions are urgently required. The emissions amplify the climate change, propelling global climate pacts, including the Paris Agreement, which tries to restrict the global warming through massive emissions cuts, including the decarbonisation of transport worldwide (Kern and Jess, 2021). These targets should be met through concerted activities to shift towards low-carbon transport systems and energy-effective transport systems across the globe.

The effects of freight and passenger transportation on the environment are not confined to GHG emissions and include air pollution, traffic congestion, and unnecessary consumption of energy. The Traditional transport fuels have negative health effects and environmental damage, particularly when they are exposed in the fast-urbanising areas that are typical of the Global South (Li and Tang 2017). The high rates of urbanisation and growing volumes of trade in such economies add to the environmental pressure, as the growing urban freight and passenger traffic cause more congestion and increased energy use (Timilsina et al. 2025). In most developing countries, there is the challenge of inadequacy of finance and infrastructure that is oriented towards sustainable mobility (Anozie 2024). It is increasingly acknowledged that integrated solutions are required to accommodate modal changes, electrification, and better logistics in order to minimize emissions and provide more equity and economic growth (Winkler et al. 2023). The environmental imprint of the transport sector therefore calls out dire and structural changes in order to be in line with the objectives of sustainable development.

Dcarbonising transport is one of the priorities of climate resilience and sustainable urbanisation as highlighted in international frameworks, such as the Paris Agreement and major UN Sustainable Development Goals (SDG 9: Industry, Innovation and Infrastructure; SDG 11: Sustainable Cities and Communities; SDG 12: Responsible Consumption and Production; SDG 13: Climate Action) (Herman 2021; Kern and Jess 2021). The worldwide trend to low-carbon transport systems combines technology development (electric vehicles and green hydrogen) and the policy tool (carbon trading schemes) to reduce emissions (Timilsina et al. 2025; Li and Tang 2017; Elegbeleye et al. 2025). Nevertheless, the lack of financing in the Global South disrupts the process, and it requires increased investments and specific policy frameworks to speed up the green transition (Anozie 2024). Transport transition in the developing economies is vital to the climate goals and the attainment of sustainable development; therefore, there is an urgent need to act swiftly and combine the environment, social and economic aspects.

5 Structural Barriers to Sustainable Transport in the Global South

An enormous lack of infrastructure impedes the adoption of sustainable transport in the Global South, as it compromises the attempts at green mobility. The developing countries are characterised by a lack of rail systems, and they overly depend on poorly maintained roads that are not always sufficient to meet the needs of transport (Okyere et al. 2019). The absence of extensive charging networks of electric vehicles (EVs) also places an additional limitation on the adoption of cleaner modes of transport, which is a significant logistical and operational burden to the stakeholders seeking to adopt EVs on a large scale (Dadashzada et al. 2024). Such physical infrastructure lapses not only escalate reliance on fossil-fuel-based road transport, but also make emissions worse by creating inefficiencies and congestion (Ashraf and Idrisi 2024). This is because the potential advantages of multimodal logistics systems are constrained due to poor integration between the various forms of transport, which reduces sustainability potential to critical levels. These infrastructure issues demand a significant amount of investment, planning and coordination so as to develop efficient and sustainable transport corridors that will suit the specific socoa-economic factors that enable developing countries.

Another key impediment to sustainable transport transitions in the Global South is that of financial constraints. Initial expenses of developing green infrastructure such as EV fleets, charging stations, modernised urban transit systems, are also above the reach of domestic funds, and many countries have a lack of capital mobilisation (Hmamed et al. 2025). The accessibility to foreign climate funds is limited, which increases the risks involved in investments, and there is a lack of confidence in the economic prospects in the country and the unwillingness to involve the private sector (Anozie 2024). Challenges encountered by developing countries in raising and properly implementing climate funds are often based on a weak fiscal infrastructure and governance complexities (Ibrahim 2024). Such financial constraints delay the spread of technology and kill sustainable mobility innovation. This, therefore, implies that to overcome such constraints, specialised financial measures, mitigation of risks, and enhancement of international climate finance flows should be prepared specifically to the low-income economies.

There are also institutional and governance issues that impede transport sustainability in the developing contexts. The laxity of the regulations and the disjointed governance systems dispense policy coherence and incentives to adopt green technologies (Hmamed et al. 2025). Urbanisation speed is a factor that leads to dependency on informal transport systems and logistics networks that are common in most cities of the Global South, which are characterised by low-level regulatory regulation and low safety standards (Okyere et al. 2019). This informal reliance on road transport maintains high-emission behaviour that restricts efficiency gains in the formal sector (Sovacool and Griffiths 2019). The inability of institutions to balance between various levels of government also impedes the production of the comprehensive transport policies in accordance with the global climatic and development objectives. These institutional barriers require capacity building, urban planning and governance reforms that could help enable transitions to sustainable and resilient transport ecosystems.

6 Technological Innovations and Sustainable Logistics Solutions

The transport industry is also experiencing a high level of electric mobility that presents a great avenue in the decarbonisation of freight and passenger logistics. Electric buses and electric freight cars are being implemented more frequently so as to decrease the emission and enhance energy efficiency in the transport systems in the city and suburbs (Kashem et al. 2024; Demir et al. 2022). Such cars can help to reduce carbon footprints significantly particularly when used in conjunction with renewable energy sources. Although it faces infrastructure difficulties and the initial investment is high, the technological advances have turned electric mobility into a more practical and cheaper option to extend it to developing and developed markets (Muchenje 2024). A phased transition of diesel-drilled fleets to electric substitutes helps to improve the quality of air in the locality and is consistent with international climate obligations, highlighting the strategic significance of increasing the proportion of electric vehicles in freight and public transport.

Coupled with electric mobility, the use of low carbon fuels e.g. hydrogen, biofuels, and alternative renewable fuels is an important contributor to make logistics sustainable. Green hydrogen, which is generated by electrolyzing renewable energy, provides an emission-free alternative to heavy freight and transportation over long distances, which directly can hardly be electrified (Nagy and Szentesi 2024). These efforts are complemented by biofuels and other alternative fuels that offer drop-in substitutes of the traditional fossil fuels to lower the lifecycle greenhouse gas emissions (Aktas et al. 2017). There is improvement in the viability and scalability of these fuels due to technological advances in the area of fuel synthesis, storage and distribution infrastructure. Their inclusion in the logistics networks facilitates the diversification of the energy mix which enhances resilience and decarbonisation objectives thereby aiding the provision of greener freight and passenger transport solutions.

Online innovation and modal have become disruptive innovations that will result in operational efficiency and emission reductions in logistics. The introduction of AI, smart freight systems, and route management technologies makes it possible to make decisions based on the data, minimise idle periods, enhance load consolidation, and reduce unproductive movements (Demir et al. 2022; Muchenje 2024). These digital technologies make supply chains more transparent and responsive to improve the sustainability of transport operations (Lin 2008). Besides, promoting changes in modalities toward rail freight, intermodal transport, and inland waterways utilises more efficient transport options and lower-emission options (Generalov 2024). Through these technological and modal innovation, the transport and logistics systems will be able to have a substantial reduction in their carbon footprint as well as increase their efficiency and competitiveness in further complicated global supply chains.

7 Policy Frameworks, Financing Mechanisms, and Governance for Green Transport

Good policy frameworks restricting the emission regulation, fuel efficiency standards, and the sustainable urban mobility policy are strong factors towards effective green transport transitions in the Global South. Emission regulations are important tools to control the level of pollution and the promotion of cleaner technologies in vehicles, whereas fuel efficiency standards encourage innovation and decreasing the use of fossil fuels (Wu et al. 2025). The policies of sustainable urban mobility encourage the use of public transport, non-motorised mode of transport and low-emission areas in order to minimize congestion and enhance urban air quality. Governance bodies are critical when setting and implementing such policies so that they are in line with climate objectives and societal interests. These policy instruments will have to be used to drive transport sectors towards decarbonisation and resource efficiency to support climate pledges including the Paris Agreement and Sustainable Development Goals (SDG 7, 11, 13). The implementation of transport policies through a complete and strong regulatory framework has been shown to show more sustainable mobility transitions in countries than those with unsustainable or weak regulatory frameworks (Okeke et al. 2024).

The funding systems also play a vital role in green transport transformation particularly in emerging economies where funding scarcity is a major barrier to the adoption of innovation. The green bonds, blended finance, and development bank investments are examples of climate financing instruments that offer essential funding sources of low-carbon infrastructure and technologies (Birindelli and Palea 2022; Rhouma et al. 2025). Green bonds appeal to environment-oriented investors by putting capital on projects of sustainability, and blended finance is a mixture of both government and non-government funds to reduce risk and open up more significant investments. Development banks also will strategically facilitate green transport projects by providing concessional loans, guarantees as well as technical support to initiate these projects. Incentives provided by policy and risk mitigation plans boost investor trust and lower the cost of financing, which will speed up the implementation of green transport solutions. Quite on top of these opportunities, there are still issues such as regulatory gaps, the local financial capacity, and the necessity of standardised green finance criteria to increase transparency and accountability (Sule et al. 2024).

Governance structures, like public- private partnerships (PPP), and regional cooperation, offer complementary routes to successful sustainable transport delivery. PPPs capitalise on the efficiency and innovation of the private sector and share risk and responsibility with the public agencies, which helps to develop infrastructure and implement technologies at large scale (Mukherjee et al. 2024). On a large scale, regional and international collaboration facilitates the transfer of knowledge and capacity building and the sharing of technology, which is essential to overcome institutional constraints in the nation and promote harmonised standards (Faijue et al. 2024). Cross border infrastructure development which promotes intermodal connectivity and clean logistics networks is also supported by multilateral collaborations and financial arrangements. Effective governance systems entrench stakeholder involvement and dynamic management to address the new challenges and prospects in green transport. Combined, the three policies, innovative financing, and joint governance can have a synergistic effect to increase sustainable transitions to transport and streamline environmental goals with economic growth and social inclusiveness (Hulko et al. 2025).

8 Methodology

Electronic databases such as Google scholar, Scopus, Web of Science and Science Direct were used to access the information needed to investigate drivers of sustainability, dilemmas and models that are responsive to Global South realities. Keywords such as Green transport, Sustainable logistics, Global South, Low-carbon infrastructure, Sustainable supply chains, Climate-resilient mobility. were used in isolation or in combination to search for data. Peer reviewed journal articles that are based on empirical data were prioritised for reliability of findings in this chapter. The chapter was based on English published sources that were related to this topic. There was no year limit for the searched items that were sourced from the published articles. The published sources that were generated from the search process were further filtered based on their relevance to come up with those that are relevant and most appropriate to the chapter. Policy documents, scholarly research, international reports and other grey literature were consulted for this chapter. As such, this this chapter was based on the assessment of existing reports and claims rather than fresh empirical data from the field.

9 Findings

To present the findings of this chapter, two case studies have been identified and analysed. The section starts with a case study on China.

10 Case Study 1: China: Electrification of Urban Transport and Green Freight Systems

The country has become a world leader in the urban transport electrification and green freight system implementation due to the determination of the government, the technological advancements, and the acute environmental issues. Swift urbanisation and heavy air pollution, especially in the large cities, stimulated the strategic intervention in order to decrease the transport emissions (Aijaz and Ahmad 2021). Shenzhen is a positive example of this leadership with a fleet of fully electric buses, the largest in the world, which significantly reduces the emissions of the city and the quality of the air (Fried et al. 2023). The extensive use of electric cabs and delivery trucks in the city also highlights the intent of China to make its urban mobility electric. These have been complemented by widespread investments in charging systems, making it the largest EV charging network in the world, with the help of which the scalability of electric vehicles (EVs) is provided across the cities (Aijaz and Ahmad 2021). The strong infrastructure supply will tackle the usual challenges that prevent adoption of EV and will ease the everyday activities of the public and freight transportation systems in the fast-growing urban centers in China.

Besides electrification of passenger transport, green freight and logistics have also become a priority in China so as to limit goods movement within the cities. The government is working hard to encourage urban freight vehicles that use electricity to substitute trucks that use diesel as their fuel, which are some of the significant contributors of air pollution and Carbon dioxide emissions in urban areas (Zheng et al. 2024). The implementation of digital logistics platforms with high transportation scheduling and route optimisation technologies has been established to streamline freight operations, saving the company on unnecessary mileage and fuel use (Centobelli et al. 2020). They are not only efficient in operational aspects but also enhance data-based management approaches to urban freight systems that can act as a model of sustainable urban logistics (Sustainable Urban Logistics: Concepts, Methods and Information Systems 2014). Notably, the green logistics innovations play a role in alleviating the congestion of infrastructure in the cities of China and lowering carbon footprints. The concept of electrification and digitalisation of the freight transport is a vivid example of how many aspects China applied to the problem of urban transport emission and economic productivity.

The complex support of policies and the high national goals are the two key pillars of China to achieve green transport transformation. The government has established significant subsidies on EV production and purchase, which has increased the rapid market penetration and domestic innovation (Nicolaides et al. 2018). Carbon neutrality by 2060 is prioritized in the national policies with green mobility being considered within the context of larger urban policy frameworks and environmental policies (Centobelli et al. 2020). The funding mechanisms and the use of public–private partnerships has still brought the capital necessary to fund the enormous infrastructure development, such as charging networks and the use of green vehicles, they are an active approach on the part of the state to achieve sustainability (Zheng et al. 2024). The Chinese example shows that appropriately coordinated policy frameworks, infrastructural investment, and technological innovation all contribute to sustainable transport transition acceleration. Its experiences are pertinent in offering information to the Global South nations that have to find a balance between the need to ensure rapid urban development and the need to reduce emissions and at the same time promote economic growth (Schulte and Ny 2018).

11 Case Study 2. India: Digital Logistics Platforms and Rail-Based Freight Decarbonization

The fast economic growth in India has provoked the increasing transportation demand with the result of rising emissions and inefficiency of freight logistics. The government has been keen on addressing these issues by modernising the systems of logistics by coming up with the concept of the Dedicated Freight Corridors (DFC). The purpose of these electrified rail freight corridors is to replace freight transportation by road with an increased capacity and less reliance on fossil fuels, which explicitly reduce the carbon footprint of freight movements (Kumar and Anbanandam 2019). DFC program is of essence in helping India achieve its sustainability goals by creating rail as a low-carbon mode of transport of long-haul freight. This policy government is accompanied by policy frameworks such as the National Logistics Policy that incorporates sustainability objectives and implements solutions to logistical inefficiencies, facilitating multimodal transport solutions to optimise the total freight movement in terms of cost and emission (Kumar and Anbanandam 2019).

In addition to the physical infrastructure development, India is fast becoming digitalised in the realm of logistics with the use of the Unified Logistics Interface Platform (ULIP). The ULIP and other digital programs also use modern data analytics to manage freight flows with a greater level of efficiency through increasing transparency, eliminating bottlenecks, and optimising supply chain synchronisation (Wang et al. 2024). This digitalization also makes it easier to optimise routes, monitor them in real time, predict demand, which altogether minimises the time spent idle by vehicles and ultimately avoids unnecessary fuel use (Yang and Lin 2023). In addition, network freight platforms form a collaborative ecosystem between logistics service providers, freight shippers, and carriers to share resources and align their efforts to achieve better sustainability outcomes (Wang et al. 2024). Such innovations indicate how digital tools can revolutionise freight efficiency and at the same time reduce emissions to support the overall environmental agenda.

Electric mobility plans also support the green freight and urban transport transformation in India. The Faster Adoption and Manufacturing of Electric Vehicles (FAME) program promotes the use of electric buses, two- and three-wheelers, which will help decrease fossil fuel consumption in the urban freight and passenger transportation (Wang et al. 2024). The National Logistics Policy policy reforms establish an enabling regulatory framework because they encourage sustainable practices and environmental objectives become immediately integrated into transport planning (Kumar and Anbanandam 2019). The interplay of committed rail freight corridors, digital freight platforms, electric mobility, and harmonised policy structures is indicative of the potential to lower transport emissions through the collaboration of technological advancement and institutional backing and enhance operational performance. It is a valuable paradigm to the Global South and is an example of how the change in logistics can be sustainable even in the context of the increasing economic and transportation needs.

12 Discussion

The Multi-Level Perspective (MLP) of sustainability transitions adopted in the chapter finds a lot of relevance to the Kanyepe et al. (2023) framework that proposes that green shifts in transport are realised in the interplay between niches, regimes and landscapes, especially in Global South situations where fast urbanisation and climate changes are being encountered. This is similar to the results of Larbi et al. (2021) who implemented MLP to urban mobility in Curitiba and Freiburg in which niche innovations such as electric buses disrupt fossil-fuels regimes under the influence of landscape forces such as the Paris Agreement. This is expanded by Sovacool and Griffiths (2019) in the Global South by highlighting cultural obstacles that cement high-emission activities, but the niche experiments of digital logistics can disrupt them as seen in the analysis of policy voids in the chapter. Khan et al. (2019) also advocate this correspondence by postulating that MLP also indicates path dependencies in emerging economies, where landscape pressures such as SDGs allow niche breakthroughs in low-carbon infrastructure. Therefore, the MLP implementation of the chapter highlights practical niches in terms of transport in Global South, which supports the ideas of the scholars in terms of regime changes that cannot be repeated and should be context-specific. Muchenje (2024) adds to the latter by incorporating MLP with the African-based empirical knowledge of logistics, confirming the multi-scale nature of the focus of the chapter.

As it is applied in the chapter, the Ecological Modernisation Theory (EMT) appeals to Bugden (2022), who perceives the technological innovation such as EVs and smart systems as the ability to decouple economic growth and emissions in resource-constrained contexts. This mirrors Frijns et al. (2000) who critiques but modifies EMT to developing countries where they emphasize institutional change and green financing to address the threats of superficial adoption in Global South transport. EMT lens in the chapter is consistent with the article by Khan et al. (2019), which shows that the policy incentives and PPPs can encourage sustainable logistics through financial barriers, similar to the case of electrification in China. In addition, Hall provides support to this by associating EMT with supply chain greening where digital technologies streamline freight in informal economies, directly corroborating the chapter on the emission reductions through innovation.

Ntuli et al. (2024) confirm the usefulness of EMT in South African settings when it comes to dealing with road transport dominance using tech transfers. On the whole, these researchers confirm the MLP-EMT integrated approach of the chapter as a strong direction in the development and decarbonisation of the logistics of the Global South. The infrastructure, financial and institutional barriers mentioned in the chapter overlap with Okyere et al. (2019), which record regime lock-ins such as poor rail integration in African freight, which hinders modal shifts according to MLP dynamics. Anozie (2024) is even more aligned, as it is a critique of the weak aspects of climate finance, which contribute to the vulnerability of the Global South, which is consistent with the chapter, which is asking to adopt blended financing. This is supported by Hmamed et al. (2025) in his case study on the automotive sector, which describes the poor governance that fragments the implementation of the green tech, just like the institutional capacity that was emphasised in the chapter. Sovacool and Griffiths (2019) contribute to the understanding to add detail by stating that informal economies preserve high-emission regimes but transitions can be unlocked through the introduction of policy measures to support the chapter in its qualitative aspect. Dadashzada et al. (2024) also support the lack of EV charging as obstacles to scalability. These alignments validate the desk review in the chapter as being empirically based and encouraging scholars to adopt holistic barrier mitigation strategies to suit emerging contexts.

The chapter provides insights on electric mobility and digital logistics, which are similar to Kashem et al. (2024), who support renewable-powered EVs to Global South resilience, which is in line with niche innovations under MLP. Demir et al. (2022) endorse the emission reduction of route optimisation, which is reflective of the ULIP platforms in India as regime opponents. Nagy and Szentesi (2024) apply this to green hydrogen of heavy freights, supporting the emphasis on low-carbon fuels in the chapter in the context of landscape decarbonisation pressures. Muchenje (2024) incorporates them with EMT and advocates PPPs of scalable supply chains, like in the case study of China. Generalov (2024) highlights changes to rail, which can complement the efficiency gains in non-formal environments. Together, these authors support the 40–60% emission reductions by 2050 pathways (blended finance, regional ties) of the chapter as a way to make the theory a practical endeavour toward Global South practice.

13 Conclusion and Recommendations

This chapter throws light on the acute necessity of green transformations in transport and logistics throughout the Global South, and how institutional weaknesses such as insufficiencies of infrastructure, lack of financial resources, and landscape demands of climate commitments and urbanisation block change. The main implications highlight the possibility of a 40–60% reduction in emissions by 2050 using the niche technologies of electric mobility, digital freight platforms, and low-carbon fuels, which are evident in the case of China and India. The combination of the Multi-Level Perspective and the Ecological Modernisation Theory puts these dynamics into perspective, as regimes are entrenched in fossil fuel addiction and niches provide disruptive possibilities in the wake of global sustainability agendas. This qualitative desk review addresses a critical literature gap on context-specific approaches, the focus on the integrated investment between the public and the private, regional partnerships, and capacity-building to build resilient supply chains. The analysis has identified practical pathways that combine economic development and decarbonisation, and recommend policymakers to focus on low-cost, scalable options to respond to the needs of informal economies and fast trade. Such lessons not only support the urgency of ensuring that transport systems meet the requirements of the Paris Agreement and SDGs, but also make the Global South an early-stage experiment with fair-minded, innovation-oriented shifts that redefine the concepts of global logistics.

The policymakers and other interested parties should undertake specific measures based on the findings of the chapter to hasten the realisation of sustainable transport. Governments must be on the frontline fostering mixed financing schemes, which integrate national funding with global climate funds, which open up investment in EV charging services and railways to surmount capital constraints and scale electrification, as China did with infrastructural development. Transport policy should require digital logistics solutions to optimise routes and consolidate loads that reduce idle times and emissions in urban freight and meet the realities of the informal sector that can be seen in South African environments. The regional organisations should also promote cross-border cooperation in terms of common green facilities, including harmonised biofuels and hydrogen standards, which will increase trade efficiency and shock resistance in the world market. Capacity-building interventions require urgent implementation with training of regulators and operators to close institutional gaps to implement emission standards and modal shifts off-road in favour of rail. To encourage the low-carbon innovations, the privately owned companies need incentives such as electric fleet subsidies which reduce the costs of operation in the long run. The international benefactors have a duty of ensuring that they focus on technology transfers that are appropriate to resource constrained environments, and such accessibility to smart systems is equitable.

References

Aijaz I, Ahmad A (2021) Electric vehicles for environmental sustainability (pp 131–145). Springer. https://doi.org/10.1007/978-3-030-80702-3_8

Aktas E, Bloemhof JM, Fransoo JC, Günther H-O, Jammernegg W (2017) Green logistics solutions. Flex Serv Manuf J 30(3):363–365. https://doi.org/10.1007/s10696-017-9301-y

Anozie C (2024) Left behind: funding climate action in the global south. Texas A & M Law Review, 11(2), 333–363. https://doi.org/10.37419/lr.v11.i2.2

Ashraf A, Idrisi MJ (2024) Smart and sustainable public transportation—a need of developing countries. Int J Netw Distrib Comput 12(1):144–152. https://doi.org/10.1007/s44227-024-00023-2

Birindelli G, Palea V (2022) To green or not to green? How CSR mechanisms at the governance level affect the likelihood of banks pursuing green product strategies. Corp GovAnce: Int J Bus Soc 23(1):219–242. https://doi.org/10.1108/cg-09-2021-0349

Bugden D (2022) Technology, decoupling, and ecological crisis: examining ecological modernization theory through patent data. Environ Sociol, ahead-of-print (ahead-of-print), 228–241. https://doi.org/10.1080/23251042.2021.2021604

Centobelli P, Cerchione R, Esposito E, Shashi S (2020) Evaluating environmental sustainability strategies in freight transport and logistics industry. Bus Strateg Environ 29(3):1563–1574. https://doi.org/10.1002/bse.2453

Dadashzada I, Garza-Reyes JA, Roy Ghatak R, Gonzalez-Aleu F (2024) An investigation into the major barriers to the adoption of electric vehicles in last-mile deliveries for sustainable transport. Int J Logist Manag 36(3):906–938. https://doi.org/10.1108/ijlm-01-2024-0007

Demir E, Syntetos A, Van Woensel T (2022) Last mile logistics: Research trends and needs. IMA J Manag Math 33(4):549–561. https://doi.org/10.1093/imaman/dpac006

Elegbeleye I, Oguntona O, Elegbeleye F (2025) Green hydrogen: pathway to net zero green house gas emission and global climate change mitigation. Hydrogen 6(2):29. https://doi.org/10.3390/hydrogen6020029

Faijue DD, Segui AO, Shringarpure K, Razavi A, Hasan N, Dar O, Manikam L (2024) Constructing a One Health governance architecture: a systematic review and analysis of governance mechanisms for One Health. Eur J Pub Health 34(6):1086–1094. https://doi.org/10.1093/eurpub/ckae124

Ferrer ALC, Thomé AMT (2023) Carbon emissions in transportation: a synthesis framework. Sustain 15(11):8475. https://doi.org/10.3390/su15118475

Fried T, Goodchild A, Browne M, Sanchez-Diaz I (2023) Seeking equity and justice in urban freight: where to look? Transp Rev 44(1):191–212. https://doi.org/10.1080/01441647.2023.2247165

Frijns J, Phuong PT, Mol APJ (2000) Developing countries: ecological modernisation theory and industrialising economies: the case of Viet Nam. Environ Polit 9(1):257–292. https://doi.org/10.1080/09644010008414519

Generalov O (2024) Analysis of modern trends and opportunities in the logistics channels of energy products producers. Balt J Econ Stud, 10(1), 39–43. https://doi.org/10.30525/2256-0742/2024-10-1-39-43

Herman KS (2021) Green growth and innovation in the Global South: a systematic literature review. Innovation and Development 13(1):43–69. https://doi.org/10.1080/2157930x.2021.1909821

Hmamed H, Cherrafi A, Shokri A, Garza-Reyes JA, Foster M (2025) Barriers to low-carbon technology adoption: insights from the automotive manufacturing sector in emerging economies. Bus Strateg Environ. https://doi.org/10.1002/bse.70337

Hulkó G, Kálmán J, Lapsánszky A (2025) Financial supervision for the green transition: comparative insights from the <scp>EU</scp>, hungary, and Singapore. Thunderbird Int Bus Rev. https://doi.org/10.1002/tie.70039

Ibrahim A (2024) Transitioning emerging economies toward sustainability by implementing targeted policy interventions. ESTIDAMAA, 2024, 53–59. https://doi.org/10.70470/estidamaa/2024/008

Kanyepe J, Alfaneta M, Zizhou B (2023) Green logistics adoption and operational performance within the road freight sector: the moderating role of management commitment. Logist Res 16(1):1–14. https://doi.org/10.23773/2023_12

Kashem MA, Shamsuddoha M, Nasir T (2024) Sustainable transportation solutions for intelligent mobility: a focus on renewable energy and technological advancements for electric vehicles (EVs) and flying cars. Futur Transp 4(3):874–890. https://doi.org/10.3390/futuretransp4030042

Kern C, Jess A (2021) Reducing global greenhouse gas emissions to meet climate targets—a comprehensive quantification and reasonable options. Energies 14(17):5260. https://doi.org/10.3390/en14175260

Khan SAR, Sharif A, Golpîra H, Kumar A (2019) A green ideology in Asian emerging economies: From environmental policy and sustainable development. Sustain Dev 27(6):1063–1075. https://doi.org/10.1002/sd.1958

Kumar A, Anbanandam R (2019) A flexible policy framework for analysing multimodal freight transportation system in India: SAP–LAP and efficient IRP method. Glob J Flex Syst Manag 21(1):35–52. https://doi.org/10.1007/s40171-019-00226-5

Larbi M, Kellett J, Palazzo E (2022, June) Urban sustainability transitions in the Global South: a case study of Curitiba and Accra. Urban Forum 33(2):223–244. https://doi.org/10.1007/s12132-021-09438-4

Li X-Y, Tang B-J (2017) Incorporating the transport sector into carbon emission trading scheme: an overview and outlook. Nat Hazards 88(2):683–698. https://doi.org/10.1007/s11069-017-2886-3

Lin C-Y (2008) Determinants of the adoption of technological innovations by logistics service providers in China. Int J Technol Manag Sustain Dev 7(1):19–38. https://doi.org/10.1386/ijtm.7.1.19_1

Muchenje KC (2024) Green logistics and supply chain management (pp. 33–61). Igi Global. https://doi.org/10.4018/979-8-3693-3755-4.ch002

Mukherjee A, Owen R, Scott JM, Lyon F (2024) Financing green innovation startups: a systematic literature review on early-stage SME funding. Venture Capital, ahead-of-print(ahead-of-print), 1–27. https://doi.org/10.1080/13691066.2024.2410730

Nagy G, Szentesi S (2024) Green logistics: transforming supply chains for a sustainable future. Adv Logist Syst-Theory Pract, 18(3), 29–42. https://doi.org/10.32971/als.2024.026

Nicolaides D, Cebon D, Miles J (2018) Prospects for electrification of road freight. IEEE Syst J 12(2):1838–1849. https://doi.org/10.1109/jsyst.2017.2691408

Ntuli MN, Eloka-Eboka AC, Mwangi FM, Ewim DRE, Dioha MO (2024) Energy sustainability and carbon dioxide emissions mitigation options for South Africa's road transport sector. Bull Natl Res Cent 48(1):37. https://doi.org/10.1186/s42269-024-01192-4

Okeke N, Bakare O, Achumie G (2024) Integrating policy incentives and risk management for effective green finance in emerging markets. Int J Front Sci Technol Res, 7(1), 076–088. https://doi.org/10.53294/ijfstr.2024.7.1.0050

Okyere S, Yang JQ, Aning KS, Zhan B (2019) Review of sustainable multimodal freight transportation system in African developing countries: evidence from Ghana. Int J Eng Res Afr 41:155–174. https://doi.org/10.4028/www.scientific.net/jera.41.155

Rhouma A, Daher B, Vrachioli M, Mohtar R, Gil JM (2025) Financing the Water-Energy-Food-Ecosystem Nexus project: challenges, opportunities, and pathways for sustainable investment. Front Sustain Resour Manag, 4. https://doi.org/10.3389/fsrma.2025.1590161

Schulte J, Ny H (2018) Electric road systems: strategic stepping stone on the way towards sustainable freight transport? Sustain 10(4):1148. https://doi.org/10.3390/su10041148

Sovacool BK, Griffiths S (2019) The cultural barriers to a low-carbon future: A review of six mobility and energy transitions across 28 countries. Renew Sustain Energy Rev 119:109569. https://doi.org/10.1016/j.rser.2019.109569

Sule A, Eyo-Udo N, Onukwulu E, Agho M, Azubuike C (2024) Green finance solutions for banking to combat climate change and promote sustainability. Gulf J Adv Bus Res, 2(6), 376–410. https://doi.org/10.51594/gjabr.v2i6.54

Sustainable Urban Logistics: Concepts, Methods and Information Systems (2014) Springer Berlin Heidelberg. https://doi.org/10.1007/978-3-642-31788-0

Timilsina RR, Zhang J, Rahut DB, Patradool K, Sonobe T (2025) Global drive toward net-zero emissions and sustainability via electric vehicles: an integrative critical review. Energy, Ecol Environ 10(2):125–144. https://doi.org/10.1007/s40974-024-00351-7

Wang G, Hu X, Wang T, Liu J, Feng S, Wang C (2024) Tripartite evolutionary game analysis of a logistics service supply chain cooperation mechanism for network freight platforms. Int J Intell Syst 2024:1–17. https://doi.org/10.1155/2024/4820877

Weber H, Weber M (2020) When means of implementation meet ecological modernization theory: a critical frame for thinking about the sustainable development goals initiative. World Dev 136:105129. https://doi.org/10.1016/j.worlddev.2020.105129

Winkler L, Pearce D, Nelson J, Babacan O (2023) The effect of sustainable mobility transition policies on cumulative urban transport emissions and energy demand. Nat Commun, 14(1). https://doi.org/10.1038/s41467-023-37728-x

Wu H, Du J, Rasool Y (2025) Navigating environmental concerns: unveiling the role of economic governance, energy transition, and population aging on transport-based CO2 emissions in China. Energies 18(7):1748. https://doi.org/10.3390/en18071748

Yang C-S, Lin MS-M (2023) The impact of digitalization and digital logistics platform adoption on organizational performance in maritime logistics of Taiwan. Marit Policy Manag 51(8):1884–1901. https://doi.org/10.1080/03088839.2023.2234911

Zheng S, Yang H, Hu H, Liu C, Shen Y, Zheng C (2024) Station placement for sustainable urban metro freight systems using complex network theory. Sustain 16(11):4370. https://doi.org/10.3390/su16114370

Climate Impacts of Transport Systems

Munyaradzi Chibaro, James Kanyepe, Wilbert Manyanga, Masimba Evis Manyangara, and Unity Maringosi

Abstract The chapters' content addresses the contribution of transport and logistics to climate change by evaluating how these systems produce greenhouse gases (and other harmful environmental factors) and are exposed to climate related threats. In this respect, it provides an overview of transport systems from the perspective of national and international climate policies, discusses the urgency for mitigation efforts with regard to developing nations and reviews the dual role of transportation and logistics as contributors to climate change but also as sectors most susceptible to climate-related risks. Aviation, road and marine shipping together comprise approximately a quarter of all energy related GHG emissions resulting from increased demand for travel (passenger mobility) and cargo movement. While the primary concern has been CO_2 emissions, various forms of pollution generated by transportation activity have negative effects on human health and environmental quality. This chapter also focuses on the vulnerability of transportation infrastructure to natural disasters such as flooding, storm surges, extreme temperatures, rising sea levels and storms which may cause disruptions to supply chains, damage or destroy critical components of transportation infrastructure and create financial uncertainty. Transportation infrastructure especially airports, seaports, rail lines and roads is becoming increasingly vulnerable to loss of operation due to climate related events and long term degradation of its condition; therefore there is a pressing need for planning and maintaining climate resilient infrastructure. As part of the evaluation of climate

M. Chibaro (✉) · J. Kanyepe
Department of Logistics and Supply Chain Management, University of Botswana, Gaborone, Botswana
e-mail: munyachibaro@gmail.com

W. Manyanga
Department of Marketing, Logistics and Sport Management, Faculty of Commerce, Human Sciences and Education, Namibia University of Science and Technology, Windhoek, Namibia

M. E. Manyangara
Department of Marketing, Chinhoyi University of Technology, Chinhoyi, Zimbabwe

U. Maringosi
Department of Education, Great Zimbabwe University, Masvingo, Zimbabwe

J. Kanyepe and I. Chirisa (eds.), *Advancing the Green Agenda in the Transport and Logistics Sector in Global South Countries*, World Sustainability Series,
https://doi.org/10.1007/978-3-032-31374-4_2

governance structures applicable to transportation systems nationally and internationally, this chapter assesses policy tools such as emission limits, carbon taxes, modal substitution policies and investments in alternative energy. Additionally, it outlines the roles of global agreements and sector specific initiatives to guide mitigation strategies while acknowledging differences in responsibility and capability among developed versus developing countries. The chapter specifically highlights the dilemma of promoting economic development, improving access to transportation services and reducing greenhouse gas emissions in rapidly urbanizing areas. The chapter asserts that transformation of transportation systems is vital to achieving global goals for climate protection and sustainable development. It stresses the importance of financing mechanisms and technological assistance along with coherent policies necessary to facilitate transition to low-carbon, climate-resilient transportation systems in the Global South.

Keywords Climate impacts · Transport systems · Climate-resilient transport transitions · Policy frameworks

1 Background to the Study

Climate Change is currently one of the largest global issues we face today. This issue will affect how our economy works; whether or not the earth can sustain itself environmentally; and the health, welfare and quality of life of people all over the world. The transport sector is among the biggest contributors to worldwide GHG emission through using fossil fuel and through an ever-increasing demand for passenger and freight movement. Worldwide, transportation produces about 15–16% of the total greenhouse gases produced globally which makes it the second highest producing sector after electricity production (Statista 2025). As a result of growing international trade; urbanization and increasing vehicle ownership especially in emerging markets, emissions from transport have grown greatly since 1990. Specifically, road transport contributes almost three quarters of the sector's emissions (UNECE 2020).

Beginning with the international context, it is becoming increasingly evident at a global scale that transportation systems can be viewed as contributing agents in climate change, but also that they may be one of the most vulnerable sectors to the impacts of climate change (Huber et al. 2022). The various types of climate-related events or hazards such as increased sea levels, increased frequency and intensity of extreme weather events, such as floods and heatwaves, and the impact of storms on critical infrastructure associated with transportation modes (roads, railways, ports and airports), could create significant disruptions to global supply chain activities and increase the cost associated with maintaining and adapting to climate related changes (Guild et al. 2024). As a result of these issues, international response mechanisms have been developed and implemented through global frameworks such as the Paris Agreement. The Paris Agreement provides an opportunity for countries to implement economy-wide decarbonization strategies, as well as provide a framework for transitioning existing transportation modes toward low carbon mobility solutions, electric mobility solutions and modal shift opportunities toward rail and marine transportation. In addition to the Paris Agreement, other global agencies such as the International Maritime Organization (IMO) and the International Civil Aviation Organization (ICAO) are working to develop regulatory policies regarding greenhouse gas emissions from international air and oceanic transportation. Historically, these two sectors were considered to fall beyond national emission inventory boundaries. Collectively, these global frameworks demonstrate the need for transformational efforts across all modes of transportation to help mitigate climate change while promoting sustainable economic development and continued connectivity (Burbidge et al. 2024).

From a local or regional viewpoint, developing nations particularly those located within Africa encounter an extremely difficult situation (Onwudiwe 2023). One side of the coin is that transport is necessary for both regional trade and global market access; on the other side of the coin is that expanding transportation facilities often results in higher emissions, land use alteration, and environmental damage. This problem is quite serious in the rapidly urbanizing/motorizing developing world with little investment in sustainable public transit options. Additionally, most African states have neither the monetary nor technical resources available to develop very-large-scale, low carbon transport options and thus continue to rely upon older, less efficient vehicles and fossil fuel-based logistics.

The Southern African Development Community (SADC) region exemplifies the interplay between the need for economic development and environmental sustainability. The SADC region's economy relies heavily upon land-transport corridors that link landlocked countries like Zambia, Zimbabwe, and Botswana to ports in South Africa, Mozambique, and Tanzania. Consequently, efficient transportation systems are crucial for the movement of goods from southern Africa into international markets. To this end, SADC has emphasized the role of transportation integration via policies including the Protocol on Transport, Communications and Meteorology and the Regional Infrastructure Development Master Plan, which encourage the development of multi-modal transportation networks and cross border corridors to support economic co-operation among member states.

Southern Africa is considered one of the world's most climate-vulnerable areas and it is where fuel-dependent systems embed environmental risks into the very structure of economic expansion.

This raises a critical question: to what extent is the prevailing model of mobility-driven development inherently unsustainable? Much of the policy discourse assumes that environmental impacts can be mitigated through technological improvements, such as cleaner fuels or electric vehicles. While these interventions are important, they risk reinforcing what has been described as a "technological optimism bias," where efficiency gains are expected to offset increasing demand (Geels 2012). In practice, such gains are often undermined by rebound effects, whereby improvements in efficiency lower the cost of mobility and lead to increased consumption. This suggests that technological solutions alone are unlikely to resolve the environmental contradictions of mobility systems.

The environmental implications of transport are not limited to emissions but include broader ecological and social impacts. Urban air pollution, driven in part by transport emissions, poses significant public health risks, particularly in rapidly growing cities. In Lagos, for example, the dominance of informal transport systems contributes to high levels of congestion and pollution, reflecting both infrastructural deficits and regulatory limitations (Adeniji 2019). Similarly, Delhi's persistent air quality crisis highlights the cumulative effects of motorisation, weak enforcement, and fragmented policy responses (UNEP 2021). These cases illustrate that environmental degradation is not simply a by-product of growth, but a consequence of governance failures and institutional constraints that shape how mobility systems evolve.

From a political economy perspective, the persistence of environmentally harmful mobility systems can be understood as a form of structural lock-in. Investments in road infrastructure, vehicle manufacturing, and fossil fuel supply chains create vested interests that reinforce existing systems and resist transition. This aligns with the socio-technical transition literature, which emphasises that dominant mobility regimes are stabilised by institutional, economic, and behavioural factors (Geels 2012). As a result, even when more sustainable alternatives are available, they often struggle to scale due to path dependency and institutional inertia.

The concept of environmental externalities provides further insight into these dynamics. The costs associated with transport such as emissions, congestion, and health impacts are rarely fully internalised in pricing or policy frameworks (Litman 2023). This leads to a systematic undervaluation of environmental harm and over-investment in high-emission transport modes. In the Global South, these distortions are exacerbated by limited regulatory capacity and competing development priorities, where immediate economic gains often take precedence over long-term environmental considerations.

Importantly, as explained by Hersh (2026), the environmental impacts of mobility systems are deeply intertwined with issues of inequality. Low-income populations are disproportionately exposed to environmental risks, often residing in areas with high levels of pollution while having limited access to reliable transport. This creates what can be described as an "environmental justice paradox," where those who contribute least to emissions bear the greatest burden of their consequences. At the same time, policies aimed at reducing environmental impacts such as fuel taxes or vehicle restrictions can disproportionately affect these groups if not carefully designed. This highlights the need to critically interrogate not only the environmental performance of mobility systems, but also their distributional effects.

Ultimately, the environmental implications of mobility systems reveal a fundamental tension at the heart of development policy. The assumption that mobility expansion can simultaneously drive economic growth and environmental sustainability is increasingly contested. A more critical perspective suggests that achieving sustainable mobility requires not just incremental improvements, but a reconfiguration of the underlying development model shifting from mobility-intensive systems toward accessibility-oriented and low-carbon alternatives. This transition is not merely technical but political, requiring changes in governance, investment priorities, and societal expectations around mobility (Munuhwa 2024).

2 Trade-Offs in Mobility Systems

Economic Growth versus Environmental Sustainability

Infrastructure-led development has historically been central to economic growth strategies, particularly in the Global South, where investments in transport networks are viewed as catalysts for trade, industrialisation, and regional integration (Creutzig et al. 2015). Expanding road infrastructure, ports, and logistics systems can reduce travel times, lower transaction costs, and enhance connectivity between markets, thereby supporting economic productivity and competitiveness. As such, mobility expansion is often positioned as a prerequisite for development. However, this growth model is frequently underpinned by carbon-intensive transport systems, particularly those reliant on road-based mobility and fossil fuels. While such systems deliver short-term economic benefits, they also generate significant environmental costs, including increased greenhouse gas emissions, air pollution, and resource depletion. A key issue lies in the phenomenon of induced demand, whereby improvements in road capacity initially reduce congestion but subsequently encourage greater vehicle use, ultimately offsetting efficiency gains and reinforcing car dependency (Karimi et al. 2022).

From a critical perspective, this raises questions about the long-term sustainability of mobility-driven growth models. The assumption that economic growth and environmental sustainability can be simultaneously achieved through incremental efficiency improvements is increasingly contested. Instead, evidence suggests that current development trajectories may be environmentally unsustainable and economically inefficient over time, as the costs of environmental degradation including health impacts, climate risks, and productivity losses are not fully accounted for (Martins 2025; Burger 2025; Hidayati 2025). This highlights the need to move beyond growth models that are inherently mobility-intensive and carbon-dependent, toward approaches that prioritise accessibility, efficiency, and low-carbon innovation.

Social Inclusion versus Environmental Efficiency

Efforts to improve the environmental performance of transport systems often focus on increasing efficiency and reducing emissions through policy instruments such as congestion pricing, fuel taxation, and vehicle restrictions. While these measures can be effective in reducing environmental impacts, they may also generate unintended social consequences, particularly for low-income populations.

The case of Bogotá's TransMilenio Bus Rapid Transit (BRT) system illustrates this tension. While the system has been widely recognised for improving transport efficiency and reducing emissions, concerns have been raised regarding fare affordability and accessibility for lower-income users (Hersh 2026; Cervero 2013). As fares increase to sustain system operations, some users may be priced out, limiting their access to employment and services. This demonstrates that improvements in environmental efficiency do not automatically translate into socially inclusive outcomes.

Policies aimed at reducing emissions can have regressive effects if they disproportionately burden those with limited financial resources, a concern increasingly highlighted in recent mobility and climate justice literature. Measures such as fuel taxation, congestion pricing, or restrictions on older vehicles may contribute to environmental objectives, yet they can simultaneously increase transport costs for low-income households, who are more likely to depend on less efficient and more affordable modes of transport (Hidayati 2025; Burger 2025). In many Global South cities, informal transport systems such as minibuses and motorcycle taxis play a critical role in maintaining accessibility where formal systems are insufficient or absent. Although these systems are often associated with higher emissions and operational inefficiencies, their removal or restriction without viable and affordable alternatives risks exacerbating social exclusion and limiting access to essential opportunities (Martins 2025). This highlights a fundamental tension between environmental efficiency and social inclusion that cannot be resolved through purely technical or regulatory interventions.

This tension underscores the importance of adopting a just transition perspective in transport policy, which emphasises the need to align environmental sustainability with social equity. Recent research argues that sustainable mobility transitions must be designed in ways that ensure affordability, accessibility, and fairness, particularly for vulnerable populations (Caggiani and Prencipe 2025; Chen et al. 2025). Without such an approach, policies intended to reduce emissions may inadvertently reinforce existing inequalities, undermining both their social legitimacy and long-term effectiveness. Consequently, achieving sustainable mobility requires moving beyond narrow efficiency-driven frameworks toward more inclusive and context-sensitive policy approaches that explicitly address distributional impacts.

Short-Term versus Long-Term Sustainability

A further critical trade-off lies between short-term development priorities and long-term sustainability objectives, a tension that is increasingly highlighted in recent urban mobility and governance literature. Governments in the Global South often face immediate pressures to alleviate congestion, improve mobility, and stimulate economic growth, leading to a preference for infrastructure investments that deliver rapid and visible results. In this context, road expansion is frequently prioritised due to its short-term political and economic benefits, including reduced travel times and improved connectivity. However, recent research suggests that such approaches reflect a broader tendency toward short-termism in infrastructure planning, where immediate gains are prioritised over long-term sustainability considerations (Caggiani and Prencipe 2025; Burger 2025).

Yet, these investments frequently reinforce unsustainable mobility patterns. By prioritising road-based development, policymakers risk entrenching car-dependent systems that are difficult to reverse, a phenomenon increasingly understood through the lens of path dependency and socio-technical lock-in (Chen et al. 2025). Once established, these systems shape urban form, travel behaviour, and investment priorities, creating structural barriers to the adoption of low-carbon alternatives. Emerging studies argue that this lock-in effect is particularly pronounced in rapidly urbanising regions, where early infrastructure decisions have long-lasting implications for environmental sustainability and social equity (Martins 2025).

In contrast, sustainable mobility solutions—such as mass public transport systems, non-motorised transport infrastructure, and integrated land-use planning—offer more resilient and environmentally sustainable pathways. These approaches are increasingly recognised as essential for reducing emissions and improving accessibility, particularly in dense urban environments. However, they require long-term planning, substantial financial investment, and strong institutional coordination, which can be difficult to achieve in resource-constrained settings (Chen et al. 2025). Moreover, their benefits often materialise over extended time horizons, making them less politically attractive in contexts characterised by short electoral cycles and immediate development pressures (Caggiani and Prencipe 2025).

From a critical standpoint, this trade-off exposes the limitations of reactive, short-term policy approaches that prioritise immediate outcomes over systemic transformation. While such strategies may deliver short-term improvements, they risk entrenching unsustainable mobility systems that undermine long-term development goals. Recent literature increasingly calls for a shift toward strategic, forward-looking planning that aligns short-term interventions with long-term sustainability objectives, emphasising the need for integrated governance and policy coherence (Burger 2025; Martins 2025).

Taken together, these trade-offs highlight that mobility systems are inherently characterised by competing priorities that cannot be fully reconciled. Economic growth, social inclusion, and environmental sustainability are deeply interconnected, yet often operate in tension. Rather than seeking to eliminate these trade-offs, policymakers must develop the capacity to navigate and manage them explicitly, adopting context-sensitive strategies that balance competing objectives while supporting long-term, sustainable development pathways (Chen et al. 2025).

3 Development Pathways and Mobility Outcomes

Different development trajectories in the Global South shape mobility demand, accessibility patterns, and environmental outcomes in fundamentally different ways. These pathways are not merely technical choices but reflect broader socio-economic priorities, institutional capacities, and political economies. As such, they play a decisive role in determining whether mobility systems reinforce unsustainable patterns or support more inclusive and low-carbon development.

Car-dependent Development

Car-dependent development, evident in cities such as Johannesburg, represents a pathway characterised by rapid motorisation, low-density urban expansion, and a strong reliance on private vehicles (UN-Habitat 2021). This model is often associated with rising incomes and is frequently perceived as a marker of economic progress and modernisation. Governments may reinforce this trajectory through investments in road infrastructure, fuel subsidies, and policies that prioritise private vehicle use. However, the outcomes of this pathway are increasingly recognised as problematic. Car-dependent systems are associated with high levels of greenhouse gas emissions, traffic congestion, and air pollution, as well as significant resource consumption. Moreover, they tend to exacerbate social inequality, as access to mobility becomes contingent on the ability to afford private vehicles. In spatial terms, this pathway is closely linked to urban sprawl, which increases travel distances and further entrenches reliance on motorised transport (Burger 2025).

From a critical perspective, car-dependent development creates a form of structural lock-in, where infrastructure investments, land-use patterns, and behavioural norms reinforce each other, making transitions to more sustainable systems difficult and costly. This raises important questions about the long-term viability of replicating such models in the Global South, particularly given their environmental and social consequences.

Public Transport-Oriented Development

In contrast, public transport-oriented development offers a more sustainable and inclusive pathway, prioritising investment in high-capacity, efficient, and accessible transport systems. Cities such as Bogotá and Curitiba have demonstrated the potential of Bus Rapid Transit (BRT) systems to provide cost-effective alternatives to private vehicle use, improving accessibility while reducing emissions (Cervero 2013). This approach is typically combined with transit-oriented urban planning, which promotes compact, mixed-use development around transport corridors. By reducing travel distances and encouraging modal shifts toward public and non-motorised transport, this pathway can significantly lower environmental impacts while enhancing accessibility for a broader population.

However, while public transport-oriented systems are often presented as best-practice models, their implementation is not without challenges. They require strong institutional coordination, long-term planning, and sustained investment, which may be difficult to achieve in resource-constrained contexts. Additionally, issues such as overcrowding, service reliability, and affordability can limit their effectiveness. As seen in Bogotá, even successful systems such as TransMilenio have faced criticism related to rising fares and declining service quality, raising concerns about long-term sustainability and inclusivity. From a socio-technical perspective, the success of this pathway depends not only on infrastructure but also on governance, policy alignment, and user behaviour. Simply replicating models from other contexts without adapting them to local conditions can lead to suboptimal outcomes (Martins 2025).

Informal and Hybrid Mobility Systems

A distinctive feature of many Global South cities is the prevalence of informal and hybrid mobility systems, such as matatus in Nairobi, minibuses in Lagos, and motorcycle taxis in Kampala. These systems play a critical role in providing flexible, demand-responsive, and often affordable transport services, particularly in areas underserved by formal infrastructure (Salon and Gulyani 2019). From an accessibility perspective, informal transport systems are highly effective in filling gaps in mobility provision, enabling millions of people to access employment and services. They also contribute to local economies by generating employment and supporting small-scale entrepreneurship.

However, these systems are often associated with significant challenges, including safety concerns, regulatory gaps, operational inefficiencies, and environmental impacts due to ageing vehicle fleets and inefficient routing. As a result, they are frequently viewed as transitional or problematic within formal planning frameworks. A more critical perspective suggests that this framing is overly simplistic. Rather than viewing informal systems as anomalies to be eliminated, they can be understood as adaptive responses to institutional and infrastructural limitations (Caggiani et al. 2025). The key challenge is therefore not removal but integration and improvement. This includes formalising operations, upgrading vehicle fleets, improving regulation, and incorporating digital technologies to enhance efficiency and sustainability. From a transition perspective, informal systems represent both a challenge and an opportunity. While they contribute to environmental pressures, they also provide a foundation upon which more inclusive and context-sensitive mobility systems can be built.

4 Comparative Insights

Taken together, these development pathways highlight that mobility outcomes are not predetermined but shaped by policy choices, institutional capacity, and socio-economic conditions. Car-dependent models tend to prioritise short-term economic gains but generate long-term environmental and social costs. Public transport-oriented approaches offer more sustainable outcomes but require strong governance and long-term commitment. Informal systems, while often overlooked, play a crucial role in ensuring accessibility and must be integrated into broader mobility strategies. Crucially, these pathways are not mutually exclusive. Many cities exhibit hybrid systems that combine elements of each approach, reflecting the complexity of real-world mobility systems. The challenge for policymakers is therefore not to adopt a single model, but to develop context-sensitive strategies that leverage the strengths of different pathways while mitigating their limitations.

5 Global South Case Studies of Integrated Mobility

The case studies of Bogotá, Lagos, and Nairobi provide a valuable empirical foundation for examining the mobility–development–environment nexus within the Global South. Collectively, they capture diverse mobility pathways, ranging from formal, planned systems to informal and hybrid arrangements. While their inclusion is appropriate and strengthens the applied relevance of the chapter, their current use remains largely illustrative rather than fully analytical (Table 1).

The case studies of Bogotá, Lagos, and Nairobi highlight that there is no single optimal model of urban mobility in the Global South; rather, each system reflects context-specific trade-offs shaped by institutional capacity, economic constraints, and social needs. Recent literature increasingly emphasises that mobility systems must be understood as socio-technical configurations embedded within local governance structures and development trajectories, rather than transferable "best practice" models (Chen et al. 2025; Burger 2025). Bogotá demonstrates the potential of formal, planned systems to improve efficiency and reduce emissions through high-capacity public transport, yet it also reveals that such gains are often fragile and contingent. Issues such as affordability, overcrowding, and system saturation suggest that efficiency improvements do not automatically translate into equitable or sustainable outcomes, particularly as demand grows over time (Caggiani and Prencipe 2025). In contrast, Lagos and Nairobi illustrate the critical role of informal and hybrid mobility systems in providing widespread accessibility, particularly for low-income populations. These systems are highly adaptive and demand-responsive, but they frequently operate within weak regulatory frameworks, resulting in low efficiency, safety concerns, and significant environmental externalities (Martins 2025). Collectively, the cases demonstrate that while formal systems tend to prioritise efficiency and environmental performance, informal systems prioritise accessibility, often at the expense of sustainability, reinforcing the inherent tensions within mobility systems.

A key takeaway from the comparative analysis is that mobility systems in the Global South are structurally shaped by institutional limitations, governance challenges, and uneven development patterns, and must therefore be analysed through a lens of trade-offs rather than idealised models. Bogotá reflects a partial socio-technical transition toward more sustainable mobility, yet one that remains incomplete and socially uneven, highlighting the limits of infrastructure-led solutions in the absence of sustained governance capacity (Chen et al. 2025). Lagos represents a system where institutional constraints sustain informality as the dominant mode, illustrating how accessibility can be maintained even in the absence of formalised systems, but at the cost of environmental and operational efficiency (Martins 2025). Nairobi, meanwhile, exemplifies a hybrid trajectory, where informal systems persist alongside emerging efforts at regulation, digitalisation, and integration, reflecting ongoing attempts to reconcile flexibility with system efficiency (Burger 2025). These cases collectively suggest that sustainable mobility cannot be achieved through the wholesale adoption of formalised systems alone, but requires context-sensitive strategies that integrate accessibility, governance reform, and environmental objectives.

Table 1 Global South Case studies of Integrated mobility

Dimension	Bogotá (Colombia)	Lagos (Nigeria)	Nairobi (Kenya)
Mobility model	Formal, public transport-oriented (BRT—TransMilenio)	Predominantly informal (danfos, minibuses)	Hybrid system (matatus + emerging formal/digital systems)
Development pathway	Public transport-led, planned urban mobility	Informal, demand-driven mobility system	Transitional, combining informal and formal elements
Economic efficiency	High capacity, reduced travel times, improved system efficiency	Low efficiency due to congestion and fragmented operations	Moderate efficiency, improved by partial regulation and digitalisation
Social inclusion (Accessibility)	Improved access but affordability concerns for low-income users	High accessibility due to flexible and affordable services	High accessibility, especially for low-income populations
Environmental sustainability	Reduced emissions relative to private vehicles but still constrained by scale and congestion	High emissions due to ageing vehicles and inefficient operations	Mixed environmental outcomes; informal systems contribute to emissions
Environmental externalities	Partially managed through structured system	High levels of congestion, air pollution, and unregulated emissions	Significant externalities, though some improvements through regulation
Governance & institutional capacity	Relatively strong governance and coordinated planning	Weak regulatory frameworks and enforcement capacity	Moderate governance; ongoing reforms and policy interventions
Socio-Technical transition	Attempted transition toward structured, low-emission transport system	Limited transition; persistence of informal regime	Hybrid transition with gradual integration of informal and formal systems
Key strengths	Scalable, cost-effective public transport model; improved efficiency	Flexible, accessible, and demand-responsive mobility system	Adaptive system combining accessibility with emerging innovation
Key challenges	Overcrowding, affordability, long-term sustainability concerns	Environmental degradation, inefficiency, lack of regulation	Safety, emissions, regulatory inconsistencies
Overall Trade-off	Efficiency vs equity (affordability and capacity constraints)	Accessibility vs environmental sustainability	Accessibility vs efficiency and environmental performance

Ultimately, recent research argues that effective mobility policy must move beyond binary distinctions between formal and informal systems, focusing instead on how different models can be combined and adapted within specific institutional and socio-economic contexts to deliver more inclusive and sustainable outcomes (Caggiani and Prencipe 2025; Chen et al. 2025).

6 Policy Implications

Policy responses must balance environmental sustainability with social inclusion and economic development, recognising that mobility systems operate within complex and often competing priorities (Chikwere et al. 2026). Rather than treating these objectives as separate, effective policy design requires an integrated approach that explicitly acknowledges and manages trade-offs. In the context of the Global South, this is particularly challenging due to resource constraints, institutional limitations, and the need to deliver both immediate and long-term development outcomes.

A central policy priority is the promotion of low-carbon transport systems. This involves a strategic shift away from fossil fuel-dependent mobility toward more sustainable alternatives, including high-capacity public transport systems, electrified vehicle fleets, and non-motorised transport infrastructure (IPCC 2022). Investment in public transit such as Bus Rapid Transit (BRT), metro systems, and rail networks can significantly reduce emissions while improving accessibility. Electrification of transport, particularly when supported by cleaner energy sources, offers further potential to decarbonise mobility. However, from a critical perspective, these interventions must go beyond technological substitution. Without addressing underlying demand patterns and urban form, there is a risk that electrification may simply replace one form of high-consumption mobility with another, rather than fundamentally reducing environmental pressures.

Equally important is ensuring that sustainability policies do not exacerbate social inequalities. Policies aimed at reducing emissions such as fuel pricing, congestion charging, or vehicle restrictions can disproportionately affect low-income populations if not carefully designed (Litman 2023). As such, affordability and accessibility must be central considerations in policy development. This may involve targeted subsidies, differentiated pricing mechanisms, or investment in affordable public transport options. Adopting a just transition approach is critical to ensuring that environmental improvements do not come at the expense of social inclusion. In this regard, accessibility not simply mobility should serve as a guiding principle, ensuring that all population groups can access essential services and opportunities.

The integration of informal transport systems represents another key policy challenge and opportunity. In many Global South cities, informal systems such as minibuses, shared taxis, and motorcycle transport provide essential mobility services where formal systems are inadequate (Salon & Gulyani 2019). While these systems are often associated with inefficiencies and environmental concerns, their elimination is neither practical nor desirable. Instead, policy should focus on formalisation and integration, including regulatory frameworks, fleet modernisation, and the incorporation of digital technologies to improve efficiency and reduce emissions. This approach recognises the role of informal systems in supporting accessibility while seeking to enhance their environmental and operational performance.

Strengthening governance and institutional capacity is fundamental to the successful implementation of sustainable mobility policies. Effective transport planning requires coordination across multiple sectors, including transport, urban development, and environmental management. However, in many contexts, institutional fragmentation and limited technical capacity hinder integrated planning efforts. The development of metropolitan transport authorities improved regulatory frameworks, and enhanced inter-agency coordination can support more coherent and effective policy implementation (World Bank 2020). Furthermore, governance reforms must address issues of accountability and transparency to ensure that policy decisions reflect broader public interests rather than narrow institutional or political priorities.

In addition, the adoption of data-driven approaches offers significant potential to improve the efficiency and responsiveness of mobility systems. Advances in digital technologies enable real-time monitoring of transport networks, improved demand forecasting, and more effective resource allocation. Intelligent transport systems, integrated ticketing platforms, and mobility-as-a-service (MaaS) models can enhance user experience while optimising system performance. However, the benefits of digitalisation depend on appropriate regulatory frameworks and data governance structures to ensure equity, privacy, and accessibility (Litman, 2023).

Finally, a critical insight is that policy interventions must be context specific. There is no universal model of sustainable mobility that can be directly transferred across different settings. Policies must be tailored to local socio-economic conditions, institutional capacities, and mobility needs. This requires moving beyond the replication of "best practice" models toward more adaptive and flexible approaches that reflect local realities. Sustainable mobility policy in the Global South requires a shift from fragmented and reactive interventions toward integrated, long-term strategies that align environmental sustainability, social equity, and economic development. Rather than attempting to eliminate trade-offs, policymakers must develop the capacity to navigate them strategically, ensuring that mobility systems contribute to inclusive and resilient development pathways.

7 Limitations of the Study

This chapter is based on a qualitative narrative literature review and therefore relies heavily on existing secondary sources rather than primary empirical data. As a result, the findings are interpretive and may be influenced by the scope and availability of the selected literature. In addition, the study focuses primarily on Global South contexts, which may limit the generalisability of the findings to developed economies (Snyder 2019). The chapter also adopts a broad conceptual perspective, meaning that some country-specific institutional and socio-political dynamics may not have been explored in sufficient depth.

8 Areas for Further Research

Future research should incorporate empirical case studies and primary data collection to provide deeper insights into how mobility trade-offs are experienced within specific urban and regional contexts (Martins 2025; Chen et al. 2025). Comparative studies across different Global South cities could further examine how governance structures, institutional capacity, and socio-economic conditions shape sustainable mobility transitions (Caggiani and Prencipe 2025). Additional research is also needed on the role of emerging technologies, digital mobility platforms, and electric transport systems in promoting inclusive and low-carbon mobility (Delaere et al. 2024). Furthermore, future studies should explore the long-term social justice implications of sustainable mobility policies, particularly their effects on vulnerable and low-income populations (Hidayati 2025; Litman 2023).

9 Conclusion

This chapter has demonstrated that mobility systems occupy a dual and often contradictory role within development processes, functioning simultaneously as enablers of economic growth and social inclusion while also acting as significant sources of environmental pressure. In addressing Objective 1, the analysis has shown that while mobility facilitates market integration, productivity, and access to essential services, its benefits are unevenly distributed, with accessibility constraints continuing to marginalise low-income and spatially disadvantaged populations. In relation to Objective 2, the chapter has highlighted that prevailing mobility systems particularly those characterised by rapid motorisation and infrastructure-led expansion generate substantial environmental trade-offs, including rising emissions, air pollution, and unsustainable land-use patterns. Crucially, these tensions are not incidental but structural, embedded within the dominant models of development and reinforced by institutional, economic, and behavioural dynamics. As such, they cannot be fully resolved but must be explicitly recognised and strategically managed. This requires moving beyond narrow, mobility-centric paradigms toward approaches that prioritise accessibility, efficiency, and equity, alongside a reorientation of policy frameworks toward integrated, long-term, and context-sensitive interventions. Ultimately, achieving sustainable mobility in the Global South depends on the capacity of policymakers to navigate these interdependencies, balancing competing objectives in ways that support inclusive development while remaining within ecological limits.

References

Adeniji K (2019) Transport challenges in Nigeria. Lagos Transp Rev

Banister D (2008) The sustainable mobility paradigm. Transp Policy 15(2):73–80

Burbidge R, Paling C, Dunk RM (2024) A systematic review of adaption to climate change impacts in the aviation sector. Transp Rev 44(1):8–33

Burger K (2025) Towards equitable, smart, and sustainable urban mobility: Governance archetypes and their relations. Transp Res Part D.

Caggiani L, Prencipe LP (2025) Development Trends of Sustainable Mobility. Sustain 17(14):6409. https://doi.org/10.3390/su17146409

Cervero R (2013) Bus rapid transit (BRT): An efficient and competitive mode of public transport. J Public Transp 16(1):1–30

Chen Y, Li C, Wang W, Zhang Y, Chen XM, Gao Z (2025) The landscape, trends, challenges, and opportunities of sustainable mobility and transport. npj Sustainable Mobility and Transport.

Chikwere D, Munuhwa S, Chibaro M (2026) Transport and logistics in Africa: Policy and future direction. In: Munuhwa S (ed) The changing terrain of transport and logistics in Zimbabwe: Technology, sustainability and resilience management. Springer Nature Switzerland, pp 279–299

Creutzig F, Jochem P, Edelenbosch OY, Mattauch L, van Vuuren DP, McCollum D, Minx J (2015) Transport: A roadblock to climate change mitigation? Science 350(6263):911–912

Datta S (2012) The impact of improved highways on Indian firms. J Dev Econ 99(1):46–57

Delaere H, Basu S, Macharis C, Keseru I (2024) Barriers and opportunities for developing, implementing and operating inclusive digital mobility services. Eur Transp Res Rev 16(1):67

Geels FW (2012) A socio-technical analysis of low-carbon transitions. Res Policy 41(6):991–1001

Guild R, Wang X, Quijón PA (2024) Climate change impacts on coastal ecosystems. Environ Res Climate 3(4):042006

Hersh M (2026) Mobility, inclusion and exclusion. In Mobility of visually impaired people: Fundamentals and ICT assistive technologies (pp. 587–608). Springer Nature Switzerland

Hidayati I (2025) Assessing justice in sustainable mobility transitions. J Urban Aff

Huber D, Viere T, Nemoto EH, Jaroudi I, Korbee D, Fournier G (2022) Climate and environmental impacts of automated minibuses in future public transportation. Transp Res Part D Transp Environ 102:103160

Intergovernmental Panel on Climate Change (2022) Climate change 2022: Mitigation of climate change. Cambridge University Press

Karimi H, Ghadirifaraz B, Shetab Boushehri SN, Hosseininasab SM, Rafiei N (2022) Reducing traffic congestion and increasing sustainability in special urban areas through one-way traffic reconfiguration. Transportation, 49(1), 37–60

Larrañaga AM, de Dios Ortúzar J, Oviedo D, Ruiz-Padillo A (2025) Active mobility, equity and inclusion. Lat Am Transp Stud 3:100040

Litman T (2023) Evaluating transportation equity. Vic Transp Policy Inst.

Martins JO (2025) Bridging policy and practice in sustainable urban mobility in African cities. Front Sustain Cities

Munuhwa S (Ed.) (2024) Contemporary solutions for sustainable transportation practices. IGI Global

Onwudiwe NN (2023) Review on climate change impacts on air quality in Nigeria. In: Climate change impacts on Nigeria: environment and sustainable development, pp 217–232

Salon D, Gulyani S (2019) Mobility, poverty, and accessibility in African cities. World Bank Policy Res Work Pap

Snyder H (2019) Literature review as a research methodology: An overview and guidelines. J Bus Res 104:333–339

Statista (2025) Global infrastructure. Statista. https://www.statista.com/study/44345/infrastructure-worldwide/

United Nations (2015) Transforming our world: The 2030 Agenda for Sustainable Development

United Nations Environment Programme (2021) Global environment outlook 6
UN-Habitat (2021) Cities and climate change
World Bank (2020) Transforming transportation: The path to sustainable mobility
United Nations Economic Commission for Europe (UNECE) (2020) Facilitation of international road transport. Working Party on Road Transport, 115th Session. Geneva, United Nations. https://unece.org/fileadmin/DAM/trans/doc/2020/sc1/ECE-TRANS-SC1-2020-1e_01.pdf

Mobility, Development and Environmental Trade-Offs

Shakerod Munuhwa, Tendai Silvaziso Mandere, and David Chikwere

Abstract Mobility systems are central to economic development and social inclusion, yet they are also a major source of environmental degradation, particularly in rapidly urbanising regions of the Global South. This chapter examines the mobility, development and environment nexus, focusing on trade-offs between economic growth, equitable access, and environmental sustainability. Drawing on an integrated conceptual framework and supported by case studies from cities such as Bogotá, Lagos, and Nairobi, the chapter demonstrates how prevailing development pathways shape mobility demand and environmental outcomes. It argues that car-dependent models risk locking countries into unsustainable trajectories, while alternative approaches centred on accessibility and public transport offer more inclusive and low-carbon pathways. However, such transitions require strong governance, institutional capacity, and context-sensitive policy design. The chapter concludes that sustainable mobility in the Global South depends on explicitly managing trade-offs rather than attempting to eliminate them.

Keywords Sustainable mobility · Transport–Development nexus · Environmental externalities · Inclusive accessibility

S. Munuhwa (✉)
Department of Management, Glasgow School for Business and Society, Glasgow Caledonian University, Glasgow, UK
e-mail: Shakerod.munuhwa@gcu.ac.uk

T. S. Mandere
Department of Business Management, Faculty of Commerce, Ba Isago University, Gaborone, Botswana

D. Chikwere
Leeds Business School, Leeds Beckett University, Leeds, UK
e-mail: D.chikwere@leedsbeckett.ac.uk

J. Kanyepe and I. Chirisa (eds.), *Advancing the Green Agenda in the Transport and Logistics Sector in Global South Countries*, World Sustainability Series,
https://doi.org/10.1007/978-3-032-31374-4_3

1 Introduction

Mobility systems play a critical role in shaping economic development, social inclusion, and spatial transformation; however, contemporary research increasingly challenges the assumption that mobility expansion inherently produces positive developmental outcomes. While transport and logistics networks facilitate economic activity by enabling the movement of goods, labour, and information, their contribution to development is contingent on how they are structured and governed. Recent studies emphasise that mobility systems are deeply embedded within broader urban and economic systems, where rapid urbanisation and population growth intensify both opportunities and constraints (Chen et al. 2025; Caggiani and Prencipe 2025). In the Global South, mobility has become central to development policy, yet this has often translated into infrastructure-led strategies that prioritise expansion over systemic sustainability, raising concerns about long-term resilience and equity.

At the same time, mobility is widely framed as a key determinant of social inclusion, yet this framing is increasingly contested. Recent literature highlights that access to transport does not automatically translate into equitable access to opportunities, particularly in cities characterised by spatial inequality and fragmented infrastructure systems (Martins 2025; Hidayati 2025). In fact, mobility systems can reinforce exclusion when access is shaped by income, geography, and service availability. This has led to a growing shift toward an accessibility-based perspective, which prioritises the ability to reach opportunities rather than movement itself. From this viewpoint, conventional mobility planning often centred on travel speed and infrastructure provision may fail to address deeper structural inequalities embedded within urban systems.

Despite its developmental importance, the environmental implications of mobility systems have become increasingly central to academic and policy debates. Transport remains a major contributor to global emissions, with recent studies highlighting that urban transport systems are under growing pressure due to rising demand, energy consumption, and congestion (Chen et al. 2025). Crucially, the environmental impact of mobility is not limited to emissions but extends to broader issues of resource use, land consumption, and public health. Emerging research argues that current mobility systems are structurally misaligned with sustainability objectives, as they continue to prioritise high-consumption, car-oriented models that are difficult to decarbonise. Furthermore, the persistence of such systems reflects deeper governance and planning challenges, where short-term development priorities often outweigh long-term environmental considerations.

The coexistence of economic benefits and environmental costs underscores the inherently contested nature of mobility systems. Rather than representing a straightforward pathway to development, mobility systems are increasingly understood as sites of competing priorities and structural trade-offs. Recent research highlights those technological solutions alone such as electrification are insufficient to address these challenges without broader systemic changes in urban form, governance, and behaviour (Caggiani and Prencipe 2025). At the same time, policies aimed at improving environmental performance may generate unintended social consequences if they fail to account for issues of affordability and accessibility. This reinforces the argument that mobility systems must be analysed not as neutral infrastructures, but as socio-technical systems shaped by power, policy, and institutional dynamics.

Against this backdrop, this chapter addresses two core objectives. First, it critically analyses the relationship between mobility systems, economic development, and social inclusion in Global South contexts. Second, it evaluates the environmental trade-offs associated with different mobility and development pathways. By drawing on recent debates in sustainable mobility and urban development, the chapter seeks to move beyond simplified narratives of mobility as a driver of growth, offering instead a more critical and nuanced understanding of how mobility systems can be designed and governed to balance economic efficiency, social equity, and environmental sustainability.

2 Definition of Key Terms

Mobility refers to the ability of people and goods to move efficiently within and between locations. It encompasses not only physical transport systems (e.g., roads, rail, public transit) but also the accessibility, affordability, and reliability of these systems. In this chapter, mobility should be understood as both a driver of economic activity and a determinant of social inclusion (Hidayati 2025).

Sustainability in the context of transport and logistics refers to the capacity to meet current mobility needs without compromising the ability of future generations to meet theirs. It involves balancing three interrelated dimensions: environmental protection (e.g., reducing emissions and resource use), economic viability (e.g., supporting growth and efficiency), and social equity (e.g., ensuring inclusive access to mobility services) (Caggiani and Prencipe 2025; Burger 2025).

Trade-offs describe the competing priorities and tensions that arise when pursuing mobility, development, and environmental objectives simultaneously. For example, policies that promote economic growth through expanded transport infrastructure may increase environmental degradation, while environmentally focused interventions may impose social or economic costs. Understanding these trade-offs is central to analysing policy choices and development pathways in the Global South (Chen et al. 2025).

3 Objectives

- To analyse the relationship between mobility systems, economic development, and social inclusion in the context of the Global South.

- To evaluate the environmental trade-offs associated with different mobility and development pathways, particularly in relation to economic growth, accessibility, and sustainability.

4 Methodology

The chapter adopts a qualitative and integrative research approach, drawing on a comprehensive review and synthesis of existing literature across transport studies, development studies, and environmental sustainability. The methodology is primarily based on a narrative literature review, which enables the identification of key themes, theoretical perspectives, and empirical findings relevant to the mobility–development–environment nexus in the Global South. The study included peer-reviewed journal articles, books, policy documents, and institutional reports relevant to mobility, transport, development, and environmental sustainability in the Global South. Priority was given to literature published between 2010 and 2026, although seminal studies were also considered. Both theoretical and empirical studies written in English were included.

The next section focuses on the conceptual framework that underpins the study.

5 Conceptual Framework

The analysis is grounded in an integrated conceptual framework combining the transport–development nexus, sustainability theory, environmental externalities, and socio-technical transitions as shown in Fig. 1.

The transport-development nexus conceptualises mobility as both a driver and an outcome of development (Burger 2025). On the one hand, improvements in transport and logistics systems enhance connectivity, reduce transaction costs, and facilitate the movement of goods, services, and labour, thereby stimulating economic activity and supporting regional and national development. On the other hand, development processes such as rising incomes, urbanisation, and industrialisation generate increased demand for mobility, as individuals and firms require greater movement to access opportunities and markets. This creates a dynamic and self-reinforcing relationship in which mobility and development continuously shape one another. However, this feedback loop can also intensify environmental pressures, particularly where mobility expansion is based on carbon-intensive systems.

To critically evaluate these dynamics, the framework adopts the three pillars of sustainable development—economic efficiency, social equity, and environmental sustainability (United Nations 2015). Economic efficiency emphasises the role of mobility in enhancing productivity, competitiveness, and growth, while social equity focuses on ensuring that transport systems provide inclusive and affordable access to opportunities. Environmental sustainability, in contrast, highlights the need to

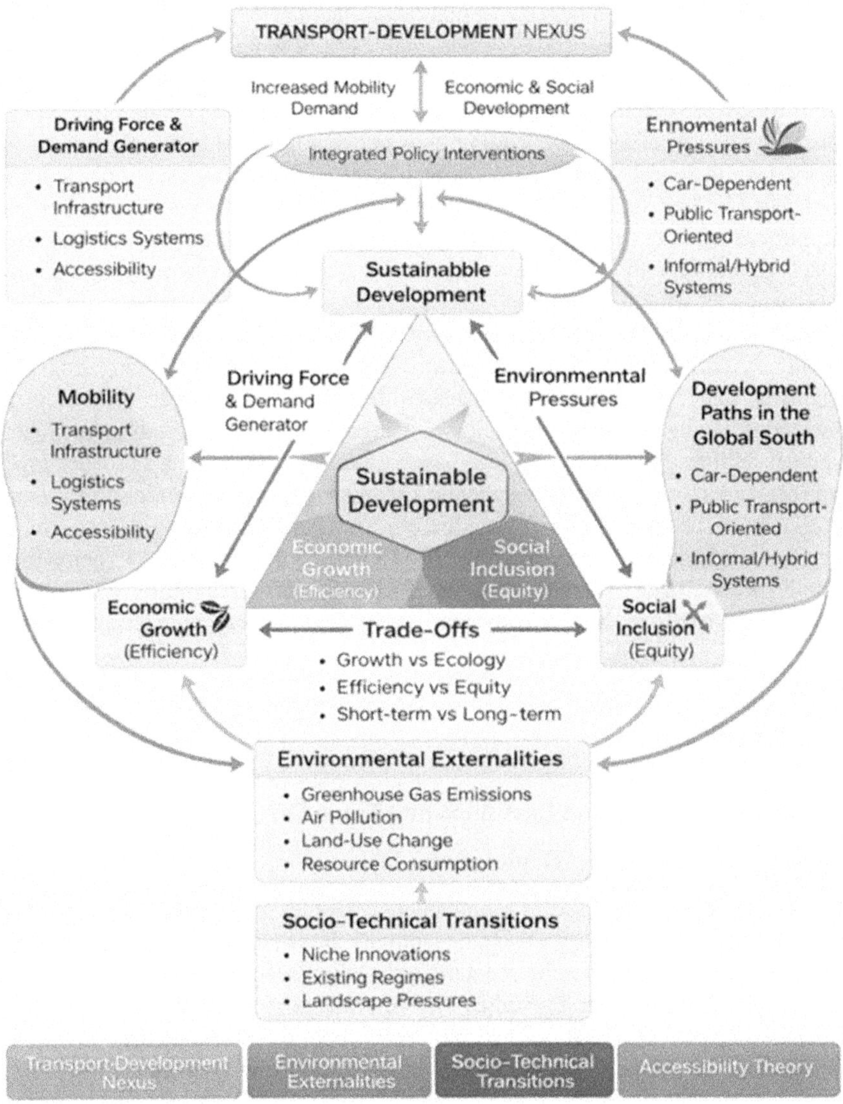

Fig. 1 The integrated framework

minimise negative ecological impacts such as emissions, pollution, and resource depletion. Although these dimensions are conceptually interdependent, in practice they often exist in tension. For example, policies that prioritise economic growth through infrastructure expansion may increase environmental degradation, while environmentally focused interventions may impose costs that limit accessibility for vulnerable populations.

The concept of environmental externalities further deepens this analysis by highlighting how the true social and environmental costs of mobility such as greenhouse gas emissions, congestion, and air pollution are frequently not fully internalised within transport systems (Litman 2023). As a result, market signals often fail to reflect the full cost of mobility, leading to overconsumption of high-emission transport modes and underinvestment in sustainable alternatives. This issue is particularly pronounced in the Global South, where regulatory frameworks may be weaker and institutional capacity to manage externalities is limited. Consequently, the negative impacts of mobility are often disproportionately borne by low-income populations, exacerbating existing inequalities.

Finally, socio-technical transition theory provides a dynamic perspective on how mobility systems evolve over time (Geels 2012). Rather than viewing transport systems as static, this approach emphasises the interaction between technological innovations (such as electric vehicles and digital platforms), established socio-technical regimes (such as fossil fuel-based transport systems), and broader landscape pressures (including climate change, urbanisation, and economic shifts). Transitions toward sustainable mobility are therefore not purely technological but require coordinated changes in infrastructure, institutions, policies, and user behaviour. This perspective underscores the complexity of transforming mobility systems and highlights the need for long-term, systemic approaches to achieving sustainability.

6 Literature Review

Mobility, Development and Environmental Trade-Offs

Mobility as a Driver of Development and Inclusion

Mobility is widely recognised as a fundamental enabler of economic development. Efficient transport and logistics systems facilitate the movement of goods, services, labour, and information, thereby reducing transaction costs, improving market efficiency, and enhancing overall productivity (Larrañaga et al. 2025). By connecting producers to markets, workers to employment opportunities, and firms to supply chains, mobility systems play a central role in supporting economic growth and regional integration. In many Global South contexts, investments in transport infrastructure are therefore closely linked to broader development strategies, particularly those aimed at industrialisation, trade expansion, and urban economic transformation.

Empirical evidence demonstrates the significant developmental impact of improved mobility systems. For instance, Ethiopia's large-scale investment in road infrastructure has enhanced rural accessibility, enabling farmers to participate more effectively in markets and improving access to essential services such as healthcare and education (World Bank 2020). Similarly, India's Golden Quadrilateral highway project strengthened connectivity between major economic centres, contributing to increased industrial activity, reduced travel times, and regional economic growth

(Delaere et al. 2024). These examples illustrate how strategic transport investments can stimulate both local and national development by integrating previously disconnected regions into wider economic systems.

Martind (2025) emphasised that beyond economic outcomes, mobility also plays a critical role in shaping social inclusion. Access to reliable and affordable transport systems determines individuals' ability to participate in economic and social life, influencing access to employment, education, healthcare, and social networks. In this sense, mobility is not only about movement but also about enabling opportunities and improving quality of life. Well-functioning transport systems can therefore contribute to poverty reduction and social mobility by expanding access to essential services and economic opportunities.

However, the benefits of mobility are not evenly distributed. In many Global South cities, structural inequalities in urban form, infrastructure provision, and income levels result in significant disparities in access to transport. Low-income populations, particularly those living in peripheral or informal settlements, often face long travel times, high transport costs, and limited-service availability. In cities such as Nairobi, these constraints restrict access to employment opportunities and essential services, reinforcing cycles of poverty and exclusion (Salon and Gulyani 2019). Informal transport systems may partially address these gaps, but they often come with trade-offs in terms of cost, safety, and environmental performance.

These dynamics highlight the critical distinction between mobility and accessibility. While mobility refers to the ability to move, accessibility captures the ability to reach desired destinations and opportunities. From a development perspective, accessibility is a more meaningful measure, as increased movement does not necessarily translate into improved access. In some cases, mobility expansion such as increased reliance on private vehicles may even exacerbate inequality by privileging those who can afford it while marginalising others (Larrañaga et al. 2025).

7 Environmental Implications of Mobility Systems

Transport systems are a major contributor to environmental degradation, accounting for a substantial and growing share of global greenhouse gas emissions (IPCC 2022). While mobility expansion is frequently framed as a necessary condition for economic development, these framing obscures the extent to which current transport systems are structurally dependent on carbon-intensive energy sources. In the Global South, rapid urbanisation, rising incomes, and expanding logistics networks have accelerated motorisation, intensifying emissions and environmental pressures (Creutzig et al. 2015). However, the issue is not merely the scale of mobility growth, but the form it takes. Development pathways that prioritise road-based, fossil fuel-dependent systems embed environmental risks into the very structure of economic expansion.

This raises a critical question: to what extent is the prevailing model of mobility-driven development inherently unsustainable? Much of the policy discourse assumes that environmental impacts can be mitigated through technological improvements, such as cleaner fuels or electric vehicles. While these interventions are important, they risk reinforcing what has been described as a "technological optimism bias," where efficiency gains are expected to offset increasing demand (Geels 2012). In practice, such gains are often undermined by rebound effects, whereby improvements in efficiency lower the cost of mobility and lead to increased consumption. This suggests that technological solutions alone are unlikely to resolve the environmental contradictions of mobility systems.

The environmental implications of transport are not limited to emissions but include broader ecological and social impacts. Urban air pollution, driven in part by transport emissions, poses significant public health risks, particularly in rapidly growing cities. In Lagos, for example, the dominance of informal transport systems contributes to high levels of congestion and pollution, reflecting both infrastructural deficits and regulatory limitations (Adeniji 2019). Similarly, Delhi's persistent air quality crisis highlights the cumulative effects of motorisation, weak enforcement, and fragmented policy responses (UNEP 2021). These cases illustrate that environmental degradation is not simply a by-product of growth, but a consequence of governance failures and institutional constraints that shape how mobility systems evolve.

From a political economy perspective, the persistence of environmentally harmful mobility systems can be understood as a form of structural lock-in. Investments in road infrastructure, vehicle manufacturing, and fossil fuel supply chains create vested interests that reinforce existing systems and resist transition. This aligns with the socio-technical transition literature, which emphasises that dominant mobility regimes are stabilised by institutional, economic, and behavioural factors (Geels 2012). As a result, even when more sustainable alternatives are available, they often struggle to scale due to path dependency and institutional inertia.

The concept of environmental externalities provides further insight into these dynamics. The costs associated with transport such as emissions, congestion, and health impacts are rarely fully internalised in pricing or policy frameworks (Litman 2023). This leads to a systematic undervaluation of environmental harm and over-investment in high-emission transport modes. In the Global South, these distortions are exacerbated by limited regulatory capacity and competing development priorities, where immediate economic gains often take precedence over long-term environmental considerations.

Importantly, as explained by Hersh (2026), the environmental impacts of mobility systems are deeply intertwined with issues of inequality. Low-income populations are disproportionately exposed to environmental risks, often residing in areas with high levels of pollution while having limited access to reliable transport. This creates what can be described as an "environmental justice paradox," where those who contribute least to emissions bear the greatest burden of their consequences. At the same time, policies aimed at reducing environmental impacts such as fuel taxes or vehicle restrictions can disproportionately affect these groups if not carefully designed. This highlights the need to critically interrogate not only the environmental performance of mobility systems, but also their distributional effects.

Ultimately, the environmental implications of mobility systems reveal a fundamental tension at the heart of development policy. The assumption that mobility expansion can simultaneously drive economic growth and environmental sustainability is increasingly contested. A more critical perspective suggests that achieving sustainable mobility requires not just incremental improvements, but a reconfiguration of the underlying development model shifting from mobility-intensive systems toward accessibility-oriented and low-carbon alternatives. This transition is not merely technical but political, requiring changes in governance, investment priorities, and societal expectations around mobility (Munuhwa 2024).

8 Trade-Offs in Mobility Systems

Economic Growth versus Environmental Sustainability

Infrastructure-led development has historically been central to economic growth strategies, particularly in the Global South, where investments in transport networks are viewed as catalysts for trade, industrialisation, and regional integration (Creutzig et al. 2015). Expanding road infrastructure, ports, and logistics systems can reduce travel times, lower transaction costs, and enhance connectivity between markets, thereby supporting economic productivity and competitiveness. As such, mobility expansion is often positioned as a prerequisite for development. However, this growth model is frequently underpinned by carbon-intensive transport systems, particularly those reliant on road-based mobility and fossil fuels. While such systems deliver short-term economic benefits, they also generate significant environmental costs, including increased greenhouse gas emissions, air pollution, and resource depletion. A key issue lies in the phenomenon of induced demand, whereby improvements in road capacity initially reduce congestion but subsequently encourage greater vehicle use, ultimately offsetting efficiency gains and reinforcing car dependency (Karimi et al. 2022).

From a critical perspective, this raises questions about the long-term sustainability of mobility-driven growth models. The assumption that economic growth and environmental sustainability can be simultaneously achieved through incremental efficiency improvements is increasingly contested. Instead, evidence suggests that current development trajectories may be environmentally unsustainable and economically inefficient over time, as the costs of environmental degradation including health impacts, climate risks, and productivity losses are not fully accounted for (Martins 2025; Burger 2025; Hidayati 2025). This highlights the need to move beyond growth models that are inherently mobility-intensive and carbon-dependent, toward approaches that prioritise accessibility, efficiency, and low-carbon innovation.

Social Inclusion Versus Environmental Efficiency

Efforts to improve the environmental performance of transport systems often focus on increasing efficiency and reducing emissions through policy instruments such as congestion pricing, fuel taxation, and vehicle restrictions. While these measures can be effective in reducing environmental impacts, they may also generate unintended social consequences, particularly for low-income populations.

The case of Bogotá's TransMilenio Bus Rapid Transit (BRT) system illustrates this tension. While the system has been widely recognised for improving transport efficiency and reducing emissions, concerns have been raised regarding fare affordability and accessibility for lower-income users (Hersh 2026; Cervero 2013). As fares increase to sustain system operations, some users may be priced out, limiting their access to employment and services. This demonstrates that improvements in environmental efficiency do not automatically translate into socially inclusive outcomes.

Policies aimed at reducing emissions can have regressive effects if they disproportionately burden those with limited financial resources, a concern increasingly highlighted in recent mobility and climate justice literature. Measures such as fuel taxation, congestion pricing, or restrictions on older vehicles may contribute to environmental objectives, yet they can simultaneously increase transport costs for low-income households, who are more likely to depend on less efficient and more affordable modes of transport (Hidayati, 2025; Burger, 2025). In many Global South cities, informal transport systems such as minibuses and motorcycle taxis play a critical role in maintaining accessibility where formal systems are insufficient or absent. Although these systems are often associated with higher emissions and operational inefficiencies, their removal or restriction without viable and affordable alternatives risks exacerbating social exclusion and limiting access to essential opportunities (Martins, 2025). This highlights a fundamental tension between environmental efficiency and social inclusion that cannot be resolved through purely technical or regulatory interventions.

This tension underscores the importance of adopting a just transition perspective in transport policy, which emphasises the need to align environmental sustainability with social equity. Recent research argues that sustainable mobility transitions must be designed in ways that ensure affordability, accessibility, and fairness, particularly for vulnerable populations (Caggiani & Prencipe, 2025; Chen et al., 2025). Without such an approach, policies intended to reduce emissions may inadvertently reinforce existing inequalities, undermining both their social legitimacy and long-term effectiveness. Consequently, achieving sustainable mobility requires moving beyond narrow efficiency-driven frameworks toward more inclusive and context-sensitive policy approaches that explicitly address distributional impacts.

Short-Term Versus Long-Term Sustainability

A further critical trade-off lies between short-term development priorities and long-term sustainability objectives, a tension that is increasingly highlighted in recent urban mobility and governance literature. Governments in the Global South often face immediate pressures to alleviate congestion, improve mobility, and stimulate economic growth, leading to a preference for infrastructure investments that deliver rapid and visible results. In this context, road expansion is frequently prioritised due to its short-term political and economic benefits, including reduced travel times and improved connectivity. However, recent research suggests that such approaches reflect a broader tendency toward short-termism in infrastructure planning, where immediate gains are prioritised over long-term sustainability considerations (Caggiani and Prencipe 2025; Burger 2025).

Yet, these investments frequently reinforce unsustainable mobility patterns. By prioritising road-based development, policymakers risk entrenching car-dependent systems that are difficult to reverse, a phenomenon increasingly understood through the lens of path dependency and socio-technical lock-in (Chen et al. 2025). Once established, these systems shape urban form, travel behaviour, and investment priorities, creating structural barriers to the adoption of low-carbon alternatives. Emerging studies argue that this lock-in effect is particularly pronounced in rapidly urbanising regions, where early infrastructure decisions have long-lasting implications for environmental sustainability and social equity (Martins 2025).

In contrast, sustainable mobility solutions—such as mass public transport systems, non-motorised transport infrastructure, and integrated land-use planning—offer more resilient and environmentally sustainable pathways. These approaches are increasingly recognised as essential for reducing emissions and improving accessibility, particularly in dense urban environments. However, they require long-term planning, substantial financial investment, and strong institutional coordination, which can be difficult to achieve in resource-constrained settings (Chen et al. 2025). Moreover, their benefits often materialise over extended time horizons, making them less politically attractive in contexts characterised by short electoral cycles and immediate development pressures (Caggiani and Prencipe 2025).

From a critical standpoint, this trade-off exposes the limitations of reactive, short-term policy approaches that prioritise immediate outcomes over systemic transformation. While such strategies may deliver short-term improvements, they risk entrenching unsustainable mobility systems that undermine long-term development goals. Recent literature increasingly calls for a shift toward strategic, forward-looking planning that aligns short-term interventions with long-term sustainability objectives, emphasising the need for integrated governance and policy coherence (Burger 2025; Martins 2025).

Taken together, these trade-offs highlight that mobility systems are inherently characterised by competing priorities that cannot be fully reconciled. Economic growth, social inclusion, and environmental sustainability are deeply interconnected, yet often operate in tension. Rather than seeking to eliminate these trade-offs, policymakers must develop the capacity to navigate and manage them explicitly, adopting context-sensitive strategies that balance competing objectives while supporting long-term, sustainable development pathways (Chen et al. 2025).

9 Development Pathways and Mobility Outcomes

Different development trajectories in the Global South shape mobility demand, accessibility patterns, and environmental outcomes in fundamentally different ways. These pathways are not merely technical choices but reflect broader socio-economic priorities, institutional capacities, and political economies. As such, they play a decisive role in determining whether mobility systems reinforce unsustainable patterns or support more inclusive and low-carbon development.

Car-dependent Development.

Car-dependent development, evident in cities such as Johannesburg, represents a pathway characterised by rapid motorisation, low-density urban expansion, and a strong reliance on private vehicles (UN-Habitat 2021). This model is often associated with rising incomes and is frequently perceived as a marker of economic progress and modernisation. Governments may reinforce this trajectory through investments in road infrastructure, fuel subsidies, and policies that prioritise private vehicle use. However, the outcomes of this pathway are increasingly recognised as problematic. Car-dependent systems are associated with high levels of greenhouse gas emissions, traffic congestion, and air pollution, as well as significant resource consumption. Moreover, they tend to exacerbate social inequality, as access to mobility becomes contingent on the ability to afford private vehicles. In spatial terms, this pathway is closely linked to urban sprawl, which increases travel distances and further entrenches reliance on motorised transport (Burger 2025).

From a critical perspective, car-dependent development creates a form of structural lock-in, where infrastructure investments, land-use patterns, and behavioural norms reinforce each other, making transitions to more sustainable systems difficult and costly. This raises important questions about the long-term viability of replicating such models in the Global South, particularly given their environmental and social consequences.

Public Transport-Oriented Development

In contrast, public transport-oriented development offers a more sustainable and inclusive pathway, prioritising investment in high-capacity, efficient, and accessible transport systems. Cities such as Bogotá and Curitiba have demonstrated the potential of Bus Rapid Transit (BRT) systems to provide cost-effective alternatives to private vehicle use, improving accessibility while reducing emissions (Cervero 2013). This approach is typically combined with transit-oriented urban planning, which promotes compact, mixed-use development around transport corridors. By reducing travel distances and encouraging modal shifts toward public and non-motorised transport, this pathway can significantly lower environmental impacts while enhancing accessibility for a broader population.

However, while public transport-oriented systems are often presented as best-practice models, their implementation is not without challenges. They require strong institutional coordination, long-term planning, and sustained investment, which may be difficult to achieve in resource-constrained contexts. Additionally, issues such as overcrowding, service reliability, and affordability can limit their effectiveness. As seen in Bogotá, even successful systems such as TransMilenio have faced criticism related to rising fares and declining service quality, raising concerns about long-term sustainability and inclusivity. From a socio-technical perspective, the success of this pathway depends not only on infrastructure but also on governance, policy alignment, and user behaviour. Simply replicating models from other contexts without adapting them to local conditions can lead to suboptimal outcomes (Martins 2025).

Informal and Hybrid Mobility Systems

A distinctive feature of many Global South cities is the prevalence of informal and hybrid mobility systems, such as matatus in Nairobi, minibuses in Lagos, and motorcycle taxis in Kampala. These systems play a critical role in providing flexible, demand-responsive, and often affordable transport services, particularly in areas underserved by formal infrastructure (Salon and Gulyani 2019). From an accessibility perspective, informal transport systems are highly effective in filling gaps in mobility provision, enabling millions of people to access employment and services. They also contribute to local economies by generating employment and supporting small-scale entrepreneurship.

However, these systems are often associated with significant challenges, including safety concerns, regulatory gaps, operational inefficiencies, and environmental impacts due to ageing vehicle fleets and inefficient routing. As a result, they are frequently viewed as transitional or problematic within formal planning frameworks. A more critical perspective suggests that this framing is overly simplistic. Rather than viewing informal systems as anomalies to be eliminated, they can be understood as adaptive responses to institutional and infrastructural limitations (Caggiani et al. 2025). The key challenge is therefore not removal but integration and improvement. This includes formalising operations, upgrading vehicle fleets, improving regulation, and incorporating digital technologies to enhance efficiency and sustainability. From a transition perspective, informal systems represent both a challenge and an opportunity. While they contribute to environmental pressures, they also provide a foundation upon which more inclusive and context-sensitive mobility systems can be built.

10 Comparative Insights

Taken together, these development pathways highlight that mobility outcomes are not predetermined but shaped by policy choices, institutional capacity, and socio-economic conditions. Car-dependent models tend to prioritise short-term economic gains but generate long-term environmental and social costs. Public transport-oriented approaches offer more sustainable outcomes but require strong governance and long-term commitment. Informal systems, while often overlooked, play a crucial role in ensuring accessibility and must be integrated into broader mobility strategies. Crucially, these pathways are not mutually exclusive. Many cities exhibit hybrid systems that combine elements of each approach, reflecting the complexity of real-world mobility systems. The challenge for policymakers is therefore not to adopt a single model, but to develop context-sensitive strategies that leverage the strengths of different pathways while mitigating their limitations.

11 Global South Case Studies of Integrated Mobility

The case studies of Bogotá, Lagos, and Nairobi provide a valuable empirical foundation for examining the mobility–development–environment nexus within the Global South. Collectively, they capture diverse mobility pathways, ranging from formal, planned systems to informal and hybrid arrangements. While their inclusion is appropriate and strengthens the applied relevance of the chapter, their current use remains largely illustrative rather than fully analytical (Table 1).

The case studies of Bogotá, Lagos, and Nairobi highlight that there is no single optimal model of urban mobility in the Global South; rather, each system reflects context-specific trade-offs shaped by institutional capacity, economic constraints, and social needs. Recent literature increasingly emphasises that mobility systems must be understood as socio-technical configurations embedded within local governance structures and development trajectories, rather than transferable "best practice" models (Chen et al. 2025; Burger 2025). Bogotá demonstrates the potential of formal, planned systems to improve efficiency and reduce emissions through high-capacity public transport, yet it also reveals that such gains are often fragile and contingent. Issues such as affordability, overcrowding, and system saturation suggest that efficiency improvements do not automatically translate into equitable or sustainable outcomes, particularly as demand grows over time (Caggiani and Prencipe 2025). In contrast, Lagos and Nairobi illustrate the critical role of informal and hybrid mobility systems in providing widespread accessibility, particularly for low-income populations. These systems are highly adaptive and demand-responsive, but they frequently operate within weak regulatory frameworks, resulting in low efficiency, safety concerns, and significant environmental externalities (Martins 2025). Collectively, the cases demonstrate that while formal systems tend to prioritise efficiency and environmental performance, informal systems prioritise accessibility, often at the expense of sustainability, reinforcing the inherent tensions within mobility systems.

A key takeaway from the comparative analysis is that mobility systems in the Global South are structurally shaped by institutional limitations, governance challenges, and uneven development patterns, and must therefore be analysed through a lens of trade-offs rather than idealised models. Bogotá reflects a partial socio-technical transition toward more sustainable mobility, yet one that remains incomplete and socially uneven, highlighting the limits of infrastructure-led solutions in the absence of sustained governance capacity (Chen et al. 2025). Lagos represents a system where institutional constraints sustain informality as the dominant mode, illustrating how accessibility can be maintained even in the absence of formalised systems, but at the cost of environmental and operational efficiency (Martins 2025).

Table 1 Global south case studies of integrated mobility

Dimension	Bogotá (Colombia)	Lagos (Nigeria)	Nairobi (Kenya)
Mobility model	Formal, public transport-oriented (BRT – TransMilenio)	Predominantly informal (danfos, minibuses)	Hybrid system (matatus + emerging formal/digital systems)
Development pathway	Public transport-led, planned urban mobility	Informal, demand-driven mobility system	Transitional, combining informal and formal elements
Economic efficiency	High capacity, reduced travel times, improved system efficiency	Low efficiency due to congestion and fragmented operations	Moderate efficiency, improved by partial regulation and digitalisation
Social inclusion (Accessibility)	Improved access but affordability concerns for low-income users	High accessibility due to flexible and affordable services	High accessibility, especially for low-income populations
Environmental sustainability	Reduced emissions relative to private vehicles but still constrained by scale and congestion	High emissions due to ageing vehicles and inefficient operations	Mixed environmental outcomes; informal systems contribute to emissions
Environmental externalities	Partially managed through structured system	High levels of congestion, air pollution, and unregulated emissions	Significant externalities, though some improvements through regulation
Governance & institutional capacity	Relatively strong governance and coordinated planning	Weak regulatory frameworks and enforcement capacity	Moderate governance; ongoing reforms and policy interventions
Socio-Technical transition	Attempted transition toward structured, low-emission transport system	Limited transition; persistence of informal regime	Hybrid transition with gradual integration of informal and formal systems
Key Strengths	Scalable, cost-effective public transport model; improved efficiency	Flexible, accessible, and demand-responsive mobility system	Adaptive system combining accessibility with emerging innovation
Key challenges	Overcrowding, affordability, long-term sustainability concerns	Environmental degradation, inefficiency, lack of regulation	Safety, emissions, regulatory inconsistencies
Overall Trade-off	Efficiency versus equity (affordability and capacity constraints)	Accessibility versus environmental sustainability	Accessibility versus efficiency and environmental performance

Nairobi, meanwhile, exemplifies a hybrid trajectory, where informal systems persist alongside emerging efforts at regulation, digitalisation, and integration, reflecting ongoing attempts to reconcile flexibility with system efficiency (Burger 2025). These cases collectively suggest that sustainable mobility cannot be achieved through the wholesale adoption of formalised systems alone, but requires context-sensitive strategies that integrate accessibility, governance reform, and environmental objectives. Ultimately, recent research argues that effective mobility policy must move beyond binary distinctions between formal and informal systems, focusing instead on how different models can be combined and adapted within specific institutional and socio-economic contexts to deliver more inclusive and sustainable outcomes (Caggiani and Prencipe 2025; Chen et al. 2025).

12 Policy Implications

Policy responses must balance environmental sustainability with social inclusion and economic development, recognising that mobility systems operate within complex and often competing priorities (Chikwere et al. 2026). Rather than treating these objectives as separate, effective policy design requires an integrated approach that explicitly acknowledges and manages trade-offs. In the context of the Global South, this is particularly challenging due to resource constraints, institutional limitations, and the need to deliver both immediate and long-term development outcomes.

A central policy priority is the promotion of low-carbon transport systems. This involves a strategic shift away from fossil fuel-dependent mobility toward more sustainable alternatives, including high-capacity public transport systems, electrified vehicle fleets, and non-motorised transport infrastructure (IPCC 2022). Investment in public transit such as Bus Rapid Transit (BRT), metro systems, and rail networks can significantly reduce emissions while improving accessibility. Electrification of transport, particularly when supported by cleaner energy sources, offers further potential to decarbonise mobility. However, from a critical perspective, these interventions must go beyond technological substitution. Without addressing underlying demand patterns and urban form, there is a risk that electrification may simply replace one form of high-consumption mobility with another, rather than fundamentally reducing environmental pressures.

Equally important is ensuring that sustainability policies do not exacerbate social inequalities. Policies aimed at reducing emissions such as fuel pricing, congestion charging, or vehicle restrictions can disproportionately affect low-income populations if not carefully designed (Litman 2023). As such, affordability and accessibility must be central considerations in policy development. This may involve targeted subsidies, differentiated pricing mechanisms, or investment in affordable public transport options. Adopting a just transition approach is critical to ensuring that environmental improvements do not come at the expense of social inclusion. In this regard, accessibility not simply mobility should serve as a guiding principle, ensuring that all population groups can access essential services and opportunities.

The integration of informal transport systems represents another key policy challenge and opportunity. In many Global South cities, informal systems such as minibuses, shared taxis, and motorcycle transport provide essential mobility services where formal systems are inadequate (Salon and Gulyani 2019). While these systems are often associated with inefficiencies and environmental concerns, their elimination is neither practical nor desirable. Instead, policy should focus on formalisation and integration, including regulatory frameworks, fleet modernisation, and the incorporation of digital technologies to improve efficiency and reduce emissions. This approach recognises the role of informal systems in supporting accessibility while seeking to enhance their environmental and operational performance.

Strengthening governance and institutional capacity is fundamental to the successful implementation of sustainable mobility policies. Effective transport planning requires coordination across multiple sectors, including transport, urban development, and environmental management. However, in many contexts, institutional fragmentation and limited technical capacity hinder integrated planning efforts. The development of metropolitan transport authorities improved regulatory frameworks, and enhanced inter-agency coordination can support more coherent and effective policy implementation (World Bank 2020). Furthermore, governance reforms must address issues of accountability and transparency to ensure that policy decisions reflect broader public interests rather than narrow institutional or political priorities.

In addition, the adoption of data-driven approaches offers significant potential to improve the efficiency and responsiveness of mobility systems. Advances in digital technologies enable real-time monitoring of transport networks, improved demand forecasting, and more effective resource allocation. Intelligent transport systems, integrated ticketing platforms, and mobility-as-a-service (MaaS) models can enhance user experience while optimising system performance. However, the benefits of digitalisation depend on appropriate regulatory frameworks and data governance structures to ensure equity, privacy, and accessibility (Litman 2023).

Finally, a critical insight is that policy interventions must be context specific. There is no universal model of sustainable mobility that can be directly transferred across different settings. Policies must be tailored to local socio-economic conditions, institutional capacities, and mobility needs. This requires moving beyond the replication of "best practice" models toward more adaptive and flexible approaches that reflect local realities. Sustainable mobility policy in the Global South requires a shift from fragmented and reactive interventions toward integrated, long-term strategies that align environmental sustainability, social equity, and economic development. Rather than attempting to eliminate trade-offs, policymakers must develop the capacity to navigate them strategically, ensuring that mobility systems contribute to inclusive and resilient development pathways.

13 Limitations of the Study

This chapter is based on a qualitative narrative literature review and therefore relies heavily on existing secondary sources rather than primary empirical data. As a result, the findings are interpretive and may be influenced by the scope and availability of the selected literature. In addition, the study focuses primarily on Global South contexts, which may limit the generalisability of the findings to developed economies (Snyder 2019). The chapter also adopts a broad conceptual perspective, meaning that some country-specific institutional and socio-political dynamics may not have been explored in sufficient depth.

14 Areas for Further Research

Future research should incorporate empirical case studies and primary data collection to provide deeper insights into how mobility trade-offs are experienced within specific urban and regional contexts (Martins 2025; Chen et al. 2025). Comparative studies across different Global South cities could further examine how governance structures, institutional capacity, and socio-economic conditions shape sustainable mobility transitions (Caggiani and Prencipe 2025). Additional research is also needed on the role of emerging technologies, digital mobility platforms, and electric transport systems in promoting inclusive and low-carbon mobility (Delaere et al. 2024). Furthermore, future studies should explore the long-term social justice implications of sustainable mobility policies, particularly their effects on vulnerable and low-income populations (Hidayati 2025; Litman 2023).

15 Conclusion

This chapter has demonstrated that mobility systems occupy a dual and often contradictory role within development processes, functioning simultaneously as enablers of economic growth and social inclusion while also acting as significant sources of environmental pressure. In addressing Objective 1, the analysis has shown that while mobility facilitates market integration, productivity, and access to essential services, its benefits are unevenly distributed, with accessibility constraints continuing to marginalise low-income and spatially disadvantaged populations. In relation to Objective 2, the chapter has highlighted that prevailing mobility systems particularly those characterised by rapid motorisation and infrastructure-led expansion generate substantial environmental trade-offs, including rising emissions, air pollution, and unsustainable land-use patterns. Crucially, these tensions are not incidental but structural, embedded within the dominant models of development and reinforced by institutional, economic, and behavioural dynamics. As such, they cannot be fully resolved but must be explicitly recognised and strategically managed. This requires moving beyond narrow, mobility-centric paradigms toward approaches that prioritise accessibility, efficiency, and equity, alongside a reorientation of policy frameworks toward integrated, long-term, and context-sensitive interventions. Ultimately, achieving sustainable mobility in the Global South depends on the capacity of policymakers to navigate these interdependencies, balancing competing objectives in ways that support inclusive development while remaining within ecological limits.

References

Adeniji K (2019) Transport challenges in Nigeria. Lagos Transport Review

Banister D (2008) The sustainable mobility paradigm. Transp Policy 15(2):73–80

Burger K (2025) Towards equitable, smart, and sustainable urban mobility: Governance archetypes and their relations. Transportation Research Part D.

Caggiani L, Prencipe LP (2025) Development Trends of Sustainable Mobility. Sustainability 17(14):6409. https://doi.org/10.3390/su17146409

Cervero R (2013) Bus rapid transit (BRT): An efficient and competitive mode of public transport. J Public Transp 16(1):1–30

Chen Y, Li C, Wang W, Zhang Y, Chen XM, Gao Z (2025) The landscape, trends, challenges, and opportunities of sustainable mobility and transport. npj Sustain Mobil Transp

Chikwere D, Munuhwa S, Chibaro M (2026) Transport and logistics in Africa: Policy and future direction. In: Munuhwa S (ed) The changing terrain of transport and logistics in Zimbabwe: Technology, sustainability and resilience management. Springer Nature Switzerland, pp 279–299

Creutzig F, Jochem P, Edelenbosch OY, Mattauch L, van Vuuren DP, McCollum D, Minx J (2015) Transport: A roadblock to climate change mitigation? Sci 350(6263):911–912

Datta S (2012) The impact of improved highways on Indian firms. J Dev Econ 99(1):46–57

Delaere H, Basu S, Macharis C, Keseru I (2024) Barriers and opportunities for developing, implementing and operating inclusive digital mobility services. Eur Transp Res Rev 16(1):67

Geels FW (2012) A socio-technical analysis of low-carbon transitions. Res Policy 41(6):991–1001

Hersh M (2026) Mobility, inclusion and exclusion. In Mobility of visually impaired people: Fundamentals and ICT assistive technologies (pp. 587–608). Springer Nature Switzerland

Hidayati I (2025) Assessing justice in sustainable mobility transitions. J Urban Aff

Intergovernmental Panel on Climate Change (2022) Climate change 2022: Mitigation of climate change. Cambridge University Press

Karimi H, Ghadirifaraz B, Shetab Boushehri SN, Hosseininasab SM, Rafiei N (2022) Reducing traffic congestion and increasing sustainability in special urban areas through one-way traffic reconfiguration. Transportation, 49(1), 37–60

Larrañaga AM, de Dios Ortúzar J, Oviedo D, Ruiz-Padillo A (2025) Active mobility, equity and inclusion. Latin American Transport Studies 3:100040

Litman T (2023) Evaluating transportation equity. Vic Transp Policy Inst

Martins JO (2025) Bridging policy and practice in sustainable urban mobility in African cities. Front Sustain Cities

Munuhwa S (Ed.) (2024) Contemporary solutions for sustainable transportation practices. IGI Global

Salon D, Gulyani S (2019) Mobility, poverty, and accessibility in African cities. World Bank Policy Res Work Pap

Snyder H (2019) Literature review as a research methodology: An overview and guidelines. J Bus Res 104:333–339

United Nations (2015) Transforming our world: The 2030 agenda for sustainable development
United Nations Environment Programme. (2021). Global environment outlook 6.
UN-Habitat (2021) Cities and climate change
World Bank (2020) Transforming transportation: The path to sustainable mobility.

Conceptual Frameworks for Green Transport Transition

Cleopatra Mtetwa and King Christopher Muchenje

Abstract Transport systems significantly contribute to greenhouse gas emissions, urban congestion, air pollution and resource depletion, particularly in rapidly urbanising Global South contexts. However, existing approaches to sustainable transport transitions often fail to address the interconnected technological, institutional, social and environmental dimensions of mobility systems. This chapter examines how Socio-Technical Transitions Theory, Systems Thinking, the Triple Bottom Line (TBL) and Circular Economy Theory collectively explain green transport transitions in developing regions. The study adopted a structured narrative review approach using peer-reviewed literature, policy reports and conceptual studies sourced from databases including Scopus, Web of Science, ScienceDirect, Emerald and Google Scholar. Thematic synthesis and critical conceptual analysis were used to identify key patterns and theoretical linkages shaping sustainable transport transitions. The analysis revealed that sustainable transport transitions are complex and adaptive processes shaped by interactions between technological innovation, governance systems, infrastructure conditions and social practices. Socio-technical transitions theory explains regime transformation and innovation dynamics, systems thinking highlights interdependencies and feedback loops, TBL evaluates environmental, economic and social sustainability outcomes, while circular economy theory emphasizes resource efficiency and lifecycle sustainability. The findings further show that successful green transport transitions in the Global South require context-sensitive policies that integrate informal transport systems, strengthen institutional coordination and support inclusive and low-carbon mobility pathways. The chapter concludes that integrating these theoretical perspectives provides a more comprehensive framework for understanding and guiding sustainable transport transitions in developing

C. Mtetwa (✉)
Fleet Management Department, Mega-Size College, Gaborone, Botswana
e-mail: cmtetwa@megasize.ac.bw

K. C. Muchenje
Faculty of Commerce, Department of Business Management, Ba Isago University, Gaborone, Botswana
e-mail: king.muchenje@baisago.ac.bw

J. Kanyepe and I. Chirisa (eds.), *Advancing the Green Agenda in the Transport and Logistics Sector in Global South Countries*, World Sustainability Series,
https://doi.org/10.1007/978-3-032-31374-4_4

regions. The study contributes both theoretically and practically by offering an integrated analytical framework and policy-relevant insights for promoting inclusive, resilient and sustainable mobility systems.

Keywords Green transport transitions · Sustainable mobility · Socio-technical transitions · Systems thinking · Triple bottom line · Circular economy · Global south

1 Introduction

The global transportation industry is central to social mobility, trade facilitation, and economic growth. It enables the movement of people and goods, supports regional integration, and enhances productivity. However, it is also a major contributor to environmental degradation and greenhouse gas (GHG) emissions due to its heavy reliance on fossil fuels. The transport sector accounts for a significant share of global carbon emissions, making it a key focus of climate mitigation efforts (Turan et al. 2024). Achieving targets under frameworks such as the Paris Agreement therefore requires a fundamental shift toward sustainable transport systems (Jelti et al. 2023; Ruggieri et al. 2021).

Transport decarbonisation is one of the most complex challenges facing developing economies. It sits at the intersection of reducing emissions and ensuring affordable, inclusive mobility for growing populations. The International Transport Forum and the United Nations Environment Programme estimate that transport accounts for about 23% of global energy-related CO_2 emissions, with freight emissions expected to rise significantly by 2050 under business-as-usual scenarios due to urbanisation and trade growth. Consequently, transport decarbonisation has become a central global policy priority. Despite increasing attention to sustainable mobility, implementation remains challenging, particularly in the Global South. Transport systems in these regions combine formal infrastructure with extensive informal transport networks and operate under financial, institutional, and infrastructural constraints. As a result, effective strategies must reflect local socio-economic realities rather than rely on models from industrialised contexts (Parnell and Pieterse 2023; Behrens et al. 2025).

Existing research often examines transport transitions in isolation, focusing on technology, policy, or infrastructure separately. However, sustainable transport transformation is inherently systemic, involving interactions between technology, governance, economic incentives, social practices, and environmental pressures. This has led to growing calls for integrated analytical frameworks that capture these interdependencies (Caldarola et al. 2023). To address this gap, this chapter develops an integrated framework combining Socio-Technical Transitions Theory, Systems Thinking, the Triple Bottom Line (TBL), and Circular Economy Theory. Socio-Technical Transitions Theory explains co-evolution between technology, institutions, and social practices. Systems Thinking highlights interdependencies and feedback loops within transport systems. Circular Economy Theory emphasises resource efficiency and closed-loop systems, while the TBL framework focuses on balancing environmental, social, and economic outcomes.

By integrating these perspectives, the chapter provides a comprehensive lens for analysing green transport transitions in the Global South. It enables a more holistic understanding of how technological, institutional, economic, and environmental factors interact in shaping sustainable mobility systems and contributes to sustainability transitions literature through a structured approach to complex transport change processes.

2 Background to the Study

The Global South, including Africa, Asia, Latin America, and the Caribbean, faces distinctive transport challenges that shape green transport transitions. Rapid urbanisation is increasing mobility demand, with the United Nations projecting that around 60% of the global population will live in urban areas by 2030, largely in developing regions. This growth is accompanied by rising motorisation, often outpacing infrastructure development as private vehicle ownership increases faster than public transport systems can adapt (Kumar and Barrett 2024). Transport systems in many Global South cities differ from those in high-income countries due to the prevalence of informal and shared mobility services, including minibus taxis, motorcycle taxis, auto-rickshaws, and shared cars. These services account for 50–80% of passenger transport in many sub-Saharan African cities and are essential for low-income mobility (Behrens et al. 2025; Diaz Olvera et al. 2023). While they address mobility gaps and provide livelihoods, they also raise environmental, safety, and governance concerns (Rizzo, 2024; Klopp and Cavoli 2023).

The coexistence of formal and informal systems complicates green transport transitions. Policies designed for highly regulated environments often fail where informal governance dominates. For example, electrifying motorcycle taxis requires not only technology and infrastructure but also affordable financing, adaptable charging systems, and regulations that accommodate existing mobility practices (Collett and Hirmer 2024; Ehebrecht et al. 2022). Environmental pressures are intensifying. Transport emissions are rising due to freight growth and expanding vehicle fleets, with freight transport in emerging economies increasing by 4–6% annually and vehicle ownership in some countries by 10–15% per year. Transport-related air pollution contributes to 4.2 million premature deaths annually, disproportionately affecting urban poor populations. Climate change further increases vulnerability through extreme weather impacts on infrastructure (IPCC 2023; World Bank 2025).

Despite these pressures, transport policy must also address development needs. Mobility is essential for access to jobs, education, and healthcare, yet low-income households often spend 15–25% of income on transport (Lucas 2023; Venter 2024). Informal transport also remains a key source of employment, particularly for youth in Africa and Southeast Asia (Rizzo 2024). Poorly designed green transition policies may therefore increase inequality or reduce access.

Recent initiatives, such as Uganda's Just Transition Framework, highlight the importance of balancing environmental goals with socio-economic impacts. The framework considers how reforms like electric motorcycles affect jobs, income, and mobility access, emphasising inclusive transition pathways (Newell and Mulvaney 2023; Sovacool et al. 2025).

Infrastructure constraints remain a major barrier. Limited and unreliable electricity supply restricts electric mobility adoption, particularly where users lack access to home or workplace charging (Collett and Hirmer 2024; Börjesson and Eliasson 2026). In addition, weak road networks and inefficient ports and border systems increase transport costs and reduce logistics performance (World Bank 2024; Raballand and Teravaninthorn 2023).

Institutional capacity constraints further limit implementation. Many transport agencies face weak coordination, limited expertise, and fragmented governance, while informal systems operate through associations rather than formal regulation (Meadowcroft 2024; Marsden and Reardon 2023; Behrens et al. 2025). These complexities highlight the need for integrated analytical approaches. Socio-technical transitions theory explains interactions between innovation and systems (Geels 2022), systems thinking captures interdependencies (Meadows 2022), the Triple Bottom Line balances sustainability dimensions (Elkington 2024), and circular economy theory emphasises resource efficiency (Kirchherr et al. 2023). However, each provides only a partial explanation when applied in isolation. Therefore, rather than relying on a single lens, this study adopts an integrated framework that combines these perspectives to reflect the complex realities of transport systems in the Global South (Dingil et al. 2021; Damer et al. 2026).

3 Research Objectives

This chapter was guided by the following research objectives:

1. To synthesise major theoretical perspectives on green transport transitions into an integrated analytical framework suitable for examining sustainable mobility transformations in Global South contexts.
2. To analyse the mechanisms through which technological innovation, institutional capacity, governance arrangements, policy coherence, and socio-economic conditions interact to facilitate or constrain green transport transitions in developing countries.
3. To assess the applicability of the integrated framework through comparative case study analysis of green transport initiatives implemented in emerging economies.
4. To develop context-sensitive policy recommendations that support green transport transitions while balancing environmental sustainability, social equity, and broader developmental priorities.

To provide a theoretical foundation for these objectives, this chapter adopts an integrated sustainability transitions perspective. It combines Socio-Technical Transitions Theory, Systems Thinking, the Triple Bottom Line (TBL), and Circular Economy Theory to capture the multidimensional nature of transport transformation. Socio-Technical Transitions Theory explains how technological innovation, policy, and social practices interact to transform transport systems over time (Geels 2022). Systems Thinking complements this by emphasising interconnections across infrastructure, governance, economic structures, and behaviour, highlighting feedback loops and system interdependencies (Meadows 2022). The Triple Bottom Line framework broadens the analysis by balancing environmental, social, and economic sustainability (Elkington 2024). It underscores the need to reduce emissions, improve energy efficiency, and ensure equitable access to mobility, particularly in developing regions where transport is closely tied to inclusion and economic participation.

Circular Economy Theory strengthens the framework by promoting resource efficiency, waste reduction, and extended product life cycles through recycling, remanufacturing, and sustainable logistics (Kirchherr et al. 2023). In transport systems, it supports reduced material use, improved vehicle efficiency, and more sustainable supply chains. Together, these perspectives provide a comprehensive lens for analysing green transport transitions. Such transitions aim not only to reduce environmental impacts but also to support inclusive and equitable mobility. Achieving sustainable transport systems therefore requires integrated approaches that balance environmental, economic, and social objectives (Sovacool et al. 2025).

4 Conceptual Model for Green Transport Transitions

A conceptual model is developed to synthesise four theoretical perspectives into a unified analytical framework for understanding green transport transitions in the Global South (Fig. 1).

The conceptual framework presents green transport transitions as a dynamic, multi-level process shaped by interactions between technological innovation, institutional governance, socio-economic conditions, and sustainability imperatives. Rather than treating theories separately, it integrates them into a unified structure explaining how transport systems evolve toward low-carbon, resource-efficient, and socially inclusive mobility. This is particularly relevant in Global South contexts characterised by infrastructure deficits, institutional constraints, informality, and uneven development (Sovacool et al. 2025; Coenen et al. 2023).

The framework is organised into four interrelated dimensions: transition drivers, system interactions, sustainability evaluation, and resource efficiency, derived from Socio-Technical Transitions Theory, Systems Thinking, the Triple Bottom Line (TBL), and Circular Economy Theory.

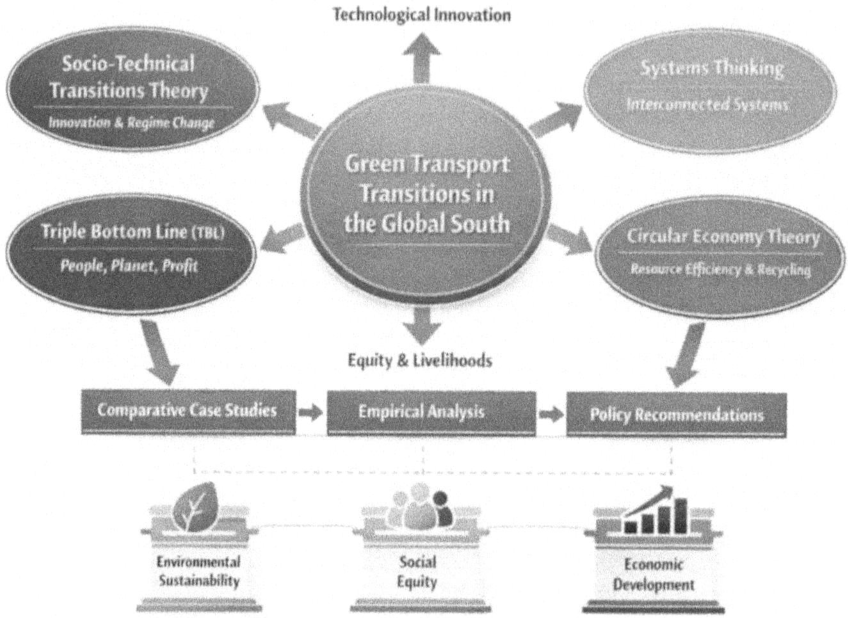

Fig. 1 Integrated conceptual framework for green transport transitions. (*Source* Synthesized by Authors)

5 Socio-Technical Transitions as the Driver of Change

Socio-Technical Transitions Theory explains how innovation and behavioural change emerge within transport systems. Technologies such as electric vehicles, shared mobility, and intelligent transport systems develop as niche innovations that may challenge existing regimes. However, their diffusion depends on institutions, infrastructure, markets, and user practices (Geels 2022; Kumar and Barrett 2024). This perspective highlights transition as an evolutionary and contested process shaped by both innovation and regime resistance.

6 Systems Thinking as the Integrative Mechanism

Systems Thinking frames transport as a complex adaptive system with interdependent elements, including infrastructure, governance, energy systems, markets, and user behaviour (Meadows 2022). Innovations generate feedback loops and system-wide effects; for example, electric mobility increases electricity demand and reshapes regulation and urban planning. These interactions may support or undermine sustainability outcomes, including equity and infrastructure capacity.

7 Triple Bottom Line as the Evaluation Layer

The Triple Bottom Line evaluates whether transitions achieve environmental, economic, and social goals (Elkington 2024; Ozanne et al. 2023). Environmental sustainability focuses on emissions reduction, economic viability on affordability and investment, and social inclusion on equitable access. Trade-offs are common; for example, electric mobility may reduce emissions but raise affordability challenges. TBL therefore clarifies synergies and tensions in policy design.

8 Circular Economy as the Resource-Efficiency Layer

Circular Economy Theory promotes recycling, remanufacturing, shared use, and sustainable logistics to reduce material consumption and waste (Kirchherr et al. 2023; Coenen et al. 2023). It replaces linear production models with closed-loop systems that extend product lifecycles and improve resource efficiency. In transport systems, it supports decarbonisation while reducing material demand and improving sustainability over time.

9 Contribution of the Framework

The framework contributes by integrating technological, institutional, economic, and sustainability dimensions into a single analytical model. It combines transition processes with sustainability evaluation and applies a context-sensitive lens suited to Global South conditions. It also identifies leverage points for inclusive and low-carbon transport transitions.

10 Research Methodology

This study adopts a structured narrative review to develop an integrated conceptual framework for green transport transitions. The approach is appropriate for theory-building, focusing on synthesising existing literature rather than collecting primary data. Literature was sourced from Google Scholar, Scopus, Web of Science, Emerald Insight, and ScienceDirect to ensure interdisciplinary coverage across transport, sustainability, governance, and development studies.

Search terms included green transport transitions, sustainable mobility, socio-technical transitions, systems thinking, Triple Bottom Line, circular economy, green logistics, and Global South transport systems. Literature published between 2018 and 2026 was prioritised, alongside key foundational works. Inclusion criteria comprised peer-reviewed articles, academic books, policy reports, and studies addressing transport sustainability and relevant theoretical frameworks, particularly in Global South contexts. Exclusion criteria included non-peer-reviewed sources, unrelated technical studies, duplicate records, non-English publications, and narrowly technical engineering studies lacking policy relevance.

The literature was analysed using thematic synthesis, identifying patterns across four themes: technological innovation, governance and institutions, sustainability outcomes, and circular economy mechanisms. Comparative analysis revealed convergences, tensions, and gaps, informing the development of the conceptual framework. The study is conceptual and does not involve empirical data collection or hypothesis testing. It instead generates insights through critical synthesis of existing literature. Limitations include potential language bias, publication bias, and the lack of empirical validation. Additionally, contextual differences across Global South regions may limit generalisability. Despite these limitations, the structured review enhances transparency and coherence, providing a strong foundation for the proposed framework and supporting policy-relevant insights for sustainable transport transitions.

11 Literature Review

11.1 Socio-Technical Transitions Theory

Socio-technical transitions theory emerged in the late twentieth century from evolutionary economics, institutional theory, and science and technology studies. It explains how large-scale systems transform through interactions among technological innovation, institutions, governance, user practices, and cultural values (Geels 2022). In this study, socio-technical transitions are defined as long-term co-evolutionary processes in which technologies, policies, institutions, and social practices interact to shift transport systems toward sustainable mobility.

Unlike technology-centred approaches, the theory emphasises that innovation is embedded in political, economic, and social contexts. Transport systems are central to transition research due to their strong institutional embedding in infrastructure, urban planning, and everyday mobility practices (Schwanen 2023). Existing regimes are shaped by path dependency and lock-in effects, including automobile-oriented planning, long infrastructure lifecycles, fossil fuel dependence, and incumbent industries such as automotive and road construction (Geels and Turnheim 2023). These factors slow systemic change even under environmental pressure. Urban mobility transitions are therefore highly political, involving competing interests between incumbent actors and sustainability coalitions advocating low-carbon alternatives such as public transport, cycling, walking, and shared mobility (Marsden and Reardon 2023).

Although the theory provides a strong analytical foundation, it has been criticised for Eurocentric assumptions and limited attention to inequality, informality, and governance complexity (Avelino 2023). Much of the early work reflects high-income contexts with formalised transport systems, limiting its direct applicability in the Global South, where transport is shaped by infrastructure deficits, fragmented governance, and informality. In these contexts, transport systems combine formal and informal modes such as minibus taxis, motorcycle taxis, and shared transport (Behrens et al. 2025). Rather than system failure, informality represents an adaptive response to structural constraints (Rizzo 2024). Examples include matatus in Kenya and boda-bodas in Uganda, which provide essential mobility and livelihoods through decentralised governance arrangements.

The interaction between formal and informal systems is central to green transport transitions. Evidence from Bus Rapid Transit (BRT) in Lagos shows that informal operators were partially integrated rather than displaced, reflecting adaptation, negotiation, and coexistence rather than linear replacement. Recent scholarship also emphasises justice and equity, as green transport policies may create distributional impacts affecting affordability, access, and livelihoods (Sovacool et al. 2025). This has strengthened "just transition" perspectives that prioritise inclusive and equitable mobility outcomes. Overall, socio-technical transitions theory is most effective when integrated with complementary frameworks addressing institutional complexity, equity, and resource constraints.

11.2 Systems Thinking

Systems thinking is an analytical approach that examines relationships, feedback loops, and interdependencies within complex systems rather than isolated components (Meadows 2022). In this study, it conceptualises transport as a complex adaptive system made up of interconnected infrastructural, institutional, economic, environmental, and behavioural elements.

Transport systems are embedded within wider urban and economic systems. Roads, rail, ports, energy networks, and digital systems interact dynamically rather than functioning independently (Dingil et al. 2021). This perspective allows for a holistic understanding of how transport outcomes emerge from multiple interacting subsystems. Systems thinking is particularly useful for understanding unintended consequences. For example, road expansion may initially reduce congestion but later generate induced demand that restores traffic levels. Similarly, public transport investments may improve accessibility but also contribute to land value increases and displacement effects. These outcomes demonstrate that transport interventions produce long-term system-wide effects rather than isolated impacts.

In developing contexts, systems interactions are especially complex due to infrastructure constraints and institutional fragmentation. For instance, electricity shortages in South Africa limit electric mobility adoption, while port inefficiencies in West Africa affect regional trade flows. These examples show that transport sustainability cannot be separated from broader infrastructure systems. Systems thinking also helps explain interactions between formal and informal transport. Informal systems respond dynamically to fuel prices, road conditions, enforcement practices, and demand fluctuations. At the same time, they influence congestion patterns and accessibility outcomes, creating continuous feedback loops within urban mobility systems.

A further contribution of systems thinking is its emphasis on resilience. In many developing regions, transport resilience depends not only on infrastructure robustness but also on institutional flexibility and informal adaptation. During disruptions such as fuel shortages, informal transport systems often provide critical continuity in mobility services. However, systems thinking has limitations. It may underemphasise political power, inequality, and institutional conflicts (Meadowcroft 2024). As a result, it is most effective when combined with governance-focused and equity-oriented frameworks.

11.3 Triple Bottom Line

The Triple Bottom Line (TBL) framework was developed by Elkington (1994) to expand sustainability assessment beyond financial performance. In this study, it is defined as a framework that evaluates transport systems across three dimensions: environmental sustainability, economic viability, and social equity (Elkington 2024). TBL has become widely used in transport planning because it broadens traditional evaluation methods that focus mainly on cost and efficiency. Environmental indicators include emissions and pollution, economic indicators include affordability and investment efficiency, and social indicators include accessibility, safety, and inclusion (Damer et al. 2026).

A key strength of TBL is its ability to identify trade-offs. For example, electric vehicles reduce emissions but may remain unaffordable for low-income users. Similarly, public transport systems may improve equity and environmental outcomes but require significant public subsidies. TBL therefore supports balanced decision-making rather than single-objective optimisation. However, the framework has limitations. In practice, economic objectives often dominate policy decisions, while environmental and social goals may be secondary (Elkington 2024). It may also oversimplify sustainability by reducing complex social issues into measurable indicators.

In Global South contexts, TBL must be adapted to reflect mobility realities shaped by informal transport dependence and unequal access to services (Lucas 2023; Venter 2024). Informal systems such as matatus and motorcycle taxis play essential economic and mobility roles and must be considered in sustainability assessments. TBL therefore provides a useful evaluative tool but requires integration with institutional and systems-based frameworks to fully capture transport sustainability dynamics.

11.4 Circular Economy Theory

Circular economy theory emerged in response to the environmental limitations of linear production systems characterised by "take–make–dispose" models (Kirchherr et al. 2023). It is defined as a regenerative system that minimises waste and extends product lifecycles through reuse, repair, refurbishment, remanufacturing, and recycling. Within transport systems, circular economy principles are increasingly relevant due to high material consumption and environmental impacts associated with vehicle production, infrastructure development, and energy use (Di Ruocco 2025). The approach draws on industrial ecology, sustainable resource management, and ecological economics.

A key application is vehicle lifecycle management. Modular design, recyclable materials, and remanufacturing processes extend vehicle lifespans and reduce waste. Electric mobility further highlights circularity challenges, particularly regarding battery production and critical mineral use. Battery recycling and second-life applications, such as energy storage systems, are therefore central to sustainable electric mobility transitions. However, circular economy implementation faces governance and infrastructure constraints, especially in developing regions. Limited recycling infrastructure and weak regulatory systems restrict full-scale circularity. In some cases, informal recycling systems operate under unsafe conditions, raising environmental justice concerns (Kirchherr et al. 2023).

At the same time, informal actors play an important role in resource recovery. Waste pickers and informal recyclers contribute significantly to material reuse in countries such as India, Brazil, and South Africa (Di Ruocco 2025). This suggests that circular economy strategies in the Global South should integrate rather than exclude informal systems. Circular economy approaches also enhance supply chain resilience by reducing dependence on imported materials and promoting local resource recovery. In transport systems, this supports domestic refurbishment industries and employment creation. Despite its benefits, circular economy theory must be applied cautiously, as it may overlook governance inequalities and labour conditions if used in isolation. It is therefore most effective when combined with socio-technical, systems, and equity-based frameworks.

11.5 Results and Discussion

There is substantial evidence in the literature that sustainable transportation systems are designed to meet present mobility needs without compromising the ability of future generations to meet their own needs (Glazener and Khreis 2019; Gudmundsson et al. 2016). In this study, sustainable transport refers to transport systems that promote environmentally responsible, economically viable, and socially inclusive mobility. Such systems support low-emission mobility solutions, including electric vehicles, public transportation, active mobility, and energy-efficient infrastructure.

Sustainable transport also incorporates innovations such as intelligent traffic management systems, integrated mobility planning, and digital transport technologies. Importantly, these systems seek to ensure accessibility, affordability, safety, and mobility equity for all citizens regardless of income level, age, gender, or physical ability (Gudmundsson et al. 2016; Kanyepe et al. 2022; Mehar et al. 2014).

Despite these potential benefits, the transition toward sustainable transport remains complex and contested. Existing transport systems are deeply embedded within economic structures, urban development patterns, governance arrangements, and everyday mobility practices. As a result, sustainability transitions often encounter resistance arising from institutional inertia, infrastructural lock-ins, fragmented governance, and socio-economic inequalities. The literature identifies several barriers to sustainable mobility transitions, including the absence of universally accepted definitions of sustainable mobility, continued dependence on private vehicles, limited institutional coordination, inadequate financing mechanisms, and pressures associated with rapid urbanisation and population growth (Gudmundsson et al. 2016; Kennedy et al. 2005; Litman and Burwell 2006). In many Global South cities, these challenges are intensified by weak infrastructure, high levels of transport informality, and limited state capacity.

Although technological innovation is frequently presented as a solution to transport sustainability challenges, the literature increasingly argues that technological change alone cannot guarantee sustainable outcomes. Developments such as electric vehicles, autonomous mobility systems, and intelligent transport infrastructure may improve efficiency and reduce environmental impacts, yet their effectiveness depends on governance arrangements, infrastructure availability, affordability, and public acceptance (Bleviss 2021). For example, electric mobility transitions in South Africa are constrained by electricity shortages and unstable energy supply systems, demonstrating that technological transitions are shaped by broader infrastructural and institutional conditions.

Similarly, investments in modern public transport systems in cities such as Lagos and Nairobi have had to coexist with dominant informal transport networks rather than replace them entirely. These examples suggest that sustainable transport transitions are multidimensional processes requiring institutional adaptation, stakeholder collaboration, and context-sensitive policy design. Against this background, the present analysis examines how four conceptual frameworks- Socio-Technical Transitions Theory, Systems Thinking, the Triple Bottom Line (TBL), and Circular Economy Theory contribute to understanding green transport transitions in Global South environments. Rather than treating these theories as separate explanatory approaches, the analysis synthesises them to demonstrate how they collectively explain technological transformation, institutional complexity, sustainability trade-offs, and resource-efficiency strategies within transport systems.

11.6 Socio-Technical Transitions Theory

Socio-technical transitions theory provides an important analytical lens for understanding how transport systems evolve through interactions among technological innovation, governance systems, institutional structures, and social practices (Geels 2022). The theory recognises that transport transitions are not purely technical processes but also involve broader changes in behaviour, regulation, infrastructure, and power relations. However, transport systems in many developing regions differ significantly from the highly formalised systems on which much of the original transitions literature was based.

In many Global South cities, formal and informal transport systems coexist and interact (Behrens et al. 2025). Informal mobility systems such as matatus in Kenya, danfo minibuses in Nigeria, boda-boda motorcycle taxis in Uganda, and auto-rickshaws in India provide essential mobility services for millions of urban residents. These systems operate through distinct institutional arrangements and social practices, creating strong path dependencies within urban mobility systems (Rizzo 2024). Rather than representing temporary or marginal transport modes, informal systems often function as dominant transport regimes that compensate for the limitations of formal public transport.

The analysis therefore suggests that socio-technical transitions theory must be adapted to reflect the realities of Global South mobility systems. Early transitions research largely assumed that sustainability transitions occur through formal regulatory systems and technologically advanced infrastructure. However, evidence from cities such as Lagos demonstrates that transition pathways are often negotiated through hybrid governance arrangements involving both formal authorities and informal operators. For instance, the implementation of Bus Rapid Transit (BRT) systems in Lagos initially sought to replace informal danfo operators, yet policymakers ultimately had to incorporate informal operators into the evolving mobility structure. This case illustrates that sustainable transport transitions in developing contexts are more likely to succeed through collaboration and institutional integration than through attempts to eliminate informality.

The literature also identifies important limitations within socio-technical transitions theory. Critics argue that the theory has historically focused heavily on technological innovation while giving insufficient attention to issues of social justice, inequality, and political power (Avelino 2023; Sovacool et al. 2025). Policies such as congestion pricing, electric vehicle subsidies, and urban redevelopment initiatives may unintentionally disadvantage lower-income populations if social equity considerations are not integrated into transition planning. Consequently, recent scholarship increasingly incorporates just transition perspectives that emphasise equitable mobility access, livelihood protection, and inclusive governance.

11.7 Systems Thinking

Systems thinking conceptualises transport as a complex adaptive system composed of interconnected infrastructural, institutional, environmental, and socio-economic components (Meadows 2022). Unlike linear planning approaches that analyse transport problems in isolation, systems thinking emphasises interdependencies, feedback loops, and unintended consequences within mobility systems. This perspective is particularly relevant in Global South contexts where transport systems are closely linked to broader infrastructure limitations, governance challenges, and development pressures.

The findings demonstrate that sustainable mobility transitions depend not only on transport technologies but also on the functioning of interconnected systems such as electricity supply, digital infrastructure, logistics networks, and urban planning. Electric mobility initiatives, for example, require stable electricity infrastructure, accessible charging systems, and effective maintenance networks. In South Africa, electricity shortages and load-shedding have constrained electric vehicle adoption despite strong policy commitments toward low-carbon transport. Similarly, poor road conditions in many African countries increase vehicle operating costs, fuel consumption, and logistics inefficiencies, illustrating how infrastructure weaknesses in one sector can undermine the sustainability of the broader transport system.

Systems thinking also highlights how policy interventions may generate unintended social and economic consequences. Investments in formal transport infrastructure, such as BRT systems, can alter passenger flows and disrupt the livelihoods of informal transport operators. Likewise, expanding road capacity may initially reduce congestion but often encourages additional private vehicle use over time, ultimately reproducing congestion through induced demand. These examples demonstrate the importance of understanding transport systems holistically rather than evaluating interventions through narrow technical indicators alone.

Another important contribution of systems thinking lies in its ability to analyse interactions between formal and informal mobility systems. Informal transport networks operate as adaptive systems influenced by infrastructure quality, fuel prices, policing practices, urban spatial patterns, and passenger demand. Changes in transport regulation or infrastructure can therefore create ripple effects throughout the wider mobility ecosystem. Systems thinking enables policymakers to identify leverage points such as institutional coordination, stakeholder collaboration, integrated planning, and improved data systems that can strengthen transport resilience and sustainability outcomes.

However, the literature also identifies limitations associated with systems thinking approaches. Critics argue that systems models sometimes prioritise technical optimisation while underestimating political conflict, institutional inequalities, and power relations that shape transport governance (Meadowcroft 2024). In resource-constrained settings, limited financial capacity, inadequate data systems, and fragmented institutions may further restrict the implementation of systems-based planning models. Consequently, systems thinking is most effective when combined with approaches that explicitly address governance complexity, social equity, and institutional adaptation.

11.8 Triple Bottom Line (TBL)

The Triple Bottom Line framework evaluates sustainability according to three interconnected dimensions: environmental sustainability, economic viability, and social equity (Elkington 2024). Within transport studies, the framework provides a multi-dimensional approach for assessing whether mobility systems contribute to broader sustainable development objectives. Traditional transport planning has often prioritised economic efficiency through indicators such as travel time savings, infrastructure costs, and logistics performance. The TBL framework broadens this perspective by recognising that transport systems also generate significant environmental and social impacts.

The environmental dimension of transport sustainability is increasingly important in developing countries where expanding vehicle fleets and freight demand contribute to rising greenhouse gas emissions and urban air pollution (International Energy Agency 2025). Many Global South cities experience severe air quality problems linked to ageing vehicle fleets, traffic congestion, and inefficient fuel systems. In cities such as Delhi and Lagos, transport-related emissions have become major public health concerns, disproportionately affecting low-income populations living near congested transport corridors. Consequently, transport sustainability policies increasingly emphasise cleaner fuels, public transportation systems, non-motorised mobility, and low-emission technologies.

The social dimension of the TBL framework highlights the relationship between transport and social inclusion. Access to affordable and reliable mobility influences access to employment, healthcare, education, and economic opportunities (Lucas, 2023; Venter 2024). In many developing cities, low-income households spend substantial portions of their income on transportation while also facing long travel times and inadequate mobility infrastructure. Informal transport systems often fill these accessibility gaps by providing flexible and affordable mobility services in underserved areas. However, sustainability policies that seek to formalise transport systems without protecting informal-sector livelihoods may unintentionally increase unemployment and social exclusion.

The economic dimension focuses on transport efficiency, logistics performance, and the contribution of mobility systems to economic development. Efficient transport infrastructure supports trade, investment, supply chain reliability, and regional integration, whereas poor infrastructure and regulatory inefficiencies increase operational costs and reduce economic competitiveness (World Bank 2024). Nevertheless, tensions frequently emerge between environmental goals, social equity, and economic efficiency. For example, stricter emission regulations may increase vehicle costs for operators and consumers, while preserving informal transport employment may slow transitions toward cleaner technologies.

The findings therefore suggest that sustainable transport policies in the Global South require context-sensitive approaches capable of balancing environmental protection, economic growth, and social inclusion simultaneously. The TBL framework is valuable because it exposes the trade-offs associated with sustainability transitions and highlights the importance of inclusive policy design. However, critics argue that TBL approaches sometimes assume that sustainability dimensions can be equally achieved despite the political and institutional tensions shaping real-world transport governance. The framework is therefore strengthened when combined with systems thinking and socio-technical transitions perspectives.

11.9 Circular Economy Theory

Circular economy theory emphasises resource efficiency, waste reduction, reuse, recycling, and regenerative production systems as alternatives to conventional linear economic models based on extraction, consumption, and disposal (Kirchherr et al. 2023). Within transport systems, circular economy approaches focus on extending vehicle lifecycles, recovering valuable materials, reducing waste generation, and improving resource productivity throughout transport supply chains.

The transition toward electric mobility has increased the relevance of circular economy strategies because electric vehicle batteries contain critical minerals such as lithium, cobalt, and nickel. Battery recycling, remanufacturing, refurbishment, and second-life applications are increasingly recognised as essential components of sustainable transport systems (Di Ruocco 2025). Used electric vehicle batteries, for example, can be repurposed for stationary energy storage systems that support renewable energy integration and electricity grid stability. These approaches reduce dependence on primary resource extraction while improving resource efficiency and supply chain resilience.

The literature also highlights important links between circular economy practices and supply chain resilience. Circular supply chains reduce dependence on imported raw materials by promoting local recovery, recycling, and remanufacturing processes. During global disruptions such as the COVID-19 pandemic, organisations with diversified and circular supply systems demonstrated greater operational flexibility and resilience. This suggests that circular economy strategies contribute not only to environmental sustainability but also to economic stability and industrial resilience.

In Global South contexts, however, circular economy implementation is strongly shaped by informal-sector dynamics and infrastructure limitations. Waste collection and recycling systems in many developing countries rely heavily on informal waste pickers, scrap dealers, repair technicians, and small-scale recyclers. In countries such as Ghana, India, and South Africa, informal recycling networks recover substantial quantities of reusable materials while creating employment opportunities for low-income populations. However, these activities often occur under hazardous working conditions with limited social protection and weak environmental regulation.

The findings therefore suggest that circular economy strategies in developing regions should not seek to eliminate informal systems but rather integrate them into more formalised and socially protected resource recovery structures. This requires supportive governance systems, investment in recycling infrastructure, and policies that recognise informal workers as legitimate stakeholders within circular transport systems. At the same time, infrastructure deficits in many developing countries may limit the immediate implementation of advanced circular economy technologies. Gradual and context-sensitive approaches are therefore necessary.

11.10 Synthesis of Frameworks

The synthesis of the four conceptual frameworks demonstrates that no single theory adequately captures the complexity of green transport transitions in the Global South. Socio-Technical Transitions Theory explains how technological innovation, institutional change, and mobility practices interact to shape transport transformation. Systems Thinking highlights interdependencies and systemic interactions across transport, infrastructure, governance, and socio-economic systems. The Triple Bottom Line framework evaluates environmental, social, and economic sustainability outcomes, while Circular Economy Theory contributes a resource-efficiency and lifecycle sustainability perspective.

Importantly, these frameworks are complementary rather than competing approaches. Socio-technical transitions theory explains how change emerges, systems thinking explains how change spreads across interconnected systems, the TBL framework evaluates whether transition outcomes are sustainable and socially inclusive, and circular economy theory ensures that transitions remain resource-efficient and resilient. Integrating these perspectives therefore produces a more comprehensive analytical framework capable of addressing technological innovation, institutional complexity, sustainability trade-offs, resource management, and social justice simultaneously.

This integrated perspective is particularly valuable in Global South contexts characterised by transport informality, institutional fragmentation, infrastructure deficits, and rapid urbanisation. Sustainable transport transitions in such settings cannot rely solely on technological innovation or infrastructure investment. Instead, they require adaptive governance systems, inclusive planning approaches, stakeholder collaboration, and context-sensitive policy interventions that recognise the interconnected nature of environmental sustainability, social equity, and economic development.

11.11 Further Research

Although this chapter contributes to understanding conceptual frameworks for green transport transitions in Global South contexts, several areas require further investigation. Future research should examine the role of informal transport systems in green mobility transitions. Many cities in Africa, Asia, and Latin America rely heavily on informal transport services such as minibus taxis, motorbike taxis, and auto-rickshaws. Despite their importance in providing affordable mobility and employment opportunities, these systems remain underrepresented in sustainable transport research. Further studies should therefore explore how informal transport systems can be integrated into formal transport planning frameworks while promoting environmental sustainability, operational efficiency, and social equity.

In addition, future research should investigate governance models that support collaboration between formal authorities and informal transport operators. Greater attention is also needed on how policy interventions can improve service quality, safety, and environmental performance without undermining the livelihoods of those who depend on the informal transport sector. Such research would contribute to more inclusive and context-sensitive approaches to sustainable transport transitions in developing regions.

11.12 Managerial and Policy Implications

The findings of this study generate several important policy implications for policymakers, transport planners, infrastructure managers, and development agencies seeking to advance sustainable and inclusive transport systems in Global South contexts. These implications emerge directly from the integrated conceptual framework developed in this study, which combines Socio-Technical Transitions Theory, Systems Thinking, the Triple Bottom Line (TBL), and Circular Economy Theory. Together, these perspectives explain how technological innovation, institutional dynamics, sustainability objectives, and resource-efficiency strategies interact within green transport transitions. Consequently, the policy recommendations are presented as interconnected strategies rather than isolated interventions.

11.13 Adopt an Integrated and Multi-Level Transport Planning Approaches

The integrated framework demonstrates that transport transitions are multidimensional processes shaped by interactions among infrastructure systems, governance structures, technological innovation, and socio-economic conditions. Drawing particularly from Systems Thinking and the Triple Bottom Line framework, the findings suggest that policymakers should move beyond fragmented and sector-specific planning approaches toward integrated transport systems that simultaneously address environmental sustainability, social inclusion, and economic development.

In practice, this requires coordination across transport, land use, energy, housing, and digital infrastructure sectors. Investments in public transport, non-motorised mobility, and transit-oriented urban development should therefore align with broader urban planning strategies aimed at reducing congestion, emissions, and spatial inequality. The framework also highlights that transport interventions often generate ripple effects across interconnected systems. For example, electric mobility transitions depend not only on vehicle technology but also on electricity reliability, charging infrastructure, and institutional coordination. Policymakers should therefore adopt multi-level governance approaches that encourage collaboration among national governments, municipalities, transport authorities, energy providers, and private-sector actors.

The Triple Bottom Line perspective further suggests that transport policies should be evaluated according to their environmental, social, and economic impacts simultaneously. Policies focused exclusively on emission reduction may unintentionally undermine affordability or accessibility if equity considerations are overlooked. Integrated planning frameworks can therefore help policymakers identify synergies and manage trade-offs between sustainability objectives.

11.14 Recognise and Integrate Informal Transport Systems

One of the key insights emerging from the integration of Socio-Technical Transitions Theory and Systems Thinking is that informal transport systems are essential components of mobility networks in many Global South cities. Operators such as minibus taxis, motorcycle taxis, and auto-rickshaw services are deeply embedded within urban transport systems and play an important role in providing affordable and accessible mobility.

The findings therefore suggest that sustainable transport transitions should not seek to eliminate informal systems through purely technocratic modernisation strategies. Instead, policymakers should develop inclusive governance mechanisms that integrate informal operators into broader transport planning processes. Experiences from cities such as Lagos, where informal danfo operators were partially incorporated into Bus Rapid Transit systems, demonstrate that collaborative governance arrangements are often more effective than exclusionary approaches.

This implication reflects the framework's emphasis on institutional adaptation and inclusive transition pathways. Informal transport operators possess extensive knowledge of local mobility patterns and service demands, particularly in underserved communities. Formalising partnerships through licensing reforms, route integration, skills development programmes, and social protection measures can improve service quality while protecting livelihoods. Such approaches also support just transition principles by ensuring that sustainability policies do not disproportionately disadvantage vulnerable populations.

11.15 Promote Technological Innovation Through Context-Sensitive Approaches

The conceptual framework demonstrates that technological innovation alone is insufficient to achieve sustainable transport transitions. Socio-Technical Transitions Theory highlights that technological change must be supported by institutional reform, behavioural adaptation, infrastructure investment, and governance capacity. Similarly, Systems Thinking emphasises that technologies operate within interconnected systems shaped by infrastructure quality, energy availability, financing structures, and social acceptance.

Policymakers should therefore promote green transport technologies such as electric vehicles, renewable energy integration, intelligent transport systems, and smart mobility platforms through coordinated and context-sensitive strategies. Incentives including subsidies, concessional financing, public–private partnerships, and tax reforms may accelerate adoption; however, these measures must align with local infrastructural realities and institutional capacity. For example, electric vehicle adoption in many African countries remains constrained by unreliable electricity systems and limited charging infrastructure. Investments in electric mobility should therefore be integrated with broader energy-sector reforms and renewable energy expansion. Likewise, intelligent transport systems require digital infrastructure, reliable data systems, and institutional capacity for implementation and monitoring. The framework therefore suggests that policymakers should avoid isolated technology-driven interventions and instead support broader systemic innovation strategies.

11.16 Strengthen Institutional Capacity, Coordination, and Data Systems

The Systems Thinking dimension of the framework highlights that transport sustainability depends heavily on institutional coordination, information flows, and adaptive governance capacity. Fragmented institutional arrangements, weak regulatory enforcement, and limited technical expertise often undermine sustainable transport planning in developing contexts. The findings therefore suggest that strengthening institutional capacity should be treated as a central component of green transport transitions rather than a secondary administrative issue. Governments should invest in integrated transport authorities, inter-agency coordination mechanisms, and professional capacity development to improve policy coherence and implementation effectiveness.

The framework also demonstrates the importance of transport data systems for evidence-based decision-making. Investments in geographic information systems (GIS), predictive modelling, transport monitoring platforms, and digital mobility databases can help policymakers identify systemic risks, anticipate unintended consequences, and evaluate sustainability outcomes. Improved data systems are particularly important for understanding informal mobility patterns, which are often poorly captured within conventional transport planning frameworks.

Stronger institutional capacity also enhances resilience by improving governments' ability to respond to disruptions, adapt policies over time, and coordinate infrastructure investments across interconnected sectors. This reflects the framework's emphasis on adaptive and iterative transition processes rather than fixed policy pathways.

11.17 Embed Circular Economy Principles Within Transport Policy

The Circular Economy component of the framework highlights the importance of resource efficiency, lifecycle sustainability, and supply chain resilience within transport transitions. The findings suggest that policymakers should integrate circular economy principles into transport planning through strategies that promote vehicle reuse, remanufacturing, recycling, repair systems, and sustainable material recovery. Particular attention should be given to electric vehicle battery management because battery technologies depend on critical minerals such as lithium, cobalt, and nickel. Policies supporting battery recycling infrastructure, second-life battery applications, and reverse logistics systems can reduce environmental pressures while improving resource security and supply chain resilience.

In many developing regions, circular economy strategies should also recognise the contribution of informal recycling networks and waste recovery systems. Informal waste pickers, repair technicians, and small-scale recyclers already play an important role in material recovery processes. Policymakers should therefore integrate these actors into formal sustainability initiatives through supportive regulation, skills training, occupational protection, and infrastructure investment.

This approach reflects the framework's emphasis on inclusive sustainability transitions. Circular economy policies that ignore informal systems may undermine livelihoods and reduce policy effectiveness, whereas context-sensitive strategies can simultaneously promote environmental sustainability, employment creation, and local economic resilience.

11.18 Develop Gradual and Adaptive Transition Pathways

A central contribution of the integrated conceptual framework is its recognition that transport transitions are dynamic, nonlinear, and context-dependent processes. Socio-Technical Transitions Theory explains that transformation occurs incrementally through interactions among innovations, institutions, and social practices, while Systems Thinking highlights feedback loops, unintended consequences, and adaptive responses within complex transport systems.

The findings therefore suggest that policymakers should avoid rapid or universal policy transfers based on models developed in high-income countries. Instead, transition strategies should be gradual, flexible, and responsive to local socio-economic and institutional realities. Phased vehicle electrification programmes, incremental public transport expansion, adaptive regulatory reforms, and pilot-based experimentation may provide more feasible and socially sustainable pathways in resource-constrained environments. This adaptive approach is particularly important in cities characterised by rapid urbanisation, infrastructure deficits, institutional fragmentation, and large informal economies. Sustainable transport transitions in such contexts require iterative learning processes, continuous stakeholder engagement, and policy flexibility capable of responding to emerging challenges and unintended outcomes.

Overall, the integrated conceptual framework developed in this study demonstrates that green transport transitions cannot be understood or governed through isolated technological, environmental, or economic interventions. Effective policy responses require multidimensional strategies that address institutional dynamics, systemic interdependencies, sustainability trade-offs, and resource-efficiency objectives simultaneously. By linking technological innovation with social inclusion, governance reform, and circular resource management, the framework provides a comprehensive basis for designing equitable and context-sensitive transport policies in the Global South.

12 Conclusion

This chapter explored conceptual frameworks for green transport transitions in Global South contexts by synthesising Socio-Technical Transitions Theory, Systems Thinking, the Triple Bottom Line (TBL), and Circular Economy Theory. The analysis shows that sustainable transport requires more than technological innovation; it also demands coordinated change across governance, institutions, infrastructure, and social practices. Each framework offers a distinct perspective: Socio-Technical Transitions Theory explains innovation and institutional change, Systems Thinking highlights interdependencies and feedback loops, the TBL framework addresses environmental, social, and economic sustainability, and Circular Economy Theory focuses on resource efficiency and resilience. However, their application in the Global South requires adaptation to local realities, including infrastructure constraints, institutional complexity, and development priorities.

The chapter further shows that sustainable transport transitions require integrated, inclusive, and context-sensitive approaches. Policymakers and practitioners must engage stakeholders, consider local institutional conditions, and embed equity in planning and implementation to ensure balanced sustainability outcomes. Overall, integrating these perspectives provides a comprehensive foundation for understanding green transport transitions in developing regions. This approach supports the design of transport systems that are environmentally sustainable, socially equitable, and economically resilient, contributing to more inclusive urban mobility futures.

References

African Development Bank (2024) African infrastructure development index 2024. Abidjan: African Development Bank

Avelino F (2023) Power in sustainability transitions: A systematic review. Environ Innov Soc Trans 48:100745

Behrens R, McCormick D, Mfinanga D (2025) Paratransit in African cities: Operations, regulation and reform. Transp Rev 45(2):145–168

Bleviss DL (2021) Transportation is critical to reducing greenhouse gas emissions in the United States. Wiley Interdiscip Rev: Energy Environ 10(2):e390

Börjesson M, Eliasson J (2026) Cost-efficient scenarios for reaching transportation climate targets. Transp Res Part a: Policy Pract 185:104201

Caldarola B, Mazzilli D, Napolitano L, Patelli A, Sbardella A (2023) Economic complexity and the sustainability transition: a review of data, methods, and literature. arXiv. https://doi.org/10.48550/arXiv.2308.07172

Coenen L et al (2023) Advancing circular economy in the Global South: Challenges and opportunities. J Ind Ecol 27(4):612–628

Collett KA, Hirmer SA (2024) Electric motorcycles in sub-Saharan Africa: A socio-technical transition perspective. Energy Res Soc Sci 98:103012

Damer L, Weiss D, Gruber J (2026) Barriers to fuel cell cargo bike adoption in urban logistics: A multi-level transition analysis. Front Sustain Cities 7:892345

Di Ruocco I (2025) Supply chain resilience and just transitions in the Global South. Contesti. Città, Territori, Progetti 1:70–85

Diaz Olvera L, Plat D, Pochet P (2023) Looking for the obvious: Motorcycle taxi services in sub-Saharan African cities. J Transp Geogr 108:103567

Dingil AE, Rupi F, Esztergár-Kiss D (2021) An integrative review of socio-technical factors influencing travel decision-making and urban transport performance. Sustain 13(18):10158

Ehebrecht D, Heinrichs D, Lenz B (2022) Motorcycle-taxis in sub-Saharan Africa: A review of their characteristics and implications for urban transport. J Transp Geogr 96:103189

Elkington J (2024) 25 years of triple bottom line: A retrospective. Calif Manage Rev 66(3):5–25

Geels FW (2022) Socio-technical transitions to sustainability: A review of criticisms and elaborations of the multi-level perspective. Curr Opin Environ Sustain 59:101234

Geels FW, Turnheim B (2023) The great reconfiguration: A socio-technical analysis of low-carbon transitions in UK electricity, heat, and mobility. Cambridge University Press, Cambridge

Geissdoerfer M, Savaget P, Bocken NMP, Hultink EJ (2023) The circular economy: A new sustainability paradigm? J Clean Prod 345:131046

Glazener A, Khreis H (2019) Transforming our cities: Best practices towards clean air and active transportation. Curr Opin Environ Sustain 6(1):22–37

Gudmundsson H, Hall RP, Marsden G, Zietsman J (2016) Sustainable transportation: Indicators, frameworks, and performance management. Springer, Berlin
Intergovernmental Panel on Climate Change (2023) Climate change 2023: Impacts, adaptation and vulnerability. Cambridge University Press, Cambridge
International Energy Agency (2025) Global EV outlook 2025. IEA, Paris
International Transport Forum (2025) ITF transport outlook 2025. OECD Publishing, Paris
Jelti F, Allouhi A, Aoul KAT (2023) Transition paths towards a sustainable transportation system: A literature review. Sustain 15(21):15457
Kanyepe J, Tukuta M, Chirisa I (2022) Urban transportation and sustainable development in African cities. Urban Forum 33(2):215–234
Kennedy C, Miller E, Shalaby A, Maclean H, Coleman J (2005) The four pillars of sustainable urban transportation. Transp Rev 25(4):393–414
Kirchherr J, Yang N-H, Schulze F (2023) Breaking the cycles of carbon lock-in: A systematic review of circular economy business models. J Ind Ecol 27(3):456–472
Kivimaa P, Kern F (2024) Creative destruction or mere niche creation? Innovation policy mixes for sustainability transitions. Res Policy 53(2):104789
Klopp JM, Cavoli C (2023) The paratransit puzzle: Mapping and understanding informal transport in African cities. Transp Rev 43(4):567–589
Kumar A, Barrett F (2024) Stuck in traffic: Urban transport in Asia's developing cities. Asian Dev Rev 41(1):78–102
Litman T, Burwell D (2006) Issues in sustainable transportation. Int J Glob Environ
Lucas K (2023) Transport and social exclusion: A review of evidence and policy implications. Transp Policy 125:156–168
Marsden G, Reardon L (2023) Governance of transitions: A review of transport policy integration. J Transp Geogr 108:103589
Meadowcroft J (2024) Planning for sustainability transitions: The role of the state. Policy Sci 57(2):189–208
Meadows DH (2022) Thinking in systems: A primer, 2nd edn. Chelsea Green Publishing, White River Junction, VT
Mehar S, Zeadally S, Rémy G, Senouci SM (2014) Sustainable transportation management system for a fleet of electric vehicles. IEEE Trans Intell Transp Syst 16(3):1401–1414. https://doi.org/10.1109/TITS.2014.2367099
Messner L, Mark-Herbert C (2026) Bus rapid transit and paratransit interactions: A socio-technical transition theory approach. In Evidence and action on the UNSDGs: Reviews and case studies (pp. 145–172). Bingley: Emerald Publishing
National Planning Authority Uganda & African Development Bank. (2025) Uganda just transition framework: Transport sector screening and assessment. Kampala: NPA
Newell P, Mulvaney D (2023) The political economy of the 'just transition.' Geogr J 189(2):256–268
Oxford Infrastructure Analytics (2025) Ghana climate-resilient infrastructure roadmap: Systems modelling for adaptation planning. OIA, Oxford
Ozanne L, Ballantine P, Mitchell J (2023) Sustainable consumption and the triple bottom line: A review. J Clean Prod 372:133664
Parnell S, Pieterse E (2023) Africa's urban revolution: Sustainability transitions in the 21st century. Zed Books, London
Papadis E, Tsatsaronis G (2020) Challenges in the decarbonization of the energy sector
Raballand G, Teravaninthorn S (2023) Transport prices and costs in Africa: A review of the main corridors, 2nd edn. World Bank, Washington, DC
Rizzo M (2024) Taken for a ride: The politics of public transport in urban Africa. Pluto Press, London
Ruggieri A et al (2021) Energy transition in sustainable transport: Concepts, policies and methodologies. Environ Sci Pollut Res
Schulze F, Bening CR (2024) Circular economy and climate change: A systematic review of synergies and trade-offs. J Clean Prod 415:137890

Schwanen T (2023) Low-carbon mobility in the Global South: A socio-technical transitions perspective. J Transp Geogr 106:103512
Slaper TF, Hall TJ (2023) The triple bottom line: What is it and how does it work? Indiana Business Review 98(2):4–12
Slocat (2024).Transport and climate change global status report 2024. Paris: SLOCAT Partnership
Smith A, Raven R (2024) What is protective space? Reconsidering niches in transitions to sustainability. Res Policy 53(4):104812
Sovacool BK, Turnheim B, Hook A, Martiskainen M (2025) Dispossessed by decarbonisation: Analysing the distributional impacts of low-carbon transitions. Energy Res Soc Sci 110:103456
Turan B, Hemmelmayr V, Larsen A, Puchinger J (2024) Transition towards sustainable mobility: The role of transport optimization. CEJOR 32:435–456
United Nations Department of Economic and Social Affairs (2024) World urbanization prospects: The 2024 revision. United Nations, New York
United Nations Environment Programme (2024) Emissions gap report 2024. Nairobi: UNEP
Venter C (2024) Transport poverty in African cities: Measurement, dynamics, and policy implications. J Transp Geogr 112:103678
Williams B, Hummelbrunner R (2024) Systems concepts in action: A practitioner's toolkit, 2nd edn. Stanford University Press, Stanford, CA
World Bank (2024) Connecting to compete: Trade logistics in the global economy 2024. World Bank, Washington, DC
World Bank (2025) The transport infrastructure lifecycle framework: A systems approach to resilient infrastructure. World Bank, Washington, DC
World Health Organization (2024) Ambient air pollution: Health impact assessment. WHO, Geneva

Infrastructure Lock-In and Path Dependency

Cleopatra Mtetwa, Mandere Tendai Silvaziso, Bunu Sandra, and Moyo Vasma

Abstract This chapter discusses how infrastructure lock-in and path dependency are significant obstacles to sustainable and low-carbon urban growth, especially in the fast-urbanising regions of the Global South. The study applies path dependency theory, socio-technical transitions theory, systems thinking and just transition perspectives in order to develop an integrated conceptual framework for explaining how historical infrastructure investments, institutional arrangements, governance systems and socio-behavioural practices reinforce carbon-intensive development pathways. The chapter finds that infrastructure lock-in is a multi-dimensional socio-technical process sustained through interactions between transport systems, energy infrastructure, land-use planning, financial systems and governance structures, based on a conceptual qualitative research design using an interdisciplinary literature review. The research also emphasizes the key, but often underappreciated, role of informal transport systems as hybrid infrastructures that can both underpin and undermine high-carbon urban paths. The findings imply that successful sustainability transitions require integrated and context-sensitive policy interventions that combine institutional transformation, strategic planning, financial innovation and behavioral adaptation. This chapter builds on the literature to move away from formal infrastructure systems, to highlight the importance of governance dynamics, hybrid infrastructures and coordinated policy intervention pathways as central to achieving resilient, inclusive and low-carbon urban futures in the Global South.

C. Mtetwa (✉)
Fleet Management Department, Mega-Size College, Gaborone, Botswana
e-mail: cmtetwa@megasize.ac.bw

M. T. Silvaziso · B. Sandra
Business Management Department, BA ISAGO University, Gaborone, Botswana
e-mail: tendai.mandere@baisago.ac.bw

B. Sandra
e-mail: Sandra.bunu@baisago.ac.bw

M. Vasma
Business Department, Gaborone University College, Gaborone, Botswana
e-mail: Vasma.Moyo@guc.ac.bw

J. Kanyepe and I. Chirisa (eds.), *Advancing the Green Agenda in the Transport and Logistics Sector in Global South Countries*, World Sustainability Series,
https://doi.org/10.1007/978-3-032-31374-4_5

Keywords Infrastructure lock-in · Path dependency · Sustainability transitions · Socio-technical systems · Low-carbon development · Informal transport systems · Urban sustainability · Governance · Global south · Sustainable infrastructure

1 Introduction

Infrastructure systems are central to economic development, urbanization and social well-being, while also influencing long-term environmental outcomes. As infrastructure consists of durable physical assets, institutional arrangements and behavioural patterns, it can constrain opportunities for more sustainable, low-carbon development pathways. This phenomenon, commonly referred to as infrastructure lock-in, arises when existing systems limit flexibility for low-carbon transitions (Seto et al. 2021; Goldstein et al. 2023a, b, c). While infrastructure development is a key driver of urban growth and economic progress, established systems often generate structural inertia, reinforcing carbon-intensive and resource-dependent development trajectories. Early investments, coupled with entrenched socio-technical and institutional arrangements, create self-reinforcing pathways that are difficult and costly to reverse (Unruh and Carrillo-Hermosilla 2006; Goldstein et al. 2023a, b, c). These dynamics hinder the adoption of more sustainable alternatives, perpetuating dominant patterns of energy consumption and urban expansion.

The concept of infrastructure lock-in is closely linked to path dependency, a theoretical framework that explains how historical decisions shape present and future choices. Path dependency emerges from feedback mechanisms, network effects and institutional commitments, which privilege incumbent technologies and practices, making change increasingly difficult over time (Arthur 2022d, a, b, c; Goldstein et al. 2023a, b, c). In socio-technical systems such as transport, energy and urban infrastructure, these effects are particularly pronounced due to the long lifespan of capital assets, complex governance structures and entrenched institutional practices. Consequently, infrastructure systems often exhibit persistent high-carbon emissions and limited adaptability.

These dynamics are especially significant in rapidly urbanizing regions of the Global South. Cities in Africa, Asia and Latin America are investing heavily in infrastructure to meet population growth and development needs. Without deliberate integration of sustainability and low-carbon strategies, such investments risk embedding carbon-intensive systems for decades (Seto et al. 2021). Moreover, the increasing prevalence of stranded assets infrastructure that loses economic value prematurely due to policy changes, market shifts or technological advances poses fiscal risks for governments, particularly where projects are debt-financed (Bolton et al. 2021). Misaligned infrastructure investments can therefore undermine both climate mitigation efforts and long-term economic resilience.

To examine these dynamics, this chapter adopts an integrative conceptual framework that combines path dependency theory, sustainability transitions research, systems thinking and just transition principles. Path dependency provides insight into how early infrastructure decisions constrain future options (Arthur 2022d, a, b, c; Goldstein et al. 2023a, b, c). Sustainability transitions research highlights the socio-technical interactions that reinforce incumbent regimes (Geels 2018). Systems thinking emphasizes feedback loops and interdependencies that contribute to lock-in, while just transition frameworks foreground equity, governance and inclusivity in transformation processes (Seto et al. 2021). Together, these viewpoints provide a thorough comprehension of infrastructure lock-in and offer strategies for creating resilient, low-carbon and socially inclusive infrastructure systems in the Global South.

This study makes three key contributions to the literature on infrastructure change. First, it synthesizes path dependency theory, socio-technical transitions theory, systems thinking and just transition principles into one conceptual framework to analyze infrastructure lock-in in fast urbanising regions. Although past research has frequently treated these perspectives in isolation, this chapter illustrates how their relationship impacts long-term infrastructure trajectories. Secondly, the study contributes to the current lock-in literature by explicitly including informal transit systems and hybrid infrastructures as reinforcing and possibly disruptive forces within socio-technical regimes. This is especially true in the Global South where informal transportation systems continue to play a key role in urban access yet remain mostly absent from sustainability transition frameworks. Third, the chapter pushes forward policy-oriented studies by identifying leverage points for intervention that link governance reform, integrated planning and socio-technical adaptation, so offering a more context-sensitive framework for low-carbon infrastructure transitions in developing countries.

2 Background to the Study

Traditionally, infrastructure development has been considered as an enabler of economic expansion, industrialisation and social progress. Connectivity, market access and standards of living in developed and developing economies are possible due to investments in roads, railroads, ports, electrical grids and urban services (Abdullah et al. 2024a, b, c, d, e; Zhang et al. 2025a, b). Infrastructure is also crucial in determining access to healthcare, education, employment and public services, and consequently in shaping broader patterns of social inclusion and urban development.

However, mounting evidence shows that many infrastructure systems are major contributors to environmental deterioration and greenhouse gas emissions (IPCC 2022; IEA 2023a). Carbon intensive transport systems, fossil fuel based electricity networks and expanding metropolitan forms maintain unsustainable patterns of energy use and land usage. Once these systems are rolled out, they provide long-term lock-in effects, which stifle the adoption of other low-carbon options. Infrastructure systems thus not only enable progress, but also determine future possibilities through the embedding of specific technological, institutional and behavioral trajectories.

Infrastructure lock-in is caused by a number of interconnected causes. Historical investment decisions have generally favoured centralized energy systems, road-based mobility and industrial growth for short-term economic goals. These expenditures were integrated in legal frameworks, engineering standards, financial systems and social expectations throughout time (Breetz et al. 2022). Such processes create positive feedback loops that entrench the systems of incumbents and increase the economic and political costs of transition.

The continued existence of high-carbon infrastructure also poses enormous economic hazards. As climate policies tighten and low-carbon technologies become more competitive, carbon-intensive assets may lose economic value before the end of their operational life. These stranded assets pose financial concerns for governments, investors and infrastructure operators, especially in developing nations where infrastructure expansion is often debt-financed (World Bank 2023). Infrastructure planning must therefore progressively reconcile development aspirations with long-term environmental sustainability and economic resilience.

The research on infrastructure lock-in and sustainability transition has grown substantially, yet critical gaps remain. Most of the existing research on transitions has been concerned with technological transformations in formal infrastructure sectors, with very little emphasis paid to informal infrastructure systems that dominate mobility and service provision in many Global South cities. In addition, past studies have generally focused on the technological, institutional or behavioural elements in isolation, rather than examining how these dimensions interact across sectors and governance scales.

This chapter addresses these shortcomings by creating an integrated analytical framework that considers infrastructure lock-in as a multi-dimensional socio-technical process driven by the interaction of formal and informal systems, governance structures, financial arrangements and behavioral practices. Special focus is given to the function of informal transport systems as hybrid infrastructures that both promote and potentially disrupt carbon-intensive development patterns.

3 Research Objectives

This study was guided by the following research objectives:

1. To analyse the mechanisms of path dependency and structural inertia within existing infrastructure systems.
2. To assess the financial and economic risks associated with long-lived, carbon-intensive infrastructure.
3. To evaluate policy interventions aimed at disrupting unsustainable infrastructure pathways.
4. To examine the window of opportunity for strategic intervention in rapidly urbanising regions of the Global South.

In alignment with the chapter's focus on infrastructure lock-in and path dependency, the study adopts an integrated theoretical framework that draws on path dependency theory, socio-technical transitions theory and systems thinking. This interdisciplinary perspective is well-suited to analysing how long-lived infrastructure systems generate structural inertia, thereby constraining the capacity of societies to transition toward sustainable and low-carbon development pathways (Geels 2018; Arthur 2022d, a, b, c).

Path dependency theory provides a foundational lens for understanding how historical decisions related to infrastructure investment, urban design and technological systems create self-reinforcing trajectories. Over time, these trajectories embed economic, institutional and behavioural patterns that privilege existing systems, limiting the adoption of emerging sustainable alternatives (Arthur 2022d, a, b, c; Unruh 2000). Within the context of infrastructure lock-in, this perspective helps explain the persistence of carbon-intensive systems such as road-dominated transport networks and fossil fuel-based energy systems despite growing environmental concerns and policy pressures for change (Goldstein et al. 2023a, b, c).

Collectively, these objectives establish a comprehensive framework for examining how infrastructure lock-in and path dependency shape development trajectories. They highlight the interplay between structural inertia, policy constraints and financial risks in reinforcing unsustainable pathways. At the same time, they provide a basis for identifying targeted, context-sensitive interventions capable of disrupting lock-in dynamics, encouraging the uptake of low-carbon alternatives and fostering sustainable infrastructure development, particularly within increasingly urbanizing countries of the Global South (Sovacool et al. 2025; Seto et al. 2016).

4 Conceptual Model

This chapter presents a conceptual model of infrastructure lock-in and path dependency that explains how historical decisions, systemic dynamics and policy environments interact to shape the long-term evolution of infrastructure systems (Fig. 1). Drawing on key insights from sustainability transitions and path dependency literature, the model conceptualises infrastructure development as a socio-technical process embedded within interdependent institutional, technological and behavioural structures (Goldstein et al. 2023a, b, c; Kaya et al. 2025).

At the foundation of the conceptual model are historical and structural drivers, which establish the initial conditions shaping infrastructure trajectories. These include legacy investments in road networks and fossil fuel-based energy systems, urban planning approaches that promote low-density, automobile-dependent development and institutional frameworks governing infrastructure provision. Economic incentives, such as subsidies for carbon-intensive industries, further reinforce these trajectories. Once embedded, these drivers produce development patterns that are difficult to reverse, locking in specific technological configurations and spatial forms

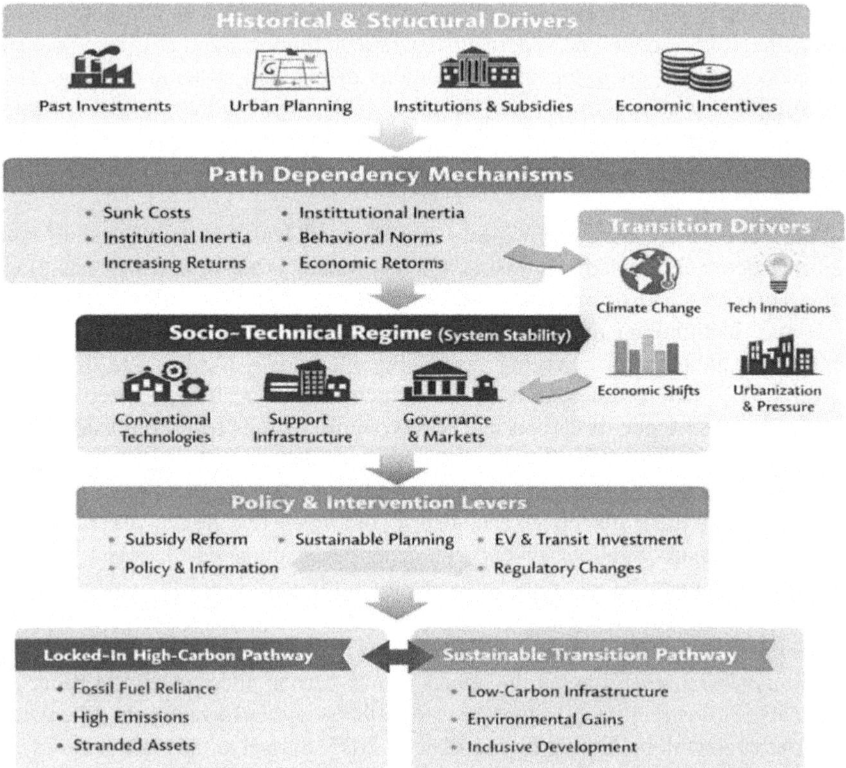

Fig. 1 Infrastructure lock-in and path dependency model. (*Source* Developed by the authors)

within the broader economy (Seto et al. 2016). In rapidly urbanising contexts, such early decisions exert enduring effects on long-term sustainability outcomes.

These foundational drivers give rise to path dependency mechanisms, which function as reinforcing processes that entrench infrastructure lock-in. Mechanisms such as increasing returns to adoption enhance the relative attractiveness of existing systems over time, while high sunk costs and long asset lifespans discourage replacement or transformation. Institutional inertia and inflexible policy frameworks further constrain adaptive capacity, while behavioural and social norms such as dependence on private vehicles sustain demand for established systems. Collectively, these dynamics generate structural inertia, making transitions away from carbon-intensive infrastructure increasingly difficult (Arthur, 2022d, a, b, c; Unruh and Carrillo-Hermosilla 2006). This highlights the cumulative and self-reinforcing nature of lock-in processes.

At the centre of the model is the socio-technical regime, representing the dominant configuration of technologies, institutions and practices that stabilise existing systems. This regime encompasses established technologies (e.g., internal combustion engines), supporting infrastructure (such as road networks and fuel distribution systems), governance arrangements and market structures that sustain demand. Its stability is reinforced through continuous interactions among key actors, including governments, firms and users, which collectively uphold prevailing carbon-intensive development pathways (Geels and Schot 2021). Consequently, even where alternative technologies exist, the entrenched nature of the socio-technical regime limits their large-scale adoption.

The model also incorporates the influence of external pressures and transition drivers, which challenge the stability of existing systems. These pressures include global climate commitments, such as the Paris Agreement, technological advancements in renewable energy and electric mobility, macroeconomic shifts and rapid urbanisation. Increasing environmental and social sustainability concerns further intensify demand for systemic change. Such factors can provide "windows of opportunity" that destabilise present regimes and allow for transitions to more sustainable configurations (IPCC 2022; Sovacool et al. 2025). However, the presence of these pressures alone is insufficient to drive transformation in the absence of enabling policy frameworks.

To facilitate systemic change, the model emphasises the role of policy and intervention levers as critical mediating variables capable of disrupting lock-in dynamics. These include strategic policy sequencing that aligns short-term actions with long-term sustainability objectives; subsidy reforms and sunset clauses aimed at phasing out support for carbon-intensive sectors; and climate-aligned investment screening mechanisms prioritising low-carbon infrastructure. Additionally, integrated land-use and transport planning, alongside institutional reforms, enhances the capacity for coordinated and sustained transformation. These interventions play a decisive role in determining whether infrastructure systems remain locked into unsustainable pathways or transition toward more sustainable alternatives (Seto et al., 2016; Organisation for Economic Co-operation and Development [OECD], 2023).

The interaction of these components ultimately produces two contrasting development outcomes. In the absence of effective intervention, infrastructure systems are likely to remain on a locked-in, high-carbon trajectory characterised by continued reliance on fossil fuels, rising emissions, environmental degradation and increased risks of stranded assets and fiscal instability. Conversely, timely and well-coordinated interventions can enable a sustainable transition pathway, marked by the adoption of low-carbon infrastructure, reduced emissions, enhanced economic resilience and improved social inclusivity. These divergent outcomes highlight the crucial role of governance and policy in defining infrastructure futures (United Nations Environment Programme [UNEP], 2024).

Overall, the conceptual model demonstrates that infrastructure lock-in emerges from the interaction of historical decisions, reinforcing mechanisms and institutional dynamics, rather than from isolated factors. It illustrates that, without deliberate and proactive intervention, development trajectories tend to reproduce carbon-intensive systems. By leveraging appropriate policy instruments and acting within critical windows of opportunity particularly in rapidly urbanising regions of the Global South policymakers can redirect these trajectories toward sustainable outcomes. As a result, the model provides a strong analytical foundation for determining leverage points and designing integrated strategies to overcome lock-in and support long-term sustainability transitions.

5 Research Methodology

The study followed a conceptual qualitative research design, based on a thorough evaluation and synthesis of inter-disciplinary literature on infrastructure lock-in, route dependency and sustainability transitions. The conceptual approach was selected because the study aimed at a critical analysis and integration of existing theoretical and empirical knowledge in order to develop an improved analytical framework to explain how infrastructural, institutional and socio-technical dynamics reinforce carbon-intense development pathways, especially in fast urbanising regions of the Global South. The study used an organized and transparent review method to discover recurring themes, conceptual gaps and potential intervention pathways across the many strands of academia, unlike solely descriptive literature reviews.

The literature review process was composed of four consecutive stages: identification, screening, eligibility assessment and topic synthesis. Systematic searches of major academic databases (e.g. Google Scholar, Scopus, Web of Science, Emerald Insight and ScienceDirect) were conducted to identify relevant literature. Search strings combined key ideas using Boolean operators, including "infrastructure lock-in", "carbon lock-in", "path dependency", "socio-technical transitions", "stranded assets", "urban sustainability", "informal transport systems", and "low-carbon infrastructure". We also manually searched reference lists and policy papers to find influential research that were not identified through database searches.

Inclusion criteria were developed to increase rigor and relevance. Sources were selected if they: (1) were peer-reviewed journal articles, scholarly books or institutional reports; (2) focused on infrastructure systems, sustainability transitions or path dependency; (3) provided theoretical, empirical or policy insights relevant to the research objectives; and (4) were published primarily between 2015 and 2026 to ensure contemporary relevance. Where appropriate seminal previous works that are basic to route dependency and socio-technical transitions theory were also maintained. Institutional publications of organizations like Intergovernmental Panel on Climate Change, World Bank, United Nations Human Settlements Programme and International Energy Agency were included for their policy importance and global authority. We excluded duplicate data, non-English publications, opinion articles without analysis and studies not directly relevant to infrastructure or sustainability transitions.

The collected literature was submitted to a structured theme analysis following screening and appraisal of eligibility. To systematically extract the relevant information from each source, a data extraction matrix was developed. This matrix captured publication details, geographical focus, theoretical framework, infrastructure sector examined, methodological approach, lock-in mechanisms identified, policy interventions and main findings. Thematic coding was then utilized to find patterns and conceptual linkages in the literature. Special emphasis was placed on: (1) lock-in mechanisms of structures and institutions, (2) socio-technical reinforcement processes, (3) informal infrastructure dynamics in the Global South, (4) governance and policy interventions, and (5) cross-sectoral sustainable transitions.

The study adopted a comparative synthesis approach instead of a merely narrative evaluation to enhance analytical transparency and conceptual rigor. This helped to identify convergences, tensions and gaps between rival scholarly positions. The synthesis process has also contributed to the conceptual model of the chapter, especially regarding the inclusion of informal transport systems and policy intervention paths as neglected features in previous infrastructure lock-in frameworks.

No primary empirical data were obtained; yet the methodical and interdisciplinary nature of the review increases the dependability and analytical depth of the conclusions. There are still some limits. Publication bias may also favour research reporting successful transition interventions over failed cases in the literature. Further, the results cannot be experimentally extrapolated without further case-study validation, given the conceptual nature of the study. Despite these limitations, the technique offers a strong framework for both the analysis of infrastructure lock-in as a multi-dimensional socio-technical phenomenon and the identification of strategic leverage points for sustainable and equitable infrastructure transitions in the Global South.

6 Literature Review

6.1 Mechanisms of Infrastructure Lock-In

Infrastructure lock-in arises from the interaction of historical investments, sunk costs, institutional rigidity and socio-behavioral norms, which collectively generate self-reinforcing patterns that constrain the adoption of alternative, low-carbon pathways (Goldstein et al. 2023a, b, c; Zhang et al. 2025a, b). Historical capital expenditures in fixed assets such as highways, power plants and water networks create sunk costs that make transitioning to superior low-carbon alternatives economically challenging. These sunk costs are reinforced by institutional lock-in, wherein regulations, planning norms and budgetary procedures are aligned with legacy systems and by socio-behavioral norms that influence expectations regarding urban mobility and lifestyle patterns.

For example, road-centric urban planning shapes the physical form of cities while reinforcing private vehicle dependence by allocating disproportionate space and fiscal resources to roads and parking infrastructure. This approach discourages public transit use, walking and cycling, thereby perpetuating higher energy consumption and carbon emissions (Banister 2022; Lucas and Wang 2023). Moreover, such spatial and infrastructural biases influence land markets and housing patterns, promoting sprawling urban growth that further entrenches car dependence and complicates subsequent efforts to implement high-capacity transit systems.

Similarly, centralized fossil fuel-based energy systems establish dependencies across extraction, refining, and distribution processes that are technically, economically and politically difficult to replace (Liu et al. 2024; Kaya and Mbembe 2025). Long asset lifespans, entrenched financing mechanisms and regulatory frameworks designed for centralized generation create structural inertia that constrains the uptake of decentralized renewable energy and demand-side innovations. Persistent fossil fuel subsidies, utility monopolies and regulatory barriers further reinforce high-carbon energy pathways (International Energy Agency [IEA], 2023b; Stern and Stiglitz 2024; Caldecott 2026).

These mechanisms exemplify path dependency, whereby early infrastructure and policy decisions generate positive feedback loops that increase the cost and complexity of alternative approaches. Cities with extensive highway networks and private vehicle infrastructure often face prohibitive costs and technical challenges when retrofitting comprehensive public transit or cycling networks (Seto et al. 2021; Delbosc and Currie 2024). Likewise, energy markets dominated by fossil fuels maintain subsidies, regulatory advantages and institutional inertia, slowing the adoption of renewable technologies (IEA 2023a; Kaya and Mbembe 2025; Pollitt and Sha, 2026).

Lock-in is further reinforced through multi-level socio-technical interactions, as technologies, institutions, markets and user behaviors co-evolve to stabilize high-carbon regimes over extended periods (Geels, 2011; Foxon et al. 2021; Meadowcroft and Rosenbloom 2023). Consequently, interventions that target a single dimension such as technology without addressing complementary domains like finance, regulation or user behavior often fail to disrupt entrenched pathways. For instance, the introduction of electric vehicles (EVs) without adequate charging infrastructure, supportive urban planning and aligned pricing signals may yield only marginal emissions reductions while leaving broader socio-technical systems largely unchanged (Pacini and Bauknecht, 2025; Herman et al. 2025). In some cases, EV deployment without integrated planning can exacerbate social inequities, as charging infrastructure is disproportionately installed in affluent neighborhoods.

The co-evolutionary nature of lock-in also shapes user expectations and routines. Once populations internalize norms such as the primacy of private car use, political and institutional support for alternative mobility options remains weak, slowing policy innovation and reinforcing high-carbon pathways (Ekers et al., 2024; Hult and Hu, 2025). Similarly, utility reforms that do not reconfigure pricing structures or investment incentives tend to preserve incumbent behaviors rather than accelerate renewable energy transitions (Caldecott and Van der Veer, 2025; Meadowcroft and Langhelle, 2026). Addressing infrastructure lock-in therefore requires multi-dimensional strategies that align technology, policy, finance and behavior. Effective approaches go beyond the deployment of new technologies to include institutional reforms, incentive redesign and cultural shifts in infrastructure use.

6.2 *Socio-Technical Perspectives on Transitions*

Socio-technical transitions theory emphasizes that infrastructure systems are embedded regimes, stabilized by interdependent networks of technologies, governance frameworks, market structures, professional norms and social practices (Herman et al. 2025). These regimes are not merely technical assemblages; they represent configurations of actors, institutions, regulations and cultural expectations that collectively shape the evolution of infrastructure. Consequently, infrastructure systems such as transport, energy, and water networks are sustained not solely by technology but by enduring patterns of investment, regulation, organizational routines and user behaviors that reinforce existing trajectories.

Innovations emerging outside dominant systems, including electric vehicles (EVs), decentralized energy storage and generation, micromobility solutions and data-driven urban management tools, often encounter significant institutional, regulatory and market barriers that constrain their scalability. For example, EV deployment may be slowed by misalignments among vehicle technologies, grid capacity, parking infrastructure and regulatory standards (Pacini and Bauknecht 2025). Similarly, decentralized renewable energy systems face challenges from legacy utility monopolies, fossil fuel subsidies and financing mechanisms designed around large-scale centralized generation (International Energy Agency [IEA] 2023a; Stern and Stiglitz 2024). These barriers reflect the broader socio-technical lock-in that privileges incumbent systems over emerging alternatives.

The multi-level perspective (MLP) is an effective paradigm for understanding how fundamental sustainability transformations occur. The MLP conceptualizes socio-technical development as the interaction of three analytical levels:

1. **Niche innovations**- Protected spaces where radical novelties are developed and refined outside dominant market and regulatory pressures. Examples include experimental technologies (e.g., battery storage, hydrogen fuels), innovative business models (e.g., mobility-as-a-service) and community-driven solutions (e.g., localized microgrids). Actors at this level experiment, learn and mature alternatives that can eventually challenge the status quo (Sovacool et al. 2025; Ahlborg and Hammar 2025).
2. **Socio-technical regimes**- The dominant configuration of infrastructure, institutions, norms, and markets that structure how services are produced and consumed. Regimes are inherently conservative, embodying vested interests, standardized practices and embedded incentives that resist disruptive change (Geels 2011; Foxon et al. 2021; Herman et al. 2025a, b).
3. **Socio-technical landscapes**- The broader exogenous environment, including macro-social trends, cultural values, political pressures, climate commitments, demographic shifts, and global economic conditions. Landscape pressures are typically slow-moving but exert significant directional influence on both niches and regimes (Seto et al. 2021; Sovacool et al. 2025).

The interaction among these levels determines whether low-carbon pathways can be adopted and scaled. When landscape pressures such as climate policy targets, volatile fuel prices and public demand for resilience align with maturing niche innovations and vulnerabilities in regime stability, windows of opportunity may emerge for transformative change (Hoffmann 2025; Meadowcroft and Langhelle 2026). For instance, stringent emissions regulations, rising urbanization and heightened environmental awareness have amplified demand for sustainable mobility and decentralized renewable energy, providing niches with leverage to challenge incumbent regimes (Kern and Rogge 2025; Watson 2025).

Socio-technical perspectives highlight that transitions are non-linear, contingent and multi-dimensional. Transformations do not proceed through smooth, predictable progressions but occur via punctuated, iterative adjustments shaped by power relations, institutional adaptation, technological learning and shifting public preferences. Transitions involve co-evolutionary dynamics, where technological change interacts with regulatory reform, market restructuring, cultural shifts and reconfigured user practices (Meadowcroft and Rosenbloom 2023; Lucas and Wang 2023). Given this complexity, successful transformations require coordinated interventions across multiple domains. Technical solutions alone are insufficient; policy innovation, institutional reform, finance restructuring and behavioral adaptation must be pursued in concert. Urban managers and policymakers must therefore address:

- **Physical infrastructure**-Designing and investing in systems that facilitate low-carbon options (e.g., transit networks, distributed energy resources, charging infrastructure) rather than reinforcing legacy configurations (Percastre Gómez et al. 2024; Curtis and Scheurer 2026).
- **Governance and regulatory frameworks**-Reforming codes, standards and institutional mandates to lower barriers for niche adoption and align incentives with sustainability goals (Hodson and Marvin 2021; Hoffmann 2025).
- **Financing mechanisms**- Innovating funding models, including climate finance instruments, blended public–private investment, and life-cycle cost assessments, to de-risk sustainable infrastructure investments (Caldecott and Van der Veer 2025; Pollitt and Sha 2026).
- **Social and behavioral adaptation**- Cultivating user practices and societal expectations that support sustainable transitions through education, participatory planning, pricing instruments, and incentive structures (Watson 2025).

Without integrated strategies, piecemeal interventions such as deploying EVs without grid modernization or equitable charging access risk reinforcing existing socio-technical configurations rather than transforming them (Pacini and Bauknecht 2025; Herman et al. 2025a, b). Conversely, approaches that align technological innovation with enabling policies, supportive markets, and positive cultural narratives can cultivate pathways that break entrenched lock-ins and accelerate just, low-carbon transitions (Bulkeley and Castán Broto 2025; Meadowcroft and Langhelle 2026).

6.3 *Informal Transport Systems and Hybrid Infrastructures*

Many cities in Africa, Asia and Latin America rely heavily on informal transportation networks such as minibuses, shared taxis, motorcycle taxis, and paratransit. Despite their importance, they are typically overlooked in mainstream sustainability research and policy discussions (Chen et al., 2025a, b, c, d; Abdullah et al. 2024; International Association of Public Transport [UITP] 2025). Unlike highly regulated formal transit modes, such as buses and metro systems, informal services emerge organically in response to unmet mobility demand and often operate outside comprehensive regulatory oversight, rendering them both essential and challenging to govern within conventional planning frameworks.

These systems provide critical mobility and livelihoods for millions of urban residents, particularly low-income populations, women and workers in peri-urban or informal settlement areas underserved by formal transport networks (Cervero and Golub, 2023; Pojani and Stead 2015). Their operational flexibility characterized by on-demand routing, frequent stops and low fare thresholds enables rapid adaptation to demand fluctuations, traffic patterns and the spatial complexity of informal settlements. This flexibility often results in higher accessibility and reduced first-/last-mile barriers compared with fixed-route systems, particularly in contexts where formal planning capacities are limited or infrastructure investment lags behind population growth (Watson 2025; Hult and Hu, 2025).

However, informal systems frequently contribute to carbon-intensive transport pathways. Vehicles used in minibuses, shared taxis and motorcycle taxis are often older, poorly maintained, and mechanically inefficient, leading to higher emissions per passenger-kilometer relative to optimized formal buses or rail (Lucas and Wang 2023; Mudzonga et al. 2025). Operating outside strict emission standards and idling in congested corridors, these vehicles exacerbate urban air pollution, greenhouse gas emissions and noise, while also increasing congestion and safety risks for pedestrians and cyclists (Delbosc and Currie 2024; Meadowcroft and Rosenbloom 2023).

These dynamics illustrate how informal transport can both reinforce and challenge infrastructure lock-in. On one hand, the prevalence of outdated and carbon-intensive operations can entrench high-emission mobility patterns, particularly when informal services are integrated into traffic flows without systemic planning. On the other hand, their embeddedness in daily mobility and local economies makes them critical nodes for sustainability interventions, particularly when supported by enabling policies and technology (Seto et al. 2021; Bulkeley and Castán Broto 2025).

Evidence from Global South contexts highlights pathways for reorienting informal systems toward low-carbon, equitable outcomes. In Kenya, pilot programs integrating electric minibuses into route-planning platforms have reduced local emissions, fuel costs and operator expenses while expanding access to underserved neighborhoods (Abdullah et al. 2024a, b, c, d, e; Mudzonga et al. 2025). In South Africa, participatory planning initiatives involving minibus taxi associations in formal transport processes have led to negotiated service contracts, joint scheduling frameworks and coordinated ticketing systems, demonstrating that collaboration rather than exclusion enhances both sustainability and service reliability (Cervero and Golub 2023; Pojani and Stead 2015).

Brazil provides another illustrative example through "Sustainable Mobility Hubs." In several metropolitan regions, informal van and moto-taxi networks have been incorporated into multimodal integration schemes linking them with rapid transit, bike lanes and pedestrian corridors, supported by digital payment systems and incentive-based fleet upgrades. These interventions reduce operational redundancies, lower emissions and improve safety, highlighting that informal mobility is not inherently unsustainable but becomes so when excluded from planning, financing and regulatory frameworks (Watson 2025).

Research further indicates that policy frameworks and institutional reforms are pivotal in this transformation. Formal recognition of informal operators through licensing and cooperative structures can create entry points for training, vehicle upgrade subsidies and emission standards that enhance service quality and align operations with urban climate strategies (Hodson and Marvin 2021; Hoffmann 2025). Additionally, digital platforms that aggregate informal service data allow municipalities to better understand demand patterns and coordinate infrastructure investments, such as safer pick-up/drop-off facilities and shared electric vehicle charging infrastructure where feasible (Pacini and Bauknecht 2025; Meadowcroft and Langhelle 2026).

In conclusion, informal transportation systems in the Global South are multifaceted socio-technical phenomena. They fill critical mobility and livelihood gaps left by formal networks, yet under current conditions, they may contribute to high-carbon pathways and infrastructure lock-in. When integrated into broader low-carbon strategies through context-sensitive planning, regulatory reform, financial incentives and technological adaptation, these systems can act as vectors of sustainable mobility transformation rather than obstacles to it.

6.4 Policy, Governance and Strategic Interventions

Policy and governance are central levers for breaking high-carbon infrastructure lock-in and enabling low-carbon transitions (Grasso and Delatin Rodrigues 2025; Seto et al. 2021). High-carbon regimes are not sustained solely by technology or finance; they are reinforced by institutional rules, bureaucratic practices and regulatory frameworks that privilege legacy investments and constrain experimentation with sustainable alternatives. Consequently, institutional reform is essential to overcome path dependencies. Such reforms include revising zoning laws, building codes, utility regulations and procurement procedures to create enabling conditions for low-carbon technologies, sustainable mobility solutions and renewable energy systems (Hodson and Marvin 2021; Kern and Rogge 2025). Without addressing these structural barriers, even technically feasible interventions risk marginalization or underutilization.

Effective governance requires multi-level coordination. Fragmented sectoral approaches often reinforce lock-in, as transport, energy, land-use and water sectors operate under divergent mandates and investment priorities. Aligning policies across sectors ensures that strategic investments such as renewable energy infrastructure, electric mobility networks, or compact urban development are mutually reinforcing rather than contradictory (Percastre Gómez et al. 2024; Curtis and Scheurer 2026). Adaptive regulatory systems are particularly critical, allowing policymakers to respond to technological innovations, shifting socio-economic conditions and emergent mobility or energy patterns without rigidly adhering to outdated norms (Hoffmann 2025; Meadowcroft and Langhelle 2026).

Strategic policy interventions can redirect capital and planning priorities toward sustainable infrastructure while phasing out high-carbon assets. Subsidy reforms that reduce support for fossil fuels and incentivize renewable energy or low-emission transport can shift market signals toward low-carbon alternatives (Caldecott and Van der Veer 2025; Stern and Stiglitz 2024). Investment screening frameworks that incorporate life-cycle carbon costs, social benefits and resilience criteria can ensure that public and private funds are directed toward sustainable projects. Urban land-use integration and transit-oriented development (TOD) can reshape mobility patterns by densifying cities around sustainable transit hubs, reducing reliance on private vehicles, limiting urban sprawl and improving accessibility (Pacini and Bauknecht, 2025; Kaya and Mbembe 2025).

Capacity-building initiatives complement these interventions by strengthening technical expertise, data-driven decision-making and stakeholder engagement among municipal planners, policymakers, and utility operators. Training in spatial analysis, energy modeling, climate risk assessment and participatory governance equips actors to identify reinforcing lock-in mechanisms, anticipate systemic barriers and implement adaptive strategies (Meadowcroft and Rosenbloom 2023; Bulkeley and Castán Broto 2025). Data-driven tools such as mobility modeling platforms, GIS-based land-use analysis, and integrated energy dashboards can reveal vulnerabilities in existing systems and highlight leverage points for low-carbon interventions (Watson 2025).

In the Global South, governance strategies must be context-sensitive, reflecting the complexity of informal transport, rapid urban growth and pronounced socio-economic inequalities. (Abdullah et al. 2024; Chen et al. 2025a, b, c, d; UITP 2025). Policies that integrate informal systems through licensing, route coordination, electrification incentives and cooperative structures can leverage their operational flexibility while aligning them with sustainable mobility objectives (Mudzonga et al. 2025; Cervero and Golub 2023). Adaptive urban planning must also account for high-density informal settlements, land tenure complexities and unequal access to infrastructure to ensure that low-carbon transitions are equitable, socially inclusive, and resilient.

Overall, governance and policy function as multi-dimensional levers. By reforming institutions, aligning sectoral strategies, incentivizing sustainable investments and building adaptive management capacity, policymakers can reduce the persistence of high-carbon lock-in, enhance resilience and create pathways for long-term low-carbon urban development in rapidly urbanizing contexts (Seto et al. 2021; Grasso and Delatin Rodrigues 2025; Pacini and Bauknecht 2025).

6.5 Cross-Sectoral Implications and Integrated Approaches

Infrastructure lock-in and path dependency extend across multiple interdependent sectors, producing systemic effects that reinforce high-carbon urban trajectories. Transport infrastructure, for example, shapes energy demand by incentivizing particular mobility patterns. Sprawling, road-oriented urban layouts favor private vehicle use and increase fuel consumption, whereas dense, transit-oriented designs reduce energy intensity per passenger-kilometer (Banister 2022; Lucas and Wang 2023). Conversely, energy systems influence urban growth patterns. Centralized, fossil-fuel-based electricity grids favor large-scale urbanization in areas with reliable supply, whereas decentralized renewable energy solutions enable more flexible, compact urban settlements (Liu et al. 2024; Kaya and Mbembe 2025). Land-use and spatial planning decisions further interact with these sectors: zoning that prioritizes low-density development or separates residential and commercial areas increases travel distances, reinforces private vehicle dependence and amplifies emissions (Seto et al. 2021; Delbosc and Currie 2024).

Financial and investment frameworks represent additional sources of path dependency. Short-term project evaluation metrics, risk-averse investment practices and entrenched financing mechanisms favor conventional high-carbon infrastructure projects over low-carbon alternatives, even when the latter offer greater long-term benefits (Caldecott and Van der Veer 2025; Stern and Stiglitz 2024). Subsidy structures supporting fossil fuels, regulatory regimes limiting competition in electricity or transport markets and public procurement norms that prioritize proven technologies collectively entrench existing systems, making transformative interventions costly and politically challenging (Hodson and Marvin 2021; Pacini and Bauknecht 2025).

Recent research emphasizes the importance of integrated, multi-dimensional approaches to overcoming cross-sectoral lock-in. Technical, policy, financial and social dimensions must be addressed concurrently to avoid piecemeal interventions that fail to shift high-carbon trajectories. Integrated infrastructure planning that aligns energy, transport and land-use strategies can reduce redundancies, optimize connectivity and enhance system efficiency (Sovacool et al. 2025; Zhang et al. 2025a, b). Complementary policy reforms including adaptive regulatory frameworks, climate-aligned procurement and urban design codes establish institutional conditions necessary for low-carbon transitions (Hoffmann 2025; Meadowcroft and Langhelle 2026). Financial mechanisms, such as blended public–private investment, climate bonds and life-cycle cost assessments, can redirect capital flows toward sustainable projects and de-risk emerging technologies (Caldecott 2026; Pollitt and Sha of Cambridge 2026).

Equally critical is social engagement and participatory governance, which ensures that interventions are context-sensitive, equitable and responsive to diverse urban populations, including those reliant on informal transport networks (Abdullah et al. 2024; Mudzonga et al. 2025). Engaging local stakeholders facilitates co-design of mobility solutions, integrates informal systems into formal planning, and builds social legitimacy for transformative investments. Collectively, these integrated strategies are essential for realizing multi-dimensional urban sustainability in rapidly urbanizing regions. By simultaneously addressing sectoral interdependencies, institutional rigidities, financing biases and behavioral patterns, cities can break high-carbon lock-in, foster resilience, and enable transitions toward low-carbon, inclusive, and adaptable urban futures (Seto et al. 2021; Sovacool et al. 2025; Zhang et al. 2025a, b).

7 Results and Discussion

7.1 Structural Persistence as a Multi-Dimensional Process

One significant result from the analysis is that infrastructural persistence should be seen as a multi-dimensional socio-technical process rather than a consequence of long-lived physical assets alone. In existing research, lock-in is often conceived in terms of technological inertia and sunk capital inputs. However, the findings of this study suggest that structural persistence is reproduced at the same time through spatial planning systems, institutional mandates, funding structures and everyday movement behaviors.

The analysis shows that interdependent development pathways are generated by transport infrastructure, fossil fuel-based energy systems and urban spatial patterns, which become increasingly hard to modify over time. The road-oriented urban growth leads to a greater dependency on the private car, but also has an influence on land use patterns, commuting cultures, the priorities of infrastructure finance and the political expectations about the provision of mobility. In this manner, infrastructure lock-in is perpetuated via physical systems and social activities.

This finding contributes theoretically by expanding standard path dependence viewpoints beyond infrastructure assets themselves to the wider socio-spatial systems within which these assets operate. Therefore, infrastructural persistence in fast-growing urban contexts is rooted in the co-evolution of urban form, governance structures and behavioural routines. The analysis also shows that informal transportation systems exacerbate this persistence and reveals gaps in existing lock-in frameworks. Traditional theories of transition tend to overemphasize formal infrastructure systems and underestimate the role of hybrid and informal infrastructures. But for many cities in the Global South, minibuses, motorbike taxis and paratransit networks are dominant, not peripheral, mobility infrastructures. Their integration into daily travel patterns means that they actively shape emissions trajectories, accessibility and urban development consequences. This study implies that formal infrastructure analysis is not sufficient to explain infrastructure lock-in in the Global South.

7.2 Socio-Technical Reinforcement Beyond Technology

The synthesis also provides an important insight: carbon-intensive infrastructure systems survive because reinforcing processes act simultaneously across institutional, economic and behavioural dimensions. The results indicate that technological alternatives alone cannot break the grip of ingrained socio-technical regimes when regulatory systems, planning standards and market incentives are still aligned with incumbent systems.

The paper shows how governance institutions often unintentionally repeat lock-in through planning procedures, investment evaluation frameworks, and sectoral fragmentation. For instance, under transport planning frameworks that focus on road growth, public resources are often still being directed towards vehicle-oriented infrastructure, even when sustainable mobility options are available. Financing mechanisms, utility laws and risk assessment models continue to favour incumbent technologies over decentralized renewable alternatives, thereby continuing reliance on centralized fossil fuel infrastructure in energy systems. Thus, a fundamental theoretical contribution of this study is the discovery of "reinforcing governance loops" as a key mechanism for sustaining infrastructure lock-in. Such loops arise when institutional regulations, funding systems and professional practices repeatedly recreate situations that are conducive to current infrastructural regimes. This adds to socio-technical transition theory by showing that institutional persistence may be more stubborn to change than technology itself.

The findings further demonstrate that socio-behavioural adaptation plays a far larger role in maintaining lock-in than is commonly recognized in the literature on infrastructure transitions. Over time, people acquire accustomed to the idea of owning their own private vehicle, or having services delivered by a central agency or informal mobility systems. This leads to political and societal resistance to radical change. This means infrastructure transitions are not only technological transitions but societal transitions requiring realignment of behaviour, institutions and culture.

7.3 Hybrid Infrastructure Systems and Cross-Sectoral Lock-in

The study also shows how infrastructure lock-in comes about through cross-sectoral feedback mechanisms between transport, energy, land-use planning and financial institutions. Instead of operating separately, these sectors reinforce one other in ways that sustain high-carbon urban trajectories. A significant discovery from the investigation is that official and informal infrastructures co-produce urban sustainability results. Current transition frameworks often consider informal systems as transitional or inefficient elements that should be eventually codified or replaced. However, the findings contest this notion by demonstrating that informal transport systems also act as adaptable mobility solutions, livelihood systems and integrated socio-economic infrastructures.

This study contributes to the understanding of "hybrid infrastructure regimes" where formal and informal systems are not dichotomous but interact dynamically. As such, in many Global South contexts, sustainable transition paths may emerge not from the abolition of informal systems but from their purposeful incorporation into low-carbon planning frameworks. The analysis also concludes that cross-sectoral lock-in raises the danger of stranded assets and fiscal vulnerability. Investments in road-dominated transport systems, infrastructure for fossil fuel use and spreading urban development patterns lead to long-term dependence which may become economically unsustainable under pressures for decarbonization. In debt-dependent countries, these risks are amplified as governments may remain financially connected to carbon-intensive assets, even when such systems have proven environmentally or economically inefficient.

The results so also highlight infrastructure lock-in as a challenge to governance and budgetary sustainability, as well as an environmental challenge. This contributes to the previous arguments on sustainability transitions by explicitly connecting infrastructure persistence with macroeconomic fragility and public financial risk.

7.4 *Policy Intervention Pathways*

The study suggests that successful intervention paths are less dependent on discrete technical developments and more on coordinated institutional and governance transformation. One major finding from the study is that policy interventions are most effective when executed as integrated policy mixtures that simultaneously address infrastructure, finance, governance and social behavior. The results suggest that piecemeal interventions such as electric vehicle roll-out without supported land-use planning, renewable energy integration or behavioural incentives tend to reproduce rather than disrupt current high-carbon systems. Conversely, measures combining subsidy reform, transit-oriented development, participatory governance and institutional coordination produce reinforcing conditions that support lasting transitions.

Importantly, the study identifies informal infrastructure systems as potential leverage points for transformation rather than barriers to sustainability. Examples of this include the electrification of informal transport fleets, integrated fare systems, multimodal mobility hubs and participatory planning processes demonstrating how hybrid infrastructures can allow adaptable low-carbon transitions when incorporated in enabling governance frameworks. This is one of the major theoretical contributions of the chapter. Current socio-technical transition models tend to focus on niche technological innovation outside dominant regimes. The outcomes of this study, however, imply that in the Global South, transformation may also result from the reconfiguration and upgrading of existing informal systems rather than complete technology replacement.

The analysis also underlines the relevance of timing and sequencing in policy intervention. Rapidly urbanising regions are key "windows of opportunity" when infrastructure systems are relatively flexible before the full fixation of long-term capital investments and institutional structures. Thus, early action has a disproportionate influence on future trajectories of urban sustainability. In general, the results indicate that infrastructure lock-in is not inevitable. But historical investments and institutional inertia impose significant limits, and coordinated and context-sensitive interventions can help guide development trajectories towards low-carbon, resilient and socially inclusive results. The study consequently contributes to the scholarship on infrastructure transitions by highlighting the role of hybrid infrastructures, governance dynamics and integrated intervention techniques in constructing sustainable urban futures in the Global South.

7.5 Further Research

While this chapter presents a thorough conceptual foundation for understanding infrastructure lock-in, path dependency and sustainability transitions, several avenues warrant further investigation to enhance both theoretical insights and practical application, particularly in Global South contexts. Future research should prioritise the analysis of cross-sectoral interactions within lock-in dynamics, recognizing that infrastructure systems are inherently interdependent. For instance, transport networks shape energy demand, urban planning influences mobility patterns and financial and investment policies direct capital flows, creating complex feedback loops that reinforce or constrain development trajectories. Empirical and modeling studies that capture these interconnections can provide critical guidance for sequencing policies, prioritizing interventions and designing integrated strategies that promote low-carbon and resilient infrastructure. By elucidating the interactions between physical systems, institutional arrangements, financial mechanisms and socio-behavioral factors, such research can inform evidence-based policy-making and encourage the creation of context-sensitive, sustainable routes in rapidly urbanizing locations (Seto et al. 2021; Grasso and Delatin Rodrigues 2025).

7.6 Managerial and Policy Implications

The results of this study underscore that addressing infrastructure lock-in and advancing sustainable urban development requires multi-dimensional, coordinated interventions. Effective strategies must simultaneously target physical infrastructure, institutional arrangements, socio-technical systems and behavioral patterns to redirect high-carbon pathways toward low-carbon, resilient, and inclusive outcomes.

7.7 Integrated Infrastructure Planning

Adopting integrated infrastructure planning approaches is essential for mitigating path dependency and reinforcing systemic flexibility (Seto et al. 2021; Kaya et al. 2025). Conventional siloed approaches that treat transport, energy and urban development independently tend to entrench high-carbon trajectories. In contrast, integrated planning aligns land-use strategies, multi-modal transport networks and energy transition objectives, ensuring interventions are mutually reinforcing. For example, coupling electric mobility with renewable energy infrastructure and compact urban design reduces emissions while enhancing accessibility and social inclusivity. Cross-sector coordination between transport, energy, finance and urban planning agencies is therefore critical to maximize the effectiveness of these interventions.

7.8 Engagement of Informal Infrastructure and Stakeholders

Engaging informal infrastructure and stakeholders is another key strategy, particularly in the Global South, where minibus operators, paratransit services, decentralized energy providers and local communities play central roles in urban mobility and service provision (Abdullah et al. 2024; Mudzonga et al. 2025). Integrating these actors into planning, regulatory frameworks and decision-making processes improves efficiency, equity and sustainability. Engagement strategies may include participatory workshops, inclusion in formal transit schemes, co-designed incentives for adopting low-carbon technologies and support for electrification of informal fleets. Recognizing informal actors as legitimate partners leverages their operational flexibility, local knowledge and networks while minimizing the risk of social exclusion or livelihood disruption.

7.9 Technological Innovation and Green Investments

Promoting technological innovation and green investments is critical to accelerating low-carbon transitions (Grasso and Delatin Rodrigues 2025; Liu et al. 2024). Managers should support electric mobility, renewable energy systems, smart grids and digital monitoring infrastructure through public–private partnerships, innovation hubs and strategic subsidies. Investments in predictive maintenance and digital monitoring enhance infrastructure adaptability, enable early risk detection and strengthen resilience against lock-in. For example, electrifying minibus fleets in cities such as Nairobi and Cape Town illustrates how informal transport can be leveraged to achieve sustainability goals while maintaining accessibility.

7.9.1 Improving Institutional Capabilities and Data Systems

Improving institutional capabilities and data systems underpins the successful implementation of these strategies (Herman et al. 2025; Goldstein et al. 2023). Building technical expertise, enhancing regulatory oversight and fostering cross-sector coordination enable evidence-based planning and early identification of reinforcing lock-in mechanisms. Investments in real-time monitoring, predictive modeling and systems-based approaches facilitate proactive interventions, which are particularly critical in rapidly urbanizing contexts characterized by uncertainty, informal sector dynamics and emergent urban challenges.

7.9.2 Context-Sensitive Policy Alignment

Aligning policy with context-specific conditions ensures interventions are effective, equitable, and socially acceptable (Sovacool et al. 2025; Kaya et al. 2025). In the Global South, rapid urbanization, informal sector prevalence and resource constraints necessitate phased and adaptive strategies rather than uniform solutions. Context-sensitive policies leverage local capacities, address equity considerations and strategically sequence interventions. For instance, incentivizing small-scale low-emission vehicles within informal transport networks can complement broader public transit expansion, while phased electrification of energy systems supports affordability and societal acceptance.

7.9.3 Integrated Multi-Dimensional Strategies

Collectively, these strategies highlight that successful urban sustainability transitions require integrated, multi-dimensional approaches that simultaneously address infrastructure, institutional, financial and social dynamics. By combining strategic planning, stakeholder engagement, technological innovation and adaptive governance, policymakers and managers can disrupt entrenched high-carbon pathways, foster resilience and promote inclusive low-carbon urban development.

8 Conclusion

This study highlights that infrastructure lock-in and path dependency are systemic, multi-dimensional phenomena that profoundly shape urban development trajectories, particularly in the rapidly urbanizing Global South (Goldstein et al. 2023; Zhang et al. 2025a, b). Historical investments in transport networks, energy systems and urban layouts create structural inertia that is reinforced by institutional rigidity, regulatory frameworks and socio-behavioral norms, entrenching high-carbon pathways (Herman et al. 2025; Liu et al. 2024). Transitioning toward low-carbon, resilient and inclusive urban systems therefore requires integrated, context-sensitive strategies that simultaneously address physical infrastructure, governance, finance and behavioral patterns.

Effective approaches include engaging informal transport operators, promoting green technologies and investments, strengthening institutional capacity and aligning cross-sector planning to disrupt lock-in and foster adaptive, sustainable infrastructure (Abdullah et al. 2024; Grasso and Delatin Rodrigues 2025; Sovacool et al. 2025). Holistic planning and multi-stakeholder engagement, leveraging policy, technological and social innovations, are critical to ensuring interventions are equitable, responsive to rapid urbanization and sensitive to informal sector dynamics and socio-economic inequalities (Mudzonga et al. 2025; Herman et al. 2025a, b). Ultimately, infrastructure lock-in is as much a socio-technical and governance challenge as a technical one and coordinated, multi-level interventions can prevent the perpetuation of carbon-intensive systems, mitigate the risks of stranded assets and enable inclusive, low-carbon urban futures. These findings provide both theoretical insights and actionable guidance for infrastructure managers, urban planners and policymakers aiming to advance sustainable urban infrastructure pathways in the Global South (Seto et al. 2021; Sovacool et al. 2025; Zhang et al. 2025a, b).

References

Abdullah H, Chen M, UITP (2024) Informal transport systems and sustainable urban mobility. Int Transp Forum

Abdullah M, Chen Y, Rajabifard A (2024b) Informal transit systems in emerging cities: Dynamics, governance, and sustainability. Transp Policy 130:1–14. https://doi.org/10.1016/j.tranpol.2023.12.003

Abdullah M, Dias C, Muley D, Shahin M (2024c) Exploring the role of informal transport in sustainable urban mobility transitions in developing cities. Transp Policy 141:12–24. https://doi.org/10.1016/j.tranpol.2023.11.005

Abdullah S, Kumar A, Nair R (2024d) Integrating informal transport into sustainable urban mobility planning: Challenges and opportunities in low- and middle-income cities. Transp Policy 130:102–115. https://doi.org/10.1016/j.tranpol.2023.12.009

Abdullah U, Raman S, Touré F (2024) Informal transport systems and sustainable mobility transitions in African cities. J Urban Transp Soc, 12(3), 215–233

Ahlborg H, Hammar L (2025) Designing fair and politically viable transport transitions: a framework for transition assistance policies in higher-income countries. Energy Res Soc Sci 131:104470. https://doi.org/10.1016/j.erss.2025.104470

Arthur WB (1989) Competing technologies, increasing returns, and lock in by historical events. Econ J 99(394):116–131

Arthur WB (2022a) Foundations of complexity economics. Oxford University Press

Arthur WB (2022b) Foundations of path dependence in economic systems: Revisiting increasing returns and lock-in. Oxford University Press

Arthur WB (2022c) Path dependence in economic development. Oxford University Press

Arthur WB (2022d) Foundations of path dependence in economic systems. University of Michigan Press

Banister D (2022) Transport and climate change: The role of infrastructure in shaping mobility. Transport Policy Review 8(2):95–112

Bolton P, Després M, Pereira da Silva L, Samama F, Svartzman R (2021) The green swan: Central banking and financial stability in the age of climate change. Bank Int Settl
Breetz H, Mildenberger M, Stokes L (2022) The political logics of clean energy transitions. Environ Polit 31(4):541–563. https://doi.org/10.1080/09644016.2021.1952466
Bulkeley H, Castán Broto V (2025) Socio-technical systems and urban sustainability transitions: Integrating informal and formal infrastructures. Urban Stud J 62(1):45–68
Caldecott B (2026) Climate finance and economic transitions: Rethinking investment frameworks. Energy Econ Policy 14(2):84–105
Caldecott B, Van der Veer R (2025) Blended finance mechanisms for low-carbon infrastructure investment. Sustainable Finance Journal 10(4):52–70
Cervero R, Golub A (2023) Integrating informal transport into formal planning: Lessons from South Africa. Transp Integr Policy 5(1):33–59
Chen H, Mwangi T, Silva P (2025a) Hybrid transport regimes in the Global South: Informal mobility and carbon lock-in. Int J Sustain Cities 9(1):118–140
Chen Y, Abdullah M, Wu K (2025b) Understanding informal urban transport: Patterns and implications for sustainable mobility. Sustain Cities Soc 95:104670. https://doi.org/10.1016/j.scs.2025.104670
Chen Y, Li K, Hu M (2025c) Social and institutional lock-in in urban infrastructure development: Evidence from the Yangtze River Delta. Sustain 17(12):5229. https://doi.org/10.3390/su17125229
Chen Y, Zhang X, Wang L (2025d) Informal transport and sustainability transitions: Evidence from rapidly urbanizing cities. J Transp Geogr 112:103681. https://doi.org/10.1016/j.jtrangeo.2024.103681
Curtis C, Scheurer J (2026) Coordinated land-use and transport planning for resilience and sustainability. J Urban Plan Dev 152(1):1–14
Delbosc A, Currie G (2024) The spatial impacts of road-oriented development and private vehicle dependence. Transport Geography 98:103–119
Ekers M, Loftus A, Shear B (2024) Behavioral lock-ins and sustainability transitions in urban mobility. Environ Policy Gov 33(2):102–123
Foxon TJ, Hammond GP, Pearson PJG (2021). Socio-technical transitions in UK electricity: part 2—technologies and sustainability. Proc Inst Civil Eng Energy 173(3):123–136. https://doi.org/10.1680/jener.19.00076
Geels FW (2011) The multi-level perspective on sustainability transitions: Responses to seven criticisms. Environ Innov Soc Trans 1(1):24–40
Geels FW (2018) Socio-technical transitions to sustainability: A review of criticisms and elaborations of the multi-level perspective. Curr Opin Environ Sustain 33:47–55. https://doi.org/10.1016/j.cosust.2018.05.005
Geels FW, Schot J (2021) The dynamics of transitions: A socio-technical perspective. In: Geels FW (ed) Sustainability transitions: Theory and practice. Routledge, pp 45–68
Goldstein A, Unruh G, Carrillo-Hermosilla J (2023a) Infrastructure lock-in and sustainability transitions. Environ Innov Soc Trans 45:1–18
Goldstein B, Gounaridis D, Newell JP (2023b) The hidden carbon footprint of urban infrastructure. Nat Sustain 6(2):123–130. https://doi.org/10.1038/s41893-022-00980-0
Goldstein JE, Neimark B, Garvey B, Phelps J (2023c) Unlocking "lock-in" and path dependency: A review across disciplines and socio-environmental contexts. World Dev 161:106116. https://doi.org/10.1016/j.worlddev.2022.106116
Grasso M, Delatin Rodrigues C (2025) Policy instruments for low-carbon infrastructure transitions. Energy Policy 172:113429
Herman KS, Hall JK, Sovacool BK, Iskandarova M (2025a) The industrial decarbonization paradigm: Carbon lock-in or path renewal in the United Kingdom? Ecol Econ 235:108628. https://doi.org/10.1016/j.ecolecon.2025.108628
Herman KS, Sovacool BK, Kim J (2025b) Beyond technology: Socio-technical barriers to energy transitions. Energy Res Soc Sci 109:103412. https://doi.org/10.1016/j.erss.2024.103412

Hodson M, Marvin S (2021) Urban infrastructure trajectories and lock-in effects. J Urban Aff 43(5):687–707
Hoffmann C (2025) Decarbonization rush? The problem of speed in the energy transition. Econ Soc 54(4):789–810. https://doi.org/10.1080/03085147.2025.2588931
Hult C, Hu X (2025) Tracking progress towards net zero mobility: a concept for a mobility transition index. In: McNally C, Carroll P, Martinez-Pastor B, Ghosh B, Efthymiou M, Valantasis-Kanellos N (eds) Transport transitions: advancing sustainable and inclusive mobility (Lecture Notes in Mobility). Springer, pp 87–93. https://doi.org/10.1007/978-3-031-89444-2_13
IPCC (2022) Climate change 2022: mitigation of climate change. contribution of working group III to the sixth assessment report of the intergovernmental panel on climate change. Cambridge University Press. https://www.ipcc.ch/report/ar6/wg3/
International Energy Agency (2023a) World energy outlook 2023. https://www.iea.org/reports/world-energy-outlook-2023
International Energy Agency (2023b) Breakthrough agenda report 2023: road transport. IEA, IRENA & UN Climate Change High-Level Champions. https://www.iea.org/reports/breakthrough-agenda-report-2023
Kaya D, Seto K, Zhang L (2025) Urban infrastructure and path dependency in low-carbon transitions. Glob Environ Chang 73:102486
Kern F, Rogge KS, Howlett M (2025) The acceleration of low-carbon transitions: Insights, concepts, challenges, and new directions for research. Energy Res Soc Sci 121:103948. https://doi.org/10.1016/j.erss.2025.103948
Liu S, Guo Y, Wagner F, Liu H, Cui RY, Mauzerall DL (2024) Diversifying heat sources in China's urban district heating systems will reduce risk of carbon lock-in. Nat Energy 9:1021–1031. https://doi.org/10.1038/s41560-024-01560-4
Lucas K, Wang F (2023) Mobility patterns and energy demand: Transport-urban form interactions. J Transp Energy 11(3):189–207
Meadowcroft J, Longhelle A (2026) Transition of the transport sector in a mining region: a cross-impact balance analysis. Eur J Futures Res 14(7). https://doi.org/10.1186/s40309-026-00267-x
Meadowcroft J, Rosenbloom D (2023). Governing the net-zero transition: strategy, policy, and politics. Proc Natl Acad Sci 120(47):e2207727120. https://doi.org/10.1073/pnas.2207727120
Mudzonga E, Chikozho C, Phiri T (2025) Electrification of informal transport in African cities: Opportunities and governance challenges. Energy Res Soc Sci 102:103689. https://doi.org/10.1016/j.erss.2024.103689
Organisation for Economic Co-operation and Development [OECD] (2023) ITF transport outlook 2023. OECD Publishing. https://doi.org/10.1787/b6cc9ad5-en
Pacini C, Bauknecht D (2025) Steering sustainability transitions with reflexivity: Policy strategies for handling emergent lock-ins and path dependencies. Energy Res Soc Sci 96:104423. https://doi.org/10.1016/j.erss.2025.104423
Percastre Gómez A, Bonilla D, Banister D (2024) Decoupling transport-CO2 emissions: Mexico, Spain and The USA: a trend analysis. Transp Res Part D Transp Environ 137:104510. https://doi.org/10.1016/j.trd.2024.104510
Pojani D, Stead D (2015) Sustainable urban transport in the developing world: beyond megacities. Sustainability 7(6):7784–7805. https://doi.org/10.3390/su7067784
Pollitt H, Shar I (2026) A policy framework for clean energy technology adoption. npj Clean Energy 2:9. https://doi.org/10.1038/s44406-026-00022-y
Seto KC, Davis SJ, Mitchell RB, Stokes EC, Unruh G, Ürge-Vorsatz D (2016) Carbon lock-in: types, causes, and policy implications. Annu Rev Environ Resour 41:425–452. https://doi.org/10.1146/annurev-environ-110615-085934
Seto KC, Davis SJ, Mitchell RB, Stokes EC, Unruh G, Ürge-Vorsatz D (2021) Carbon lock-in: Types, causes, and policy implications. Annu Rev Environ Resour 46:425–452. https://doi.org/10.1146/annurev-environ-012220-011555
Stern N, Stiglitz JE (2024) Climate change and growth. Grantham Research Institute on Climate Change and the Environment, London School of Economics and Political Science

Sovacool BK, Kester J, Noel L (2025) Systems approaches to infrastructure transitions. Energy Res Soc Sci 71:102–124

UITP (2025) Informal mobility in global cities: Trends and implications. International Association of Public Transport

United Nations Environment Programme (2024) Emissions Gap Report 2024: no more hot air … please! With a massive gap between rhetoric and reality, countries draft new climate commitments. Nairobi: UNEP. https://doi.org/10.59117/20.500.11822/46404

Unruh GC (2000) Understanding carbon lock-in. Energy Policy 28(12):817–830. https://doi.org/10.1016/S0301-4215(00)00070-7

Watson M (2025) How theories of practice can inform transition to a decarbonised transport system. Working paper, White Rose Research Online. https://eprints.whiterose.ac.uk/74552/

Zhang H, Martínez A, Singh R (2025a) Urban path dependency and infrastructure lock-in in emerging megacities. J Urban Syst 21(1):35–59

Zhang X, Li H, Chen Y (2025b) Urban infrastructure and carbon lock-in: Understanding path dependency in low-carbon transitions. Energy Res Soc Sci 97:103034. https://doi.org/10.1016/j.erss.2023.103034

Circular Economy in Transport and Logistics: Pathways to Resource Efficiency, Reverse Logistics, and Zero-Emission Supply Chains in Developing Countries

Ernest Mugoni and Pardon J. Muzondo

Abstract The transition from linear take-make-dispose models to circular economy principles represents a paradigm shift for transport and logistics systems globally. This chapter investigates the application of circular economy principles to transport and logistics, examining three interconnected dimensions: resource efficiency optimisation, reverse logistics and life cycle management, and the establishment of zero-emission supply chains with particular focus on developing countries. Through a narrative synthesis of contemporary literature (2020–2026) drawing on 95 sources, the chapter develops a conceptual framework that integrates dynamic capability theory and institutional theory to explain how logistics firms sense, seize, and reconfigure resources to achieve circular outcomes. Dynamic capability theory provides the analytical lens for understanding how technology adoption, encompassing artificial intelligence, the Internet of Things (IoT), and blockchain serves as an organisational capability enabling firms to reconfigure supply chain processes in line with circular economy imperatives. Institutional theory complements this by explaining how regulatory environments, industry norms, and policy mandates shape the adoption of circular practices. The findings reveal that emerging technologies serve as critical enablers of circular logistics, while regulatory pressure moderates the relationship between technology adoption and sustainability outcomes. For developing countries, the chapter moves beyond generic observations to examine context-specific dynamics: the structural role of the informal sector in reverse logistics, infrastructure and grid-capacity constraints on electric vehicle uptake, the enabling potential of mobile-first digital platforms, and the strategic importance of international climate

E. Mugoni (✉) · P. J. Muzondo
Department of Supply Chain Management, Faculty of Agribusiness and Entrepreneurship, Marondera University of Agricultural Sciences and Technology (MUAST), Marondera, Zimbabwe
e-mail: ernmugoni@gmail.com

P. J. Muzondo
e-mail: pmuzondo@muast.ac.zw

J. Kanyepe and I. Chirisa (eds.), *Advancing the Green Agenda in the Transport and Logistics Sector in Global South Countries*, World Sustainability Series,
https://doi.org/10.1007/978-3-032-31374-4_6

finance mechanisms including the Climate Investment Funds' Industry Decarbonisation programme. The chapter concludes with managerial implications for logistics service providers and targeted policy recommendations for governments seeking to accelerate circular transition in transport and logistics systems in the Global South.

Keywords Circular economy · Transport logistics · Reverse logistics · Zero-emission supply chains · Developing countries · Resource efficiency · Emerging technologies · Dynamic capabilities · Institutional theory

1 Introduction

The contemporary global economy remains predominantly linear, operating on a take-make-dispose trajectory that maximises resource extraction whilst externalising environmental costs (Seroka-Stolka and Ociepa-Kubicka 2019). Transport and logistics systems, as the circulatory system of global commerce, account for approximately 40% of global CO_2 emissions and up to 70% of other pollutants in urban areas (Melkonyan-Gottschalk and Li 2026). The urgency of transitioning towards circular models has intensified as supply chains face unprecedented disruptions from geopolitical instabilities, climate events, and resource scarcity (Dzwigol et al. 2021).

The circular economy proposes a regenerative economic paradigm one that heals and rejuvenates itself, retaining products, components, and materials at their highest utility value at all times (Van Buren et al. 2016). For the transport and logistics sector, this entails rethinking how goods move, how vehicles are powered, how materials re-enter the production cycle, and how infrastructure is designed to support closed-loop systems (Beames et al. 2021). Logistics service providers occupy a strategically privileged position in this shift, acting as connectors between organisations, coordinators of material flows, and poolers of shared resources across the supply chain (Mate et al. 2025).

This chapter makes an explicit and applied use of dynamic capability theory and institutional theory to explain how these transitions occur, and why they succeed or fail in different organisational and national contexts. Dynamic capability theory (Teece et al. 1997) explains how firms develop the capacity to sense circular economy opportunities, seize them through technology adoption, and reconfigure their operational processes to sustain circular outcomes. Institutional theory (DiMaggio and Powell 1983) explains why regulatory environments, policy signals, and industry norms are not merely background conditions but active moderators of the technology-sustainability relationship. These two frameworks together generate testable and actionable propositions that go beyond descriptive accounts of technology adoption.

The chapter focuses on three interrelated dimensions of circular economy application in transport and logistics. First, it examines how circular economy principles can enhance resource efficiency across vehicle utilisation, packaging systems, and logistics infrastructure. Second, it investigates how reverse logistics and life cycle management mechanisms close material loops and recover value from end-of-life products. Third, and most distinctively, it explores pathways to zero-emission supply chains in developing countries, with sustained attention to context-specific barriers and opportunities that generic reviews routinely overlook.

The focus on developing countries is both timely and analytically necessary. These economies face distinctive structural conditions: rapid urbanisation driving logistics demand ahead of green infrastructure provision; large informal sectors that simultaneously hinder and enable circular material flows; grid-capacity and financing constraints on electric vehicle adoption; and institutional voids that weaken regulatory pressure the very moderating variable that institutional theory identifies as critical. At the same time, developing countries enjoy leapfrogging opportunities the ability to bypass carbon-intensive legacy infrastructure by adopting frontier technologies from the outset and they are increasingly targeted by international climate finance mechanisms, including the Climate Investment Funds' 2025 Industry Decarbonisation programme (Climate Investment Funds 2025). Analytically, these conditions require the framework to be adapted rather than applied wholesale from developed-economy contexts, and this chapter undertakes that adaptation explicitly.

The chapter is structured as follows. Section 2 reviews the evolution of circular economy concepts in logistics and the strategic position of logistics service providers. Section 3 sets out the three research objectives. Section 4 presents the conceptual framework. Section 5 describes the methodological approach. Section 6 synthesises the literature across the three research dimensions. Section 7 discusses the findings in relation to theory and practice. Sections 8, 9, and 10 address future research directions, managerial and policy implications, and conclusions.

2 Background to the Study

2.1 Evolution of Circular Economy Concepts in Logistics

The circular economy has deep theoretical roots, but its application to logistics is comparatively recent. Early logistics scholarship focused on green logistics, which sought to reduce the environmental impact of logistics activities through emissions reduction, energy efficiency, and the adoption of renewable energy (Seroka-Stolka and Ociepa-Kubicka 2019). Green logistics is now recognised as a constituent pathway towards the circular economy (Dzwigol et al. 2021; Jayarathna et al. 2023).

The relationship between green logistics and the circular economy involves both overlap and distinction. Green logistics focuses principally on the environmental impacts of logistics activities emissions, noise, congestion whilst the circular economy seeks a systemic transformation of the economy towards sustainable and regenerative principles (De Souza et al. 2022). Reverse logistics, which manages the flow of goods from end consumers back to manufacturers or processors, sits at the intersection of these two domains (Yu et al. 2018). Mate et al. (2025) identify persistent conceptual ambiguity in this space, noting that the logistics literature remains heavily focused on green efficiency metrics and manufacturer-centric circular models, undervaluing the systemic contribution of logistics service providers.

3 The Strategic Position of Logistics Service Providers

Logistics service providers are the connective tissue of supply chains: they have direct line-of-sight to material flows and exercise significant influence over how resources circulate through the economy (Van Buren et al. 2016). In a circular economy, where closed material loops and end-of-life product recovery are central, this positional advantage is strategically significant (Turoń and Czech 2016).

Yet logistics has traditionally been positioned as a cost centre rather than an innovation hub. When supply chains are analysed through the Supply Chain Operations Reference (SCOR) framework, the 'make' phase attracts the most innovation attention, whilst the 'deliver' and 'return' phases are associated with efficiency rather than value creation (Vegter et al. 2020). The circular economy reconfigures this hierarchy: the 'return' phase becomes a strategic value recovery function, and logistics service providers become architects of closed-loop systems rather than executors of point-to-point delivery.

4 The Developing Country Context

Developing countries present a distinctive and internally varied set of conditions for circular economy adoption in transport and logistics. These conditions cannot be reduced to a uniform narrative of 'challenges'; they require differentiation across income levels, urbanisation trajectories, institutional capacities, and sector structures.

In lower-middle-income economies such as Nigeria and Zimbabwe, infrastructure deficits are the primary constraint: unreliable electricity grids limit electric vehicle adoption; poor road conditions increase vehicle wear and fuel consumption; and limited cold-chain or digital infrastructure impedes the traceability systems on which reverse logistics depends. The informal sector is large, structurally embedded, and simultaneously a barrier to formalised circular systems and an operational backbone of waste recovery waste pickers, informal repair shops, and second-hand goods markets collectively manage volumes of material that formal systems cannot absorb (Nwaichi et al. 2025). Regulatory frameworks are often fragmented, under-enforced, or absent, weakening the institutional pressure that theory identifies as a key moderator of the technology-sustainability relationship.

In upper-middle-income developing economies such as Brazil, South Africa, and Malaysia, the picture differs. These countries have more developed logistics infrastructure, stronger regulatory institutions, and greater access to international capital markets. South Africa, for example, has a formalised waste picker cooperative movement that provides a model for informal sector integration (Mugoni et al. 2024). Brazil's national solid waste policy mandates the inclusion of waste picker cooperatives in municipal waste management systems. Malaysia has adopted national circular economy roadmaps with quantified targets. These differentiated contexts demand context-specific analysis rather than blanket 'developing country' generalisations.

The leapfrogging opportunity the ability to bypass carbon-intensive legacy infrastructure by adopting frontier technologies from the outset is real but unevenly distributed. Mobile telecommunications penetration exceeding 80% in many developing countries creates a foundation for platform-based logistics coordination. However, rural–urban digital divides mean that platform-based solutions often consolidate efficiency gains in urban logistics corridors while leaving rural and peri-urban areas underserved. The analytical implication is that leapfrogging potential must be assessed against infrastructure realities and distributional consequences, not assumed as a universal developing-country advantage.

5 Research Objectives

This chapter addresses three primary research objectives, each corresponding to a dimension of circular economy application in transport and logistics:

RO1: To investigate how circular economy principles can enhance resource efficiency in transport and logistics systems. This encompasses vehicle utilisation, load consolidation, packaging optimisation, infrastructure longevity, and the role of emerging technologies as efficiency enablers.

RO2: To examine the development of reverse logistics and life cycle management as mechanisms for closing material loops. This includes returns management, repair and refurbishment, remanufacturing, and end-of-life material recovery, with attention to both operational processes and strategic positioning.

RO3: To explore pathways for establishing zero-emission supply chains in developing countries. This addresses technological transitions, infrastructure requirements, policy frameworks, financing mechanisms, informal sector dynamics, and the distinctive socio-economic contexts that shape circular economy adoption in the Global South.

These objectives are analytically interconnected: resource efficiency provides the operational foundation for circular logistics systems; reverse logistics operationalises material loop closure; and zero-emission transitions in developing countries represent the frontier application of circular economy principles, where the enabling conditions of theory must be adapted to structural realities.

6 Conceptual Framework

The framework presented in Fig. 1 integrates Dynamic Capability Theory (Teece et al. 1997) and Institutional Theory (DiMaggio and Powell 1983) into a capability–constraint interaction model, explaining how and under what conditions emerging technologies produce circular economy outcomes in developing-country logistics systems.

Dynamic capabilities are operationalised as three technology-enabled mechanisms: sensing (IoT/AI systems detecting inefficiencies in forward and reverse flows), seizing (strategic deployment of AI optimisation and blockchain traceability), and reconfiguring (structural redesign of logistics networks around circular principles). Critically, the adoption of emerging technologies (AOET) is positioned not as a direct driver of sustainability outcomes but as the observable expression of sensing and seizing capabilities—firms that layer technology onto unchanged linear operations realise efficiency gains but not circular transformation (Mate et al. 2025).

Institutional pressures—coercive, mimetic, and normative—function as conversion filters determining whether dynamic capabilities translate into circular outcomes. Grounded in Kumari et al. (2025), who empirically confirm that regulatory pressure (RP) significantly moderates the AOET–SICE relationship ($\beta = 0.099$; $p = 0.015$) in China's supply chain, the framework positions RP not as a background

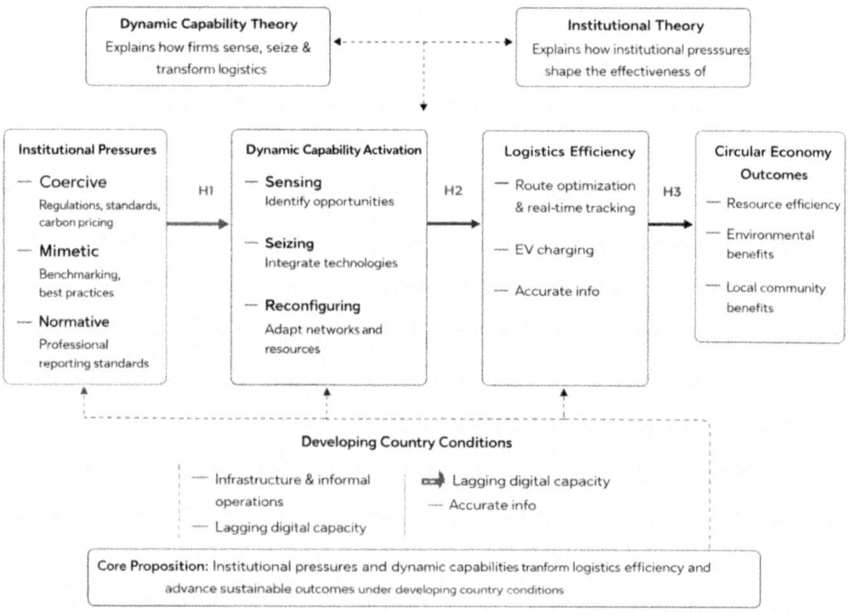

Fig. 1 Conceptual framework linking technology adoption, logistics efficiency, regulatory pressure, and circular economy sustainability (Authors 2026)

condition but as an active moderator that amplifies the effectiveness of technology adoption under strong compliance environments.

The core causal pathway—AOET → Enhanced Logistics System efficiency (ELS) → Sustainability in the Circular Economy (SICE)—is supported by Kumari et al.'s (2025) finding that ELS partially mediates the AOET–SICE relationship ($\beta = 0.193$; $p = 0.000$), confirming that logistics system transformation is the mechanism through which technology produces circular outcomes. Three developing-economy contextual moderators further condition this pathway: infrastructure constraints suppress capability activation; informal sector fragmentation interrupts reconfiguring capabilities; and leapfrogging potential can accelerate transformation where institutional conditions are enabling (Climate Investment Funds 2025).

This generates three testable hypotheses: H1 (institutional pressure conditions AOET effectiveness); H2 (dynamic capabilities expressed through AOET positively transform ELS); and H3 (improved ELS enhances SICE outcomes).

H1: Institutional pressures positively condition the effectiveness of emerging technology adoption by shaping the activation and direction of dynamic capabilities in logistics firms.

H2: Dynamic capabilities (sensing, seizing, reconfiguring), expressed through emerging technologies, positively transform logistics system efficiency.

H3: Improved logistics system efficiency driven by dynamic capabilities positively enhances circular economy sustainability outcomes.

7 Research Methodology

This chapter employs a structured narrative review methodology, drawing on 95 peer-reviewed publications and authoritative grey literature sources published between 2020 and 2026. The choice of narrative rather than formal systematic review methodology is deliberate and transparent: the research objectives span three analytically distinct domains (resource efficiency, reverse logistics, and developing country zero-emission transitions), each requiring thematic synthesis rather than the aggregation of comparable quantitative effect sizes that systematic review meta-analysis entails. A narrative approach allows the integration of empirical studies, theoretical developments, case evidence, and policy analysis across these domains in a manner that a single systematic review protocol would artificially constrain (Tranfield et al. 2003).

The review nevertheless follows structured and replicable procedures to ensure rigour and transparency.

8 Search Strategy and Database Selection

Searches were conducted across five databases: Scopus, Web of Science, ScienceDirect, SpringerLink, and Wiley Online Library. Search terms were constructed using Boolean operators to combine domain-specific terminology: circular economy concepts ("circular economy" OR "circular logistics" OR "green logistics"), transport and logistics terms ("transport" OR "logistics" OR "supply chain" OR "reverse logistics"), and developing country qualifiers ("developing countr*" OR "emerging econom*" OR "Global South" OR "Africa" OR "Asia" OR "Latin America"). Technology-specific search strings ("artificial intelligence" OR "blockchain" OR "Internet of Things") were combined with circular economy and logistics terms to capture the technology adoption literature relevant to RO1 and RO2. The temporal boundary of 2020–2026 was applied to ensure currency, with exceptions for foundational theoretical works (Teece et al. 1997; DiMaggio and Powell 1983; Barney 1991) that anchor the conceptual framework.

9 Inclusion and Exclusion Criteria

9.1 The Following Inclusion Criteria Were Applied

- **Scope**: studies must address circular economy principles in the context of transport, logistics, or supply chain management. Studies focused exclusively on manufacturing without logistics relevance were excluded.
- **Publication type**: peer-reviewed journal articles and book chapters were included. Conference proceedings were included only where the work had not been subsequently published in a journal. Grey literature from international organisations (UNIDO, Climate Investment Funds, UNCTAD) was included where it provided primary data or policy evidence not available in peer-reviewed sources.
- **Language**: English-language publications only, acknowledging the limitation that this excludes relevant research published in Spanish, Portuguese, French, Chinese, and other languages.
- **Quality assessment**: studies were assessed for methodological clarity, theoretical grounding, and relevance to the three research objectives. Studies lacking methodological transparency or theoretical basis were excluded from the synthesis, although they may be cited in passing.

The initial search and database screening returned 95 publications meeting inclusion criteria after removal of duplicates and irrelevant results. Of these, 92 (97%) were published between 2021 and 2026, reflecting the recency of scholarly attention to the intersection of circular economy and logistics. Reference list searching of key articles contributed an additional 38 sources, including foundational theoretical works and policy documents. All sources were managed using reference management software and coded according to research objective, country context, technology focus, and methodological approach to facilitate thematic synthesis.

10 Analytical Approach

Thematic synthesis was employed as the analytical method (Thomas and Harden 2008). Each source was coded against the three research objectives and the constructs of the conceptual framework (technology adoption, logistics efficiency, regulatory pressure, circular sustainability). Within each theme, the synthesis identifies: (a) the mechanisms through which relationships operate; (b) enablers and barriers documented in the empirical literature; (c) outcomes reported at firm, sector, and national levels; and (d) the contextual conditions particularly developing country institutional and infrastructure conditions that moderate these relationships. Particular analytical weight was given to studies conducted in developing country contexts and to those employing theoretical frameworks consistent with the AOET-ELS-SICE model, enabling assessment of the framework's empirical support and its boundary conditions.

11 Literature Review

11.1 Resource Efficiency in Circular Transport and Logistics Systems

11.1.1 Defining Resource Efficiency in Circular Contexts

Circular resource efficiency differs fundamentally from conventional logistics efficiency. Whereas linear efficiency focuses on cost minimisation within a given operational configuration, circular efficiency entails managing materials, energy, assets, and infrastructure across their entire life cycle to maximise value retention and eliminate waste (Zheng and Zhang 2010; Ripanti and Tjahjono 2019). This distinction has profound implications for how efficiency is measured and managed: conventional indicators such as cost per kilometre, vehicle utilisation rates, and on-time delivery must be complemented by material circulation metrics, embodied carbon accounting, and end-of-life value recovery rates (De Souza et al. 2022; Andrejć 2023). Applied through the dynamic capabilities lens, circular resource efficiency is not simply an operational outcome but a sensing and reconfiguring capability: firms must develop the organisational routines and data systems to recognise where value is being lost across the logistics life cycle, and the operational flexibility to redesign processes accordingly.

11.1.2 Vehicle Utilisation and Load Consolidation

Vehicle utilisation is a core efficiency lever with direct circular economy relevance: higher utilisation reduces the number of vehicles needed, lowering embodied emissions from vehicle manufacture and reducing fuel consumption per unit of output (Makarova et al. 2019). Digital technologies particularly AI-driven route optimisation and demand aggregation platforms are the primary enablers of utilisation improvement. Jayarathna et al. (2023) demonstrate that smart load consolidation reduces vehicle kilometres travelled by 15–30%, yielding proportional emissions reductions and congestion relief.

The circular economy perspective extends vehicle utilisation considerations to vehicle design and end-of-life management. Modular vehicle designs that facilitate component repair, parts reuse, and end-of-life material recovery align circular economy principles with efficiency logic: a vehicle designed for disassembly and remanufacturing generates circular value at each life cycle stage, not only in operational deployment (Makarova et al. 2021). Dynamic platforms that aggregate shipper demand and optimise routes represent a seizing capability in dynamic capability terms, enabling firms to achieve utilisation levels that individual operators cannot reach independently (Mate et al. 2025).

11.1.3 Packaging Optimisation and Reusable Systems

Packaging represents both a major source of logistics waste and a significant opportunity for circular innovation. Single-use packaging norms generate substantial disposal costs and material losses; circular packaging principles returnable, recyclable, and reusable systems—retain material value across multiple use cycles (Schneikart et al. 2024). Schneikart et al. (2024) report a 40–60% reduction in material usage for pharmaceutical logistics through the deployment of reusable containers integrated with Industry 4.0 tracking technologies, demonstrating the scale of potential gains.

The transition to returnable packaging systems requires corresponding logistics reconfiguration: collection points, reverse flow coordination, and container cleaning and refurbishment processes (Beames et al. 2021). Logistics service providers are uniquely positioned to manage this reconfiguration, as they control both forward and reverse flow networks. This is a reconfiguring capability in dynamic capability terms: firms that redesign their network architectures to accommodate bidirectional packaging flows create operational advantages that are difficult for competitors without circular logistics infrastructure to replicate.

11.1.4 Infrastructure Longevity and Circular Design

Transport infrastructure roads, railways, ports, and logistics terminals represents substantial embedded capital in materials and energy. Extending infrastructure lifespans through circular design principles and predictive maintenance reduces the environmental costs of replacement and protects the value of embodied resources (De Abreu et al. 2022; Yildizbasi et al. 2025). Digital technologies play a central enabling role: sensor networks monitor pavement condition and structural performance, enabling data-driven maintenance scheduling that extends infrastructure lifespans whilst minimising resource consumption (Asensio et al. 2020). Building information modelling (BIM) and geographic information systems (GIS) support circular infrastructure planning by enabling life cycle assessment at the design stage (Irizarry et al. 2013).

11.1.5 Emerging Technologies as Efficiency Enablers

AI, IoT, and blockchain are the three technologies most consistently identified in the literature as enablers of circular logistics efficiency. AI optimises routes, consolidates loads, and predicts maintenance needs, reducing empty trips and extending asset lifespans (Kokkinou and Mitas 2026). IoT enables real-time tracking of asset location, condition, and utilisation, supporting informed decisions about repair, redeployment, and retirement (Sarkis et al. 2025). Blockchain creates immutable records of material provenance and lifecycle history, enabling accountability in circular processes and facilitating secondary market transactions (Bekrar et al. 2021).

Li et al. (2025) provide the most direct empirical test of the AOET-ELS-SICE framework, finding a positive relationship between technology adoption and logistics efficiency in Chinese supply chains, with regulatory pressure moderating the strength of this relationship. Wang et al. (2025) extend this by demonstrating that logistics efficiency mediates the technology-sustainability relationship, confirming the mediating role specified in H2-H3 of the framework. Mate et al. (2025) identify an important qualification: many logistics firms adopt technology within unchanged linear operational frameworks, realising efficiency gains but not circular economy transformation. The reconfiguring capability using technology to redesign processes around circular principles rather than merely optimising existing linear ones is the decisive differentiator.

11.2 Reverse Logistics and Life Cycle Management

11.2.1 Conceptual Foundations

Reverse logistics encompasses the processes managing product and material flows from end users back to manufacturers, processors, or recovery facilities for returns, repairs, refurbishment, recycling, and disposition (Yu et al. 2018). In a circular economy, reverse logistics is not a residual function but the primary mechanism through which material loops are closed and value is recovered (Biancolin et al. 2023). The scale of reverse logistics is substantial: the Institute for Supply Management (2025) estimates that reverse logistics flows account for 17% of retail sales in the United States, representing a global market of over one trillion dollars annually.

Butt et al. (2024) identify five key reverse logistics processes in retail contexts: returns management, inspection and testing, repair and refurbishment, recycling, and disposition. High-performing firms integrate these processes with forward logistics and inventory management, ensuring that returned goods of appropriate quality are repurposed to fulfil orders rather than being written off or disposed of at discount.

11.2.2 The Five Rs Framework

The Five Rs framework returns, reselling, repairs, replacements, and recycling provides a structured taxonomy for reverse logistics decision-making (Institute for Supply Management 2025). Returns management involves assessing product condition and return eligibility. Reselling leverages residual value through secondary markets. Repairs restore functionality whilst maintaining customer relationships. Replacements honour service commitments whilst managing inventory exposure. Recycling recovers material value from products that cannot be economically repaired or resold.

Each R demands distinct logistics capabilities: returns management requires efficient intake processes; reselling requires secondary market access and pricing intelligence; repair requires technical capacity and access to spare parts supply chains; recycling requires specialist processing facilities and material recovery networks. Makarova et al. (2021) illustrate these distinctions through automotive spare parts logistics, where returned components are graded for refurbishment or repair suitability, with non-repairable items entering material recovery streams and full electronic tracking maintained throughout.

11.2.3 Repair, Refurbishment, and Remanufacturing

Repair, refurbishment, and remanufacturing represent a hierarchy of value recovery intensity. Repair restores basic functionality through component replacement or adjustment. Refurbishment restores products to near-new condition through comprehensive reconditioning. Remanufacturing disassembles products to component level, replaces worn or failed parts, and reassembles to original specifications (Makarova et al. 2018).

Each level makes distinct demands on logistics systems. Repair requires bidirectional parts flows, access to technical documentation, and quality validation. Refurbishment requires space and equipment for reconditioning, testing facilities, and work-in-progress management. Remanufacturing requires full manufacturing-scale facilities, consistent replacement parts supply chains, and comprehensive quality control processes (Mugoni et al. 2023). Logistics service providers are increasingly offering these processes as value-added services: Kuehne + Nagel's aftermarket logistics service for electronic and industrial equipment exemplifies this transition from cost centre to value creator (Institute for Supply Management 2025).

11.2.4 Life Cycle Management and Material Traceability

Life cycle management treats reverse logistics as a sustained material stewardship responsibility rather than a series of discrete transactions. It involves tracking materials through multiple reuse cycles and designing products and packaging to facilitate end-of-life recovery (Beames et al. 2021). Traceability is the enabling condition: knowing what material exists, where it is, what condition it is in, and what has been done to it is foundational to circular material management.

Emerging digital tools are transforming traceability at scale. Blockchain enables tamper-resistant records of material history, ensuring accountability in multi-stakeholder circular value chains (Chen et al. 2024). IoT provides real-time location and condition data for products in circulation. Digital product passports currently under development within the EU regulatory framework will aggregate material composition, component provenance, and recyclability information at the product level (Yildizbasi et al. 2025). Renkin and Limbourg (2025) demonstrate the analytical integration of life cycle assessment into reverse logistics network optimisation for construction and demolition waste, illustrating how traceability data enables simultaneous optimisation of cost and environmental impact.

11.2.5 Reverse Logistics Network Design

Reverse logistics network design presents fundamentally different challenges from forward network design. Return flows originate from dispersed end users rather than consolidated facilities; volumes are variable and difficult to forecast; and product conditions range from near-new to end-of-life, requiring differentiated processing pathways (Renkin and Limbourg 2025). These characteristics create a tension between collection coverage and processing efficiency: centralised processing achieves economies of scale but lengthens return transportation distances; decentralised networks minimise transport but compromise processing capacity. A hybrid model distributed collection points feeding into regional processing centres with links to centralised remanufacturing facilities is identified by De Souza et al. (2022) as the optimal structure for most contexts.

Mathematical optimisation models, such as the multi-stage model proposed by Renkin and Limbourg (2025), enable simultaneous optimisation of facility locations, inventory placement, and transportation routing across differentiated processing technologies, accommodating their respective cost and environmental profiles.

11.3 Zero-Emission Supply Chains in Developing Countries

11.3.1 The Imperative and the Context-Specific Opportunity

Developing countries face a distinctive configuration of pressures and opportunities in the transition to zero-emission logistics. Transportation emissions in these economies are growing at rates significantly exceeding the global average, driven by rapid motorisation, urbanisation, and expanding trade volumes (Mugoni et al. 2025a, b). This growth trajectory, if unaddressed, will lock in high-carbon transport infrastructure for decades. Simultaneously, the absence of entrenched legacy infrastructure creates a genuine leapfrogging opportunity: countries that are building logistics networks from a relatively early stage can potentially align new investments with zero-emission technologies from the outset (Climate Investment Funds 2025).

However, the leapfrogging opportunity is distributed unevenly across and within developing countries, and it is contingent on conditions that institutional theory identifies as critical. Four conditions determine whether leapfrogging potential is realised: (1) the availability of concessional or grant financing to offset the higher upfront capital cost of zero-emission technologies relative to conventional alternatives; (2) the institutional capacity to develop and implement enabling policy frameworks; (3) the technical capacity to operate and maintain new technologies; and (4) the infrastructure capacity grid reliability, road quality, digital connectivity that determines technology performance in practice. Where these conditions are absent or weak, leapfrogging potential remains unrealised regardless of the theoretical opportunity. This is why institutional theory's moderating variable regulatory and institutional pressure is particularly consequential in developing country analysis.

Mugoni et al. (2025a, b) examine the nexus between circular economy principles and sustainable supply chain management in developing countries, arguing that the analytical value of this nexus may be greatest precisely where infrastructure is at an early stage, since circular principles can be embedded in new investments rather than retrofitted to existing systems.

11.3.2 Electric Vehicle Transition: Context-Specific Barriers and Conditions

Electric vehicles are the primary technology pathway for zero-emission last-mile and urban logistics in most developing country contexts. Battery-powered vans, trucks, and two- and three-wheelers are increasingly cost-competitive in total cost of ownership terms in many settings (Bonsu 2020). However, the barriers to adoption are substantial and context-specific, varying systematically with income level, grid infrastructure, and supply chain configuration.

In sub-Saharan African contexts such as Zimbabwe, Nigeria, and Tanzania, the binding constraint is grid capacity and reliability. Electric vehicle charging demands steady, reliable electricity supply; where grid coverage is intermittent or voltage is unstable conditions common across urban sub-Saharan Africa depot charging becomes unreliable and en-route charging is typically unavailable (De Abreu et al. 2022). Off-grid solar charging solutions are technically feasible but add capital cost and complexity. Battery replacement and recycling present a further challenge: the battery supply chain is globally concentrated in China, Europe, and North America, meaning developing countries import batteries at high foreign exchange cost and must develop domestic end-of-life battery processing capacity to avoid creating a new waste management problem (Chen et al. 2024).

South and Southeast Asian contexts present a more heterogeneous picture. India's FAME (Faster Adoption and Manufacturing of Electric Vehicles) initiative has successfully accelerated commercial EV adoption through capital subsidies and charging infrastructure investment, demonstrating that demand-side policy can shift adoption curves even in infrastructure-constrained environments. India's large domestic battery manufacturing base, growing through government industrial policy, is progressively reducing import dependency. Vietnam and Indonesia have attracted significant EV manufacturing investment from Chinese producers, creating domestic supply chains that lower vehicle costs and improve after-sales service availability.

Latin American contexts demonstrate the viability of alternative fuel pathways where EV adoption is constrained. Brazil's ethanol and biodiesel programme the world's most developed biofuel logistics infrastructure offers a lower-carbon alternative to conventional diesel without the grid dependency of battery electric vehicles, and leverages existing fuelling infrastructure. Rwanda's electric motorcycle taxi programme illustrates that two- and three-wheeler electrification requiring less grid infrastructure than heavy vehicle charging can achieve rapid penetration in urban environments with appropriate policy support and business model innovation.

These differentiated country experiences generate an institutional theory insight: the EV transition pathway is not technologically determined but institutionally shaped. Countries with clear policy mandates (India's FAME), enabling industrial policy (Vietnam), and infrastructure investment programmes (Rwanda) demonstrate higher adoption rates than countries where institutional pressure is absent or fragmented.

11.3.3 Digital Leapfrogging and Platform-Based Logistics

Digital technologies provide developing countries with an opportunity to build logistics capabilities without traversing all intermediate developmental stages. Mobile telecommunications, cloud computing, and platform-based coordination enable logistics operations to be organised with relatively limited physical infrastructure (Sarkis et al. 2025). Platform-based logistics models digital freight matching, crowd-sourced last-mile delivery, and blockchain-enabled supply chain traceability illustrate this potential.

Kokkinou and Mitas (2026) examine AI's potential to support transport and logistics firms in developing countries, finding that AI-based route optimisation, predictive maintenance, and computer-vision-assisted product grading address multiple contextual constraints simultaneously: fuel consumption reduction is especially valuable where fuel costs constitute a high proportion of logistics costs; predictive maintenance is valuable where vehicle replacement costs are high and repair infrastructure is limited; and automated product grading addresses skills gaps in reverse logistics processing.

The enabling conditions for digital leapfrogging, however, require qualification. Mobile network penetration exceeding 80% in many developing countries provides the connectivity foundation, but rural–urban digital divides remain significant (Mugoni et al. 2023). Platform-based logistics models have primarily demonstrated value in urban freight corridors: they have been less effective in rural and peri-urban contexts where demand density is insufficient for matching algorithms to generate efficiency gains, and where network effects require critical mass that dispersed rural markets cannot provide. Digital leapfrogging, therefore, is an urban opportunity that must be complemented by targeted rural infrastructure investment to achieve equitable logistics system transformation.

11.3.4 Informal Sector Integration: A Structural Challenge and Opportunity

The informal sector's role in circular logistics in developing countries cannot be adequately captured by the generic observation that it is 'large' and 'important'. Its structural position varies significantly across contexts and requires differentiated analysis.

In sub-Saharan African cities such as Lagos, Nairobi, and Harare, informal waste pickers and scrap traders manage the majority of recyclable material flows that formal waste management systems cannot absorb. These operators possess detailed knowledge of material quality and market value, established relationships with secondary material processors, and the operational flexibility to collect materials from dispersed informal settlements that formal logistics networks do not serve. They constitute a de facto reverse logistics network that processes millions of tonnes of material annually a function that, if formalised and integrated, could dramatically expand circular economy performance (Nwaichi et al. 2025).

In South Asian cities such as Dhaka, Mumbai, and Lahore, informal electronics repair clusters perform a similar function for the consumer electronics waste stream, extending product lifespans through repair and component harvesting at a scale that formal extended producer responsibility systems in the same countries do not approach. These clusters operate on thin margins, serve low-income consumers who cannot access or afford formal repair services, and generate employment for millions of workers with limited formal skills credentials.

The integration challenge is not principally technical but institutional: how to create governance arrangements that incentivise informal operators to adopt standards and reporting practices consistent with formal circular economy systems, without displacing them from the livelihoods on which they depend. Brazil's cooperative model enshrined in the National Solid Waste Policy provides the most extensively documented policy framework for this integration, requiring municipal waste management systems to contract with waste picker cooperatives and providing cooperative members with access to training, equipment, and social protection (Mugoni et al. 2024). India's informal recycling clusters have been the subject of regulatory pilot programmes seeking to register operators, provide health and safety equipment, and link them to formal recycling offtake agreements. South Africa's Reclaimers cooperative movement provides a further model, having secured municipal contracts and formal sector partnerships in several cities.

Mugoni et al. (2023) demonstrate how digital technologies mobile payment systems, blockchain-based verification, and QR-code material tracking can facilitate informal sector integration by providing the traceability infrastructure that formal buyers require, without imposing the upfront infrastructure costs that would exclude informal operators.

11.3.5 International Climate Finance and Policy Frameworks

International climate finance is increasingly consequential for industrial decarbonisation in developing countries. The Climate Investment Funds' 2025 Industry Decarbonisation programme committed USD 1 billion in concessional financing to seven inaugural countries Brazil, Egypt, Mexico, Namibia, South Africa, Türkiye, and Uzbekistan to support clean and circular industrial technologies including green hydrogen, waste heat recovery, and low-carbon materials (Climate Investment Funds 2025). The UNIDO Industrial Decarbonisation Centre's USD 30 million accelerator, announced at COP30, targets collaborative ecosystem development, investment mobilisation, near-zero-emission materials market creation, competitiveness enhancement, and harmonised carbon accounting (UNIDO 2026).

The European Union's Carbon Border Adjustment Mechanism is reshaping trade policy incentives for developing country exporters: firms in countries without domestic carbon pricing face adjustment charges on EU exports, creating strong economic incentives to develop emissions monitoring and reduction capabilities independently of domestic regulatory pressure. Free trade agreements are progressively incorporating circular economy and environmental standards (UNCTAD 2026), embedding circular economy compliance requirements in market access conditions that developing country firms must meet to participate in global supply chains.

11.3.6 Barriers and Enablers: A Structured Assessment

Drawing on the theoretical framework, barriers and enablers can be organised by the construct of the AOET-ELS-SICE model they primarily affect:

Barriers affecting technology adoption (AOET):

- Infrastructure deficits: unreliable electricity, poor road conditions, and limited digital connectivity reduce the operational effectiveness of AI, IoT, and EV technologies (De Abreu et al. 2022)
- Capital constraints: high upfront technology costs relative to firm equity and debt capacity, particularly for SMEs without access to concessional financing
- Technical capacity gaps: limited availability of workers with skills to operate and maintain advanced technologies
- Policy uncertainty: inconsistent or absent regulatory mandates reduce the investment security that private capital requires (Panghal et al. 2024)

Barriers affecting logistics efficiency (ELS):

- Informal sector fragmentation: dispersed informal operators operating outside formal quality and reporting standards reduce system-level efficiency gains
- Coordination failures: limited data-sharing between logistics actors and between formal and informal operators

Barriers affecting circular sustainability outcomes (SICE):

- Weak extended producer responsibility enforcement: manufacturers have limited incentive to design for reverse logistics where enforcement of take-back requirements is absent
- Limited secondary market depth: thin markets for secondary materials reduce the economic viability of recycling and remanufacturing

Enablers:

- Political commitment to net-zero targets provides the regulatory and policy foundation for private sector investment
- International climate finance mechanisms provide concessional capital that offsets higher upfront costs of circular and zero-emission technologies
- Multinational buyer sustainability requirements create demand-side pull for circular supply chain performance
- Mobile technology infrastructure provides the connectivity foundation for platform-based logistics coordination
- Informal sector entrepreneurship generates operational innovation in reverse logistics at low cost and without formal sector investment

11.4 Results and Discussion

11.4.1 Synthesis of Findings

The review of literature across three dimensions resource efficiency, reverse logistics, and zero-emission developing country transitions reveals three stable themes.

First, significant resource efficiency gains vehicle utilisation improvements of 15–30%, packaging material reductions of 40–60%, and infrastructure lifespan extensions measurable in decades are achievable through the combination of smart technologies, process redesign, and circular business models (Jayarathna et al. 2023; Schneikart et al. 2024; De Abreu et al. 2022). Second, reverse logistics represents a trillion-dollar value recovery opportunity that high-performing firms are systematically converting from cost centre to competitive advantage through repair, refurbishment, and remanufacturing capabilities supported by digital traceability (Institute for Supply Management 2025; Butt et al. 2024). Third, developing country circular transitions involve genuine and context-specific opportunities alongside structural constraints that generic frameworks fail to capture.

11.4.2 Theoretical Implications

The findings support and refine the theoretical framework in several ways. First, they validate the dynamic capabilities proposition that technology adoption generates circular outcomes primarily when it is accompanied by organisational reconfiguring redesigning processes and business models to embed circularity rather than being layered onto unchanged linear operations (Mate et al. 2025). This distinction between technology adoption as a seizing capability and the subsequent reconfiguring capability is analytically important: firms that invest in AI or IoT without redesigning their operational logic around circular principles fail to realise circular outcomes even when technology deployment succeeds.

Second, the findings validate the institutional theory proposition that regulatory pressure moderates the technology-sustainability relationship (Li et al. 2025; Yildizbasi et al. 2025). The evidence from developing country contexts is particularly instructive: countries with clear, enforced regulatory mandates Brazil's solid waste policy, India's FAME initiative, the EU's Carbon Border Adjustment Mechanism demonstrate higher rates of circular technology adoption and more consistent circular outcomes than countries where regulatory frameworks are absent or unenforced. This finding reinforces the analytical priority of institutional conditions over technology availability in explaining circular economy adoption patterns.

Third, the framework requires extension to incorporate the three developing-country contextual moderators identified in Sect. 4.2 infrastructure capacity, informal sector integration, and leapfrogging potential as each significantly attenuates or amplifies the core AOET-ELS-SICE relationships in ways that the original framework, derived from Chinese manufacturing supply chains, does not anticipate. Future empirical testing should explicitly model these moderators to assess their relative strength across national contexts.

Finally, the review validates and extends the critique that circular economy research has been manufacturer-centric and undervalues the strategic contribution of logistics service providers (Mate et al. 2025; Vegter et al. 2020). The evidence from reverse logistics, packaging systems, and platform-based logistics consistently positions logistics service providers as the organisational actors with the greatest leverage over circular material flows a finding with significant implications for both theory and practice.

11.4.3 Practical Implications

For logistics service providers, the strategic imperatives are: developing reverse logistics capabilities beyond simple returns processing to create differentiated value through repair, refurbishment, and material brokerage; deploying digital technologies as reconfiguring capabilities rather than linear efficiency tools; and adopting circular business models product-as-a-service, sharing platforms, performance-based contracts that align revenue structures with circular value creation (Mate et al. 2025).

For manufacturers and retailers, the key insights are: logistics partnerships are critical enablers of circular transitions; product design decisions determine the logistical feasibility of repair, refurbishment, and recycling; packaging design determines the viability of return systems; and data sharing with logistics partners on product composition, use history, and condition is foundational to circular logistics effectiveness (Butt et al. 2024).

For policymakers, the framework identifies the following leverage points: infrastructure investment in EV charging networks, digital connectivity, and recycling facilities; regulatory instruments including extended producer responsibility, recycled content mandates, and carbon pricing, calibrated to institutional capacity; concessional finance and risk-sharing instruments to mobilise private investment; informal sector engagement strategies that provide training, cooperative structures, and market access; and international engagement with climate finance mechanisms and trade negotiations to access external resources for circular transition (UNIDO 2026; UNCTAD 2026).

12 Limitations

This chapter's findings are subject to several limitations. First, the narrative review methodology, whilst appropriate to the multi-domain scope of the research objectives, does not provide the quantitative effect size aggregation or formal bias assessment that systematic review meta-analysis offers. Second, the English-language restriction excludes potentially relevant research published in other languages, which is especially significant for developing country contexts where Spanish, Portuguese, French, and Chinese-language scholarship may provide important evidence. Third, the temporal boundary of 2020–2026, whilst ensuring recency, excludes foundational empirical work that may provide important baseline comparisons for assessing the magnitude of change over time. Fourth, the empirical base for developing country analyses is thinner than for developed economy contexts, with many developing country findings resting on single-country studies or small samples that may not generalise. Fifth, the conceptual framework, whilst theoretically grounded and empirically supported, has not been tested through primary data collection; its relationships remain propositions requiring empirical validation across diverse national contexts.

12.1 Future Research Directions

12.1.1 Logistics Service Provider Business Model Transformation

The transition from transaction-based to value-based logistics business models for circular economy participation merits sustained research attention. Key questions include: What are the organisational and relational capabilities that enable this transition? How do customer-provider relationships change when circular services replace transactional logistics? What performance measurement systems are appropriate for circular logistics service models? (Mate et al. 2025). Comparative studies across asset-owning and non-asset-owning providers, and across global and regional operators, would provide insight into how business model architecture influences circular economy engagement.

12.1.2 Informal Sector Integration Governance

Formalising the informal sector's role in circular logistics systems in developing countries requires governance research grounded in the specific institutional contexts of different countries. Key research questions include: What governance arrangements most effectively balance economic inclusion with environmental and health standards? How can digital technologies facilitate integration without imposing capital costs that informal operators cannot bear? What are the success factors distinguishing Brazil's cooperative model from less successful integration attempts in other countries? Action research approaches, involving co-design with informal sector organisations, would be particularly appropriate (Nwaichi et al. 2025; Mugoni et al. 2023).

12.1.3 Technology Adoption in Resource-Constrained Contexts

Research on circular logistics technology adoption has concentrated on developed economy contexts with mature digital infrastructure. Urgent research gaps exist around: the minimum technology stack required to achieve circular benefits in infrastructure-constrained settings; the application of frugal innovation principles to circular logistics technology design; the role of mobile-first platforms in substituting for fixed digital infrastructure; and the conditions under which technology enables informal sector integration rather than exclusion (Kokkinou and Mitas 2026; Sarkis et al. 2025).

12.1.4 Circular Economy Performance Measurement

Current circular logistics performance metrics emphasise environmental outcomes and are poorly adapted to the economic and social dimensions of sustainability that are especially salient in developing economy contexts (De Souza et al. 2022). Research is needed to develop and validate comprehensive performance measurement frameworks that: integrate environmental, economic, and social dimensions; are implementable in both formal and informal sector contexts; and can be linked to existing logistics performance measurement systems to facilitate adoption by practitioners.

12.1.5 Policy Effectiveness and Sequencing

Understanding which policy instruments are most effective for circular logistics transitions in developing countries, and in what sequence they should be deployed, requires comparative research across countries at different stages of institutional development and circular economy adoption. Key questions include: Which regulatory instruments generate the strongest technology adoption responses in weak institutional environments? How should informal sector engagement policies be sequenced relative to formal sector technology mandates? What role do international climate finance mechanisms play in overcoming institutional coordination failures? (Panghal et al. 2024; Matos and Perello-Marin 2024).

12.2 Managerial and Policy Implications

12.2.1 Implications for Logistics Managers

The transition to circular economy principles requires logistics managers to extend their strategic logic from cost reduction and service reliability to resource productivity, material circulation, and lifecycle impact management. This is a fundamental reorientation of strategic purpose, operational design, and performance measurement.

Strategic Positioning: Logistics service providers must assess where in the circular value chain they can create distinctive value: from basic returns management and recycling coordination, through repair and refurbishment services, to material brokerage and circular logistics platform operation. This positioning must be grounded in realistic assessment of existing capabilities, customer requirements, and competitive dynamics (Mate et al. 2025).

Operational Redesign: Circular logistics requires network architectures that accommodate reverse flows alongside forward flows, with appropriately located inspection, grading, repair, and material recovery facilities integrated into standard warehouse operations. Performance measurement systems must capture material circulation metrics alongside conventional logistics KPIs (De Souza et al. 2022).

Technology Investment: Technology is the enabling infrastructure of circular logistics but does not autonomously generate circular outcomes. Investment decisions should be driven by the question of how technology enables circular operational redesign rather than how it optimises existing linear processes. A phased investment approach with pilot programmes reduces risk (Sarkis et al. 2025).

Capability Development: Circular logistics requires capabilities beyond traditional logistics management: technical skills for repair and refurbishment, analytical skills for material flow optimisation, and stakeholder management skills for coordinating with informal sector operators, material recovery specialists, and sustainability-oriented customers.

12.2.2 Implications for Manufacturers and Retailers

Manufacturers and retailers must integrate logistics considerations into circular economy planning from the product design stage. Design for disassembly, component standardisation, and material labelling directly determine the logistical feasibility of repair, refurbishment, and recycling. Packaging design determines the viability of return systems. Data sharing with logistics partners on product composition, use history, and condition is foundational to effective reverse logistics management (Butt et al. 2024). Long-term logistics partnerships that align incentives around circular value recovery are preferable to transactional arrangements that optimise cost at the expense of circular performance.

12.2.3 Implications for Policymakers

For policymakers in developing countries, the framework identifies five integrated levers for accelerating circular logistics transitions:

Infrastructure investment: Targeted public investment in EV charging networks, digital connectivity infrastructure, and recycling processing facilities creates the enabling conditions for private sector circular investment. Investment should be prioritised in infrastructure that unlocks the greatest private investment leverage (Climate Investment Funds 2025).

Regulatory frameworks: Extended producer responsibility, recycled content mandates, and carbon pricing create the institutional pressure that theory identifies as the critical moderator of the technology-sustainability relationship. Regulatory design should account for the capacity constraints of developing country firms, with graduated implementation timelines and technical support programmes (Panghal et al. 2024).

Financial mechanisms: Concessional finance, guarantees, and risk-sharing instruments mobilise private sector investment in circular logistics infrastructure. Strategic engagement with international climate finance Green Climate Fund, Climate Investment Funds, UNIDO programmes can leverage domestic public investment into much larger private capital flows (UNIDO 2026).

Informal sector policy: Inclusive governance frameworks that recognise informal operators as legitimate actors in circular economy systems, provide access to training and cooperative formation support, and create market access to formal sector contracts, are more likely to achieve systemic circular economy outcomes than exclusionary formalisation approaches (Mugoni et al. 2023).

International engagement: Active participation in international climate finance negotiations, carbon accounting standard development, and trade agreement environmental provisions positions developing countries to access external resources and avoid being disadvantaged by carbon border adjustments (UNCTAD 2026).

13 Conclusion

This chapter has examined the application of circular economy principles to transport and logistics across three interconnected dimensions: resource efficiency optimisation, reverse logistics and life cycle management, and zero-emission supply chain transitions in developing countries. The analysis has been grounded in an explicit and applied integration of dynamic capability theory and institutional theory, used not as rhetorical framing but as analytical tools that generate specific propositions about the mechanisms through which circular logistics outcomes occur and the conditions that enable or obstruct them.

The resource efficiency findings demonstrate that substantial gains vehicle utilisation improvements of 15–30%, packaging material reductions of 40–60%, and extended infrastructure lifespans are achievable through the combination of digital technologies, operational redesign, and circular business models. These gains are not automatic consequences of technology adoption; they require the reconfiguring capability that dynamic capability theory identifies as the highest-order organisational competency.

The reverse logistics findings establish that trillion-dollar annual material flows can be systematically converted from cost burdens to value recovery opportunities through repair, refurbishment, and remanufacturing capabilities. Digital traceability through blockchain, IoT, and digital product passports is the foundational enabling condition. Logistics service providers are uniquely positioned to orchestrate these capabilities across supply chain networks.

The developing country analysis is the chapter's most distinctive contribution. Rather than treating developing countries as a homogeneous category defined by shared deficits, the analysis differentiates across income levels, institutional contexts, and sector structures to identify context-specific pathways: grid-independent EV charging solutions for sub-Saharan Africa; biofuel alternatives for Latin American contexts with constrained grid capacity; mobile-first platform logistics for urban corridors with high mobile penetration but limited fixed infrastructure; and cooperative governance models for informal sector integration that draw on documented successes in Brazil, India, and South Africa. In each case, institutional theory's insight that regulatory pressure is the critical moderator of the technology-sustainability relationship provides the analytical foundation for distinguishing contexts where leapfrogging potential will be realised from those where it will remain unrealised.

The chapter's core argument is that logistics service providers are not peripheral actors in the circular economy transition but its central orchestrators. They control the material flows on which circular value recovery depends, they are positioned to develop the reverse logistics capabilities that circular economy systems require, and they are the organisational link between the technology adoption that dynamic capabilities theory prescribes and the circular outcomes that sustainability imperatives demand. Recognising and developing this strategic role through deliberate business model transformation, technology investment aligned with circular operational redesign, and constructive engagement with enabling policy frameworks is the practical agenda that emerges from this analysis.

Future research should prioritise longitudinal primary data collection across developing country contexts to empirically validate the conceptual framework, with particular attention to the informal sector integration governance models and the technology adoption dynamics in infrastructure-constrained environments that this narrative review has identified as the most significant knowledge gaps. The transition to circular logistics in the Global South will not occur automatically through technology diffusion; it requires the deliberate construction of institutional conditions, organisational capabilities, and policy frameworks that this chapter has sought to map.

References

Andrejć M (2023) Measurement of logistics efficiency in circular economy contexts. Sustainability 15(3):234–251

Asensio J et al (2020) Real-time data from mobile platforms to evaluate sustainable transportation infrastructure. Nat Sustain 3:463–471

Barney J (1991) Firm resources and sustained competitive advantage. J Manag 17(1):99–120

Beames A, Claassen GDH, Akkerman R (2021) Logistics in the circular economy: challenges and opportunities. In: Strategic decision making for sustainable management of industrial networks, pp 1–14

Bekrar A et al (2021) Digitalising the closing-of-the-loop for supply chains: a transportation and blockchain perspective. Sustainability 13(5):2895

Biancolin M, Capoani L, Rotaris L (2023) Reverse logistics and circular economy: a literature review. Eur Trans Trasporti Europei 94:1–19

Bonsu N (2020) Towards a circular and low-carbon economy: insights from the transitioning to electric vehicles and net zero economy. J Clean Prod 256:120659
Bozhanova V et al (2022) Green enterprise logistics management system in circular economy. In: Green enterprise logistics, pp 350–363
Butt AS, Ali I, Govindan K (2024) The role of reverse logistics in a circular economy for achieving sustainable development goals: a multiple case study of retail firms. Prod Plan Control 35(12):1490–1502
Chen Z et al (2024) Digital transformation for safer circular lithium-ion battery supply chains: a blockchain ecosystem-data perspective. Int J Prod Res Forthcom
Climate Investment Funds (2025) CIF selects seven inaugural countries to pioneer industry decarbonisation. Climate Investment Funds, Washington, DC
De Abreu VHS et al (2022) The role of the circular economy in road transport to mitigate climate change and reduce resource depletion. Sustainability 14(14):8951
De Souza ED et al (2022) Performance evaluation of green logistics: paving the way towards circular economy. Clean Logist Supply Chain 3:100019
DiMaggio PJ, Powell WW (1983) The iron cage revisited: institutional isomorphism and collective rationality in organisational fields. Am Sociol Rev 48(2):147–160
Dzwigol H, Trushkina N, Kwilinski A (2021) Green logistics as a sustainable development concept of logistics systems in a circular economy. IntellectIsation Logist Supply Chain Manag 7–8:6–25
Institute for Supply Management (2025) Reverse logistics: unlock the benefits for your business. https://www.ism.ws/logistics/reverse-logistics/. Accessed 15 Jan 2026
Irizarry J, Karan EP, Jalaei F (2013) Integrating BIM and GIS to improve the visual monitoring of construction supply chain management. Autom Constr 31:241–254
Jayarathna CP, Agdas D, Dawes L (2023) Exploring sustainable logistics practices toward a circular economy: a value creation perspective. Bus Strateg Environ 32(1):704–720
Ji-Hyland C, White D, Khaydarov R (2025) The impact of circular economy practices on sustainable logistics performance. Int J Log Res Appl 28(10):1220–1246
Kokkinou A, Mitas O (2026) Driving the circular economy: how artificial intelligence could empower transportation and logistics companies to lead the transition. Clean Logist Supply Chain 100300
Kumari S, Qalati SA, Shaikh R, Badwy HE (2025) Connecting emerging technologies with sustainability in the circular economy: the role of efficiency in logistics system and regular pressure. Sustain Futures 10:101297
Lei Z et al (2025) Artificial intelligence and big data analytics for circular economy: recycling, remanufacturing, and reuse applications. J Clean Prod 380:135012
Li X et al (2025) Adoption of emerging technologies and sustainability in circular economy supply chains: the mediating role of logistics efficiency. Sustain Fut 10:101297
Makarova I et al (2021) The role of reverse logistics in the transition to a circular economy: case study of automotive spare parts logistics. FME Trans 49(1):78–89
Makarova I et al (2018) The role of reverse logistics in the transition to a circular economy. In: International conference on reliability and statistics in transportation and communication. Springer, Cham, pp 363–373
Makarova I et al (2019) Features of logistic terminal complexes functioning in the transition to the circular economy and digitalization. In: Modelling of the interaction of different vehicles and various transport modes. Springer, Cham, pp 415–527
Mate OA et al (2025) Logistics service providers' business models in the circular economy: a systematic literature review and panarchy theory framework. Business strategy and the environment, published online 22 Sept 2025
Matos F, Perello-Marin MR (2024) Circular economy challenges within the road freight transport: case study of Portuguese companies. Proced Comput Sci 232:2181–2190
Melkonyan-Gottschalk A, Li Y (eds) (2026) Urban logistics transformation: E-commerce and sustainable circular economy approaches. Springer, Cham

Muchenje C et al (2023) Sustainable supply chain management: reducing the environmental impact of production and distribution processes. In: Sustainable marketing, branding, and reputation management: strategies for a greener future, pp 52–66
Mugoni E, Nyagadza B, Hove PK (2023) Green reverse logistics technology impact on agricultural entrepreneurial marketing firms' operational efficiency and sustainable competitive advantage. Sustain Technol Entrep 2(2):100034
Mugoni E, Kanyepe J, Tukuta M (2024) Sustainable supply chain management practices (SSCMPs) and environmental performance: a systematic review. Sustain Technol Entrepr 3(1):100050
Mugoni E, Kanyepe J, Tukuta M (2025) Sustainable supply chain management practices and environmental performance in a developing country's petroleum industry: the mediating role of operational performance. Int Bus Logist 5(1)
Mugoni E, Shumbanhete B, Nyagadza B (2025) The nexus between circular economy (CE) and sustainable supply chain management (SSCM): a systematic literature review and mapping of future research agenda. Circ Econ Sustain 1–38
Nwaichi PI et al (2025) Technological advancements in transportation and logistics: exploring circular economy principles through bibliometric insights. Circ Econ Sustain 1–51
Oluleye G et al (2025) Artificial intelligence for circular economy: a systematic review of applications in waste management and resource recovery. Resour Conserv Recycl 195:106987
Panghal A et al (2024) Adoption barriers to green logistics in the Indian food industry: a circular economy perspective. Int Soc Sci J 74(252):519–538
Renkin C, Limbourg S (2025) Optimising reverse logistics for waste materials: a multi-stage processing and transportation model. presented at Joint orbel-NGB conference on operations research, Maastricht
Ripanti EF, Tjahjono B (2019) Unveiling the potentials of circular economy values in logistics and supply chain management. Int J Logist Manag 30(3):723–742
Sarkis J et al (2025) Transport, logistics, and emergent technology within the circular economy. J Clean Prod 147155
Schneikart G et al (2024) A roadmap towards circular economies in pharma logistics based on returnable transport items enhanced with Industry 4.0 technologies. Resour Conserv Recycl 206:107615
Seroka-Stolka O, Ociepa-Kubicka A (2019) Green logistics and circular economy. Transp Res Proced 39:471–479
Teece DJ, Pisano G, Shuen A (1997) Dynamic capabilities and strategic management. Strateg Manag J 18(7):509–533
Thomas J, Harden A (2008) Methods for the thematic synthesis of qualitative research in systematic reviews. BMC Med Res Methodol 8(1):45
Tranfield D, Denyer D, Smart P (2003) Towards a methodology for developing evidence-informed management knowledge by means of systematic review. Br J Manag 14(3):207–222
Turoń K, Czech P (2016) Circular economy in the transport industry in terms of corporate social responsibility concept. J Corp Responsib LeadShip 3(4):83–94
UNCTAD (2026) Global trade trends 2026: sustainable growth. United Nations Conference on Trade and Development, Geneva
UNIDO (2026) Industrial decarbonisation centre accelerator facility: roadmap for action. United Nations Industrial Development Organisation, Vienna
Van Buren N et al (2016) Towards a circular economy: the role of Dutch logistics industries and governments. Sustainability 8(7):647
Vegter D, van Hillegersberg J, Olthaar M (2020) Supply chains in circular business models: processes and performance objectives. Resour Conserv Recycl 162:105046
Wang Y et al (2025) Emerging technologies and circular economy performance: evidence from Chinese manufacturing supply chains. Int J Prod Econ 270:109234
Yildizbasi A et al (2025) Exploring the synergy between circular economy and emerging technologies for transportation infrastructure: a systematic literature review. J Clean Prod 144553

Yu Z, Kwilinski A, Trushkina N (2018) Reverse logistics in a system of the circular economy: theoretical aspect. Ekonomichnyi Visnyk Donbasu 4(54):29–37

Zhang A et al (2022) Special issue editorial: logistics and supply chain management in an era of circular economy. Transp Res Part e Logist Transp Rev 166:102911

Zheng L, Zhang J (2010) Research on green logistics system based on circular economy. Asian Soc Sci 6(11):116

Integrated Energy–Transport Systems for Decarbonization in the Global South: A Desktop Review

James Kanyepe, Munyaradzi Chibaro, Wilbert Manyanga, and Masimba Evis Manyangara

Abstract A detailed, systematic desk review was conducted to identify relevant information from various sources (literature, policies, and institutional reports) regarding how energy and transportation systems interact; what barriers exist in moving toward lower carbon systems; and potential policy, technology, and strategy options to support the decarbonization of transportation systems. Various physical and system-wide limitations hinder the development of electric vehicles in many developing countries. Some of these problems are due to insufficient electricity-generating capacity, relatively little use of renewable energy resources, energy security concerns, and generally too few sustainable implementation capacities for large-scale transition pathways. This limits the feasibility and environmental performance of alternative transportation modes beyond those that are currently available. In addition to the physical and technical barriers listed above, socioeconomic and institutional barriers significantly affect the pace, degree, and equity of low-carbon transitions in the transportation sector. Examples of such barriers include the widespread existence of informal public transit services, poor institutional/governance capacity, poorly coordinated and/or fragmented regulatory frameworks, and high economic costs associated with the adoption of new transportation technologies/modes. Other barriers identified include continued dependence on fossil fuels, financial constraints, poor interagency/bureaucratic coordination, and behavioral/cultural resistance to changing one's mode of transportation. Various studies have provided insights into how low-carbon transport transitions may occur. Examples include the integration of energy-transport planning, technological innovation, market-based regulation, and government interventions that create social inclusion opportunities while creating

J. Kanyepe (✉) · M. Chibaro
Department of Logistics and Supply Chain Management, University of Botswana, Gaborone, Botswana
e-mail: jameskanyepe@gmail.com

W. Manyanga
Department of Marketing, Logistics and Sport Management, Faculty of Commerce, Human Sciences and Education, Namibia University of Science and Technology, Windhoek, Namibia

M. E. Manyangara
Department of Marketing, Chinhoyi University of Technology, Chinhoyi, Zimbabwe

J. Kanyepe and I. Chirisa (eds.), *Advancing the Green Agenda in the Transport and Logistics Sector in Global South Countries*, World Sustainability Series,
https://doi.org/10.1007/978-3-032-31374-4_7

sustainable mobility systems. On the basis of the results of this study, a coordinated/policy-driven approach is necessary to successfully decarbonize transportation systems in developing countries. Low-carbon transport transitions require the alignment of energy and transport governance functions with enhanced institutional capabilities to overcome both structural and behavioral barriers within a broader developmental framework.

Keywords Transport decarbonization · Energy–transport interdependence · Low-carbon mobility · Policy strategies · Sustainable transport · Global South · Desk review

1 Introduction

The transportation sector accounts for 15% of total anthropogenic greenhouse gas (GHG) emissions, 23% of global energy-related emissions, and 8.7 Gt CO_2-eq. emissions (Ferrer and Thome 2023). Urbanization trends in the Global South have created disproportionate burdens from rapid population growth, rising motor vehicle fleets, and increased urban sprawl (Dashora et al. 2025), all of which have contributed to a growing need for investment in public transportation infrastructure. Persistent underinvestment in developing country public transit systems has fuelled growing energy demand, worsened air quality, and strained both local power grids and fuel supplies (Wu 2024). Therefore, the constrained financial resources available to governments in lower- and middle-income economies limit their ability to establish and execute low-carbon mobility plans.

Low-carbon mobility options, including electrifying vehicles, reducing emissions via alternative travel modes, and improving the efficiency of existing systems, are viewed as viable methods for mitigating climate change impacts, improving air quality in cities, and promoting long-term energy sustainability (Szczepański et al. 2025). However, the success of each of these alternatives depends on the strength, resiliency, and carbon footprint of the larger energy supply chains. For example, if widespread electricity decarbonization is not implemented simultaneously with the adoption of electric vehicles in the transportation sector, there may be few environmentally beneficial results or possibly new energy security vulnerabilities (Nekahi et al. 2025).

Owing to the interdependent nature of these two systems, multidisciplinary approaches that incorporate technological innovation, infrastructure design, and social factors are needed to successfully implement decarbonization efforts. While a vast amount of research has focused on electric vehicles, integrating renewable energy into the overall energy portfolio, and decarbonizing the transportation sector, most of this research is segmented by either specific topic areas or geographic regions. Currently, there is less research on the Global South than on other regions (Okafora et al. 2025; Pozo Burgos et al. 2026; Pranawengkapti et al. 2025). Moreover, much of the research conducted to date has focused primarily on the technical dimensions of designing electric vehicle charging infrastructure, planning for grid usage associated with electric vehicles, and potential synergies between renewable energy sources without considering the interconnectedness of energy production systems, transportation networks, government policies, and societal contexts. As a result of this narrow focus, research examining the systemic issues generated by decarbonizing the transportation sector has been limited. Additionally, institutional and financing constraints, such as continued dependence on fossil fuels, fragmented regulatory regimes related to subsidy programs, and insufficient investments in upgrading grid infrastructure, have typically received scant attention in previous research (Pozo Burgos et al. 2026).

Further impediments to joint planning exist, including inconsistencies in charging standards for electric vehicles, the required level of grid infrastructure to accommodate electric vehicles, disparities in regulatory frameworks across energy policy sectors, and unharmonized standards for policymaking. In contrast to developed economies, these challenges are especially pronounced in the developing regions of the Global South, where frequent outages and instability of electricity services, limited availability of electrical transmission and distribution infrastructure, and complex governance structures compound these difficulties. Nonetheless, they are frequently absent from the literature (Emodi et al. 2022). Thus, the present literature contains limited insight into how complementary cross-sectoral initiatives can simultaneously promote environmentally friendly and economically and socially desirable objectives while avoiding undesirable secondary consequences, such as unstable grid operation or unevenly distributed access. Resolving these challenges requires locally applicable evidence-based methodologies that combine the promotion of electric vehicles with broader energy policymaking/planning activities, identification of structural impediments, and evaluation of policy instruments capable of generating equitable and sustainable outcomes (Emodi et al. 2022; Okafora et al. 2025; Pranawengkapti et al. 2025; Hossain and Cho 2026).

2 Conceptual Framework

The conceptual framework used as a basis for this research study includes three key (interconnected) elements. Energy–Transport Interdependence, Decarbonizing Transport through Policy Measures, and Energy Security. These three elements influence the viability, efficiency, and sustainability of low-carbon urban transportation options in the Global South. In addition, it provides a basic outline of the relationships among energy supply, transport demand, and the use of policy instruments to bring about change. Therefore, alternative technologies should not be expected to accomplish goals independently. Consequently, there is a need to understand the relationships between individual components when attempting to develop comprehensive policies and strategies to meet environmental goals, foster social equity, and promote economic feasibility.

2.1 *Energy–Transport Interdependence*

Transportation and energy interdependence describe how each affects the other. Transportation requires reliable, available, and decreasingly carbon-based fuels, such as electrical energy or liquid fuels, to operate effectively (Michalski et al. 2019). However, the form, size, and operation of transportation systems also directly affect the amount of energy needed, which in turn affects the type and volume of energy that must be produced, transmitted, and stored. As a result of this dependence, decisions made regarding the transportation sector will have significant impacts on future energy system design and implementation. The degree of interdependence is even greater within the Global South because of inadequate infrastructure, unreliable grid access to electrical energy, and continued reliance on imported petroleum products (Martinez et al. 2025). Consequently, efforts toward transportation decarbonization cannot occur independently of the larger energy system. Initiatives related to transportation decarbonization, including the adoption of electric vehicles, alternative fuels, and/or electrified modes of transportation, must therefore be coordinated with the ability of the existing energy system to support them through the availability, reliability, and carbon intensity of its energy production (Lah 2017). A unified view that accounts for the interdependence of transportation and energy systems is necessary to achieve effective and long-lasting decarbonization.

2.2 *Transport Decarbonization Strategies*

Transportation decarbonization strategies involve the combination of technology and policy interventions that affect mobility carbon intensity (lah 2017) to reduce greenhouse gas emissions from mobility. The main technological strategy is to implement low-carbon transportation technologies, such as vehicle electrification, hybrid vehicles, emission standards for vehicles, and energy-efficient modes of transportation (e.g., rail and high-capacity public transit systems). They are critical for reducing greenhouse gas emissions and increasing the overall energy efficiency of transportation systems. Moreover, policy strategies provide the economic and regulatory frameworks required to move toward low-carbon mobility. Policy strategies include fiscal incentives, regulatory frameworks, market-based instruments, and institutional coordination to promote cleaner transportation. Examples of policy strategies include subsidies, emission standards, and carbon pricing mechanisms to encourage investment in low-emission technologies and discourage carbon-intensive transportation. Although policy and technology strategies contribute significantly to transport decarbonization, they cannot be implemented as individual strategies. As mentioned earlier, their effectiveness depends on coordination with energy system planning, infrastructure development, and governance capacity (Amin 2025). Thus, it is important to develop a systems approach to ensure coherence between transportation policies, energy supply systems, and institutional arrangements.

2.3 Energy Security

Energy security is defined as a country's ability to ensure that its energy needs are consistently met at affordable prices and in a sustainable manner, with a limited risk of disruption from external supplies (Strojny et al. 2023). The role of energy security in transport decarbonization dual-rôle. It facilitates transitions toward low-carbon solutions by providing reliable electricity generation capacity, grid stability, and reliable distribution infrastructure for implementing electrification strategies. However, when energy systems are fragile or heavily reliant on imported fossil fuels, the cost efficiency and reliability of low-carbon transport solutions are threatened (Oshilalu and Iboh 2025). The development of energy security through the development of renewable energy sources, diversion of sources, and improvements in energy efficiency can greatly increase the feasibility and long-term sustainability of decarbonization pathways for transport (Rabbi et al. 2022). Additionally, interventions in the transport sector, such as a modal shift toward more energy-efficient mobility systems and the use of alternative fuels, reduce the total energy demand and therefore lead to diversification of consumption patterns (Oshilalu and Iboh 2025).

3 Methodology

This study used a desk-based review methodology to examine the relationship between global energy systems and transportation decarbonization. This study identified potential obstacles to the successful implementation of decarbonization policies and technologies in the Global South. As such, desk reviews were well suited for this research, as they allow for a systematic integration of evidence from various types of literature (including gray literature), empirical studies, and policy reports. The use of multiple contexts in which the findings have been synthesized provides insight into complex sociotechnical systems that are difficult to obtain through primary data collection. Additionally, this analysis identifies common themes, knowledge deficits, and issues specific to contexts relevant to lower-middle-income countries.

3.1 Search Strategy

A thorough literature review was conducted by searching three electronic databases, namely, Scopus, Web of Science, and Google Scholar. The keywords used for each database were carefully selected to represent the multifaceted nature of this research (transport decarbonization, energy-transport interdependence, low-carbon mobility, electrification of transport, energy security, policy interventions, and the Global South). Only publications published from 2000–2025 were included to ensure that all studies published since 2000 were included, therefore capturing recent technological, policy, and infrastructure developments. Boolean operators and truncation techniques were used to increase the accuracy and scope of the search and enable the systematic collection of relevant studies.

3.2 Inclusion and Exclusion Criteria

To ensure the highest quality and most relevant data for analysis, each publication included in this review was carefully evaluated using defined selection (inclusion) and rejection (exclusion) criteria. This review focused on transportation-related decarbonization, energy systems, and/or the integration of transportation and energy systems, with a specific focus on low- and middle-income countries. In addition to academic research (theoretical/empirical), this review incorporates nonacademic literature, such as government publications (e.g., reports), technical documents (assessments/briefs), and policy-related literature, all of which are available in English. All other forms of literature outside of these time frames (i.e., 2000–2025), literature focusing solely on high-income countries, and literature without adequate detail regarding methodology/policy details were excluded from the review. By employing the abovementioned methodology, the best possible literature will be selected for inclusion in this review.

3.3 Selection Process

The first step of the literature search included an initial search of over 1200 records. After removing duplicate and irrelevant references through title and abstract screening, we narrowed the total number of potential resources to 430. Ultimately, after carefully evaluating all 430 full-text reviews against our eligibility criteria, we identified 168 relevant documents. The method involves multiple iterations with cross-references to the referenced works so that important studies are not overlooked. Through this approach, we developed a highly complete and representative sample that represents a wide range of technologies available to improve energy-transport systems, as well as policies for those in the Global South.

3.4 Data Synthesis

Thematic narrative synthesis was applied to integrate the data extracted from the selected sources. This methodology is particularly well suited for combining different types of literature (i.e., quantitative, qualitative, and policy-oriented research) related to the energy–transport nexus. Inductive analysis was used to identify themes related to the relationship between energy systems and transport decarbonization, barriers to the transition toward low-carbon modes, and how policies and technologies can support sustainable development. Conceptual mapping was used to represent the relationships between energy and transportation systems, including constraints on the electric power grid, the integration of renewable energy into those grids, and policy mechanisms. Narrative synthesis allows for the aggregation of results from diverse contexts, highlights patterns and divergence in the results, and identifies areas where there are still gaps in our understanding.

4 Results

The main purpose of this section is to perform a thematic synthesis of studies on decarbonizing transportation in the Global South. Using information from all 168 studies reviewed as part of the desk review, three key themes were identified: the connection between energy systems and transportation, barriers to transitioning to lower carbon forms of transportation, and the policies and technologies needed to support the sustainable transition to lower carbon forms of transportation. By focusing on how structural, financial, institutional, and socioeconomic factors can impact the ability of an intervention to be effective, equitable, and feasible, this analysis provides a comprehensive view of what currently exists with respect to low-carbon transportation options and where future research may be most productive.

5 Interdependence Between Energy Systems and Transport Decarbonization in the Global South

The interrelationship between energy systems and transport infrastructure has been a well-recognized theme in the global literature on transport decarbonization. The ability to decarbonize transport via the use of alternative fuels (electricity), low-emission vehicles, changes in travel mode, and increased efficiency is dependent on the energy system's capacity, reliability, and level of carbon emissions. In turn, energy systems are affected by the demand for energy in the transport sector. Consequently, limitations within an electricity system limit the potential success of any low-carbon transport-related strategy.

Grid capacity and infrastructure reliability are two key determinants of transport decarbonization. Given that infrastructure is aging and unable to accommodate increasing amounts of traffic, combined with insufficient generation capacity, inadequate transmission/distribution network capacity limits the potential for the widespread adoption of electric vehicles (EVs) and electrified public transportation systems. The lack of proactive grid modernization efforts risks creating a situation where the introduction of new technologies leads to overloading of the electrical grid, which would create blackouts and reduce operational efficiency in terms of providing energy services (Powell et al. 2024).

For example, it has been estimated that peak-hour electric vehicle charging in densely populated urban centers could lead to increases in electrical loads of 15–25% above base levels in some sub-Saharan African cities, thus creating further pressure on already strained grids. This problem is exacerbated by underinvestment in transmission infrastructure and the inability to effectively deploy smart grid technology. Thus, cooperation and coordinated planning among energy utility providers and transportation agencies are needed to implement low-carbon transport solutions that consider the physical capabilities of the existing grid infrastructure. This problem is most pronounced in rapidly urbanizing cities (Goldthau et al. 2020; Minas et al. 2024).

Furthermore, the unequal availability of urban–rural power grid coverage creates additional barriers to implementing low-carbon-based transport solutions, which may exacerbate existing social inequalities regarding accessibility and quality of service (Geurs et al. 2021). Social, economic, and institutional factors play important roles in determining the degree to which energy and transportation systems are interconnected. Urban mobility patterns, income inequality levels, informal transportation networks, and population density affect energy demand. Similarly, government capacity, regulations, and taxation policies affect the feasibility of developing low-carbon alternatives. Effective policy coordination is necessary to develop low-carbon alternatives that provide equitable benefits and political sustainability. This requires integrated planning between the energy and transportation departments, favorable regulatory frameworks, and financial incentives (Ho and Tirachini 2024).

In addition to these issues, the development of effective policy coordination for low-carbon alternatives requires the consideration of socioeconomic conditions such as affordability and accessibility so that the benefits of low-carbon alternatives are shared equitably among stakeholders. Failure to do so may result in the perpetuation of carbon lock-in behaviors. Failure to coordinate effectively may result in unintended consequences, such as reduced access to energy services, overload of energy infrastructure, and continued reliance on fossil fuels (Addy et al. 2026). Therefore, socioeconomic and institutional factors significantly impact the relationship between energy and transportation systems. Mobility patterns in urban areas, income distributions in urban areas, informal transportation networks, and population densities in urban areas affect the amount of energy consumed by these systems.

However, the feasibility of introducing lower-carbon options depends on the government's ability to make decisions about energy policy, regulatory structures, and fiscal policies (Ho and Tirachini 2024). For example, attempts to develop formalized mass transit systems in urban areas that contain large numbers of informal bus fleets may be hampered by various factors. These include limited funding sources, limited technical expertise in designing and building formalized mass transit systems, and reluctance among stakeholders involved in operating informal bus fleets because of their dependence on conventional fuel-based transportation systems (Aidam et al. 2025). Therefore, effective policy coordination is needed to facilitate decarbonization efforts. This means coordinating policy-making processes across different types of governmental organizations (energy ministry/transportation ministry), establishing regulatory environments that support decarbonization, and providing specific financial incentives to encourage investments that support it. All three of those items are essential for ensuring that there is alignment between expanding the grid, deploying renewable forms of energy, and using electric vehicles (Pranawengkapti et al. 2025). If these three items are not aligned with one another through effective policy coordination, any number of unintended consequences may arise from decarbonization efforts. Some examples of these unintended consequences include limiting access to energy services, overloading energy infrastructure, and maintaining long-term reliance on fossil fuels. In addition to these consequences being likely to occur if policy coordination does not occur effectively, it is also important to consider socioeconomic concerns such as affordability and accessibility when designing decarbonization strategies. This is because it is essential to ensure that the benefits of decarbonization are equitably distributed among all stakeholders and that they are politically sustainable (Addy et al. 2026).

Therefore, socioeconomic and institutional factors strongly influence the interdependence between energy and transport systems. Urban mobility patterns, income distributions, informal transport networks, and population density in urban areas influence the energy demand of these systems. On the other hand, how feasible it is to introduce lower-carbon options depends on governmental ability to make decisions about energy policy, regulatory structures, and fiscal policies (Ho and Tirachini 2024). For example, attempts to develop formalized mass transit systems in urban areas that contain large numbers of informal bus fleets may be hampered by a variety of factors. These include limited sources of funding, limited technical expertise in designing and building formalized mass transit systems, and reluctance on the part of stakeholders involved in operating informal bus fleets owing to their dependence on conventional-fuel-based transportation systems (Aidam et al. 2025). Therefore, effective policy coordination is required to achieve successful decarbonization, involving integrated planning between the energy and transportation ministries, supporting regulatory frameworks, and offering incentivized mechanisms. Each element is crucial for achieving alignment between extending grid capacity, deploying renewable forms of energy, and adopting electric vehicles (Pranawengkapti et al. 2025).

However, without this coordination, several negative outcomes may arise from decarbonization initiatives. Examples of negative outcomes include inequitable access to energy services, overloads of energy infrastructure and persistent carbon lock-ins. Additionally, socioeconomic aspects, such as affordability and accessibility, must be considered when defining decarbonization strategies to guarantee that the benefits of those strategies are equitably distributed among stakeholders and that they are politically sustainable (Addy et al. 2026).

6 Barriers to Implementing Low-Carbon Transport and Energy Strategies in the Global South

Despite the increasing global focus on reducing greenhouse gas emissions through decarbonization, implementing low-carbon transportation and energy strategies in the Global South is hindered by various structural, technological, financial, and institutional barriers. These barriers do not operate independently but rather interactively, shaping the feasibility, impact, and equity of regional decarbonization programs. Thus, efforts to implement decarbonization through electrification, modal shifts, and energy efficiency frequently encounter systemic barriers that reduce both the viability and the environmental impacts of low-carbon transition efforts. One of the largest barriers is fossil fuel lock-in. Fossil fuel lock-in refers to how the history of the development of transportation and energy infrastructure has largely relied on carbon-intensive technology. Shao et al. (2025) reported that in many developing nations, mobility systems continue to be heavily dependent on petroleum-based fuels. Likewise, as shown by Shao et al. (2025), electricity generation in many developing countries is still heavily dependent on fossil energy, including coal.

Structural dependence on fossil fuels is maintained by established supply chains, strong industrial interests, and long-lived capital equipment that support the continued usage of fossil fuels even after viable alternative forms of energy become available. For example, Pojani and Stead (2015) demonstrated in case studies of South Asian megacities that subsidized prices for diesel and gasoline lead to many vehicles being privately owned, leading to reduced incentives to develop public transportation or electrify transportation systems. Coal-dependent electricity systems in some sub-Saharan African nations provide another example where low-carbon electric mobility has little to no environmental benefit because much of the electrical energy generated comes from burning fossil fuels; thus, the overall emissions generated by the transportation system are likely to be greater than those generated if traditional combustion engines are used. In this way, fossil fuel lock-in provides a type of technological and institutional path dependence that creates a need for major policy, financial, and technological efforts to resolve.

In addition to structural lock-in, financing and investment hurdles represent significant obstacles to low-carbon transitions, and the development of technologies such as electric buses, charging infrastructure, and renewable energy systems requires significant capital expenditures. Frequently, these costs exceed the financial capacity of municipalities and private transportation providers in low- and middle-income countries (Song et al. 2022). Additionally, high capital expenditures combined with perceived technological risks and market uncertainties may discourage public and private sector investments in emerging low-carbon technologies (Falcone 2023). While international climate financing and concessionary lending mechanisms have been developed to address the funding gap between what is needed and what currently exists, access to these resources varies greatly among countries and regions. Furthermore, there is no guarantee of long-term financial sustainability (Kalinowski 2024). Financial barriers are often compounded by limited fiscal incentives, inadequate public–private partnerships, and a lack of financial instruments designed to assist investors in managing risks associated with integrated energy-transportation projects (Ukoba et al. 2024). Together, these financial constraints make the widespread diffusion of low-carbon technologies difficult in many Global South contexts.

In addition to financial issues, the application of decarbonization strategies is complicated by uncoordinated regulatory frameworks and divided institutional structures. In different countries, the responsibility for managing energy systems, transportation infrastructure, and city development is often distributed among separate institutions with limited interinstitutional collaboration. Therefore, policy frameworks regulating energy and transportation systems frequently lack coherence or consistency (Nwosu et al. 2025). For example, electricity regulators may focus primarily on maintaining grid reliability and managing loads without providing direct support for transitioning to electric mobility. Transportation policies may provide too few incentives for shifting transportation modes or adopting low-carbon technologies (Aidam et al. 2025). Ambiguity surrounding overlapping institutional roles and authority, along with weak compliance mechanisms, impedes policy success (Sundaram 2024). Bureaucratic resistance, inadequate technical expertise, and political instability are additional factors contributing to the delayed implementation of comprehensive decarbonization strategies and creating uncertainty for investors and stakeholders (Sun and Ahmed 2024). Therefore, enhancing interagency cooperation and institutional capacity building is a prerequisite for the successful execution of low-carbon transitions.

Social conditions and consumer behavior related to transportation play significant roles in the application of low-carbon transportation and energy strategies. Many cities in the Global South feature a high degree of dependence on informal transportation systems, limited disposable income at the household level, and restricted availability of economically accessible formal mobility services. These structural conditions define consumer behavior related to travel and energy consumption patterns. They frequently restrict the implementation of emerging alternatives to carbon-intensive energy sources (Azunre et al. 2022). Affordability constraints can exclude individuals from accessing electric mobility solutions and environmentally friendly transportation services. Public perception and habituated behavior preferences for owning personal vehicles may further impede shifts toward using mass transit or walking/biking options (Mesquita et al. 2025). Moreover, decarbonization strategies usually ignore social equity concerns, which can result in increased inequality in terms of mobility in specific communities. To address such socioeconomic challenges, policy measures that combine investment in infrastructure with subsidy schemes for target groups, inclusive urban planning processes, and awareness-raising campaigns for changing user behavior are needed. The scientific literature indicates that structural, financial, institutional, and socioeconomic barriers are strongly interdependent and reinforce each other. Hence, single policy interventions cannot contribute significantly to achieving considerable results with respect to sustainable transportation decarbonization. Rather, addressing these challenges demands holistic strategies that simultaneously cover infrastructure development, fund provision, institutional capacity strengthening, and social equity improvement.

7 Policy and Technological Strategies for Sustainable Decarbonization Pathways

Achieving sustainable decarbonization of transport through coordinated combinations of policy, technical, and institutional reform is critical for transport transition in the Global South. A key consideration in the literature is the need to address the interdependencies between energy systems and transport infrastructure, as well as the associated socioeconomic and environmental risk implications. Pathways must move beyond isolationist technological solutions and integrate multiple dimensions of sustainability in their design, namely, energy production, infrastructure development, finance, and governance. Four interconnected strategic domains—integrated energy-transport planning, technological innovation and infrastructure development, financial and market-based instruments, and social and governance interventions—are consistent throughout the reviewed literature.

Decarbonizing transport through the integration of energy and transport sector planning is a fundamental component of all sustainable decarbonization strategies. Integration involves coordinating the deployment of electricity generation capacity and extending the electricity grid with the development of transport infrastructure so that electrification is both technologically possible and environmentally acceptable (Oshilalu and Iboh 2025). In practice, this will require policymakers to deploy electric vehicles at scales that match the available electricity generating capacity, renewable energy availability, and geographically aligned charging infrastructure. Studies conducted in South Asia and Sub-Saharan Africa have demonstrated how aligning the timing of electric vehicle deployment with grid capacity and renewable energy output minimizes grid overload and improves overall energy use efficiency (Ajao et al. 2024; Saklani et al. 2024). Long-term integrated planning frameworks are typically designed to coordinate longer-term urban mobility strategies, land use planning, and climate change mitigation targets to ensure that low-carbon transport decarbonization supports wider national development goals (Stoeglehner 2020). Through such coordination, policymakers can better predict changes in electricity demand, develop effective load management strategies, and provide greater resilience for both energy and transport systems.

Technological innovation is central to supporting the transition to low-carbon forms of mobility. Advancements in vehicle electrification, renewable energy generation, grid modernization, energy storage technologies, and intelligent transportation systems have contributed to increased rates of transition to more sustainable forms of mobility (Aidam et al. 2025). Of particular significance are the opportunities provided by electrifying high-utilization forms of transport (e.g., buses, taxis, and motorbikes), resulting in large reductions in emissions under the assumption that sufficient charging infrastructure and access to a low-carbon electricity supply are available (Li et al. 2025). The integration of renewable energy into transport infrastructure (such as solar-powered charging stations or localized microgrid systems) can also increase energy security while simultaneously decreasing reliance on imported fossil fuels (Ukoba et al. 2024). Smart grids and DSMs combined with energy storage systems enable the reliable operation of electricity supplies by facilitating grid networks to accommodate fluctuations in the output of renewable energy generators. When centralized infrastructure may be constrained, distributed energy systems (DESs) and hybrid energy transport models (HETMs) represent flexible and scalable options for supporting regionalized decarbonization pathways (Eltohamy et al. 2026).

Financial mechanisms and market-based instruments are widely regarded as necessary facilitators of the widespread adoption of low-carbon transport technology. As noted in the Literature Review section, capital cost barriers remain significant obstacles to the adoption of electric vehicles, EV charging infrastructure, and renewable energy systems in many low- and middle-income countries (Matsvai 2026). Financial mechanisms developed to alleviate these capital cost barriers include subsidies for electric vehicles, conventional financing for fleet operators, tax incentives, and public–private partnerships (Mwesigwa et al. 2023; Ka et al. 2026). Evidence from case study research in Latin America and Southeast Asia illustrates how targeted subsidies for public transport electrification and renewable energy deployment accelerate technology adoption while potentially lessening social inequality issues (Guzman and Oviedo 2018; Delina and Shi 2025). Market-based instruments (MBIs), such as congestion pricing and fuel taxes, can also encourage modal shifts away from private car usage toward public transport, cycling, and walking. Therefore, MBIs can act in conjunction with technological interventions and support the decarbonization of urban transport (Veitch and Rhodes 2024).

In addition to technological innovations, financial tools and governance reforms are critical components in ensuring that decarbonization strategies achieve both effectiveness and social equity goals. Governance reforms that enhance coordination among ministries and departments, improve regulatory frameworks, and strengthen institutional capacity are commonly cited as necessary prerequisites for successfully implementing integrated policy approaches (Pojani and Stead 2018). Additionally, inclusive governance processes that involve diverse stakeholders (including local communities, private sector actors, and informal transport providers) assist in ensuring that decarbonization policies meet the needs of affected populations and generate equitable social outcomes (Bhatt 2025). Behaviorally oriented interventions, such as public awareness campaigns and mobility demand management initiatives, can also assist in fostering sustainable transport practices, thereby enhancing the legitimacy of decarbonization policies (Pojani and Stead 2018). Together, these governance/social strategies provide enabling conditions in which technical innovation, financial investment and institutional coordination can function synergistically to support resilient/equitable decarbonization pathways in the Global South.

8 Discussion

This literature review concludes that energy systems and transport decarbonization are systematically intertwined in the Global South. Thus, there are two specific points in the literature. First, low-carbon mobility transitions cannot be viewed as technology-driven technology for electric vehicles or alternative fuels. Rather, they are part of larger sociotechnical transformations encompassing electricity generation systems, infrastructure development, governance, and mobility. Second, the literature supports sustainability transition scholarship, which views decarbonization as a multidimensional process influenced by the relationships among technological innovation, institutional change, and socioeconomic dynamics (Sovacool et al. 2022; Goldthau et al. 2020). Taken together, the themes synthesized through this literature review demonstrate that transport electrification, renewable energy expansion, and governance frameworks are all interconnected aspects of the overall transition system.

First, the literature review demonstrates that electricity grid capacity and infrastructure reliability are critical enablers of the widespread adoption of transportation electrification. Many countries in the Global South have aging and underdeveloped grids, resulting in limitations in expanding and increasing the electricity supply. Additionally, several countries experience high levels of transmission and distribution losses. Therefore, the rapid expansion of electric mobility is likely to add to the existing strain on these constrained energy systems. Accordingly, studies have highlighted the need for proactive grid modernization and effective demand-side management to prevent large-scale electrification from adding to supply instability and undermining both mobility services and energy reliability (Henderson et al. 2017; Ponce-Jara et al. 2017).

Therefore, as identified in broader energy transition research, the level of preparedness of the energy system is a primary determinant of success when decarbonization outcomes are implemented (Weko and Goldthau 2022). Therefore, as a condition precedent to achieve a sustainable transition of mobility through the electrification of transport, it is necessary to develop transport electrification strategies aligned with investments in energy system resilience. Second, this literature review highlights the importance of using renewable energy to achieve the environmental benefits of electric mobility. Electric mobility can substantially reduce transport-related greenhouse gas emissions. However, realizing this benefit fully depends on whether the electricity used to charge the vehicles is produced from carbon-intensive sources. Several countries in the Global South continue to produce most of their electricity from fossil fuel-based sources. Therefore, reducing greenhouse gas emissions from electric mobility requires increased use of renewable energy. Studies have demonstrated that increasing renewable energy capacity, especially from solar and wind energy, is a critical component of decarbonizing energy systems (Li et al. 2023a, b).

Studies have also identified the technical challenges of integrating intermittent renewable energy into existing grid systems and managing their variability. To address these challenges, researchers have proposed investing in complementary technologies to enable greater flexibility in energy production and consumption. Examples include investing in smart grid infrastructure and distributed energy systems capable of stabilizing variations in the renewable energy supply (Ukoba et al. 2024). These examples illustrate the close linkages between transport decarbonization strategies and broader energy transition processes occurring at the national level within electricity systems.

Third, energy security has been shown to influence transportation decarbonization strategies in many Global South countries. For example, dependence on imported petroleum-based fuels increases exposure to volatile international oil prices and geopolitical risks associated with supply chain disruptions. Conversely, transitioning to domestically sourced renewable energy for electric mobility reduces reliance on foreign-supplied fuels. Furthermore, transitioning to domestically sourced renewable energy for electric mobility provides two benefits: greenhouse gas emission reduction and enhanced national energy security (Falcone 2023). Research has shown that countries with more diverse energy mixes and higher proportions of renewable energy are better able to scale up electric mobility options without compromising their ability to provide reliable energy supplies. Therefore, transport electrification offers opportunities to advance climate change mitigation objectives and enhance economic resilience and energy independence in individual countries.

Finally, this literature review illustrates that institutional capacity and governance coordination play important roles in influencing global decarbonization pathways. In some instances, the responsibilities for energy policy, transport planning, and urban development are dispersed across multiple government agencies, with little coordination between them. Governance fragmentation is widely recognized as a limitation on the ability of governments to implement holistic decarbonization policies. Research has shown that effective low-carbon transitions require strong regulatory frameworks that enable collaborative decision-making among government departments responsible for coordinating actions around infrastructure investment, technological deployment, and mobility planning (Pojani and Stead 2018). Enhancing institutional capacity and coordination through policy design and delivery is a vital precursor to successfully translating decarbonization objectives into sustainable mobility transitions.

Finally, this literature review highlights how social inequality and affordability constraints affect the adoption of low-carbon mobility options. Many cities in developing countries are characterized by informal modes of transport as the dominant mode due to a lack of access to affordable formal public transport or private vehicles among lower income households. Therefore, decarbonization strategies focused solely on technological innovation alone run the risk of widening preexisting social mobility disparities. This observation resonates strongly with emerging research on energy justice (Sovacool et al. 2022), which recognizes that sustainability transitions must be environmentally effective and socially inclusive. Therefore, policymakers must consider equity issues as part of their transport and energy policy frameworks if the environmental benefits of decarbonization are to be distributed equitably throughout society.

Collectively, the results of this literature review suggest that a comprehensive and integrated approach must be adopted to facilitate the decarbonization of transport in developing regions. Energy infrastructure, institutional governance, and social inequality must be considered interconnected components of the decarbonization process. Technologies such as electric vehicles represent one aspect of this transition to a low-carbon economy. However, their adoption relies on supportive energy systems and an integrated governance framework. Decarbonizing transport in developing regions requires comprehensive strategies that simultaneously focus on transforming energy systems, investing in appropriate infrastructure, improving institutional governance capacity, and fostering equitable distribution of environmental benefits throughout society.

9 Policy and Practical Implications

The findings of this study have significant implications for policymakers and for implementing decarbonized transportation strategies in developing countries. The results of this review highlight the need for a systems-based approach that recognizes the relationship between energy and transportation systems. First, the findings demonstrate the need for integration between the energy and transportation sectors through both institutional and physical/infrastructure means. Many studies have shown that if each sector implements its own decarbonization strategy independently, it will be difficult to realize the desired outcomes. One example is that there can be no meaningful electric vehicle adoption until sufficient reliable and noncarbon-emitting electricity is available. Therefore, an integrated planning process for power generation, the building of the grid, and urban mobility plans under a coordinated governance structure is needed. A mechanism for coordinating the activities of utilities generating electricity, local government departments responsible for transportation, and local government agencies responsible for urban planning could improve coordination and reduce the risk of investing in uncoordinated infrastructure.

Second, this review highlights the crucial role of upgrading and expanding electricity grids and increasing renewable energy production as necessary conditions for a sustainable transition to cleaner forms of mobility. Many developing countries in the Global South are characterized by aging and unreliable electrical distribution systems, insufficient electrical generating capacity, and high loss rates in transmitting electricity. Therefore, the large-scale introduction of electric mobility has the potential to place even greater strain on already overstressed systems. However, many studies have reported that the installation of smart grid technologies, distributed generating systems, and energy storage technologies is essential to ensure the reliability of these systems. This issue is especially pressing in fast-growing cities that face unprecedented increases in travel demand.

Third, this study's findings highlight how ongoing financial limitations continue to constrain the rate and scope of developing countries' decarbonization initiatives. Currently, many developing countries face significant capital expenditures in purchasing electric vehicles, establishing charging infrastructure, and producing renewable energy sources. These barriers to entry are exacerbated by the lack of access to affordable funding options. As such, many researchers argue that several different types of financial assistance tools, including tax incentives, concessionary financing options, and public–private partnership arrangements, could provide lower-cost investment alternatives and encourage the diffusion of new technologies. Additionally, climate finance mechanisms provided by foreign governments or multilateral organizations are increasingly seen as a key source of funding that allows developing countries to invest in the large-scale infrastructure needed to facilitate integrated energy-transportation transitions.

10 Limitations of the Study

This study contributes substantially to understanding the role of energy and transportation in sustainable development; however, several limitations affect the findings. First, because this study was based on secondary data (which include articles from peer-reviewed journals, policy documents, and gray literature) and utilized a desk-based methodology to synthesize the existing body of knowledge about sustainable transportation in the Global South, the results were shaped by the availability of relevant literature, the methods used to collect and analyze the literature, and the perspectives employed to understand the relationship between energy consumption and transportation behavior. The second limitation relates to geographic coverage. Despite the emphasis on sustainable transportation issues in the Global South, there appears to be an unequal distribution of research on these topics. For example, many studies on sustainable transportation issues have focused on major emerging economies, namely, China, India, and Brazil. Few studies have focused on transportation-related decarbonization processes in smaller developing countries (especially those in sub-Saharan Africa), and few studies have focused on this topic. Therefore, the results may not provide a comprehensive representation of the institutional, infrastructure-based, and socioeconomic characteristics of various regional contexts in the Global South.

Finally, while the use of qualitative thematic synthesis to identify common conceptual themes and relationships present within the literature allows for the integration of knowledge related to structural linkages between energy consumption and transportation activity, it does not allow for the quantification of emissions reduction efforts, assessment of the effectiveness of policies designed to promote sustainable transportation behaviors, or measurement of the performance of transportation-related infrastructure. Thus, the results of this study can best be viewed as providing analytic and conceptual contributions to understanding the process of achieving sustainable transportation practices rather than as empirical measures of sustainability progress.

11 Future Research Directions

Several opportunities exist for additional academic research to improve our understanding of sustainable transportation transitions, one of which is the empirical evaluation of integrated energy-transportation systems at the national or regional scale. More specifically, detailed case studies evaluating the interaction between the available grid capacity, potential renewable energy generation, and deployment of electric vehicles will help identify the operational challenges related to achieving integrated decarbonization. Comparative evaluations among cities/regions may also highlight institutional arrangements, governance structures, and economic conditions that affect the success of decarbonization policies, enabling the identification of more appropriate (contextual) policy approaches while highlighting common practices that can be transferred.

Another area worthy of additional study is the contribution of informal transportation systems to carbon–neutral mobility transitions. Because informal transportation is often the dominant form of transportation in many developing cities worldwide, investigating the technical, financial, and institutional pathways by which these systems can evolve toward lower-emission forms of operation is essential. Future research should focus on assessing distributional transportation decarbonization strategies. Evaluating specific strategies for disparate socioeconomic populations, including those with limited income levels and other disadvantaged populations, will enable researchers to develop more inclusive frameworks for transitioning to new modes of transportation. Finally, future research utilizing an interdisciplinary approach that combines energy system modeling techniques, urban planning theory, political/economic analysis, and social science methodologies will aid in providing a more complete understanding of the multiple scalars and intersectoral influences impacting energy transportation transition efforts.

12 Conclusion

The effectiveness of transport decarbonization strategies in the Global South depends on complex interdependencies among energy systems, transportation system infrastructure capabilities, institutional frameworks, policies governing these sectors, and socioeconomic factors. Systematic reviews of the relevant literature have identified numerous long-standing obstacles to transitioning to sustainable mobility. These include continued reliance on fossil fuels, financial constraints for implementing alternative technologies and infrastructure, fragmentation within institutions responsible for regulating transportation services, and an unequal distribution of access to transportation services across different socioeconomic populations.

Technological innovation has provided new avenues for achieving emission reductions. Increasing the availability of renewable resources for electricity generation and developing multimodal policy platforms offer potential ways to accelerate emission pathways in the transportation sector. Ultimately, this analysis underscores the necessity of an integrated approach to achieve decarbonized transportation systems. Examples of such an approach include coordinating transportation and energy policies, providing increased funding for infrastructure investments in areas related to low-carbon transportation options (e.g., electric vehicle charging stations), and integrating principles of equity into transportation planning and decision-making processes.

References

Addy MN, Aigbavboa C, Kwofie TE (2026) Urbanization and decarbonization. In: Decarbonization of building energy in developing countries. Springer Nature Switzerland, Cham, pp 267–294

Aidam GSSK, Opoku R, Adjei EA, Davis F, Oppong DK, Adu-Poku A, Aidam SG (2025) Electrified transportation for sustainable mobility in developing countries–a review of challenges and opportunities. J Ghana Inst Eng 25(1):34–44

Ajao Q, Sodiq OO, Sadeeq L (2024) Understanding the adoption and integration of autonomous EVs in Sub-Saharan Africa: challenges and technical recommendations. In: 2024 IEEE 5th international conference on electro-computing technologies for humanity (NIGERCON). IEEE, pp 1–8

Amin S (2025) Decarbonization of road transportation in India a round-robin review on low-carbon strategies and financial policies. Fut Transp 5(2):36

Azunre GA, Amponsah O, Takyi SA, Mensah H, Braimah I (2022) Urban informalities in sub-Saharan Africa (SSA): a solution for or barrier against sustainable city development. World Dev 152:105782

Bhatt S (2025) 7 Inclusive transport in Africa. Transp Serv Manag Afr 151

Chen Z, Awan U, Nassani AA, Al-Aiban KM, Zaman K (2025) Enhancing sustainable growth in the global south: the role of mineral resource management, supply chain efficiency, technology advancement, and local downstream processing. Resour Policy 100:105451

Dashora LK, Firoz CM, Shaw R (2025) Understanding urbanization, urban sustainability, and environmental transition challenges in the global south. In: Urban sustainability: navigating environmental transition challenges in the global south. Springer Nature Singapore, Singapore, pp 1–8

Delina LL, Shi L (2025) Sustainable mobility transitions in developing states: a comparison of Southeast Asia's electric mobility policies and aspirations. Sustain Fut 9:100751

Eltohamy MS, Aldawsari MS, Raouf A, Youssef H, Ahmed I, Alsenani TR, Ahmed MMR (2026) Enhancing electric grid flexibility for the integration of variable renewable energy: challenges, innovations, and future directions. IEEE Access

Emodi NV, Okereke C, Abam FI, Diemuodeke OE, Owebor K, Nnamani UA (2022) Transport sector decarbonization in the Global South: a systematic literature review. Energ Strat Rev 43:100925

Falcone PM (2023) Sustainable energy policies in developing countries: a review of challenges and opportunities. Energies 16(18):6682

Ferrer ALC, Thome AMT (2023) Carbon emissions in transportation: a synthesis framework. Sustainability 15(11):8475

Geurs KT, Niemeier D, Giannotti M (2021) The uneven geography of the accessibility and environmental quality in the global north and south: introduction to the special issue. J Transp Geogr 97:103216

Goldthau A, Eicke L, Weko S (2020) The global energy transition and the global south. In: Hafner M, Tagliapietra S (eds) The geopolitics of the global energy transition. Lecture notes in energy, vol 73. Springer, Cham. https://doi.org/10.1007/978-3-030-39066-2_14

Guzman LA, Oviedo D (2018) Accessibility, affordability and equity: assessing 'pro-poor' public transport subsidies in Bogotá. Transp Policy 68:37–51
Hassan Q, Algburi S, Sameen AZ, Salman HM, Jaszczur M (2023) A review of hybrid renewable energy systems: solar and wind-powered solutions: challenges, opportunities, and policy implications. Results Eng 20:101621
Henderson MI, Novosel D, Crow ML (2017) Electric power grid modernization trends, challenges, and opportunities. IEEE Power Energy
Ho CQ, Tirachini A (2024) Mobility-as-a-service and the role of multimodality in the sustainability of urban mobility in developing and developed countries. Transp Policy 145:161–176
Hossain MN, Cho HM (2026) Techno-economic feasibility of photovoltaic-powered electric vehicle charging stations: a global review and future outlook. Sustain Energy Res 13(1):9
Ibáñez E, Gkritza K, McCalley J, Aliprantis D, Brown R, Somani A, Wang L (2010) Interdependencies between energy and transportation systems for national long term planning. In: Sustainable and resilient critical infrastructure systems: simulation, modeling, and intelligent engineering. Springer Berlin Heidelberg, Berlin, Heidelberg, pp 53–76
Ka I, Djité AN, Chimére Diop SA, Ayetor GK, Diouf B (2026) The electric vehicle transition in emerging economies. Vehicles 8(2):37
Kalinowski T (2024) The green climate fund and private sector climate finance in the global south. Clim Policy 24(3):281–296
Lah O (2017) Decarbonizing the transportation sector: policy options, synergies, and institutions to deliver on a low-carbon stabilization pathway. Wiley Interdiscip Rev Energy Environ 6(6):e257
Li W, Ding H, Xu N, Zhang J (2023a) Toward carbon-neutral transportation electrification: a comprehensive and systematic review of eco-driving for electric vehicles. IEEE Trans Transp Electrification 10(3):6340–6360
Li Z, Hasan MM, Lu Z (2023b) Assessing financial factors for oil supply disruptions and its impact on oil supply security and transportation risks. Environ Sci Pollut Res 30(12):33695–33710
Li Y, Purwanto AJ, Shi X, Ge S (2025) Electric vehicles in developing countries: trends, challenges, and policy imperatives for a sustainable transition. Energy Sustain Dev 101827
Lup ANK, Soni V, Keenan B, Son J, Taghartapeh MR, Morato MM, Montañés RM (2023) Sustainable energy technologies for the Global South: challenges and solutions toward achieving SDG 7. Environ Sci Adv 2(4):570–585
Martínez DTS, Kreuz T, Ridens BL, Rahman RK, Sumathi SV, Ross S, Smith NR (2025) Energy transport is a cornerstone of the energy supply chain. Energy Transp Infrastruct Decarbonized Econ 7–43
Matsvai S (2026) Carbon markets: participation and benefits. Green Financ Emerg Econ Empower Small Enterp Sustain Growth Dev Ctries 109–131
Mesquita AR, de Abreu VHS, Poyares CN, Santos AS (2025) Barriers to electric vehicle adoption: a framework to accelerate the transition to sustainable mobility. Sustainability 17(18):8318
Michalski J, Poltrum M, Bünger U (2019) The role of renewable fuel supply in the transport sector in a future decarbonized energy system. Int J Hydrogen Energy 44(25):12554–12565
Minas AM, García-Freites S, Walsh C, Mukoro V, Aberilla JM, April A, Mander S (2024) Advancing sustainable development goals through energy access: lessons from the global south. Renew Sustain Energy Rev 199:114457
Mwesigwa L, Kilani M, Siemiatycki M (2023). Road transport pricing and financing in Africa. Handb Transp Pricing Financ 364–379
Ndhlovu E, Mhlanga D, Duri B (2025) Decarbonizing urban transport: an overview of electric vehicles, public transport, and sustainable infrastructure in achieving net-zero emissions. Discov Glob Soc 3(1):53
Nekahi A, Anil Kumar MR, Deng S, Li X, Petropoulos A, Nanda J, Zaghib K (2025) Toward green renewable energies and energy storage for the sustainable decarbonization and electrification of society. Electrochem Energy Rev 8(1):12
Nwankwo N, Olaniyi TK, Morgan A (2023) A just transition framework for oil rich Sub-Saharan Africa countries. Int J Sustain Energy 11(1)

Nwosu CC, Maduekwe EE, Obalum DC, Udemba JN (2025) Governance and the energy transition in developing nations: attaining sustainable development. J Policy Dev Stud 15(2):21–29
Obi JN, Ojo E, Ujah CO (2025) Decentralized renewable energy in sub-Saharan Africa: a critical review of pathways to equitable and sustainable energy transitions. Unconv Resour 100267
Okafora CL, Van Eeden J, Booysen MJ, Bruwer M (2025) The electric mobility ecosystem: a literature review of transport electrification to define research paths for Africa. J Transp Health. https://www.tandfonline.com/doi/full/10.1080/19427867.2025.2535491
Oshilalu AZ, Iboh S (2025) Integration of renewable energy technologies into transportation infrastructure to minimize carbon footprint and enhance long-term environmental resilience. Glob J Eng Technol Adv 24(03):236–251
Pojani D, Stead D (2015) Sustainable urban transport in the developing world: beyond megacities. Sustainability 7(6):7784–7805. https://doi.org/10.3390/su7067784
Pojani D, Stead D (2018) Policy design for sustainable urban transport in the Global South. Policy Design Pract 1(2):90–102. https://doi.org/10.1080/25741292.2018.1454291
Ponce-Jara MA, Ruiz E, Gil R, Sancristóbal E, Pérez-Molina C, Castro M (2017) Smart grid: assessment of the past and present in developed and developing countries. Energ Strat Rev 18:38–52
Powell J, McCafferty-Leroux A, Hilal W, Gadsden SA (2024) Smart grids: acomprehensive survey of challenges, industry applications, and future trends. Energy Rep 11:5760–5785. https://doi.org/10.1016/j.egyr.2024.05.051
Pozo-Burgos EJ, Alpala LO, Heredia-Campaña AL (2026) Electric mobility transition, intelligent digital platforms, and grid-vehicle integration models: a systematic review. World Electr Veh J 17(3):123
Pranawengkapti K, Shrestha S, Werland S, Martin E, Lah O (2025) Sector coupling: accelerating renewable energy integration in transport with electric vehicles. Sustain Earth Rev 8(1):16
Rabbi MF, Popp J, Máté D, Kovács S (2022) Energy security and energy transition to achieve carbon neutrality. Energies 15(21):8126
Saklani M, Saini DK, Yadav M, Gupta YC (2024) Navigating the challenges of EV integration and demand-side management for India's sustainable EV growth. IEEE Access 12:143767–143796
Shao Y, Yang Z, Yan Y, Yan Y, Israilova F, Khan N, Chang L (2025) Navigating Nigeria's path to sustainable energy: challenges, opportunities, and global insight. Energ Strat Rev 59:101707
Song J, He G, Wang J, Zhang P (2022) Shaping future low-carbon energy and transportation systems: digital technologies and applications. Ienergy 1(3):285–305
Sovacool BK, Newell P, Carley S, Fanzo J (2022) Equity, technological innovation and sustainable behavior in a low-carbon future. Nat Human Behav 6(3):326–337
Stoeglehner G (2020) Integrated spatial and energy planning: a means to reach sustainable development goals. Evol InstAl Econ Rev 17(2):473–486
Strojny J, Krakowiak-Bal A, Knaga J, Kacorzyk P (2023) Energy security: a conceptual overview. Energies 16(13):5042
Sun S, Ahmed S (2024) Navigating and overcoming barriers to digital energy transition for carbon neutrality in China. Energies 17(23):5888
Sundaram A (2024) The enforcement of environmental laws in developing countries: challenges and opportunities. Libr Prog-Libr Sci Inf Technol Comput 44(3)
Szczepański E, Żochowska R, Izdebski M, Jacyna M (2025) Decision-making problems in urban transport decarbonization strategies: challenges, tools, and methods. Energies 18(15):3970
Udendhran R, Mohan TR, Uthra RA, Selvakumarasamy S, Dinesh G, Mukhopadhyay M, Chakraborty P (2025) Transitioning to sustainable E-vehicle systems–global perspectives on the challenges, policies, and opportunities. J Hazard Mater Adv 17:100619
Ukoba K, Yoro KO, Eterigho-Ikelegbe O, Ibegbulam C, Jen TC (2024) Adaptation of solar energy in the Global South: prospects, challenges and opportunities. Heliyon 10(7)
Veitch E, Rhodes E (2024) A cross-country comparative analysis of congestion pricing systems: lessons for decarbonizing transportation. Case Stud Transp Policy 15:101128

Weko S, Goldthau A (2022) Bridging the low-carbon technology gap? Assessing energy initiatives for the Global South. Energy Policy 169:113192
Wu W (2024) Infrastructure investment and finance in the global south: private participation and its determinants. Public Financ Manag 23(3):99–112
Yan C, Murshed M, Ozturk I, Siddik AB, Ghardallou W, Khudoykulov K (2023) Decarbonization blueprints for developing countries: the role of energy productivity, renewable energy, and financial development in environmental improvement. Resour Policy 83:103674
Yañez-Pagans P, Martinez D, Mitnik OA, Scholl L, Vazquez A (2019) Urban transport systems in Latin America and the Caribbean: lessons and challenges. Lat Am Econ Rev 28(1):1–25

Digital Technologies and Smart Mobility Solutions: A Conceptual Framework for Urban Mobility in Developing Countries

Brave Zizhou, Fungai Nora Mukora, and Teurai Matekenya

Abstract Cities in developing nations face incessant mobility challenges including congestion, fragmented services, infrastructural fragility and unequal access to basic amenities; all of which undermine environmental sustainability and economic productivity. Smart mobility offers a potential response to these challenges. It aims to create transport systems that are efficient, low-carbon and resilient. Digital technologies such as mobile platforms, real-time data analytics, and intelligent transport systems are the key enablers of smart mobility. However, existing research on digitally enabled smart mobility solutions is largely grounded in high-income contexts, offering limited guidance for developing nations and resource-constrained settings characterised by informality, infrastructure and institutional limitations, a significant lacuna in the scholarship canon which this research, therefore, seeks to fill. This Chapter responds to this gap through developing a framework that proffers insights on how resource-constrained and developing countries can adopt digitally enabled smart mobility solutions to support efficient, low-carbon and resilient transport systems. Drawing on a structured review and synthesis of interdisciplinary literature, the paper conceptualises digital technologies as socio-technical enablers whose effects are conditioned by governance environments, context-sensitive design principles, formal-informal system integration, and implementation pathways. The resulting five-layer framework clarifies how digitally enabled smart mobility solutions interact with and compound infrastructural, economic and social constraints to shape mobility outcomes in developing countries. The paper contributes to smart mobility scholarship by moving beyond deterministic technology narratives and in consequence, theorises how digitally enabled smart mobility solutions can be designed and deployed in ways that are responsive to context specific conditions of scarcity, informality, and institutional fragility. For policymakers and practitioners,

B. Zizhou (✉)
Loughborough University, Loughborough, UK
e-mail: zizhoub@yahoo.com

F. N. Mukora · T. Matekenya
University of Zimbabwe, Harare, Zimbabwe

J. Kanyepe and I. Chirisa (eds.), *Advancing the Green Agenda in the Transport and Logistics Sector in Global South Countries*, World Sustainability Series,
https://doi.org/10.1007/978-3-032-31374-4_8

the framework helps aligning digital smart mobility innovation with sustainability, resilience, and equity goals.

Keywords Smart mobility solutions · Urban transportation · Digital technology · Developing countries · Informal transport · Conceptual framework

1 Introduction

Urban mobility systems in low and middle-income countries are facing growing pressure. As cities expand and populations surge, existing transport infrastructures are stretched beyond limits, and new investments cannot keep up with rising demand and this puts chronic pressure on cities' amenities. This trend is especially visible across Africa, Asia and Latin America where most future urban growth is expected to occur (United Nations 2018). By 2050, approximately 68% of the global population is projected to be living in urban areas (United Nations 2018). This rapid urbanisation brings mobility challenges including congestion, pollution, unreliable public transport, fragmented service provision and long travel times which all enmesh to reduce economic productivity and undermine environmental sustainability. As urban transportation challenges intensify, transforming urban mobility has become a central agenda for governments seeking to support more inclusive, efficient and sustainable urban development.

Smart mobility has gained significant attention as a solution to modern urban mobility challenges and a means to enhance citizens' quality of life. Smart mobility refers to digitally enabled, data-driven, and user-centred transport systems that integrate multiple modes and technologies to deliver efficient, sustainable, safe, and accessible mobility for people and goods (Papa and Lauwers 2015; European Commission 2013; Giffinger et al. 2007). It includes intelligent transport systems, multimodal journey planning, digital payment integration and algorithmic management of traffic flows. The appeal of this mobility systems lies in the expectation that better information and coordination can reduce inefficiencies, lower emissions and improve safety and accessibility without requiring large scale expansion of physical infrastructure (Banister 2008; Shaheen and Cohen 2013). Digital technologies form the operational foundation of this approach (Papa and Lauwers 2015). Internet of Things (IoT) sensors support real time monitoring of traffic conditions, mobile applications connect users to services and help aggregate demand, and data analytics enable predictive planning and dynamic routing (Biyik et al. 2021). Digital payment systems reduce transaction friction and support platform-based coordination (Cledou et al. 2018). In high income contexts where electricity supply is stable, connectivity is widespread, infrastructure well developed and transport operators are formalised, these tools have delivered measurable improvements in network efficiency and user experience (Smith et al. 2020; Hensher 2020). However, the transferability of these models to developing nation contexts is far less certain (Behrens et al. 2016; UN-Habitat 2020).

Mobility systems in many cities in developing nations are shaped by infrastructural fragility, constrained public budgets and extensive informal provision (Cervero and Golub 2007). Electricity and connectivity can be intermittent, regulatory oversight fragmented and institutional capacity uneven. Informal transporters including minibuses, motorcycle taxis and shared vans provide a large share of daily mobility and are deeply embedded in local labour markets and social relations (UN-Habitat 2020). Research shows that informal transport accounts for the majority of motorised trips in many African cities, often exceeding 70% and reaching up to 90% in some contexts (Cervero and Golub 2007; Behrens et al. 2016). In Latin American cities, informal modes can represent between 30 and 50% of motorised trips, and often around 40% in cities with limited formal public transport (Cervero 2013; Gómez-Lobo 2011). In such settings, digitally enabled smart mobility solutions that have been designed for high-income countries cannot be transferred and replicated directly. They fail when transplanted into contexts of scarce infrastructure, weak institutions, and informal transport systems. Instead, their success depends on how they are designed, integrated, and implemented within these specific socio-technical environments (Docherty et al. 2018; Behrens et al. 2016).

Although smart mobility solutions have the potential to reshape transport systems in developing nations, existing research remains uneven and technologically oriented. Luke et al. (2025), in their review of 608 documents on smart mobility in Africa, found that the field is fragmented, with only 14 studies focusing specifically on African transport systems. The review highlighted the majority of the literature examines discrete technological solutions such as electric vehicles, platform-based services or operator-level digitisation without linking these interventions to the broader communities and infrastructural systems in which they operate. Similarly, Mitieka et al. (2023) reviewed 3223 global publications and observed a strong emphasis on technological aspects and limited engagement with social, institutional or economic contextual dimensions. Issues such as inclusivity, safety, and quality of life remain underexplored, despite their centrality to urban mobility outcomes (Luke et al. 2025). Collectively, these reviews highlight the need for more integrated approaches that reflect the institutional fragmentation, infrastructural constraints and informal transport dominance characteristic of many developing-nation contexts.

Existing smart mobility research shows two important gaps when considered in developing-nation contexts. First, much of the foundational work on smart mobility such as Mobility-as-a-Service (Maas) frameworks (Jittrapirom et al. 2017; Smith et al. 2018; Sochor et al. 2018), data-intensive Intelligent Transport Systems (ITS) (Lyons, 2018), and broader smart mobility and smart city governance models (Giffinger et al. 2007; Docherty et al. 2018) has been developed in high-income settings characterised by stable digital infrastructure, formalised transport operators, reliable connectivity and strong institutional capacity. By contrast, many developing cities face intermittent electricity supply, unreliable connectivity, fragmented regulatory oversight and limited institutional capacity. Although some studies address these contexts, they rarely offer integrated frameworks linking digital capabilities with such constraints (Luke et al. 2025; Mitieka et al. 2023). Tan and Taeihagh (2020) highlight that digitally enabled mobility solutions depend on alignment with broader socio-economic, legal and regulatory conditions. What remains underdeveloped in the wider literature is an account of how the digitally enabled smart mobility solutions can be designed and implemented to operate within these governance settings.

Second, informal transport systems receive minimal attention in smart mobility frameworks considering they account for up to 95% of motorised trips in many African cities (Behrens et al. 2016; Cervero and Golub 2007; UN-Habitat 2020) and around 30–50% in major Latin American cities (Gómez-Lobo 2011; Hidalgo and Gutiérrez 2013). Research either overlooks informal modes entirely or treats them as obstacles to modernisation rather than examining integration possibilities. Recent studies show that MaaS implementation faces significant spatial and institutional constraints in cities where informal transport dominates, yet existing work offers limited guidance on how such systems might be effectively integrated (Ho and Tirachini 2023; Acheampong et al. 2025). Comprehensive reviews by Luke et al. (2025) and Mitieka et al. (2023) provide meta-level evidence that smart mobility research in Africa remains fragmented, technologically deterministic and lacks system level integration with limited consideration of user perspectives and socioeconomic impacts.

This paper addresses these gaps by developing a conceptual framework that explains how developing nations can adopt digital technologies to create smart mobility solutions in resource constrained contexts. The framework synthesises literature from smart mobility, digital technologies and transportation systems streams to provide an integrated perspective currently absent from existing research. Drawing on socio-technical systems theory, the framework integrates five interdependent themes including digital functional capabilities, institutional enabling conditions, design principles, informal transport integration and implementation pathways. Each theme addresses specific aspects of how digital technologies can support efficient, sustainable and resilient mobility outcomes when properly embedded within developing nation realities.

The contribution of this study is twofold. First, the framework provides an integrated conceptual model that explicitly positions informal transport integration as a central requirement for successful smart mobility implementation in developing nations. Where existing frameworks either ignore informal systems or treat them as obstacles, this framework conceptualises how digital technologies can complement rather than displace informal networks that serve the majority of mobility solutions in many developing cities. Second, the framework identifies context specific design principles and implementation pathways appropriate for resource constrained environments characterised by intermittent connectivity, limited institutional capacity and infrastructural fragility. The framework offers researchers a structured analytical lens for examining smart mobility in developing nations and provides policymakers with guidance for aligning digital innovation with inclusive and sustainable urban development.

The remainder of this paper is organised as follows. Section 2 reviews existing literature on smart mobility, digital technologies, and urban mobility in urban developing nations which includes informal transport systems. Section 3 explains the research methodology. Section 4 presents the conceptual framework for implementing digitally enabled smart mobility solutions in developing countries. Section 5 discusses how the framework addresses identified gaps and examines its limitations and Sect. 6 concludes with implications for future research and practice.

2 Literature Review

This section presents the literature review of the chapter focusing on smart mobility, digital technologies in smart mobility, urban mobility in developing nations and digital innovation under constraint.

3 Smart Mobility

Smart mobility has emerged as a central concept in urban transport scholarship. The appeal of this paradigm lies in the expectation that better information and coordination can reduce urban transport inefficiencies, lower emissions, and improve accessibility without requiring large-scale expansion of physical infrastructure (Banister 2008; Shaheen and Cohen 2013). Scholarly attention has correspondingly concentrated on the technologies through which these outcomes are pursued, including intelligent transport systems, MaaS platforms, demand-responsive transit, shared autonomous vehicles, and integrated digital payment systems (Butler et al. 2020; Szpilko et al. 2023). Research demonstrates that these technologies, when effectively deployed, can reduce congestion, lower per-trip emissions, and improve network efficiency across diverse urban settings (Son et al. 2025; Müller-Eie and Kosmidis 2023).

Despite this promise, the dominant smart mobility literature carries assumptions that limit its transferability. These assumptions are rarely made explicit but are deeply embedded in the design logic of the frameworks themselves. A further limitation of the dominant literature is its technological determinism: smart mobility outcomes are frequently attributed to the adoption of specific technologies rather than to the relational conditions such as governance arrangements, institutional capacity, and socio-technical fit, that determine whether technologies can function as intended (Tan and Taeihagh 2020). Addressing these limitations requires critical examination of what digital technologies enable, and under what conditions.

4 Digital Technologies as Functional Enablers

The digital technologies underpinning smart mobility function not as standalone solutions but as functional enablers whose effects are conditioned by the governance and institutional environments in which they are deployed (Markus and Silver 2008; Docherty et al. 2018). Intelligent transport systems operate through sensor networks, artificial intelligence, and big data analytics to improve traffic efficiency and environmental performance (Lungu, 2024; Shiva et al. 2024). Digital twin platforms create dynamically evolving models of urban transport environments, enabling real-time traffic management and environmental impact assessment (Faliagka et al. 2024). IoT-based coordination systems manage electric vehicle charging and route optimisation to reduce peak energy demand and per-trip emissions (Reis et al. 2025). Mobile platforms enable multimodal integration and demand aggregation, reducing reliance on private vehicles and supporting more accessible service provision (Cledou et al. 2018; Biyik et al. 2021).

Critically, the effectiveness of these technologies depends on conditions that extend well beyond their technical specifications. Interoperability between systems, open data standards, regulatory frameworks that accommodate platform-based service models, and institutional capacity to manage data governance all shape whether digital tools deliver their intended benefits or reproduce existing inefficiencies in digital form (Almatar 2024; Szpilko et al. 2023). Research also highlights that the digital infrastructure sustaining smart mobility including data centres, communication networks, and cloud computing platforms, generates a carbon footprint that can partially offset sustainability gains if not managed through renewable energy sourcing and energy-efficient design (Deveci et al. 2023; Golinska-Dawson and Sethanan 2023). Taken together, this body of evidence supports the conclusion that digital technology deployment is a necessary but insufficient condition for smart mobility transformation; the governance and institutional conditions mediating technology use are equally determinative of outcomes.

5 Urban Mobility in Developing Nations: Challenges and Constraints

Urban mobility in developing nations presents a configuration of challenges that is qualitatively different from the contexts in which smart mobility frameworks have been developed. Rapid urbanisation, accelerated motorisation, and constrained public investment have produced transport systems under severe stress across cities in Africa, Asia, and Latin America (UN Habitat 2020). Chronic underinvestment in formal public transport has left mobility provision fragmented, with informal operators including minibuses, motorcycle taxis, and shared vans filling the gap and providing the majority of daily trips in many cities (Cervero and Golub 2007; Behrens et al. 2016). Research indicates that informal transport accounts for 70–90%

of motorised trips in many African cities and between 30 and 50% in major Latin American urban centres (Cervero and Golub 2007; Gómez-Lobo 2011) making these systems structurally central rather than transitionally peripheral.

The institutional and infrastructural conditions of developing nation cities further complicate smart mobility adoption. Electricity supply is frequently intermittent, connectivity is uneven and unreliable, regulatory frameworks were designed for conventional bus operations and may actively inhibit platform-based service models, and institutional capacity for cross-sector coordination is limited (Tan and Taeihagh 2020; Ho and Tirachini 2023). Financial constraints restrict both public investment in digital infrastructure and private operator capacity to absorb the costs of technology transition (Jia et al. 2022). These structural conditions mean that smart mobility solutions designed for high-income, high-connectivity, formally organised transport environments cannot be transplanted into developing nation contexts without fundamental redesign. Crucially, affordability and infrastructure resilience are not merely practical concerns to be resolved during deployment, they are design imperatives that must be embedded from the outset as primary constraints shaping platform architecture, user interface logic, and service coverage decisions (Lucas 2019; Oviedo Hernandez and Titheridge 2016). Frameworks that treat these as secondary considerations systematically underestimate the gap between technological potential and operational reality in resource-constrained settings.

6 Digital Innovation Under Constraint: Emerging Evidence and Gaps

A growing body of research has begun to examine how digital innovation can be configured to function under the resource, connectivity, and institutional constraints characteristic of developing nation contexts. Lightweight IoT sensors, edge computing architectures, cloud-based coordination platforms, and smartphone-based data collection have demonstrated potential for incremental modernisation of transport systems without requiring comprehensive infrastructure overhaul (Ahmed et al. 2022; Rojek et al. 2025). Blockchain and distributed ledger technologies offer possibilities for transparent, decentralised governance of mobility platforms in contexts where institutional trust is limited (Ahmed et al. 2022; Hoang 2024). Examples such as the Trufi app in Bolivia, the Digital Matatus project in Nairobi, and the SafeTravelPH platform in the Philippines show how participatory data collection and lightweight digital coordination can produce mobility improvements that are contextually grounded and operationally sustainable (Davidson et al. 2024).

This emerging evidence also reveals that implementation approach matters as much as technology selection. Incremental, pilot-based deployment with embedded learning mechanisms produces more durable outcomes than large-scale transplantation of solutions designed elsewhere, because pilots create the institutional knowledge, stakeholder trust, and operational refinement that large-scale deployments assume but cannot generate (Sochor et al. 2018; Docherty et al. 2018). Yet this implementation dimension of how digital mobility systems should be sequenced, tested, and scaled under conditions of institutional uncertainty and infrastructural fragility receives limited theoretical attention in existing frameworks, which tend to specify what technologies should be deployed without specifying how deployment should be managed.

Taken together, this body of literature remains fragmented and insufficiently theorised. Comprehensive reviews of smart mobility research in African cities identify only a small proportion of studies that engage specifically with African transport systems, with the remainder examining discrete technological solutions without linking interventions to the broader institutional and infrastructural systems in which they operate (Luke et al. 2025). Mitieka et al. (2023), reviewing over three thousand global publications, find a strong emphasis on technological aspects and limited engagement with social, institutional, and economic contextual dimensions. Two gaps stand out clearly. First, no integrated framework currently exists that specifies how digital capabilities, institutional conditions, context-sensitive design principles, and implementation pathways interact as a system in resource-constrained environments. Second, informal transport systems which carry the majority of urban mobility in many developing cities receive minimal theoretical attention in smart mobility scholarship, either excluded entirely from frameworks or positioned as obstacles to modernisation rather than as potential partners in digital transition. These gaps motivate the conceptual framework developed in this paper.

7 Research Methodology

This study adopts a conceptual research design grounded in an integrative literature synthesis methodology. Conceptual papers develop new theoretical insights by integrating and extending existing knowledge rather than collecting primary data, and their rigour is assessed by the quality, coherence, and explanatory power of the conceptual contribution they produce (Torraco, 2005; Jaakkola 2020). Following Snyder (2019), integrative reviews are most appropriate when the purpose is to create new theoretical models and conceptual frameworks rather than to aggregate empirical evidence toward a definitive verdict on a specific research question. The methodology was therefore designed to be sufficiently transparent to enable critical assessment while retaining the interpretive flexibility required to synthesise across the heterogeneous disciplinary literatures of smart mobility, digital technology, and development studies.

The literature search was conducted across Scopus and Web of Science, supplemented by Google Scholar for grey literature and institutional reports from organisations including the World Bank, United Nations, and the Asian Development Bank. These databases were selected for their broad coverage of peer-reviewed journals across transport and mobility studies, digital technology and information systems, and development studies—the three core disciplinary domains relevant to this study. The search strategy combined terms relating to smart mobility and urban transport, digital technologies in transportation, and urban mobility in low- and middle-income country contexts. The following Boolean query was applied to titles, abstracts, and keywords: *("smart mobility" OR "digital mobility" OR "digital transport" OR "intelligent transport system*" OR "Mobility-as-a-Service" OR "urban mobility") AND ("digital technology*" OR "data" OR "platform*" OR "mobile application*") AND ("developing countries*" OR "low-and middle-income" OR "Global South" OR Africa OR "Latin America" OR "Asia").*

The search was limited to English-language publications from 2000 to 2025, reflecting the period during which digital mobility and smart transport debates emerged and consolidated.

Publications were screened against predefined inclusion criteria requiring substantive engagement with digital technologies or platform-based functions in urban mobility and offering empirical or conceptual insights relevant to low- and middle-income country settings. Studies focused exclusively on high-income country contexts without transferable conceptual relevance, purely technical optimisation studies without mobility relevance, and publications lacking sufficient methodological or conceptual grounding were excluded. To reduce the risk of omitting influential work not captured by the database queries, backward and forward citation tracking was conducted for key articles identified during initial screening, a procedure recommended for integrative reviews to strengthen the completeness of the conceptual base (Snyder 2019).

Analysis followed an iterative, interpretive synthesis process consistent with established guidance on integrative and conceptual reviews (Torraco 2005; Jaakkola 2020). After full-text reading, publications were examined to identify recurring conceptual patterns, theoretical tensions, and knowledge gaps relating to digital technologies and smart mobility in developing country contexts. Themes were derived inductively from the literature rather than imposed through a predetermined coding framework, reflecting the interpretivist orientation of the study. The aim of the review was not exhaustive bibliographic coverage but a theoretically rich and analytically transparent synthesis capable of supporting robust conceptual framework development—a standard appropriate to integrative review methodology and distinct from the evidence aggregation protocols associated with systematic reviews (Paré et al. 2015).

8 A Proposed Conceptual Framework for Adopting Digital Enabled Smart Mobility Systems in Developing Countries

The conceptual framework was developed through synthesising literature from smart mobility, digital technologies, and transportation systems, and situating these literature themes within the institutional and infrastructural realities of developing nation contexts. Five core conceptual domains were derived from the literature: digital functional capabilities, institutional enabling conditions, context-sensitive design principles, informal transport system integration, and implementation pathways. These five domains were organised into analytically distinct but interrelated layers to reflect the logical and temporal relationships through which digitally enabled smart mobility initiatives unfold in resource-constrained urban settings (Heeks 2017; Peppard and Ward 2004). The layered structure does not imply hierarchy in importance but rather clarifies the conditions under which each domain becomes relevant and operative.

The framework is organised around five interconnected layers that together explain how digitally enabled smart mobility systems are conceived, embedded, and scaled within resource-constrained urban environments. Section 4.6 synthesises how these layers interact through five theoretical mechanisms that collectively explain why the same digital technologies produce divergent outcomes across developing nation contexts.

9 Digital Technologies as Functional Enablers

The framework conceptualises digital technologies as functional enablers, drawing on affordance theory from information systems research. Markus and Silver (2008) define functional affordances as the possibilities for goal-oriented action afforded to specific user groups by technical objects. This perspective shifts analytical attention from what technologies are to what they enable users to do within particular contexts. Affordance theory emphasises that technologies do not determine outcomes but provide action possibilities that emerge through relationships between technical artefacts, users, and contextual conditions (Zammuto et al. 2007; Leonardi 2011). This relational view is especially important in developing nation settings, where the same digital technology may produce different outcomes depending on institutional arrangements, user capabilities, and infrastructural conditions. The framework therefore treats digital capabilities not as fixed properties but as functional possibilities whose realisation depends on conditions addressed in subsequent layers.

The literature has extensively catalogued the enabling technologies that support smart mobility systems. Naeem et al. (2021) similarly emphasise the role of IoT, AI, blockchain and big data as core pillars underpinning innovative mobility solutions. While these classifications are useful, they are largely derived from studies conducted in high-income contexts where reliable digital infrastructure supports deployment. Developing nation settings present distinct constraints that shape how these technologies must be configured and implemented.

Research consistently shows that data infrastructure gaps constitute a primary barrier in developing nations. The World Bank (2021) reports that 92% of the largest low- and middle-income cities lack complete transportation maps, with overall digitalisation of transport systems remaining low. This data scarcity reflects limited formal infrastructure for sensor deployment and systematic data collection. Consequently, the literature highlights alternative approaches that leverage existing assets rather than assuming extensive sensor networks. Smartphone-based data generation has emerged as one such approach, enabling the mapping of both formal and informal transport services and providing real-time information for users and planners (World Bank 2021). The Trufi app in Bolivia illustrates this model, allowing users to contribute route information and integrate buses, microbuses, and informal taxis into a unified platform (World Resources Institute 2024).

Beyond data collection, coordination platforms represent a second critical capability. Research shows that data sharing mechanisms and open data infrastructures remain underdeveloped in many developing nation contexts, limiting the effectiveness of platform-based solutions (Tiznado-Aitken et al. 2023). In cities with large informal transport sectors, MaaS implementations are spatially constrained to areas served by formal, frequent public transport, raising concerns about coverage and equity (Tiznado-Aitken et al. 2023). Transaction management capabilities form a third component. Mobile payment innovations such as M-Pesa in Kenya demonstrate how digital financial services can expand access for populations without formal banking infrastructure (World Bank 2024; Suri 2017; Jack and Suri 2011). Analytics capabilities further support risk identification, travel pattern analysis, traffic management, and behaviour change interventions. Together, these four capabilities, data generation, coordination platforms, transaction management, and analytics form the technological foundation of the framework's first layer.

10 The Enabling Environment

Digital capabilities do not translate automatically into improved mobility outcomes. Research consistently demonstrates that institutional and governance conditions determine the practical scope and effectiveness of technological interventions (Tan and Taeihagh 2020; Docherty et al. 2018). A systematic review of smart city development in developing countries shows that technology-enabled initiatives can only be realised when concurrent socioeconomic, human, legal, and regulatory reforms are instituted (Tan and Taeihagh 2020; Kummitha and Crutzen 2017). The enabling environment therefore determines what is achievable even when digital technologies make certain outcomes technically possible.

Regulatory frameworks shape implementation feasibility. Studies show that governance structures must evolve to accommodate new mobility technologies, yet many developing nation regulations were designed for conventional public transport and may unintentionally inhibit innovation (Marsden and Reardon 2017; Docherty et al. 2018). Research on Asian megacities demonstrates that institutional fragmentation such as the division of responsibilities across multiple local government units creates inconsistent data sharing practices and uneven technical capacity, preventing coordinated smart mobility implementation despite adequate technology availability (Cervero 2013). Rigid regulatory frameworks can therefore create barriers even when technologies are mature and proven elsewhere, underscoring the need for adaptive governance mechanisms (Tan and Taeihagh 2020).

Political commitment determines whether informal operators are engaged constructively or marginalised. Research on informal transport governance shows that governments often view informal operators as obstacles to modernisation rather than legitimate mobility providers (Goodfellow 2015; Cervero and Golub 2007). Yet informal transport carries a substantial share of urban travel in the Global South and provides employment for large workforces (Behrens et al. 2016). Where political framing positions informal operators as partners, digital interventions can enhance existing systems through electrification, route optimisation and rider–driver connectivity rather than displacing livelihoods (Mitullah and Omondi 2023; Acheampong et al. 2025). Studies in African and Asian cities demonstrate that digitalisation of informal transport can reduce emissions, improve safety, and support gradual formalisation when accompanied by appropriate governance arrangements (Schalekamp and Behrens 2010).

Institutional capacity underpins implementation effectiveness. Research shows that low-performing local governments in developing countries face institutional rather than infrastructural barriers, including fragmented governance, weak coordination, and limited budgetary autonomy (Cervero 2013; Sengers and Raven 2015). Tan and Taeihagh (2020) emphasise that developing countries must strengthen human capital, construct clear regulatory frameworks, and ensure digital inclusivity. Limited capacity does not preclude digital innovation but narrows viable pathways and increases the importance of modest, maintainable technological designs. The enabling environment thus determines what is practically achievable, shaping the conditions under which digital capabilities can be deployed and sustained.

11 Design Principles as Contextual Mediators

Between technological capabilities and institutional environments lie design decisions that determine how digital mobility systems perform in practice. Academic research identifies several core principles guiding smart mobility design, including accessibility, sustainability, integration, and user-centricity (Lyons 2018; Docherty et al. 2018; Hensher 2020). Cledou et al. (2018) establish an eight-dimension taxonomy for designing smart mobility services, while Bıyık et al. (2021) define the field through intelligent transport systems, open data, big data, and citizen engagement. These principles must be adapted to developing nation contexts, where infrastructure constraints and resource limitations shape implementation possibilities.

Affordability emerges as a critical design requirement in low- and middle-income settings. Studies show that transport costs disproportionately burden disadvantaged populations, with affordability acting as a major determinant of mobility access (Lucas 2019; Oviedo Hernandez and Titheridge 2016; Guerra and Cervero 2011). Where disposable incomes are limited and operator margins are narrow, even modest cost increases can exclude participation. Open-source transport applications and low-cost digital tools have been shown to reduce barriers for municipal governments and small operators, enabling modern planning and coordination functions without high capital expenditure (World Bank 2024). Technologies that minimise hardware requirements and reduce transaction friction therefore achieve higher sustained uptake in these contexts.

Resilience represents another essential design principle for resource-constrained contexts. vidence shows that investing in resilient infrastructure yields substantial long-term benefits, with the World Bank estimating a return of four dollars for every dollar invested (Hallegatte et al. 2019). Research on telecommunications resilience demonstrates that legacy mobile networks such as 2G and 3G often maintain higher continuity during extreme events than newer, more complex systems (Simo-Reigadas et al. 2015; Kwasinski 2016). Digital mobility systems must therefore continue functioning during connectivity interruptions, power outages, or physical disruptions through offline capabilities and alternative communication channels. Without these features, user trust erodes rapidly and systems fail precisely when most needed.

User-centric accessibility is equally essential for ensuring uptake across diverse populations. The World Bank's Smart Mobility Program for São Paulo developed a micro-accessibility manual providing practical design guidance to promote safety, autonomy and dignity for all users (Bellas Lamas et al. 2023). Academic research similarly emphasises that smart mobility systems must accommodate varied literacy levels, linguistic diversity, and differing degrees of technological familiarity (Lucas 2019; Oviedo Hernandez and Titheridge 2016). Clear visual cues, simplified interaction flows, and alternative communication formats support broader participation. Together, affordability, resilience, and user-centric accessibility constitute the core design requirements that determine whether digital mobility systems can operate effectively within developing nation realities.

12 Integration with Informal Transport Systems

Smart mobility solutions must be integrated with informal transport systems, which provide the majority of urban mobility in many developing nations. Research shows that informal transport accounts for 70–95% of motorised trips in African cities (Behrens et al. 2016; Cervero and Golub 2007) and 30–50% in major Latin American cities (Gómez Lobo 2011). In Africa, paratransit services constitute more than 70% of public transport supply and typically operate with limited formal integration or regulation (Behrens et al. 2016). Their exclusion from smart mobility frameworks therefore represents both an analytical gap and a practical constraint that undermines system completeness.

Integration requires collaborative governance that recognises informal operators as legitimate mobility providers rather than obstacles to modernisation. Studies of co-production and digital crowdsourcing in the Philippines demonstrate how participatory approaches can modernise informal transport without displacing operators

(Tiglao et al. 2023). The SafeTravelPH platform engages government agencies, operators and passengers in sharing real-time data on vehicle trajectories, boarding locations and occupancy levels, enabling evidence-based improvements while preserving operator autonomy (Tiglao et al. 2023). Similar approaches in African cities show that incorporating the experiential knowledge of informal operators including their understanding of local demand patterns, route dynamics, and passenger behaviour, enhances planning in ways that cannot be replicated through data extraction alone (Schalekamp and Behrens 2010; Goodfellow 2015; Schalekamp 2017).

Equitable value distribution represents a further governance challenge for digital platforms. Research on platform cooperatives highlights alternative ownership models in which workers retain control over digital infrastructure and revenue flows (Scholz 2016; Scholz and Schneider 2017). OECD case studies show how worker-owned platforms such as Up&Go in New York enable members to retain 95% of earnings, with the remainder reinvested in platform development (OECD 2023). Research on informal transport governance in Kenya similarly highlights how collective organisation can strengthen democratic decision-making and reduce dependency on external platform owners (Mitullah and Omondi 2023). Without such mechanisms, digital mobility platforms risk concentrating benefits among platform owners rather than transport providers, undermining long-term sustainability.

Data governance arrangements shape power dynamics and determine whether operators retain meaningful control over operational information. Research on data cooperatives shows how collective data governance enables users to pool data voluntarily, exercise rights collectively, and maintain oversight over data use (Delacroix and Lawrence 2019). The European Data Governance Act (Regulation (EU) 2022/868) strengthens collective data rights by establishing new categories of data intermediaries and data altruism organisations, providing a regulatory foundation for cooperative and community-led data governance models. Allowing informal operators to participate in data governance reduces dependency on single providers and mitigates exclusion risks. Studies of African transport systems show that digital coordination does not require immediate formalisation; hybrid governance arrangements can enhance scheduling, safety, and route information while preserving the flexible service patterns characteristic of informal transport (Schalekamp and Behrens 2010; Porter et al. 2018). Through these collaborative mechanisms, integration strengthens rather than displaces existing mobility ecosystems.

13 Implementation Pathways

Digital mobility systems typically evolve through iterative processes rather than single, one-off deployments. Research on MaaS shows that pilot programmes function as learning mechanisms and are most valuable when designed with a view to scaling, rather than remaining isolated demonstrations (Sochor et al. 2018; Smith et al. 2020; Hensher 2020). Pilots allow participating actors to identify unforeseen barriers, test operational models, and refine governance arrangements before wider implementation. The São Paulo Smart Mobility Program illustrates this approach: developed as a multi-year partnership between 2019 and 2022, it combined analytical work with engagement activities to strengthen planning practices, institutional coordination, and the use of technological tools for more inclusive mobility (World Bank 2022, 2023).

Early-stage implementation requires careful selection of demonstration sites and activities. Academic research emphasises that pilots are most effective when located where institutional cooperation is feasible and where visible benefits can build stakeholder confidence (Docherty et al. 2018; Marsden and Reardon 2017). Learning mechanisms should be embedded from the outset so that user feedback and operational experience inform system refinement. In São Paulo, the Smart Mobility Program supported activities such as accessibility assessments, analysis of active mobility conditions, and governance and regulatory improvements, illustrating how initial efforts can broaden to address multiple dimensions of urban mobility (World Bank 2022, 2023). Capacity building is essential for both operators and institutions; training, local technical support, and gradual knowledge transfer reduce dependence on external vendors and enhance long-term system durability (Sengers and Raven 2015; Cervero 2013).

Scaling should proceed iteratively, with flexible planning that accommodates technological change and evolving user needs (Geels 2012; Sengers et al. 2019). Evidence from smart city initiatives in developing Asia further demonstrates that operational models must be tailored to local development status, digital maturity, and governance capacity. As systems expand, analytics can inform adjustments to services and network design, while governance arrangements must remain adaptable. Tracking indicators such as accessibility, travel times, modal shares, and service reliability over time enables cities to assess progress and benchmark performance.

Evaluation should extend beyond technical efficiency to encompass broader development objectives. Studies of mobility planning in Global South cities show that social equity, accessibility, and governance considerations often take precedence over environmental factors, reflecting the priorities of rapidly urbanising contexts (Lucas 2019; Oviedo Hernandez and Titheridge 2016; Schwanen 2021). Outcomes should therefore be assessed using multidimensional frameworks that incorporate environmental impact, resilience during disruptions, distributional equity, and governance effectiveness. Such comprehensive evaluation helps ensure that digital mobility initiatives align with wider development goals rather than narrow technical optimisation.

14 Framework Synthesis

The five layers of the framework do not operate as a linear sequence of independent steps. Rather, they constitute an interdependent socio-technical system in which technological capabilities, institutional conditions, design choices, informal transport ecologies, and implementation processes mutually shape one another. Socio-technical systems theory holds that sustainable outcomes require the joint optimisation of technical and social elements, such that neither can be developed in isolation without compromising system performance (Cherns 1976). The framework is built on the premise that mobility outcomes in developing nation contexts emerge not from any single layer but from the quality of interactions between all five. Five theoretical mechanisms explain how and why these layers interact in the ways the framework proposes.

The first mechanism is ***affordance actualisation***. Drawing on affordance theory, the digital capability layer establishes that digital technologies do not determine outcomes but create action possibilities (functional affordances) that must be actualised through use within specific contexts (Markus and Silver 2008; Leonardi 2011). The enabling environment is the primary actualisation condition. Institutional arrangements (regulatory frameworks, political commitment, and governance capacity) determine which affordances are accessible, to whom, and under what conditions. A mobility platform capable of coordinating informal routes and aggregating real-time demand data provides a functional affordance. However, this affordance remains latent unless regulatory frameworks permit data sharing, political actors commit resources, and institutional capacity exists to maintain operational systems. The implication is that identical digital technologies may produce substantially different outcomes across developing nation contexts, not because of differences in the technologies themselves, but because institutional conditions shape which possibilities are realised. This mechanism explains the frequently observed gap between technological potential and actual performance in Global South smart mobility initiatives.

The second mechanism is ***contextual translation***. The design principles layer functions as a translation mechanism between the generic possibility space created by digital technologies and the specific socio-material realities of resource-constrained environments. Technologies developed and proven in high-income settings carry embedded assumptions about infrastructure reliability, user literacy, operator formalisation, and institutional capacity that do not hold in many developing nation contexts (Heeks 2017; Docherty et al. 2018). Without deliberate translation through context-sensitive design, prioritising affordability, resilience during infrastructure disruption, and accessibility across diverse user populations, functional affordances created by the digital capability layer and enabled by the enabling environment cannot be operationalised at the local level. The design principles layer therefore mediates between technical capability and practical functionality, and its absence or inadequacy is a common source of smart mobility failure in resource-constrained settings even where institutional conditions are otherwise supportive.

The third mechanism is ***socio-technical co-evolution***. The informal transport integration layer represents more than a governance arrangement for managing informal operators; it reflects the theoretical requirement that technological systems and social structures evolve together rather than one being imposed upon the other (Geels 2002; Cherns 1976). Informal transport systems in many developing nations are not merely operational networks but deeply embedded social institutions carrying accumulated knowledge of local demand patterns, route dynamics, spatial inequalities, and passenger behaviour (Cervero and Golub 2007; Behrens et al. 2016). Smart mobility systems that extract operational data from informal networks without incorporating this experiential knowledge into platform design and governance arrangements produce technically incomplete systems that cannot serve the majority of urban trips. Genuine co-evolution requires that informal operators participate not only as users of digital platforms but as co-producers of the operational logic and governance arrangements that define how platforms function. Where this co-evolutionary process is absent, failures within the informal integration layer generate system-level consequences such as reduced coverage, contested legitimacy, and continued exclusion of the populations most dependent on informal transport.

The fourth mechanism is ***institutional path shaping***. The relationship between the enabling environment and the remaining layers is not unidirectional. While institutional conditions constrain and enable the digital capability layer, shape design choices, and determine what integration approaches are feasible, digital mobility initiatives can also exert reverse influence on institutional arrangements over time. Successful pilot programmes that demonstrate the operational value of informal operator integration, for example, can shift political framing and generate evidence that supports regulatory reform. This dynamic is consistent with niche-regime interaction dynamics in socio-technical transition theory, where innovations developed within protected spaces gradually destabilise incumbent institutional arrangements and create openings for broader reform (Geels 2002; Sengers et al. 2019). The framework therefore conceptualises the enabling environment not as a fixed precondition but as a domain that is itself shaped by the outcomes of the other layers over time.

The fifth mechanism is the ***reflexive learning loop***. The implementation layer is not the terminal stage of the framework but the mechanism through which the framework becomes self-correcting. Iterative pilots, embedded evaluation, and capacity building generate empirical evidence about which digital capabilities function as expected, which institutional conditions require adjustment, which design principles require adaptation, and which integration strategies generate the intended socio-technical co-evolution. This learning feeds back to all preceding layers, enabling refinement of digital capabilities, advocacy for institutional reform, revision of design principles, and adaptation of integration strategies. Without this reflexive mechanism, the framework degenerates into a static deployment model incapable of responding to the technological uncertainty, governance volatility, and evolving informal transport dynamics characteristic of developing nation contexts.

Together, these five mechanisms explain why identical smart mobility technologies produce divergent outcomes across different developing nation cities. Variations in affordance actualisation, contextual translation quality, co-evolutionary depth,

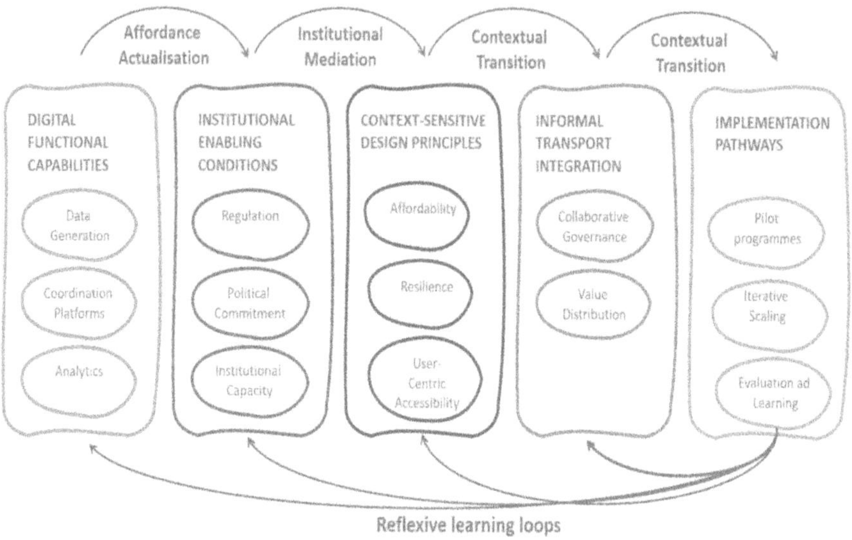

Fig. 1 A framework for adopting digital enabled smart mobility systems in developing countries (Authors' own work)

institutional path trajectories, and learning loop effectiveness account for the divergence more fully than technological differences alone. The framework thus provides not only an analytical structure for diagnosing smart mobility outcomes but a basis for designing interventions that address the systemic rather than the technological sources of failure. Figure 1 illustrates the five layers and their interactions, showing primary directional relationships, bidirectional influences, and the reflexive feedback loop that sustains system adaptation over time.

The framework generates five propositions:

Proposition 1: The governance substitution effect

In developing nation contexts where formal institutional capacity is weak, strong informal governance networks including community-based route associations, operator cooperatives, and neighbourhood mobility practices can function as substitute affordance actualisation conditions, enabling digital mobility platforms to achieve meaningful adoption even in the absence of formal regulatory support. This proposition challenges the prevailing assumption in smart mobility governance literature that institutional weakness uniformly constrains digitalisation and proposes instead that the relevant governance capacity is relational rather than formal.

Proposition 2: The formalisation paradox

Smart mobility platforms that achieve rapid formal-sector adoption without prior co-evolutionary integration of informal transport operators will produce aggregate service metric improvements that mask system-level deterioration specifically, reductions in coverage to peripheral and low-income areas, displacement of livelihood networks, and erosion of the distributed service flexibility that informal systems provide. The net effect on overall urban mobility access may therefore be negative even when headline indicators of efficiency and emissions appear positive.

Proposition 3: **Bidirectional institutional causality**

The relationship between governance conditions and smart mobility outcomes is bidirectional rather than unidirectional. While enabling institutional environments condition digital capability actualisation, successful niche implementations particularly those that demonstrate operational value through informal operator integration can progressively reshape the regulatory frameworks that initially constrained them. Consequently, waiting for full institutional readiness before deploying digital mobility solutions may itself prevent the institutional evolution necessary for sustainable transitions, as the evidence base for reform can only be generated through implementation.

Proposition 4: **Design translation and the limits of evidence transfer**

Evidence of smart mobility effectiveness generated in high-income, high-connectivity contexts does not transfer predictably to resource-constrained environments, because the functional affordances that produced measured outcomes in the original context cannot be fully actualised under different infrastructural, institutional, and social conditions. Specifically, platforms designed for stable connectivity, formalised operators, and high digital literacy will exhibit systematically lower performance in developing nation contexts not because of technological inferiority but because the contextual translation layer is absent from their design logic. Performance gaps between high-income and developing nation implementations are therefore better understood as design mismatch effects than as capacity deficits.

Proposition 5: **The equity-efficiency divergence**

Digitally enabled smart mobility implementations optimised for system-level efficiency including network coverage, average travel time reduction, and emissions performance will systematically underperform on equity dimensions, including accessibility for low-income users, coverage of peripheral areas, and protection of informal transport livelihoods, unless equity considerations are embedded as primary design constraints rather than secondary evaluation criteria. The divergence between efficiency and equity outcomes will be strongest in cities where spatial inequality is high and informal transport dependency among low-income populations is greatest.

These propositions provide direction for comparative case study research and quantitative evaluation across diverse developing nation contexts.

15 Discussion

Smart mobility scholarship has long recognised that digital technologies hold significant potential for improving urban mobility efficiency, reducing emissions, and expanding accessibility (Bıyık et al. 2021; Docherty et al. 2018). Yet evidence consistently shows that implementations in developing nation contexts fall short of these outcomes, with failures attributed variously to infrastructure deficits, institutional weakness, and limited financial capacity (Tan and Taeihagh 2020; Luke et al. 2025; Mitieka et al. 2023). This study advances a relational explanation for these persistent failures, arguing that the primary sources of underperformance are not technological but systemic arising from misalignments between digital capabilities, institutional actualisation conditions, design translation, and the co-evolutionary requirements of informal transport integration. In doing so, the study extends the socio-technical systems perspective (Cherns 1976; Geels 2002) into a domain where it has been underutilised, providing a theoretically grounded account of why the same technologies produce divergent outcomes across contexts that differ not in their access to digital tools but in the relational conditions that determine whether those tools can be made to work.

The finding that institutional conditions function as the primary actualisation mechanism for digital capabilities aligns with and extends prior work on smart city governance in developing countries (Tan and Taeihagh 2020; Kummitha and Crutzen 2017). Where this literature identifies institutional capacity as a general precondition for technology adoption, the present framework specifies the mechanisms through which this relationship operates, distinguishing between regulatory feasibility, political commitment, and governance coordination as distinct enabling functions that together determine which digital affordances can be realised and by whom (Markus and Silver 2008; Leonardi 2011). This specificity matters practically: it implies that institutional reform strategies should be differentiated rather than uniform, targeting the particular enabling function that is most constraining in a given context rather than treating institutional capacity as a single variable to be strengthened generically. The framework also challenges the assumption, implicit in much governance literature, that institutional conditions are stable preconditions for technology deployment. Drawing on niche-regime interaction dynamics (Geels 2002; Sengers et al. 2019), the framework proposes that successful implementations progressively reshape the institutional arrangements that initially constrained them, generating a co-evolutionary dynamic between governance reform and digital innovation that is absent from static precondition models.

The framework's treatment of informal transport represents its most substantive departure from prior smart mobility scholarship. Existing reviews have documented that informal systems are either ignored in smart mobility frameworks or treated as transitional phenomena to be replaced through formalisation (Luke et al. 2025; Mitieka et al. 2023; Behrens et al. 2016). This study argues that both postures misread the structural role of informal transport in developing nation mobility systems. Consistent with socio-technical co-evolution theory (Cherns 1976; Geels 2002), informal operators carry embedded operational knowledge of demand patterns, spatial inequalities, route dynamics, and passenger behaviour that formal planning instruments do not possess and that digital platforms cannot replicate through data extraction alone. The evidence from participatory digital coordination in the Philippines (Tiglao et al. 2023), cooperative and operator-led digitalisation models in African informal transport (ITDP 2022), and hybrid governance arrangements documented by Schalekamp and Behrens (2010) collectively supports this argument, demonstrating that participatory co-production of platform governance produces coverage, legitimacy and durability outcomes that displacement-oriented interventions cannot achieve. By positioning this co-evolutionary requirement as a theoretically necessary rather than practically optional design consideration, the framework extends Multi-Level Perspective scholarship (Geels 2002) into contexts where the incumbent mobility regime is itself informal; a configuration that existing transition theory has not adequately theorised.

Considered together, the relational reframing and the co-evolutionary informal integration argument produce an analytical advance that neither alone can generate. Frameworks that are sensitive to resource constraints without a theory of informal integration remain analytically incomplete, unable to engage with the systems that carry the majority of urban mobility in many developing cities (Cervero and Golub 2007). Equally, advocacy for informal integration without the institutional, design, and implementation conditions that make integration viable produces recommendations that are aspirational rather than actionable, as Tiznado-Aitken et al. (2023) demonstrate in their analysis of MaaS spatial constraints in informal transport environments. The five-layer architecture of the framework holds these two dimensions together, and the five propositions it generates specify the empirical conditions under which the framework's mechanisms are expected to operate, providing a structured basis for comparative validation across diverse developing nation contexts.

Four limitations qualify these contributions and point toward important directions for future inquiry. As a conceptual framework developed through literature synthesis rather than primary fieldwork, its propositions remain empirically unvalidated, and the conditions under which each mechanism operates as predicted across different governance structures, political economies, and levels of digital maturity are not yet known (Jaakkola 2020; Snyder 2019). The framework's deliberate generality, which enables its application across the heterogeneous contexts of the Global South, limits specific implementation guidance for individual cities, requiring contextual adaptation as a prerequisite for application rather than an optional supplement (Heeks 2017). The framework does not prescribe decision rules for resolving conflicts between objectives when they arise in practice, for instance, when affordability requirements compete with resilience investments, or when iterative scaling timelines conflict with political incentives for visible short-term outcomes. Most significantly, while the framework identifies collaborative governance and equitable value distribution as enabling conditions for informal transport integration, it does not fully theorise the political economy of resistance. The power of entrenched incumbent operators, bureaucratic incentives oriented toward formalisation rather than integration, and platform ownership structures that concentrate value among investors rather than transport providers (Scholz 2016; Scholz and Schneider 2017) represent forces that the framework acknowledges but does not model, and future theoretical work that engages explicitly with these dynamics would substantially strengthen the framework's predictive capacity.

16 Conclusion and Future Research

This study develops a five-layer conceptual framework that explains how digitally enabled smart mobility systems can be designed and implemented in developing nation contexts characterised by resource constraints, institutional fragility, and the structural dominance of informal transport networks. By integrating digital functional capabilities, institutional enabling conditions, context-sensitive design principles, informal transport system integration, and iterative implementation pathways into a theoretically grounded architecture, the framework advances two contributions to smart mobility scholarship. It provides a relational account of why identical digital technologies produce divergent outcomes across developing nation contexts, grounded in affordance theory and socio-technical systems theory, and it establishes informal transport integration as a theoretically necessary rather than practically optional component of smart mobility design, positioning the operational knowledge embedded in informal networks as an analytical and practical resource that digital platforms must incorporate rather than displace. Together, these contributions offer researchers a more precise vocabulary for diagnosing implementation failures and provide policymakers with conceptual tools for aligning digital innovation with the equity, resilience, and inclusion objectives that rapidly urbanising developing cities demand.

Future research should build on this foundation through empirical validation of the framework's five propositions across cities with varying governance structures, digital maturity, and political economies. Further work is needed on the political economy of informal transport integration, examining how entrenched interests and regulatory incentives shape the feasibility of co-evolutionary governance arrangements. Longitudinal studies tracking how informal operators adapt to digital platforms over time, and whether anticipated co-evolutionary benefits materialise in practice, would provide the empirical depth that conceptual work of this kind necessarily leaves open. Through these avenues, the framework can be refined and extended to support developing nations in leveraging digital technologies to build inclusive, resilient, and context-appropriate smart mobility systems.

References

Acheampong RA, Boateng FG, Agyemang E, Hotor Edem D (2025) Digital platforms, ICT-mediated mobility services and emerging smart mobility ecosystems in African cities: systematic evidence review of impacts and research gaps

Ahmed I, Zhang Y, Jeon G, Lin W, Khosravi MR, Qi L (2022) A blockchain- and artificial intelligence-enabled smart IoT framework for sustainable city. Int J Intell Syst 37(9):6493–6507. https://doi.org/10.1002/int.22852

Almatar KM (2024) Sustainable smart city planning: Advances in intelligent transportation and e-transportation management. Int J Transp Dev Integr, 8(1):167–177. https://doi.org/10.18280/ijtdi.080116

Banister D (2008) The sustainable mobility paradigm. Transport Policy 15(2):73–80. https://doi.org/10.1016/j.tranpol.2007.10.005

Behrens R, McCormick D, Mfinanga D (eds) (2016) Paratransit in African cities: Operations, regulation and reform. Routledge

Bellas Lamas C, Fonseca de Medeiros T, Moura dos Santos B, Pires de Souza E, Pereira Caldeira G, Di Biase L, Garcia Lippe MI (2023) São Paulo smart mobility program. World Bank Group, Washington, DC

Bıyık C, Abareshi A, Paz A, Ruiz RA, Battarra R, Rogers CD, Lizarraga C (2021) Smart mobility adoption: A review of the literature. J Open Innov: Technol, Mark, Complex 7(2):146. https://doi.org/10.3390/joitmc7020146

Butler L, Yigitcanlar T, Paz A (2020) Smart urban mobility innovations: A comprehensive review and evaluation. IEEE Access 8:196034–196049. https://doi.org/10.1109/access.2020.3034596

Cervero R (2013) Informal transport in the developing world. Oxford University Press, In The Oxford Handbook of Urban Planning

Cervero R, Golub A (2007) Informal transport: A global perspective. Transp Policy 14(6):445–457. https://doi.org/10.1016/j.tranpol.2007.04.011

Cherns A (1976) The principles of sociotechnical design. Human Relations 29(8):783–792

Cledou G, Soares E, Pinto JM (2018) A taxonomy for planning and designing smart mobility services. Gov Inf Q 35(4):577–591. https://doi.org/10.1016/j.giq.2017.11.008

Commission E (2013) Smart cities and communities—European innovation partnership. European Commission, Brussels

Davidson A, Cano Gómez D, Mazorra A, Minella Pasqual F (2024) Shared mobility and a sustainable transport future in Latin American Cities. World Resources Institute, Washington, DC

Delacroix S, Lawrence ND (2019) Bottom-up data trusts: Disturbing the 'one size fits all' approach to data governance. Int Data Priv Law 9(4):236–252

Deveci M, Gokasar I, Pamucar D, Zaidan AA, Wei W, Pedrycz W (2023) Advantage prioritization of digital carbon footprint awareness in optimized urban mobility using fuzzy Aczel Alsina based decision making. Appl Soft Comput 151:111136. https://doi.org/10.1016/j.asoc.2023.111136

Docherty I, Marsden G, Anable J (2018) The governance of smart mobility. Transp Res Part A: Policy Pract 115:114–125. https://doi.org/10.1016/j.tra.2017.09.012

Faliagka E, Christopoulou E, Ringas D, Politi T, Kostis N, Leonardos D, Tranoris C, Antonopoulos CP, Denazis S, Voros N (2024) Trends in digital twin framework architectures for smart cities: A case study in smart mobility. Sensors (Basel, Switzerland) 24(5):1665. https://doi.org/10.3390/s24051665

Geels FW (2002) Technological transitions as evolutionary reconfiguration processes: a multi-level perspective and a case-study. Res Policy 31(8–9):1257–1274

Geels FW (2012) A socio-technical analysis of low-carbon transitions: introducing the multi-level perspective into transport studies. J Transp Geogr 24:471–482

Giffinger R, Fertner C, Kramar H, Kalasek R, Pichler-Milanović N, Meijers E (2007) Smart cities: Ranking of European medium-sized cities. Centre of Regional Science, Vienna University of Technology. www.smart-cities.eu, Vienna

Golinska-Dawson P, Sethanan K (2023) Sustainable urban freight for energy-efficient smart cities—systematic literature review. Energies 16(6):2617. https://doi.org/10.3390/en16062617
Gómez-Lobo A (2011) Affordability of public transport a methodological clarification. J Transp Econ Policy (JTEP) 45(3):437–456
Goodfellow T (2015) Taming the "rogue" sector: Studying state effectiveness in Africa through informal transport politics. Comp Polit 47(2):127–147
Guerra E, Cervero R (2011) Cost of a ride: The effects of densities on fixed-guideway transit ridership and costs. J Am Plann Assoc 77(3):267–290
Hallegatte S, Rentschler J, Rozenberg J (2019) Lifelines: The resilient infrastructure opportunity. World Bank Publications
Heeks R (2017) Information and communication technology for development (ICT4D). Routledge
Hensher DA (2020) What might Covid-19 mean for mobility as a service (MaaS)? Transp Rev 40(5):551–556
Hidalgo D, Gutiérrez L (2013) BRT and BHLS around the world: Explosive growth, large positive impacts and many issues outstanding. Res Transp Econ 39(1):8–13
Ho CQ, Tirachini A (2023) Mobility-as-a-Service and the role of multimodality in the sustainability of urban mobility in developing and developed countries. Transp Policy 145:161–176. https://doi.org/10.1016/j.tranpol.2023.10.013
Hoang TV (2024) Impact of integrated artificial intelligence and internet of things technologies on smart city transformation. J Tech Educ Sci 19(1):64–73. https://doi.org/10.54644/jte.2024.1532
Institute for Transportation and Development Policy (2022) Digitalising informal transport: A review of practice in Africa. ITDP, New York
Jaakkola E (2020) Designing conceptual articles: four approaches. AMS Review 10(1):18–26. https://doi.org/10.1007/s13162-020-00161-0
Jack W, Suri T (2011) Mobile money: The economics of M-Pesa. National Bureau of Economic Research Working Paper No. 16721. https://doi.org/10.3386/w16721
Jia W, Beukes EA, Coetzee J, Van Ryneveld P (2022) Improving Paratransit in Maseru and Gaborone: Using innovative data techniques in a diagnostic approach to inform strategy. World Bank, Washington, DC
Jittrapirom P, Caiati V, Feneri AM, Ebrahimigharehbaghi S, Alonso-González MJ, Narayan J (2017) Mobility as a service: A critical review of definitions, assessments of schemes, and key challenges
Kummitha RKR, Crutzen N (2017) How do we understand smart cities? an evolutionary perspective. Cities 67:43–52. https://doi.org/10.1016/j.cities.2017.04.010
Kwasinski A (2016) Quantitative model and metrics of electrical grids' resilience evaluated at a power distribution level. Energies 9(2):93. https://doi.org/10.3390/en9020093
Leonardi PM (2011) When flexible routines meet flexible technologies: affordance constraint and the imbrication of human and material agencies1. MIS Quart 35(1):147–167. https://doi.org/10.2307/23043493
Lucas K (2019) A new evolution for transport-related social exclusion research? J Transp Geogr 81:102529. https://doi.org/10.1016/j.jtrangeo.2019.102529
Luke R, Mageto J, Twinomurinzi H, Bokaba T, Mhlongo S (2025) Smart mobility in Africa: where are we now? Cities 166:106212. https://doi.org/10.1016/j.cities.2025.106212
Lungu MA (2024) Smart urban mobility: the role of AI in alleviating traffic congestion. Proc Int Conf Bus Excell 18(1):1441–1452. https://doi.org/10.2478/picbe-2024-0118
Lyons G (2018) Getting smart about urban mobility–aligning the paradigms of smart and sustainable. Transp Res Part A Policy Pract 115:4–14. https://doi.org/10.1016/j.tra.2016.12.001
Markus ML, Silver MS (2008) A foundation for the study of IT effects: a new look at DeSanctis and Poole's concepts of structural features and spirit. J Assoc Inform Syst 9(10):609–632. https://doi.org/10.17705/1jais.00176
Marsden G, Reardon L (2017) Questions of governance: rethinking the study of transportation policy. Transp Res Part A Policy Pract 101:238–251. https://doi.org/10.1016/j.tra.2017.05.008
Mitieka D, Luke R, Twinomurinzi H, Mageto J (2023) Smart mobility in urban areas: a bibliometric review and research agenda. Sustainability 15(8):67–54. https://doi.org/10.3390/su15086754

Müller-Eie D, Kosmidis I (2023) Sustainable mobility in smart cities: A document study of mobility initiatives of mid-sized Nordic smart cities. Eur Transp Res Rev 15(1). https://doi.org/10.1186/s12544-023-00610-4
OECD (2023) Platform cooperatives and employment. OECD Publishing, Paris. https://doi.org/10.1787/3eab339f-en
Oviedo Hernandez D, Titheridge H (2016) Mobilities of the periphery: Informality access and social exclusion in the urban fringe in Colombia. J Transp Geogr 55:152–164. https://doi.org/10.1016/j.jtrangeo.2015.12.004
Papa E, Lauwers D (2015) Smart mobility: opportunity or threat to innovate places and cities. Proc Soc Behav Sci 151:295–302. http://repository.corp.at/id/eprint/36
Paré G, Trudel MC, Jaana M, Kitsiou S (2015) Synthesizing information systems knowledge: a typology of literature reviews. Inform Manage 52(2):183–199
Peppard J, Ward J (2004) Beyond strategic information systems: towards an IS capability. J Strat Inform Syst 13(2):167–194
Porter G, Tewodros A, Gorman M (2018) Mobility, transport and older people's well-being in sub-Saharan Africa: review and prospect. Geogr Transp Age 75–100
Regulation (EU) 2022/868 of the European Parliament and of the Council of 30 May 2022 on European data governance and amending Regulation (EU) 2018/1724 (Data Governance Act)
Reis MJCS, Branco F, Gupta N, Serôdio C (2025) An IoT architecture for sustainable urban mobility: Towards energy-aware and low-emission smart cities. Future Internet 17(10):457. https://doi.org/10.3390/fi17100457
Rojek I, Prokopowicz P, Piechowiak M, Kotlarz P, Náprstková N, Mikołajewski D (2025) The Impact of data analytics based on internet of things, edge computing, and artificial intelligence on energy efficiency in smart environment. Appl Sci 16(1):225. https://doi.org/10.3390/app16010225
Schalekamp H, Behrens R (2010) Engaging paratransit on public transport reform initiatives in South Africa: a critique of policy and an investigation of appropriate engagement approaches. Res Transp Econ 29(1):371–378
Schalekamp H (2017) Lessons from building paratransit operators' capacity to be partners in Cape Town's public transport reform. Res Transp Econ 62:40–48. https://doi.org/10.1016/j.tra.2017.08.002
Scholz T (2016) Platform cooperativism. Challenging the corporate sharing economy. Rosa Luxemburg Foundation, New York, p 435
Scholz T, Schneider N (2017) Ours to hack and to own: the rise of platform cooperativism, a new vision for the future of work and a fairer internet
Schwanen T (2021) Urban transport and wellbeing: a critical analysis. In: Transport in human scale cities, Edward Elgar Publishing, pp 14–26
Sengers F, Raven R (2015) Toward a spatial perspective on niche development: the case of Bus Rapid Transit. Environ Innov Soc Trans 17:166–182. https://doi.org/10.1016/j.eist.2014.12.003
Sengers F, Wieczorek AJ, Raven R (2019) Experimenting for sustainability transitions: a systematic literature review. Technol Forecast Soc Change 145:153–164. https://doi.org/10.1016/j.techfore.2016.08.031
Shaheen SA, Cohen AP (2013) Carsharing and personal vehicle services: worldwide market developments and emerging trends. Int J Sustain Transp 7(1) 5–34. https://doi.org/10.1080/15568318.2012.660103
Shiva CK, Vedik B, Manoharan G, Singh SK, Tanwar S, Jadeja R, Singh S, Polkowski Z (2024) Secure and intelligent IoT-enabled smart cities sustainable smart transportation technologies benefits challenges to urbanisation concept. IGI Global, 355–370
Simo-Reigadas J, Municio E, Morgado E, Castro EM, Martinez A, Solorzano LF, Prieto-Egido I (2015) Sharing low-cost wireless infrastructures with telecommunications operators to bring 3G services to rural communities. Comput Netw 93:245–259. https://doi.org/10.1016/j.comnet.2015.09.006

Smith G, Sochor J, Karlsson IM (2020) Public–private innovation: barriers in the case of mobility as service in West Sweden. Public Manage Rev 21(1):116–137

Smith G, Sochor J, Sarasini S (2018) Mobility as a service: comparing developments in Sweden and Finland. Res Transp Bus Manage 27:36–45. https://doi.org/10.1016/j.rtbm.2018.09.004

Snyder H (2019) Literature review as a research methodology: an overview and guidelines. J Bus Res 104:333–339. https://doi.org/10.1016/j.jbusres.2019.07.039

Son H, Jang J, Park J, Balog A, Ballantyne P, Kwon HR, Singleton A, Hwang J (2025) Leveraging advanced technologies for (smart) transportation planning: a systematic review. Sustainability 17(5):2245. https://doi.org/10.3390/su17052245

Sochor J, Arby H, Karlsson IM, Sarasini S (2018) A topological approach to mobility as a service: a proposed tool for understanding requirements and effects and for aiding the integration of societal goals. Res Transp Bus Manage 27:3–14. https://doi.org/10.1016/j.rtbm.2018.12.003

Suri T (2017) Mobile Money. Ann Rev Econ 9(1):497–520. https://doi.org/10.1146/economics.2017.9.issue-1, https://doi.org/10.1146/annurev-economics-063016-103638

Szpilko D, Naharro FJ, Lăzăroiu G, Nica E, de la Torre Gallegos A (2023) Artificial intelligence in the smart city—A literature review. Eng Manage Prod Serv 15(4):53–75. https://doi.org/10.2478/emj-2023-0028

Tan SY, Taeihagh A (2020) Smart city governance in developing countries: a systematic literature review. Sustainability 12(3):899. https://doi.org/10.3390/su12030899

Tiglao NCC, Ng ACL, Tacderas MAY, Tolentino NJY (2023) Crowdsourcing digital co-production and collaborative governance for modernizing local public transport services: the exemplar of general Santos City Philippines. Res Transp Econ 100:101328. https://doi.org/10.1016/j.retrec.2023.101328

Tiznado-Aitken I, Palm M, Farber S, Paez A (2023) The Routledge handbook of sustainable urban transport. Taylor and Francis, ISBN: 9781040395813, 1040395813

Torraco RJ (2005) Writing integrative literature reviews: guidelines and examples. Hum Resour Dev Rev 4(3):356–367. https://doi.org/10.1177/1534484305278283

UN Habitat (2020) World Cities Report 2020: The Value of Sustainable Urbanization. United Nations, Department of Economic and Social Affairs, Population Division (2018). World Urbanization Prospects: The 2018 Revision (ST/ESA/SER.A/420). United Nations, New York. https://population.un.org/wup/assets/WUP2018-Report.pdf

World Bank (2021) World Development Report 2021: Data for Better Lives. Washington, World Bank

World Bank (2023) Improving paratransit through digital innovation: Lessons from Maseru and Gaborone. World Bank Group

World Bank (2024) Kenya economic update: securing future growth—policies to support Kenya's digital transformation. Washington, World Bank Group

Zammuto RF, Griffith TL, Majchrzak A, Dougherty DJ, Faraj S (2007) Information technology and the changing fabric of organization. Organ Sci 18(5):749–762. https://doi.org/10.1287/orsc.1070.0307

Demand-Side Interventions for Low-Carbon Urban Transport in the Global South: A Behavioral and Spatial Perspective

James Kanyepe and Nyarai Kasambuwa

Abstract Rapidly expanding cities with limited resources have recognized that the primary way to achieve a transition to low-carbon transportation is through demand-side measures. This chapter evaluates the spatial, institutional, and behavioral aspects that influence transport-related greenhouse gas emissions in cities in the Global South. Specifically, this chapter seeks to address the following objectives: (1) the reasons behind the continued high emissions due to intensive carbon transportation and (2) the determinants of continuing reduced emissions from demand-side action strategies. This study was conducted via a qualitative desk-based methodology, with an overview of empirical and theoretical work published across three disciplines: transportation planning, behavioral economics, and sociotechnical transitions. The results indicate that while there are several barriers to reducing carbon emissions from transportation, including travel behavior, city form, fragmented transport infrastructure, and weak governance structures, these factors also demonstrate the ineffectiveness of many low-carbon transportation strategies when implemented in isolation. The results also confirm that a low-carbon transition requires multiple integrated strategies that address both individual and societal changes in travel behavior simultaneously, transforming urban structures and increasing the level of institutional cooperation. Therefore, this chapter identifies potential research areas for scholars, planning professionals, and government officials who wish to implement the most efficient ways to develop low-carbon transportation systems by positioning demand-side interventions within a larger social, technical, and governance context.

Keywords Low-carbon urban transport · Spatial and land-use · Demand-side transport · Behavioral change · Modal shift · Land use planning · Telecommuting · Flight consolidation

---

J. Kanyepe (✉)
Department of Logistics and Supply Chain Management, Faculty of Business, Economics and Law, University of Botswana, Gaborone, Botswana
e-mail: jameskanyepe@gmail.com

N. Kasambuwa
Academy of Teaching and Learning, Chinhoyi University of Technology, Chinhoyi, Zimbabwe

J. Kanyepe and I. Chirisa (eds.), *Advancing the Green Agenda in the Transport and Logistics Sector in Global South Countries*, World Sustainability Series,
https://doi.org/10.1007/978-3-032-31374-4_9

1 Introduction

Transport systems are responsible for a significant and increasing proportion of greenhouse gas emissions. This trend is particularly prevalent in many rapidly developing urban areas in Global South countries (Azhar et al. 2024; Azhar et al., 2024; Toledo & La Rovere 2018). As noted by Jenn et al. (2025), supply-side policies that address vehicle emissions via technology substitutions, including electric vehicles, improved fuel efficiencies, and alternative forms of energy production, have been used historically to reduce carbon dioxide. However, focusing exclusively on supplying new technologies may cause policymakers to ignore how structural or behavioral factors drive individuals' use of their personal transportation systems. Urbanization in the Global South is growing rapidly. Cities such as those mentioned above have a wide variety of fragmented land uses and transportation systems. They also increasingly rely on personal vehicles for travel. As such, they experience structural "lock-ins" that tend to maintain high-carbon modes of transportation. Additionally, new roads can create more demand for driving, and suburban or low-density developments can make it difficult to provide viable mass transit options. Thus, emissions continue to grow as more environmentally friendly vehicle technologies become available.

Brun et al. (2024) noted that while demand-side initiatives can be used to reduce "unnecessary" travel, shift the mode of travel from high- to low-carbon transportation options, transform the urban form, and restructure the way in which goods flow through an increasingly complex network of logistics, they differ from supply-side solutions, as demand-side strategies address the "underlying determinants of behavior and spatial organization" rather than just modifying the technology or fuel source for a vehicle (Model 2023). The demand-side strategy focuses on understanding how travel behaviors are formed through individual decision-making processes, how a city is designed, and how goods move within an ever-increasing and complex logistics network. Travel choices are influenced by factors such as habits, social norms, limited information processing capabilities (bounded rationality), and a preference for maintaining the current way of doing things (status quo bias) (Riggs 2022). Individuals typically maintain their long-standing mobility patterns (even after alternative travel modes become viable). Jankowski (2018), Mwale et al. (2022), and Karjalainen et al. (2023) have shown that the built environment constrains the variety of practical transportation options through its physical design attributes (density, mixed land use, connectivity), thus reinforcing this behavior. Geels (2019) further expanded the theoretical scope for analyzing structural barriers to systemic change by developing path-dependent relationships among infrastructure, institutions, cultural values/meanings, and technologies.

The need to transform demand in transportation is especially true in the Global South, where the rate of urbanization creates both great obstacles and tremendous opportunities (Addy et al. 2026). While increased motorization could result in the developing world experiencing the same pattern of transportation development that led to the high levels of emissions experienced by developed nations, many of the cities in developing countries are still in their early stages of expanding their physical infrastructures, thus providing a window of opportunity to implement compact city models, transit-oriented models, and integrated freight models (Jain & Garg 2025), which will increase accessibility and equity while reducing environmental impacts (Fatorachian & Kazemi 2025) and provide greater economic efficiency (Kanyepe et al. 2022). Although there is increasing awareness of this issue among researchers and policymakers, there is still a lack of understanding regarding how best to address this challenge because research on the topic has been primarily discipline-based and fragmented.

Research on modal choice behavior has typically ignored or downplayed spatial planning relationships. Conversely, studies examining urban design have often neglected to examine either the behavioral consistency that occurs when individuals use their transportation behaviors over time or the influence of cultural values on how people travel. Similarly, research on freight decarbonization has often focused solely on optimizing technology to reduce greenhouse gas emissions from freight activity and has rarely examined whether the reduction in emissions can occur systemically by changing the demands placed on freight infrastructure. Therefore, a theoretically integrated examination of behavioral mechanisms, land-use configurations, institutional factors, and the potential emission consequences of demand-side policy approaches is needed. This chapter aims to address this gap through a systematic qualitative review of all relevant academic theories and empirical evidence. Additionally, the specific objectives of this study include (1) what drives consistent carbon-intensive usage of personal and commercial transportation and (2) what influences continued lower levels of CO_2 emissions resulting from changes in the demand for transportation. The synthesis of findings from multiple disciplines presented here provides a comprehensive understanding of systems-level transportation decarbonization that transcends technological solutions.

The primary claim of this research is that for effective climate mitigation in transportation, it is essential to transform how we currently move around (mobility demand) and simultaneously innovate technologies. Cultural values, city layouts, and government policies must evolve simultaneously as new energy systems develop. If these changes do not occur simultaneously, electric vehicles could simply replace internal combustion engine vehicles without altering the current levels of travel demand and the continued growth of space. However, if behavior and land use planning changes are included as part of policy design, then transportation systems can make the necessary changes to move along lower-emission paths while being fiscally viable and socially equitable. This chapter provides additional insight into the larger body of literature on sustainable transitions globally by emphasizing that managing demand is one of the most important pillars of transitioning to a less carbon-based system. This highlights that behavior and physical space transformations are critical components of long-term decarbonization plans.

2 Literature Review

2.1 Behavioral Dynamics

The microlevel perspective of behavioral economics illustrates how people continue to make high-emission travel choices when they have low-carbon options available (Watabe & Yamabe-Ledoux 2023). Inertia is created by bounded rationality, habit formation, status quo bias, and social norms in travelers' modal choices (Gao et al. 2022). A traveler's decision-making process is established through routine activities, cultural influences, and the environment; therefore, travelers do not rely on cost or travel time alone when making decisions (Rana & Sadiq 2024). Identifying these microlevel behaviors will assist in the development of interventions (e.g., nudging, incentives) that may influence an individual's mode of transportation by considering both their habits and social perceptions.

2.2 Spatial and Land-Use Structures

At the macro level of study, the theoretical framework for urban planning views a person's behavior in an urban environment through the lens of larger-scale spatial and infrastructure settings (Uduokhai et al. 2023); that is, urban densities, connectivity, integrated land uses, and accessibilities can all impact the practicality and appeal of using lower carbon emission transportation options. Urban areas characterized by expansive development, featuring low-density residential zones and a clear separation of various land uses, tend to prioritize automobile travel over walking, cycling, and public transportation (Jain & Garg 2025). Conversely, urban designs that are dense, residential, employment focused, mixed use, and transit oriented promote shifts in transportation modes and reduce travel distances (Yahia et al. 2026). This study primarily investigates the reciprocal relationship between individual behavior and the urban environment, as city planners design physical spaces that either constrain or facilitate sustainable transportation choices.

2.3 Demand-Side Strategies

At the macro level, demand-side interventions target reductions in travel demand, modal shifts to lower-carbon options, and improved efficiency in urban and freight systems (Creutzig 2016; Yi et al. 2025). Demand-side interventions are multi-dimensional and integrate behavioral, spatial, and institutional considerations to achieve long-term sustainability. For example, travel demand management, behavioral nudges, and incentives encourage shifts from private vehicles to sustainable modes (such as public transport, cycling, and walking). However, the effectiveness of travel demand management depends on the alignment between infrastructure and the mitigation of habitual car-use behaviors (Javaid et al. 2020; Anagnostopoulou et al. 2020; Patil et al. 2022).

Similarly, mixed-use urban forms reduce trip lengths and enhance access to sustainable transport, whereas transit-oriented development integrates residential, commercial, and employment zones (Ayub & Hou 2026; Zain et al. 2025). Zapata-Campos et al. (2022) explained that there is a need to balance densification with equitable access, particularly in rapidly urbanizing and informal settlements. Numerous scholars have reported that teleworking and digital communication can decrease peak-hour travel and vehicle distances. However, their effectiveness is contingent on variables such as job type, infrastructure, and organizational culture (Caraiani et al. 2023; Motte-Baumvol & Schwanen 2024; Stefaniec et al. 2024; Motte-Baumvol & Schwanen 2024). The integration of these practices with transportation and urban policies further enhances their impact on emission reduction.

Kanyepe et al. (2022) argued that well-coordinated urban logistics (including shipment consolidation, route optimization, and off-peak deliveries) can significantly reduce the number of vehicle kilometers traveled, thus lowering emissions. Behavioral incentives for logistics actors play a complementary role in shaping decision-making and operational practices to reinforce efficiency gains. This highlights that demand-side transformation is not a single intervention but a layered process that combines infrastructure, technology, and human behavior (Lozzi et al. 2022; Sahu et al. 2022; Mepparambath et al. 2023).

3 Methodology

A qualitative desk-based (i.e., nonexperimental) study method was employed to identify potential behavioral changes that can result in demand-side transformations in urban transportation systems within the Global South. A structured literature review was conducted through a comprehensive database search via Scopus, Web of Science, and Google Scholar. The search terms included "demand-side transport," "behavioral change," "modal shift," "land use planning," "telecommuting", and "flight consolidation." Peer-reviewed articles from academic journals were used along with theoretical foundations, and policy documents were selected to provide context. Studies were considered eligible if they addressed either behavioral or demand-side interventions to reduce emissions in low- and middle-income countries. Studies that focused solely on vehicle technology efficiencies or engineering-based efficiencies and did not address the behavioral or spatial impacts were excluded.

Thematic synthesis was used to analyze the selected literature. Thematic synthesis systematically codes the behavioral mechanisms, spatial structures, institutional contexts, policy instruments, and emission outcomes of the selected literature. The results were compared iteratively to refine the identification of relationships and patterns among the studies. The findings were synthesized into an explanatory framework that connects individual behavior, urban form, and governance structures. The analytic methods employed in this study are informed by theories of behavioral economics, urban planning, and sociotechnical transition to allow for multiscale analyses of behaviors, land use configurations, and path dependencies at multiple levels of scale. While the analytical methods employed in this study rely on secondary data sources and have inherent limitations related to publication bias, geographic imbalances, and differences in measurement techniques, the results provide a theoretically supported basis for understanding the factors driving carbon-intensive forms of mobility and assisting in clarifying the conditions required for successful demand-side interventions in resource-constrained urban areas.

4 Results

The analysis of the reviewed literature reveals that urban mobility patterns in Global South cities are shaped by multiple interrelated factors spanning the behavioral, spatial, institutional, sociotechnical, and socioeconomic domains. These factors contribute to both the persistence of carbon-intensive mobility and the potential effectiveness of demand-side interventions aimed at reducing transportation-related emissions. The findings are organized around two overarching themes that directly address the study's objectives.

5 Drivers of Persistent Carbon-Intensive Mobility in the Global South

Research indicates that individuals' behavior plays a significant role in shaping their travel decisions. Across all three studies, it is evident that participants' travel choices are heavily influenced by habitual travel routines, perceived convenience, and social norms related to various modes of transportation. Each of these factors supports an individual's travel habits on the basis of what they view as convenient for them. Consequently, making changes in travel behavior could be difficult and require a great deal of time (Li et al. 2024). According to Ranjan and Sinha (2025), in most urban locations across the developing world, personal vehicles are perceived to offer greater reliability, flexibility, and comfort than public transportation. Perceptions related to personal vehicles are further reinforced by concerns regarding safety, loss of control over travel time, and convenience due to door-to-door services.

In addition, in many cultures around the world, owning a personal vehicle is equated with higher social status, independence, and wealth. Such social interpretations increase the probability of an individual utilizing a personal vehicle. Social interpretations of automobile use reinforce existing perceptions of convenience, reliability, and comfort. Thus, when combined, behavioral and cultural barriers present a substantial barrier to a shift away from an automobile-dominated system. Although technological advancements (e.g., electric vehicles or "cleaner" buses) exist, they cannot independently facilitate a systemic transition in travel behavior.

Each of the reviewed studies demonstrated that participants' travel decisions were heavily influenced by social norms. For example, Li et al. (2024) noted that in many countries, car ownership is viewed as a measure of success and affluence. In addition to behavioral factors, the literature highlights the significant influence of urban spatial structures on people's mobility. Many cities in the Global South are characterized by low-density development, fragmented land use, and limited integration between residential areas, employment centers, and essential services (Follmann et al. 2023). These spatial arrangements increase travel distances and reduce the practicality of walking, cycling, and using public transportation (Raza et al. 2022). Furthermore, public transport networks in many rapidly growing cities remain limited in coverage, particularly in peripheral urban areas (Venter et al. 2019). Active mobility infrastructure, including safe sidewalks and cycling lanes, is frequently inadequate (Aderibigbe et al. 2024). Consequently, private vehicles and informal transportation systems are often the most practical options for urban residents. Consequently, even when low-emission technologies are introduced, the spatial configuration of cities continues to generate high travel demand and reinforces reliance on carbon-intensive mobility.

The literature also identifies two additional significant ways in which the characteristics of institutional and governance frameworks support the persistence of high-emission mobility in developing regions. First, the literature shows that there has been an emphasis on expanding roads and designing automobiles within much of the transportation planning done in Global South Cities. This emphasis has typically come from pressures from politics, economics, and rapid urbanization (Thondoo et al. 2020) rather than from a well-planned approach to reducing the environmental impacts associated with these activities. Governments are increasingly adopting policies to promote sustainable transportation, such as investing in public transportation, cycling facilities, and low-emission vehicles.

However, because the government structure tends to be fragmented and transportation planners often lack adequate capacity to develop and implement new plans and programs, implementing sustainable transportation policies can be difficult (Colombo & Dijk 2023). Furthermore, poor communication between land-use and transportation planners creates additional obstacles to the implementation of effective sustainable transportation policies (Poku-Boansi 2022). These institutional factors lead to "policy lock-in" and "infrastructure lock-in," wherein historical decisions concerning funding and regulation constrain future decision-making processes. Owing to prior determinations regarding the regulation of transportation systems within a region, current systems predominantly favor automobiles over alternative modes of transportation, even when such alternatives are available (Haarstad et al. 2022).

A second way that institutional and governance framework characteristics contribute to continued high emission mobility in developing regions involves what can be called "sociotechnical path dependence." Sociotechnical systems refer to the interactions between technology and society, as they affect each other. Transport systems exemplify sociotechnical systems, characterized by the dynamic interplay of technological development (e.g., the construction of new highways), governance arrangements (e.g., legal agreements between various levels of government), technological advancements (e.g., the creation of new vehicle types), and social practices (e.g., behaviors associated with using different vehicle types) (Mesdaghi et al. 2022).

Over time, all these components interact with each other and become highly interdependent. Once a system becomes highly dependent on its various components, it becomes very resistant to changes. For example, introducing low-carbon technologies into a transportation system (such as electric buses) does not necessarily lead to a large decrease in the total emissions generated by the transportation sector. Low-carbon technologies may decrease emissions at the individual vehicle level; however, the overall reduction in emissions depends on whether the underlying pattern of development of cities, how people travel throughout cities, and the existing network of transportation infrastructure also change (Cinderby et al. 2024). Therefore, path dependence indicates the continuation of high-emission transportation systems.

Furthermore, both economic and equity issues influence the ability of citizens in developing countries to adopt low-carbon transportation methods. Several studies have indicated that low-carbon technologies (i.e., electric vehicles) require large upfront expenditures. Such expenditures include the cost of purchasing a new vehicle and building supporting infrastructure, such as charging stations (Shamsuddoha & Nasir 2025). Consequently, affluent individuals with adequate income levels are inclined to invest in low-carbon technologies. In contrast, most urban residents lack access to these technologies because of their comparatively lower incomes. Moreover, many citizens lack access to affordable and reliable public transportation. Public transportation is essential for providing citizens with mobility options beyond walking and driving personal vehicles alone. When public transportation is inaccessible, lower-income citizens are forced to choose between walking and driving private vehicles to meet their daily needs (Pojani & Stead 2015). Economic inequities combine with behavioral, spatial, and institutional barriers to maintain preexisting mobility inequities.

6 Determinants of Sustained Emission Reductions Through Demand-Side Interventions

This study investigated the persistence of sustainable carbon-reducing mobility practices among urban residents in resource-constrained cities. Additionally, it analyzes the conditions under which demand-based measures can effectively result in enduring reductions in greenhouse gas emissions associated with the transportation of people and goods within urban areas. The results demonstrate that the success of reducing greenhouse gas emissions through their effects on transportation-related emissions depends on how well these interventions can be integrated. This includes, but is not limited to, the degree to which behavioral changes occur (in terms of changing travel modes), the spatial structure of an urban environment (e.g., density), the level of institutional capacity available for implementing interventions, and the socioeconomic characteristics of the population being served.

The literature clearly demonstrates that behavioral interventions, including travel mode-shifting programs, educational outreach initiatives, and incentives aimed at promoting low-emission travel behaviors, have the potential to influence both individual and organizational travel patterns. These interventions can also affect the total number of vehicle kilometers traveled (VKT) (Bakker et al. 2019). Studies conducted in cities such as Nairobi indicate that the use of targeted behavioral interventions within high-density corridors, where public transportation is easily accessible, can lead to a significant decline in reliance on private vehicles (Kruger et al. 2021). Additionally, commuter incentives and reduced fares in Johannesburg have localized opportunities for changes in travel behavior (Guzman & Hessel 2022). However, as illustrated by the studies reviewed here, behavioral interventions alone rarely generate long-lasting impacts on GHG reduction. Habits surrounding travel behavior choices, perceived convenience, and perceived safety risks frequently lead to continued reliance on private vehicles (Venter et al. 2019). Consequently, behavioral interventions are generally most effective when supported by infrastructure and urban design.

The urban spatial structure is also a key factor influencing the effectiveness of demand-based interventions. For example, compact urban forms and mixed-use developments tend to decrease travel distance requirements, improve service accessibility, and facilitate the use of lower-carbon modes of transportation (Creutzig et al. 2024). Consequently, alternative forms of transportation (i.e., public transportation, walking, and cycling) have become viable alternatives to private vehicles. An example of this would include evidence from Lagos, Nigeria, indicating that combining residential and commercial development adjacent to public transportation hubs increases transit usage while decreasing reliance on private vehicles (Thondoo et al. 2020). However, sprawling urban forms with separate land uses limits the ability of demand-based interventions to be successful, as they maintain long trip lengths and therefore high travel demands (Jain & Garg 2025). Thus, there is a strong relationship between a city and the success of demand-based interventions.

Further research is needed to explore ways to minimize freight- and logistics-related emissions. Fragmented supply chains, unorganized informal delivery networks, inefficient routing methods, and poor information sharing all contribute to increased transportation emission levels (Sahu et al. 2022). Demand-side strategies (such as urban consolidation centers, route optimization technologies, shared logistics platforms, and off-peak deliveries) have proven capable of reducing the amount of VKT traveled and fuel consumed during transportation activities (Arvianto et al. 2021). Urban logistics initiatives in cities such as Johannesburg and Nairobi have shown that the coordination of last-mile delivery systems can significantly mitigate freight-related emissions in densely populated urban areas. Moreover, the efficacy of demand-side interventions is contingent on institutional capacity and governance.

Cities in the Global South frequently encounter obstacles in executing comprehensive mobility strategies because of constrained financial resources, inadequate coordination among departments and organizations, and conflicting development priorities (Haley et al. 2020). For demand-side policies to be implemented effectively, there must be alignment in land-use planning, investments in transportation infrastructure, and regulatory frameworks (Wu et al. 2021). In the absence of such coordination, individual demand-side initiatives are likely to yield only temporary emission reductions, as financial resources may be depleted owing to budgetary constraints.

Additionally, socioeconomic factors strongly affect both the adoption and long-term sustainability of demand-side interventions. To be adopted by diverse urban populations, low-carbon mobility options must be cost-effective, accessible, and reliable (Lee et al. 2024). When interventions do not consider socioeconomic differences, they can lead to low adoption rates of demand-side interventions, resulting in limited emission reductions and exacerbating existing socioeconomic inequities (Sovacool et al. 2022). However, when interventions address issues such as the affordability of low-cost public transportation for low-income residents, safety for pedestrians/walkers/bicyclists, and the expansion of cycling networks so that a greater portion of residents feel comfortable using them as a means of getting around, demand-side interventions can not only help reduce emissions but also create a more inclusive urban mobility system (Hook et al. 2025).

7 Discussion

The study revealed that the transition to sustainable urban mobility in developing countries is both complex and systematic. The persistent carbon footprint from urban mobility cannot be explained exclusively by a lack of technology but also by the interdependent effects of institutional, spatial, behavioral, and socioeconomic factors. The design of an urban transportation system influences users' mobility behaviors, which are influenced by their social and economic status. Consequently, institutional, infrastructural, spatial, and social factors limit the development of new low-carbon urban transportation solutions. While technological innovations such as electric vehicles or clean public transportation systems have the potential to contribute to a reduction in greenhouse gas emissions through changes in the transportation modes used by consumers, they will not achieve major reductions in emissions unless deeper structural and behavioral changes occur.

One common thread throughout much of the research reviewed was the role of behavioral/cultural inertia in influencing how people make travel decisions. Travel decisions are typically driven by a combination of habitual routines, convenience perceptions, and social norms surrounding automobile use. Many of the studies reviewed have indicated that commuters tend to choose transportation modes on the basis of perceived flexibility, comfort, and safety, which frequently results in the choice of automobiles over other forms of transportation (Li et al. 2024; Ranjan & Sinha 2025). Automobiles represent an important symbol of social status and independence among many city dwellers (Nasrin & Chowdhury 2024), reinforcing the preference for automobiles. Consequently, changing travel behavior is difficult because of the deep-seated habits and attitudes that exist within the population. This means that while low-carbon technologies may be available, it does not mean that travel behavior will change. Therefore, demand-side interventions must address the social and psychological barriers that prevent individuals from making different travel choices.

However, the dynamic nature of individual behaviors does not provide complete insight into why carbon-based mobility behaviors persist. The urban structure provides additional evidence of how urban spatial design affects travel demand and mobility. Many cities in the developing world are characterized by physical attributes of low-density sprawl, physically separated land use areas, and insufficient access to public transportation, creating long travel distance needs and making nonautomobile modes of transportation impractical (Raza et al. 2022; Follmann et al. 2023). Consequently, many citizens rely on personal vehicle ownership or informal forms of transportation to meet their travel needs (Venter et al. 2019). Therefore, the relationship between urban physical design and the way people travel indicates that the potential for achieving a low-carbon mobility transition is heavily influenced by the characteristics of the physical space within an area. Therefore, if one wishes to implement demand-side strategies, such as encouraging pedestrianism, cycling, or public transportation use, the degree of success may be limited until sufficient modifications are made to land use regulations to create compact, mixed-use neighborhoods and provide increased accessibility to transport corridors (Zhu & Liu 2024).

The ability to influence transportation systems through governmental institutions and regulatory frameworks can affect the rate at which cities in the Global South achieve a low-carbon mobility transition. Although there is some evidence that governance systems in much of the developing world are becoming increasingly complex in terms of transport planning, complexity may hinder the ability to effectively plan, coordinate, and prioritize transport planning efforts in developing countries. Previous research has shown that transport planning has historically focused on expanding roads and constructing automobile-oriented infrastructure, further entrenching automobile-dependent development (Haarstad et al. 2022; Thondoo et al. 2020). In addition, although many policies exist to promote low-carbon forms of mobility, their implementation is often inconsistent owing to a lack of cooperation between agencies responsible for regulating land use and investing in infrastructure (Poku-Boansi 2022; Colombo & Dijk 2023). These institutional barriers lead to "policy and infrastructure lock-in," where existing investments in automotive technologies continue to shape transport systems in ways that do not support low-carbon-based alternative forms of mobility.

Urban transportation systems also strongly depend on sociotechnical pathways. Over time, the development of transport infrastructure, governance structures, technological systems, and mobility behaviors has developed into mutually influencing elements that are highly resistant to sudden changes (Mesdaghi et al. 2022). Thus, it is not possible to implement only new technologies to fundamentally change mobility patterns because the basic structure of travel demand and spatial structure will remain unchanged. Although the introduction of electric vehicles or renewable energy-powered transport may lead to reduced emissions per vehicle, they have a very limited effect if the travel demand in a city is still very high and the existing car-oriented infrastructure has not changed (Kimuli & Kirabira 2025). This highlights the necessity of systematic measures that address the behavioral, spatial, and structural causes of intensive carbon mobility.

Both the continuity of carbon-intensive mobility behavior and the effectiveness of measures aimed at reducing this behavior by changing an individual's demands are influenced by socioeconomic factors. Low-carbon mobility options are not equally available to all people within the urban population. For example, in many resource-constrained cities, there are a wide range of barriers to adopting low-carbon mobility. The high initial investments required for electric vehicles and the necessary supporting infrastructure mean that these investments are mostly adopted by wealthier families. Therefore, lower income groups rely on conventional forms of transportation (Shamsuddoha & Nasir 2025). Furthermore, a lack of adequate public transport infrastructure and few opportunities for active mobility severely hinder disadvantaged communities' access to mobility (Pojani & Stead 2015). Therefore, an urban mobility policy that ignores social inequality can lead to only small emission reduction effects and the continuation of current unequal access to transportation. In contrast, urban mobility policies that focus on affordability and accessibility, such as subsidies for public transport, safe pedestrian infrastructure, and cycling infrastructure, can simultaneously contribute to emission reductions and a fairer distribution of access to mobility (Hook et al. 2025; Sovacool et al. 2022).

The findings indicate that significant and sustainable reductions in emissions can be achieved through demand-side measures, contingent upon their implementation within suitable institutional and infrastructural contexts. The potential to impact individual travel behavior exists with the application of behavioral programs, such as modal shifts, travel demand management programs, and public education campaigns. If feasible alternatives to motorized personal transportation exist, an individual is likely to be influenced to alter their mode of travel (Guzman & Hessel 2022). Additionally, land-use planning initiatives, such as transit-oriented development and compact city design, can both reduce travel demand and increase the practicality of using alternative lower-carbon forms of transportation (Creutzig et al. 2024).

Additionally, freight and logistics optimization strategies, such as routing coordination, urban consolidation facilities, and off-peak delivery schedules, provide additional means to decrease emissions associated with transportation by increasing the efficiency of distribution (Sahu et al. 2022; Arvianto et al. 2021). However, the long-term efficacy of these measures will depend on institutional support and policy alignment. If this does not occur, the potential to achieve significant and sustained emission reductions through demand-side measures will be reduced. Many developing world cities lack the necessary financial resources, administrative capabilities, and political stability to successfully implement and expand proven mobility projects (Wu et al. 2021; Haley et al. 2020). Thus, effective governance systems and continued policy commitment are needed to create an environment in which demand-side approaches to reducing greenhouse gas emissions from transportation can achieve significant and enduring impacts.

8 Theoretical Implications and Contributions

This chapter contributes to the body of knowledge related to sustainable urban transportation by developing an integrated analytical framework that addresses behavioral, spatial, and institutional factors specific to the Global South. Traditionally, researchers have focused primarily on supply-side technological solutions to decarbonize transportation systems (vehicle electrification, improved fuel efficiencies, and renewable energy) while ignoring the systemic barriers to travel behaviors and urban mobility patterns. Therefore, by moving forward with a demand-side focus, this chapter bridges two significant gaps in the literature and brings together the theoretical bases of behavioral economics, urban planning, and sociotechnical transition theory.

At the behavioral level, this chapter discusses how several types of cognition, including habits, social norms, preferences for the status quo, and bias, can create "behavioral lock-in" in relation to travel decision-making, thus limiting the potential of technological advancements. Spatially, this chapter examines some of the restrictions created by urban forms, land use, connectivity, and accessibility on viable travel modes. The institutional/policy dimension of this chapter demonstrates how governance systems, regulatory frameworks, and coordinating mechanisms for planning can either support or hinder the development of carbon-intensive travel patterns. Thus, this chapter develops a multilevel framework to explain why low-carbon technologies rarely result in long-term decreases in transportation-related greenhouse gas emissions.

The contribution of this chapter is also in terms of global-south-specific research: rapid urbanization, informal settlements, limited resources, and socioeconomic heterogeneity (among others) generate distinct challenges and opportunities in relation to demand-side interventions. The integration of empirical and theoretical studies from Lagos, Nairobi, Johannesburg, and Gaborone illustrates how behavioral dynamics and land use structures interact to generate contextualized emission outcomes. This study presents researchers with a conceptual and methodological foundation for studying transport decarbonization within resource-constrained urban environments and highlights the importance of systemic interdisciplinary context-sensitive approaches. Furthermore, this chapter provides methodological insights for future studies and illustrates the utility of desk-based qualitative synthesis to integrate diverse empirical evidence and theoretical constructions and identify cross-cutting patterns, causal mechanisms, and practical policy implications without reliance on primary data collection, which can be costly and time-consuming in resource-constrained environments.

9 Limitations of the Study

Although numerous studies included in this review have made significant contributions, there is evidence indicating at least four distinct limitations within this body of research. First, this research was based entirely on the existing published literature. Therefore, the literature base for this research is potentially biased toward the publication of research that reports positive findings from interventions. Interventions that report no effect (null) or negative effects are likely to remain unpublished. Consequently, the overall assessment of the effectiveness and sustainability of these interventions is skewed. Second, this study focuses on several cities in the Global South. Although these cities provide an interesting illustration of where urban development can take place in developing countries, they are unlikely to be representative of all urban areas in developing countries (DCs). Therefore, the applicability of the findings reported here is likely limited.

There are significant differences in urban forms, transport governance regimes, socioeconomic conditions, and cultural norms among Global South countries. Therefore, interventions that prove successful in one location are unlikely to be easily transferred to other locations. Third, the methodologies, emission quantification methods, and definitions of behavioral or demand-side interventions used in the studies included in this review varied significantly. These variations make direct comparisons difficult and complicate attempts to synthesize the findings into a coherent overview of the literature. In addition, differences in the quality of the data collected, reporting standards, and metrics used to measure outcomes may also impact the interpretation of the findings from each study in terms of the efficacy and sustainability of each type of intervention. Fourth, the research has concentrated on two main mobility sectors, passenger and freight transport systems, while excluding certain transportation modes, such as nonmotorized modes, informal public transportation services (e.g., jitneys), emerging forms of microscale transportation (micromobility), and rural–urban transportation connections. Consequently, some aspects of demand-side transformation may be overlooked. Finally, relying solely on a desk-based review of the existing published literature does not allow real-time behavioral changes or the dynamic impacts of emerging technology and/or policy interventions to be captured.

10 Conclusions

This chapter shows that demand-side measures are needed to achieve a sustainable reduction in greenhouse gas emissions from the transport sector in urban areas, particularly in developing countries. Rapid urbanization is occurring in many cities of the Global South and has created major new challenges, notably, rapid population growth and limited resources. Urban transport systems exhibit high levels of carbon intensity owing to technological barriers (e.g., lack of alternatives), behavioral inertia, institutional constraints, and physical/urban structures. While technology will continue to evolve (e.g., e-vehicles and LEP), this evolution is insufficient to transform behavior. To unlock the true potential of low-carbon transportation technologies, effective transformations on the demand side require multidimensional, integrated approaches that align behavioral habits with spatial arrangements and institutional frameworks to create conditions for sustainable mobility.

Urban transportation in developing countries is shaped by contextual considerations. The compactness of cities, mix of uses on parcels of land, and proximity to public transit all contribute to making it easier to implement behavioral-based interventions while improving logistical efficiency through methods such as freight consolidation and route optimization. Over time, institutional backing (including coordinated policies, enforcement of regulations, and financial rewards) will be important for maintaining sustainability. Nonetheless, unless the intervention concurrently addresses economic disparities among consumers, there is likely to be unequal acceptance and limited, long-term success. Consequently, the research conducted herein contributes to the theoretical and practical understanding of the relationships among behavior, space, and institutions in achieving greenhouse gas reductions through transportation in the developing world. It develops a conceptual model to guide the design of resource-constrained urban settings and can assist transportation professionals, planners, and policymakers in developing countries in designing equitable and environmentally sustainable solutions for their communities.

In conclusion, demand management measures are not merely supplementary to technology; they are essential for establishing sustainable urban transportation systems. Only by aligning efforts to change behavior, plan cities, and govern them in conjunction with technological advancements can we sustainably reduce emissions, enhance accessibility, and promote social justice in urban transportation systems in developing countries. Integrating these three elements is a vital initial step toward balancing urban growth, transportation needs, and climate protection goals in the world's least developed regions.

References

Addy MN, Aigbavboa C, Kwofie TE (2026) Urbanization and Decarbonization. In: Decarbonization of building energy in developing countries, pp 267–294. Springer Nature Switzerland, Cham

Aderibigbe OO, Fadare SO, Gumbo T (2024) Transport situation in the Global South: Insights from Nigeria, South Africa and India. In: emerging technologies for smart cities. Springer, Cham. https://doi.org/10.1007/978-3-031-66943-9_3

Anagnostopoulou E, Urbančič J, Bothos E, Magoutas B, Bradesko L, Schrammel J, Mentzas G (2020) From mobility patterns to behavioral change: leveraging travel behavior and personality profiles to nudge for sustainable transportation. J Intell Inf Syst 54(1):157–178

Angelini Q (2025) Interdependencies of traffic flow modeling, travel behavior, and transportation infrastructure.Travel Behav, Transp Infrastruct (August 5, 2025)

Arvianto A, Sopha BM, Asih AMS, Imron MA (2021) City logistics challenges and innovative solutions in developed and developing economies: A systematic literature review. Int J Eng Bus Manag 13:18479790211039724

Ayub S, Hou Y (2026) Impact of the built environment on urban mobility patterns and advanced transport dynamics: a systematic review. Transp Res Interdiscip Perspect 36:101842

Azhar U, Yaseen S, Arif M, Babar M, Sagir M (2024) Emission of greenhouse gases from transportation. In: Advances and technology development in greenhouse gases: Emission, capture and conversion. Elsevier, pp 147–163

Babayomi OO, Dahoro DA, Zhang Z (2022) Affordable clean energy transition in developing countries: Pathways and technologies. Iscience, 25(5)

Bakker S, Haq G, Peet K, Gota S, Medimorec N, Yiu A, Rogers J (2019) Low-carbon quick wins: Integrating short-term sustainable transport options in climate policy in low-income countries. Sustainability 11(16):4369

Brun K, Allison T, Kurz R, Wygant K (eds) (2024) Energy transport infrastructure for a decarbonized economy. Elsevier

Caraiani C, Lungu CI, Dascalu C, Stoian CA (2023) The impact of telework on organisational performance, behavior, and culture: evidence from business services industry based on employees' perceptions. Econ Res-Ekon Istraživanja, 36(2)

Cinderby S, Haq G, Opiyo R, Muhoza C, Ngabirano A, Wasike Y, Cambridge H (2024) Inclusive climate resilient transport challenges in Africa. Cities 146:104740

Colombo CM, Dijk M (2023) Understanding the policy integration challenges of sustainable urban mobility in the context of rapid decarbonization. Eur J Risk Regul 14(3):583–606

Creutzig F (2016) Evolving narratives of low-carbon futures in transportation. Transp Rev 36(3):341–360

Creutzig F, Roy J, Minx J (2024) Demand-side climate change mitigation: where do we stand and where do we go? Environ Res Lett 19(4):040201

Das DK, Aiyetan AO, Mostafa MMH (2026) Advancing smart cities in Africa: Barriers, potentials, and strategic pathways for sustainable urban transformation. Smart Cities 9(2):38

Faiyetole AA, Adewumi VA (2024) Urban expansion and transportation interaction: Evidence from Akure, southwestern Nigeria. Environ Plan B: Urban Anal City Sci 51(1):57–74

Fatorachian H, Kazemi H (2025) Sustainable optimization strategies for on-demand transportation systems: Enhancing efficiency and reducing energy use. Sustain Environ 11(1):2464388

Ferreira MC, Dias TG (2024) Exploring perceptions of comfort, security and safety in different modes of transport: a comparative study. In: Transport research arena conference, pp 640–646. Springer Nature Switzerland, Cham

Follmann A, Kennedy L, Pfeffer K, Wu F (2023) Peri-urban transformation in the Global South: a comparative sociospatial analytics approach. Reg Stud 57(3):447–461

Gao K, Shao M, Axhausen KW, Sun L, Tu H, Wang Y (2022) Inertia effects of past behavior in commuting modal shift behavior: interactions, variations and implications for demand estimation. Transportation 49(4):1063–1097

Gao Y, Zhu J (2022) Characteristics, Impacts and trends of urban transportation. Encyclopedia 2(2):1168–1182

Geels FW (2019) Sociotechnical transitions to sustainability: a review of criticisms and elaboration Multilevelulti-level perspective. Curr Opin Environ Sustain 39:187–201

Göransson J, Andersson H (2023) Factors that make public transport systems attractive: a review of travel preferences and travel mode choices. Eur Transp Res Rev 15(1):32

Guzman LA, Hessel P (2022) The effects of public transport subsidies for lower-income users on public transport use: A quasi experimental study. Transp Policy 126:215–224

Haarstad H, Sareen S, Kandt J, Coenen L, Cook M (2022) Beyond automobility? Lock-in of past failures in low-carbon urban mobility innovations. Energy Policy 166:113002

Haley B, Gaede J, Winfield M, Love P (2020) From utility demand side management to low-carbon transitions: Opportunities and challenges for energy efficiency governance in a new era. Energy Res Soc Sci 59:101312

Hook H, Durán-Rodas D, Jamal S, Schwanen T (2025) Evaluating initiatives to improve transport justice. Transp Res Part D: Transp Environ 142:104719
Jain A, Garg C (2025) Overcoming the barriers of sustainable urban development: An examination of urban morphology and transportation systems. J Transp Syst 10(3):1–12
Jankowski P (2018) Behavioral decision theory in spatial decision-making models. In: Handbook of behavioral and cognitive geography. Edward Elgar Publishing, pp 41–55
Javaid A, Khanna T, Franza M, Creutzig F (2022) Behavioral interventions change individual transport choices but have a limited impact on transport mode split. Evidence from a systematic review
Javaid A, Creutzig F, Bamberg S (2020) Determinants of low-carbon transport mode adoption: systematic review of reviews. Environ Res Lett 15(10):103002
Jenn A, Chakraborty A, Hardman S, Hoogland K, Sugihara C, Tal G, Refa N (2025) Supply-side challenges and research needs on the road to 100% zero-emissions vehicle sales. Progress in Energy 7(2):022002
Kanyepe J, Tukuta M, Chirisa I (2022) Residential density and traffic congestion in the harare metropolitan region. Eur Transp (87):1–13. https://doi.org/10.48295/ET.2022.87.4
Karjalainen LE, Tiitu M, Lyytimäki J, Helminen V, Tapio P, Tuominen A, Paloniemi R (2023) Going carless in different urban fabrics: socio demographics of household car ownership. Transportation 50(1):107–142
Khalaj F, Alidoust S, Pojani D (2023) Mapping urban transport-land use interactions worldwide, a state of practice. Handb Transp Land Use:147–166
Kimuli I, Kirabira JB (2025) Low-carbon urban transportation: Optimizing mechanical systems for sustainable electric bus and BRT deployment in Kampala. Transp Eng:100417
Krüger F, Titz A, Arndt R, Groß F, Mehrbach F, Pajung V, Wimmer L (2021) The Bus Rapid Transit (BRT) in Dar es Salaam: A pilot study on critical infrastructure, sustainable urban development and livelihoods. Sustainability 13(3):1058
Kuma C, Gideon S, Kinyui N (2025) Rethinking automobile dependency in Sub-Sahara toward sca: Toward sustainable urban planning in Cameroon. Int J Transp Eng Technol 11:47–63
Lättman K, Otsuka N (2024) Sustainable development of urban mobility through active travel and public transport. Sustainability 16(2):534
Leal Filho W, Hassen TB, Matandirotya N, Ng A (2025) Empty promises: Some requirements for a successful implementation of decarbonization strategies in developing countries. Sci Total Environ 977:179409
Lee G, Kim S, Koo J, Choo S (2024) Exploring psychological factors influencing the adoption of sustainable public transit considering preference heterogeneity. Sustainability 16(18):7924
Li S, Wang B, Zhou H (2024) Decarbonizing passenger transportation in developing countries: Lessons and perspectives1. Reg Sci Urban Econ 107:103977
Lozzi G, Iannaccone G, Maltese I, Gatta V, Marcucci E, Lozzi R (2022) On-demand logistics: Solutions, barriers, and enablers. Sustainability 14(15):9465
Mepparambath RM, Cheah L, Zegras PC, Alho AR, Sakai T (2023) Evaluating the impact of urban consolidation center and off-hour deliveries on freight flows to a retail district using agent-based simulation. Transp Res Rec 2677(2):264–281
Mesdaghi B, Ghorbani A, de Bruijne M (2022) Institutional dependencies in climate adaptation of transport infrastructures: an Institutional Network Analysis approach. Environ Sci Policy 127:120–136
Motte-Baumvol B, Schwanen T (2024) Telework, travel times, and peak hour avoidance in England: An overview using travel times across five weekdays. Travel Behav Soc 34:100668
Mwale M, Luke R, Pisa N (2022) Factors that affect travel behavior in developing cities: A methodological review. Transp Res Interdiscip Perspect 16:100683
Nasrin S, Chowdhury S (2024) Exploring transport mobility issues and adaptive behavior of women in a developing country. Transp Res Interdiscip Perspect 23:100991

Nogueira M, Dias F, Santos V (2023) Sustainable mobility choices: Exploring the impact of consumers' values, attitudes, perceived behavioral control and subjective norms on the likelihood to choose sustainable mobility options. J Consum Behav 22(2):511–528

Patil DS, Bailey A, Yadav UN, George S, Helbich M, Ettema D, Ashok L (2022) Contextual factors influencing the urban mobility infrastructure interventions and policies for older adults in low-and middle-income countries: a realist review. BMC Public Health 22(1):1489

Pojani D, Stead D (2015) Sustainable urban transport in the developing world: beyond megacities. Sustainability 7(6):7784–7805

Poku-Boansi M (2022) Public transport policy and governance in African Cities: An introduction. In: Acheampong RA, Lucas K, Poku-Boansi M, Uzondu C (eds) Transport and mobility futures in Urban Africa. The urban book series. Springer, Cham. https://doi.org/10.1007/978-3-031-17327-1_11

Rammert A (2025) Transforming mobility governance: challenges faced in Germany. Discov Sustain 6(1):576

Rana M, Sadiq H (2024) Computational modeling and simulation techniques for managing rail-urban interface constraints in metropolitan transportation systems. Am J Sch Res Innov 3(02):141–178

Ranjan R, Sinha S (2025) A systematic review of mode choice behavior in urban transportation with emphasis on individual preferences and influencing factor. Discover Cities 2(1):98

Raza A, Safdar M, Zhong M, Hunt JD (2022) Analyzing spatial location preference of urban activities with mode-dependent accessibility using integrated land use–transport models. Land 11(8):1139

Riggs W (2022) Behavioral economics and social nudges in sustainable travel. In: Innovations in transport. Edward Elgar Publishing, pp 89–110

Sahu PK, Pani A, Santos G (2022) Freight traffic impacts and logistics inefficiencies in India: Policy interventions and solution concepts for sustainable city logistics. Transp Dev Econ 8(2):31

Shamsuddoha M, Nasir T (2025) The road ahead for hybrid or electric vehicles in developing countries: market growth, infrastructure, and policy needs. World Electr Veh J 16(3):180

Sovacool BK, Newell P, Carley S, Fanzo J (2022) Equity, technological innovation and sustainable behavior in a low-carbon future. Nat Hum Behav 6(3):326–337

Stefaniec A, Brazil W, Whitney W, Zhang W, Colleary B, Caulfield B (2024) Examining the long-term reduction in commuting emissions from working from home. Transp Res Part D: Transp Environ 127:104063

Thondoo M, Marquet O, Marquez S, Nieuwenhuijsen MJ (2020) Small cities, big needs: Urban transport planning in cities of developing countries. J Transp Health 19:100944

Toledo AL, La Rovere EL (2018) Urban mobility and greenhouse gas emissions: status, public policies, and scenarios in a developing economy city, Natal, Brazil. Sustainability 10(11):3995

Uduokhai DO, Okafor MI, Sanusi AN, Garba BMP (2023) Systems-based analysis of urban mobility, land-use patterns, and sustainable city growth dynamics. Gyanshauryam, Int Sci Ref Res J 6(2):498–515

Van Acker V, Goodwin P, Witlox F (2016) Key research themes on travel behavior, lifestyle, and sustainable urban mobility. Int J Sustain Transp 10(1):25–32

Venter C, Mahendra A, Hidalgo D (2019) From mobility to access for all: Expanding urban transportation choices in the global south. World Resources Institute, Washington, DC, pp 1–48

Wang J, He W, Huang J (2025) Toward a new normal? Deconstructing travel behavior resilience in the urban transport system. Transport Reviews:1–27

Wang X, Dong X, Zhang Z, Wang Y (2024) Transportation carbon reduction technologies: A review of fundamentals, application, and performance. J Traffic Transp Eng (Engl Ed) 11(6):1340–1377

Watabe A, Yamabe-Ledoux AM (2023) Low-carbon lifestyles beyond decarboTowardson: Toward a more creative use of the carbon footprinting method. Sustainability 15(5):4681

Wu Y, Shan J, Choguill CL (2021) Combining behavioral interventions with market forces in the implementation of land use planning in China: A theoretical framework embedded with nudge. Land Use Policy 108:105569

Yahia O, Chohan AH, Arar M, Mangi MY (2026) Enhancing neighborhood functionality through transit-oriented design, a study of sustainable development, proximity, and mobility. Front Built Environ 11:1689752

Yap JBH, Chua CY, Skitmore M (2021) Toward sustainable mobility with transit-oriented development (TOD): understanding greater Kuala Lumpur. Plan Pract Res 36(3):314–336

Yi Y, Sun ZY, Fu BA, Tong WY, Huang RS (2025) Accelerating toward sustainability: Policy and technology dynamic assessments in China's road transport sector. Sustainability 17(8):3668

Zain SNM, Ismail WIFW, Hamzah AH (2025) Evaluating transit-oriented development in Malaysia: A comprehensive analysis of benefits, drawbacks and implementation challenges. Int J Res Innov Appl Sci 10:574–587

Zapata Campos MJ, Kain JH, Oloko M, Stenberg J, Zapata P (2022) Urban qualities and residents' strategies in compact global south cities: The case of Havana. J Housing Built Environ 37(1):529–551

Zhu L, Liu P (2024) How demand-side mitigation can help shape effective climate policies. Nat Clim Change 14(11):1117–1118. https://doi.org/10.1038/s41558-024-02148-2

Decarbonizing Hard-To-Abate Transport Sectors: Aviation, Maritime Shipping, and Green Freight Corridors in a Net-Zero Transition

Innocent Chirisa, Takudzwa Mharadze, Tinashe Magande, and Africa Muneti

Abstract Decarbonizing challenging-to-abate modes of transport such as aviation and maritime shipping is an essential component of achieving net-zero globally as such transport modes produce large amounts of emissions that are crucial to international trade. Despite encouraging developments, such as the implementation of international regulation, including MARPOL Annex VI, and development of new energy carriers, like ammonia and hydrogen, such as technological lock-ins, infrastructure deficits, and fragmented governance impede swift transitions. In this qualitative research, interactions between alternative fuels, policy mixes, and digital innovations to facilitate decarbonization paths will be explored. Data used in the analysis were collected through a thorough desk review of peer-reviewed articles, policy papers, and other reports published within the field and located in various databases, including Scopus and Google Scholar. Key themes were identified using textual analysis in three case studies: EU FuelEU Maritime Regulation, electrification of coastal ships in Greece, and creation of digital twins at Busan New Port. Analysis showed that cohesive policy mixes on multiple levels, as seen in the case of EU regulations, stimulate standardization of fuels and help overcome international IMO lags; innovative niches, such as green corridors and coastal ship electrification, create zero-emission transportation systems through integration of infrastructure; and and digital tools such as port digital twins optimise logistics, slashing emissions up to

I. Chirisa (✉)
Zimbabwe Ezekiel Guti University, Bindura, Zimbabwe
e-mail: innocent.chirisa@gmail.com

I. Chirisa · T. Magande
Department of Urban and Regional Planning, University of Johannesburg, Johannesburg, South Africa

T. Mharadze
Department of Development Studies, Zimbabwe Open University, Harare, Zimbabwe

A. Muneti
Firm Central Property Group, Harare, Zimbabwe

IC University of Technology, Phnom Pehn, Cambodia

J. Kanyepe and I. Chirisa (eds.), *Advancing the Green Agenda in the Transport and Logistics Sector in Global South Countries*, World Sustainability Series,
https://doi.org/10.1007/978-3-032-31374-4_10

75% without fuel changes. The results show that there is need for synergistic technological, regulatory, and digital strategies to break the fossil fuel regimes balance trade competitiveness with environmental goals, and support developing economies through coordinated investments.

Keywords Decarbonization · Maritime shipping · Green corridors · Alternative fuels · Policy mix · Digital twins

1 Introduction

Decarbonization of non-easy-to-abate transport industries, such as maritime shipping and aviation have emerged as a key challenge in meeting global climate targets. These industries contribute a large portion of total global greenhouse emissions and are highly integrated into global systems of trade (Mallouppas and Yfantis 2021; Bows-Larkin 2014). Contrary to land-based transport, shipping and aviation have technological and infrastructural bottlenecks that do not permit swift electrification and fuel conversion. The International Maritime Organization (IMO) enhanced its regulatory framework by introducing several tools that include MARPOL Annex VI and its updated emission policy, but there are still spaces of implementation (Campara et al. 2018; Sahin 2024). Other alternative fuels like ammonia, hydrogen, and methanol are also under development to decrease lifecycle emissions but face economic and safety challenges (Karvounis et al. 2024; Lindstad et al. 2021). Concurrently, market-based instruments and the price on carbon are gaining more and more acceptance as the means to bridge the price gap between fossil fuels and low-carbon sources (Lagouvardou et al. 2023). The developments show the pressing need of combined technological and governance solutions enabling a net-zero transition.

This chapter suggests that efficient decarbonization of maritime transport requires, and cannot be limited to, technological innovation, but should also encompass coherent policy mixes and multi-level governance frameworks. Green shipping corridors may be viewed as the niche innovation that can disrupt the fossil-fuel-based regimes, and sustainability transitions theory and especially the Multi-Level Perspective explain how it can be facilitated (Geels 2018). It is found in the empirical research that to be viable transition pathways, green corridors need to be coordinated through investments in ports, fuel supply chains, and shipping companies (Song et al. 2023). As global policies to govern the maritime climate have developed slowly, regional regulatory programs (e.g., maritime climate policies of the EU) show how policy entrepreneurship can speed up change (Malmborg 2024). Nonetheless, incoherent global policies and economic contradictions can result in uneven climate measures (Li and Hu 2024). Digital innovations, such as port digital twins, also demonstrate that operational efficiency can be used to cut emissions to a considerable degree without the direct replacement of fuel (Eom et al. 2023). Collectively, these observations indicate that the technological, regulatory, and digital policies should develop in a systematic manner to break structural lock-ins.

The significant objective of this research is to analyze the role of the alternative fuels, economic tools, governance regimes, and digital infrastructure in terms of their influence on the decarbonization trajectories of the maritime transport sector. The previous studies note that marine renewables have the potential to produce significant emission cuts, although their implementation requires robust carbon pricing and favourable regulation (Harahap et al. 2023; Serra and Fancello 2020). Research also indicates that the port electrification and energy system integration is instrumental in supporting zero-emission shipping networks (Damman and Steen 2021). Moreover, an assessment of the effects of decarbonization indicates extensive environmental trade-offs in both energy and transport systems (Luderer et al. 2019). To bring together these strands of literature, the present study is aimed at having a holistic picture of the interaction between governance, technology, and infrastructure in the transition process. It can be argued that the analysis adds to the current discussions regarding the need to strike a balance between environmental ambition and economic competitiveness in the global trade (Clora and Yu 2022). Finally, the study will determine coherent pathways, which can speed up the process by which the maritime sector becomes climate neutral.

2 Conceptual Framework

Sustainability transitions theory, and specifically the Multi-Level Perspective (MLP) of decarbonisation of hard to abate sectors, e.g. aviation, maritime shipping, and green freight corridors can be subject to rigorous study. The MLP models socio-technical transitions as a relation among niche innovations, socio-technical regime, and the socio-technical landscape. Applied to maritime shipping, this theory demonstrates the appearance of innovations such as green shipping corridors in niches due to the support of such international programs as the Clydebank Declaration of the International Maritime Organisation that contributes to the emergence of green shipping routes and adoption of renewable marine fuels, such as green ammonia (Akhavan 2025; Wang et al. 2022). The barriers to these niche innovations are the incumbent regime of fossil fuel dependence and regulatory lock-ins, as the policy mix of the International Maritime Organisation, that is characterised by the lack of consistency and comprehensiveness in its incentives to new propulsion technologies, shows (Bach and Hansen 2023). The landscape level of strategic interactions, such as global climate agreements, energy transitions in other sectors, and others, put pressure on the regime to change (Urban et al. 2023; Malmborg 2024). Thus, the decarbonisation of these areas requires a systematic approach to the analysis of relationships between technologies, policies, and socio-economic processes involved in the MLP scheme.

To add to the multi-level perspective, Global Governance and Policy Mix Theory can shed some light on how the international regulatory frameworks and the mix of policies contribute to the sustainability transitions in these sectors. The international shipping, which is a highly complex multi-stakeholder industry, is governed through complex coordination, and the case of green shipping corridors can be exemplified by the fact that they rely on partnerships between ports, carriers, and governments to resolve the problem of alternative fuels, bunkering infrastructure, and logistics of operations (Bengue et al. 2024; Song et al. 2023). Empirical literature confirms that recent policy tools, although they are becoming more specific to the emission reduction objective, experience technological lock-in and do not have incentives to stimulate sustainable innovation, which increases the challenge of governance (Bach and Hansen 2023). The policy mix theory emphasises the need to have coherence, consistency, and complementary tools of regulation, which is constrained, and hence the decarbonisation progress is slow. Moreover, the presence of divergent policy entrepreneurship and rival and competing narratives at the EU level demonstrates that the political establishment influences policy adoption patterns, which either strengthen incumbent technologies or speed up clean energy alternatives such as zero-carbon fuels (Malmborg 2024). Decarbonisation therefore needs proactive policy combinations to be coordinated across several governance levels to overcome the problems of fragmentation and market failures.

3 Literature Review

This section presents the literature review of the study. The reviewed literature include alternative fuels and zero-emission technologies, economic instruments and efficiency standards, international regulatory coordination, green corridors, port electrification and infrastructure development and trade, sector coupling, and impacts on trade-dependent economies.

3.1 Alternative Fuels and Zero-Emission Technologies

Studies on alternative maritime decarbonisation fuels indicate that hydrogen and ammonia are the most promising options, yet major investments in production and bunkering infrastructure are the primary obstacles. Investigations critically examine the sustainability of hydrogen and ammonia as liquids in short sea shipping in terms of its environmental and economic sustainability, and they prove their ability to achieve IMO carbon emission reduction targets by 2030 and, when policies are in place, to reduce carbon taxation to 136–965 EUR/t and ammonia to 356–2647 EUR/t (Karvounis et al. 2024). The existing global production and storage network of ammonia has got an advantage over hydrogen that has challenges such as storage and distribution with a lower density of energy and highly volatile hydrogen (Ayvali et al. 2020). Additionally, safety issues, as well as operational limitations, demand combined risk, cost, and environmental evaluation to implement adoption policies (Kim et al. 2020). The development of ammonia engines, compression and spark ignition versions ensure avenues to realistic emission cuts, although after-treatment technologies are required to fully exploit the advantage of nitrous oxide emissions (Schwarzkopf et al. 2023).

Methanol and biofuels are part of the diversified range of maritime fuels under development to achieve decarbonisation objectives. Bio-synthetic fuels like e-methanol made using renewable energy sources have demonstrated the ability to reduce GHG emissions by more than 88 percent using life cycle assessment, but cost competitiveness is still a serious issue (Lee et al. 2024). However, the benefits of using fossil-based methanol are relatively small unless this is paired with renewable energy and carbon capture, which highlights the significance of so-called green production routes (Huang et al. 2022). Biofuels based on hydrogen and drop-in biofuels based on agricultural and forestry residues are also considered promising but need specific regional and technological alignment to make maximum climate advantages (Watanabe et al. 2022). These fuels are transitional alternatives to other options as fuel cells and hybrid solutions are also examined to enhance their efficiency and minimize lifecycle emissions to meet the IMO requirements (Zhang et al. 2024).

Electrification and synthetic e-fuels are other potential opportunities to decarbonize the maritime industry, but their viability currently depends on the type of vessels and the scope of their operation. Particularly through the use of lithium-ion technology, battery electric propulsion proves more economical and more environmentally friendly to small and medium-range ships, although energy density constraints limit its use to long-range shipping (Bei et al. 2024). E-fuels, including e-hydrogen, e-ammonia and hydrocarbon e-fuels (e-methanol and e-diesel) are synthetic fuels that rely on large quantities of renewable electricity and dual-fuel engine systems to reduce fuel supply risks and flexibility in operating systems (Lindstad et al. 2021). The most feasible approach to the implementation of the Carbon Intensity Index (CII) requirements by the IMO and the promotion of green shipping development seems to be multi-fuel pathways which include LNG, methanol, hydrogen, and ammonia options (Zhang et al. 2024). Technical, economic, and regulatory obstacles to the implementation of these various zero- and low-carbon solutions require additional research and coordinated policy frameworks to overcome these difficulties (Shi et al. 2023; Reddy et al. 2023).

3.2 Economic Instruments and Efficiency Standards

Maritime decarbonization plans include major elements of economic tools and efficiency requirements that can enable the IMO to meet its greenhouse gas reduction agenda in 2050. The carbon pricing mechanisms such as taxes and emissions trading are considered effective in the bridging of the wide cost gap between fossil fuels and alternative low-carbon marine fuels (Lagouvardou et al. 2023). These market-based measures (MBMs) promote the use of cleaner fuels and can bring revenue to facilitate the development of renewable fuel production and storage and bunkering infrastructure by making the use of fossil fuels more expensive (Metziger 2021; Tu et al. 2024).

Research indicates a need to set carbon prices above 100- 200 EUR/tCO2 to ensure that renewable marine fuels such as biomethanol and e-ammonia become economically viable, highlighting the extensive role of sound pricing policies in quickening the process of fuel transitions (Harahap et al. 2023). MBMs also affect the decision to invest into emission reduction technology, like wind-assist propulsion, by raising the economic viability of the projects under the conditions of uncertainty (Metziger 2021).

Slow steaming, hull design optimisation, and the use of hybrid propulsion are strategies that operational efficiency can play a role in fuel savings and emission reduction by itself, but cannot get the company to decarbonisation by its own (Lagouvardou et al. 2023; Spinelli et al. 2022). Energy-saving devices (ESDs) include stern hydrodynamic enhancements which have shown up to 15 per cent of fuel consumption savings but these technical solutions should be combined with alternative fuels and policy incentives to have the greatest effect (Spinelli et al. 2022). The digitalisation of ports and their enhancement, e.g., with digital twins, contribute to reducing the emissions of a voyage significantly because efficient scheduling of vessels and minimised waiting periods can be brought by the efficiency of a port that is not a vessel itself (Eom et al. 2023).

The success of economic tools and efficiency benchmarks hinges heavily on coordinated regulation akin to market incentives and regulatory systems (Li and Hu 2024). The focus on international environmental public interests and considering both the economic and social aspects supports the idea that the policy mixes that consist of carbon prices, fuel regulations, and operational efficiencies would allow the maritime sector to transition to sustainability (Bengue et al. 2024; Serra and Fancello 2020). The studies highlight the importance of coherent MBMs that will be consistent with operational strategies to encourage investments and innovations that the sector needs to achieve the ambitious emission cut goals set by the IMO in 2050 (Tu et al. 2024; Serra and Fancello 2020).

3.3 International Regulatory Coordination

The climate governance mechanisms in the world are very crucial in driving the process of decarbonising the shipping and aviation industry, which contribute nearly fifty percent of the global greenhouse emissions. The International Maritime Organisation (IMO) is the main regulator of shipping emissions, with the 2023 Revised Emission Reduction Strategy with its objectives becoming the Paris Agreement aligned (Sahin 2024). MARPOL Annex VI is one of the foundational regulatory schemes in the scope of IMO which has set the limits of sulfur emissions, nitrogen oxides, PM, and carbon dioxide emissions by marine diesel engines, and certified compliance can be monitored through such tools as the International Air Pollution Prevention (IAPP) certificate (Campara et al. 2018; Kattner et al. 2015). Such regulations have fuelled technological advancement of cleaner engine designs and fuel switching, even though there are still difficulties in balancing the global action and strengthening enforcement (Kopela 2017).

The International Civil Aviation Organisation (ICAO) in aviation is mandated with the responsibility of organising the global efforts in reducing the carbon footprint of international air transport. In contrast to shipping, the regulatory framework in the aviation industry is more inclined to base on market-based interventions like the Carbon Offsetting and Reduction Scheme to International Aviation (CORSIA), but this scheme is also criticised as unable to achieve enough emissions reduction in line with 2degC temperatures targets (Bows-Larkin 2014). The differences in the complexity of the governance of shipping and aviation have impacted policy making; the complexity of shipping operations and technologies needs a range of policy intervention options such as alternative fuels, operational efficiency, and regulatory incentives (Mallouppas and Yfantis 2021). The cooperation between these areas and global organisations remains essential in order not to have fragmented climate policies that may slow down the entire progress (Campara et al. 2018).

International regulatory coordination is also another problem as different countries and industry actors have conflicting economic and environmental interests. The combination of these interests requires government designs that pre-empt the common good of the environment and adapt to the varying national conditions and economic realities (Li and Hu 2024). The concept of common yet differentiated responsibilities highlights the attempts to make sure that developing nations will obtain the necessary technical and financial assistance in the process of transition, an important aspect due to the high intensity of carbon emissions that is traditionally associated with shipping supply chains of the global scope (Dong et al. 2022). The concept of multi-stakeholder involvement is also driven in the area of multi-stakeholder participation in the process of private governance and green shipping, whereby influential corporate entities are involved in the process with the coordinated assistance of IMO in promoting sustainability beyond the regulatory requirements (Lister 2014). Such a hybrid form of governance has the potential to increase the effectiveness and legitimacy of the decarbonisation policies.

Also, global governance structures should deal with overlaps and jurisdictions in regulations, as the coastal and port states gradually follow the unilateral steps that at times contradict international agreements (Kopela 2017). Regulatory stability, uniformity, and clarity will always be the primary focus to guarantee market confidence to the maritime stakeholders who invest in alternative fuels, energy-conserving technologies, and cleaner ships (Sahin 2024). The ongoing creation of global regulations that harmonised policy tools, and implementation schemes taking advantage of governmental interventions and market mechanism are critical in the shipping and aviation industries to achieve the goal of reducing emissions and ensuring the operations remain viable (Li and Hu 2024; Mallouppas and Yfantis 2021).

3.4 Green Corridors, Port Electrification and Infrastructure Development

Green shipping corridors as a strategic framework have received significant interest in enhancing decarbonisation of maritime shipping. Such corridors include between ports, shipping firms, cargo owners, and governments to create routes using zero-emission fuels and technologies and thus facilitate sustainable trade and lower greenhouse gas emissions (Akhavan 2025). It is important to note that green ammonia corridors have been suggested as efficient means of decarbonising mass transportation on the sea, since ammonia has carbon–neutral qualities and has already developed production and transportation infrastructure (Bengue et al. 2024). Environmental compatibility and economic viability must be prioritised when developing a corridor, and the stakeholders focus on such aspects as the availability of alternative fuels, the presence of bunkering infrastructure, and decarbonisation paths of vessels (Bengue et al. 2024). The Clydebank Declaration emphasises the importance of these corridors to accelerate the process of switching to zero-carbon maritime transport through coordination of investments and risk sharing among main stakeholders (Song et al. 2023).

In line with route decarbonisation, green port strategies and port electrification are two important elements of the emissions reduction on the regional and local levels. Research shows that electrified maritime networks, which are backed by green port projects, can significantly reduce the number of emissions since powered vessels during port calls are supported by renewable electricity, which is in turn free of any use of diesel-driven auxiliary engine (Nguyen et al. 2022; Akhavan 2025). Green ports use a set of measures such as providing shores with power, using energy-efficient logistics, and digital monitoring to better regulate their work and reduce environmental damage (Nguyen et al. 2022). As developed nations are seen to display successful fulfillment of the green port policies, the emerging economies have to be seen to gain through customised policies based on the improved models to enhance the port-city ecosystems in a sustainable way (Nguyen et al. 2022).

The development of infrastructure, in particular, smart logistics hubs, is part of the facilitation of green corridors and electrified port operation. These ports also organise the operations of cargo handling, storage, customs, and inland transportation to facilitate the marine logistics and integrate the technologies with low or zero emissions and digital solutions (Issa-Zadeh and Garay-Rondero 2025). The digitalisation of the transport industry, including the port digital twins, will help optimize the work of vessels and cargos in real-time and increase their effectiveness and minimize the carbon footprints (Xiao et al. 2025). Nevertheless, hurdles of logistics, expensive investment, and slow adoption of technology are issues that challenge extensive decarbonisation in port operations (Issa-Zadeh and Garay-Rondero 2025). In this way an IT-driven solution encompassing the modernisation of the infrastructure and regulatory incentives with stakeholder cooperation is required to scale sustainable maritime logistics.

The operationalisation of green corridors and zero-emission ports is supported by technological innovation and fleet deployment strategies. Shipping companies can gradually lower the amount of emissions by retrofitting their current fleets with green technologies, including hybrid power systems and energy-saving devices, when switching to cleaner fuels (Wu et al. 2022; Spinelli et al. 2022). The sensitivity analyses indicate that the investments in green technologies become economical in the case of rising fuel and operating costs, especially when dealing with large vessels and more well-established shipping routes (Wu et al. 2022). Moreover, certain ship designs are optimized by design to achieve the maximum possible energy efficiency in line with the objective of decarbonization, and integrated electrification, alternative fuels (Huang and Duan 2023). As a result, a synergistic approach of green corridors, the electrification of ports, and the improvement of infrastructure, is crucial to the transition of the maritime industry toward sustainability (Song et al. 2023; Akhavan 2025).

3.5 Trade, Sector Coupling, and Impacts on Trade-Dependent Economies

The process of decarbonisation has a deep impact on the global trade flow due to the fact that the maritime shipping contributes approximately 90 percent of all global trade and is closely interconnected with other industries such as energy and agriculture (Shen et al. 2021; Urban et al. 2023). This links maritime transportation to energy and commodity provisions, increasing the complexity of the transitions because the demands in marine fuel use conflict with other uses of biomass and energy resources, sometimes causing an intensification of the competition over the resources in the context of decarbonisation policies (Urban et al. 2023; Meckling 2025). Furthermore, sector coupling strategies show that disruptive technological integration in the transport, energy, and industrial sector should be undertaken but difficult, and such integration should be coordinated in the adoption of innovations to achieve the goals without the loss of efficiency in trade (Beckmann et al. 2024). Simultaneously, energy-intensive sectors such as shipping also expose themselves to trade hazards due to geopolitical aspects and policy-based factors, leading to a change in trade-related economies (Dumortier et al. 2024).

It can be seen that the economic aspects of decarbonisation of shipping are characterised by high external costs imposed on societies, with the necessity to incorporate the environmental externalities into the design of maritime policies (Meckling 2025). Models that include the effects of green house gases and trade balance emphasise the tensions between high reduction of emissions and trade competitiveness especially in areas such as Europe where decarbonization can worsen external trade balances without corresponding global action (Clora and Yu 2022). Such dynamics imply that there is a need to have unified international systems and coordinated policies, such as carbon adjustments on the borders to avoid the leakage of carbon and enforce market fairness (Clora and Yu 2022). Green industrial strategies based on policies can make use of the implementation of clean technologies to increase energy security and economic development, yet they can also lead to the occurrence of trade protectionism and disparities in global economies (Meckling 2025).

Sector coupling has become a powerful analytical instrument to study the ripple effects of decarbonisation of shipping on energy systems and resource distribution. Lifecycle assessment-based studies find that decarbonisation policies, particularly those focused on electrification and renewable integration, many times multiply electricity demands (sometimes to twice current levels) and, therefore, cause electric grids to face the challenge of absorbing coupled systems transitions (Baumgartner et al. 2021; Dong et al. 2023). Nonetheless, this integration may result in a transfer of environmental pressures that were placed on carbon to metal resource losses and freshwater ecotoxicity, and finally, a complete planning should be performed to balance the reduction of emissions with the overall sustainability results (Luderer et al. 2019). The issue of temporal variations of the renewable supply and demand is further reinforced by the necessity of sophisticated energy modeling to guarantee resilient decarbonized infrastructure to sustain trade demands (Bistline 2021).

The effects of decarbonization on economies that rely on trade are diverse and change the competition of resources and their economic sustainability. In the case of Latin America, agricultural exports are under the pressure of the global climate shocks due to disruptions in trade and changes in climate policy, which can amplify the vulnerabilities of biodiverse and water-stressed regions (Dumortier et al. 2024). In the meantime, the concerted energy and trade policy actions are still necessary to achieve a balance between decreasing emissions and economic growth in the areas (Meckling 2025; Dong et al. 2023). Altogether, the connection between trade and sectoral integration with policy implies the dependency of maritime decarbonisation and its effects on global economic stability and the sustainability of the environment (Urban et al. 2023; Clora and Yu 2022).

4 Methodology

The study used a qualitative methodology that is predominantly a desk review. The qualitative nature allowed the researchers to have a deeper investigation into the sectors that are technically and economically difficult to decarbonise. It further allowed for the evaluation of alternative fuels, efficiency standards, and international regulatory coordination under institutions. Data for the study were sourced from a broad range of material, including books, policy documents, peer-reviewed journal articles, and some credible online sources. A comprehensive review was done through a selection of these different sources to capture theoretical views and empirical findings from previously existing articles. To get access to this literature, the study used electronic search engines that include Science Direct, Scopus, and Google Scholar. This helped the researchers to ground the study in authoritative and well-established academic discourse. The study used textual analysis as a primary analytical tool for data analysis. A systematic examination of selected English documents was done using this method to identify recurring themes and patterns in literature.

5 Findings

Findings in this study are presented in form of three case studies on EU FuelEU Maritime Regulation, Greek Coastal Shipping Electrification and finally Port Digital Twin–Busan New Port. The chapter commences with the EU FuelEU Maritime Regulation.

Case Study 1: EU FuelEU Maritime Regulation (Global Governance and Policy Mix)

The EU FuelEU Maritime Regulation is an example of how regional regulation can hasten decarbonization in hard-to-abate industries where global bodies such as the International Maritime Organization (IMO) have not been making fast progress. A policy window was opened after the Paris Agreement and it made the EU develop binding maritime transport fuel standards, which previously had relied on voluntary strategies (Malmborg 2024). The regulation is indicative of the scrupulous political bargaining and institutional pressure, as the main policy entrepreneurs, such as environmental NGOs, including Transport & Environment, urged the adoption of strict zero-carbon fuel requirements, in contrast to the more moderate, technology-neutral position of the European Commission (Malmborg 2024). This tension identified the existence of conflicts between climate coalitions and existing players in the industry, indicating the multi-layered nature of policy entrepreneurship under the multi-level governance system in the EU. The EU tries to redefine the international shipping market by establishing new standards of advanced fuels on its territory, and uses its economic power to become the first one to encourage decarbonization, at the time when IMO regulations are delayed (Bach and Hansen 2023).

Considering the global governance and policy mix, the FuelEU Maritime Regulation indicates the significance of consistency in the policy mix and an integrated instrument in facilitating sustainability transitions. The policy of the EU is a combination of the market-based solutions, the standard of emission intensity, and the encouragement of alternative fuels as an effort to reconcile the ambitions of the environment with the realities of industrialisation (Malmborg 2024). Nevertheless, the interactions of various governance actors pose consistent threats, including the incompatibility of the interests of the member states and the lobbying of industries that weaken the rigor of the regulations (Jancic 2016; Malmborg 2024). The case highlights the importance of regional governance structures as institutional pressure on the global actors such as the IMO to be more progressive and towards the goals of climate. However, the ensuing policy fragmentation will lead to uncertainty in the global regulations, and the unequal global standards may slow down the transition pace (Bach and Hansen 2023). In this way, the case demonstrates the importance of consistent and stable policy combinations in breaking technological lock-ins and decarbonising the maritime industry faster.

The implications in terms of trade also depict the regulatory and governance challenges of FuelEU Maritime. The regulation also has an impact on the fuel selection and operation of the shipping companies in the waters of the EU, essentially stretching the EU regulatory scope to the international suppliers and global freight routes (Malmborg 2024). This extraterritorial effect has the power to change the maritime competitiveness, which may provoke trade disputes with other non-EU partners that might regard strict regulations as obstacles (Jancic 2016). However, the economic strength of the EU allows it to be a norm-setter regarding maritime governance in the world and advance decarbonisation outside the union (Vogler and Stephan 2007). The case is therefore illustrative of empowering a global shift through regional policy entrepreneurship placed in a multi-level and multi-actor governance environment, where it is also able to shape environmental standards and trade dynamics (Malmborg 2024; Bach and Hansen 2023). In general, FuelEU Maritime firmly believes in the theoretical applicability of the global governance and policy mix frameworks in the context of decarbonization of the sector.

Case Study 2: Greek Coastal Shipping Electrification (Technological Niches and Green Corridors)

The case of the Greek electrification of coastal shipping shows that niche innovation, such as the idea of electric ferries in the case, can be an important catalyst to the restructuring of the current maritime networks. According to research conducted by Karountzos et al. (2023), the parts of the Greek coastal shipping network have a high potential to be converted into the zero-emission electric sub-network, with the help of port hubs in strategic locations and sophisticated spatial planning instruments (ibid.). Such a shift is representative of the green corridors as a safe haven where electric shipping technologies may be cultivated before broader expansion. It is important to note here that green corridors have the ability to absorb renewable energy sources and tie together island economies which not only reflects technological innovation but also the realignment of socio-technical systems. This process is conceptualised by sustainable transitions theory, in particular, the Multi-Level Perspective (MLP), whose components of interaction are niche and regime, in which the electric ferry niche and the established maritime regime based on fossil fuels are the respective mode and sector (Geels 2018; 2024).

Among the key enablers of electrification in Greek coastal shipping, infrastructure preparedness can be distinguished. The case illustrates that the port hubs that want to be functional energy centers that can accommodate electric ferry charging as well as integrating into the local renewable energy infrastructure will also serve to bound the sector between the maritime transport system and electricity grids (Akhavan 2025; Damman and Steen 2021). Furthermore, tools of spatial planning that rely on Geographic Information Systems (GIS) contribute to the determination of optimal paths and clusters where zero-emission sub-networks can flourish and one must consider accessibility and the presence of renewable resources (Karountzos et al. 2023). This type of integration highlights how technological innovation interacts with organisational change as a part of the larger regime which points to specific policy tools to support infrastructure investments and operational changes (Bergek et al. 2021).

Electrification of the Greek coastal shipping has far reaching regional economic consequences, particularly to the islands economies which rely on sea transport for their livelihood. Electric ferry green corridors can become a stimulus of the sustainable economic development because they allow minimising operating expenses, pollutants, and increase energy autonomy with local renewable resources (Karountzos et al. 2023). The transition is also another factor to transform socioeconomic pathways in coastal areas through creating new business models, green employment, and tourism gains associated with cleaner transport solutions (Akhavan 2025). Placed in the context of the MLP, the case points to the vital niche-regime processes in which the new electric shipping networks engage with old maritime systems demonstrating one of the possible pathways of transitioning to the experimentation phase to potential mainstream adoption (Geels 2024).

Case Study 3: Port Digital Twin–Busan New Port (Logistics Optimization and Freight Corridors)

The Port Digital Twin used in Busan New Port is a good example of how digital innovation can become an enormous boost to the efficiency of port logistics and emission reduction without modifying fuel types. This digital twin system uses real-time data aggregation and sophisticated scheduling algorithms to optimise the vessel arrival and departure times where in a tested case scenario the overall carbon emissions related to ports was reduced by more than 75% of the traditional operations (Eom et al. 2023). The system allows making proactive decisions through simulating interactions of ships, cargo handling, and port infrastructure to prevent congestion and unnecessary wait times. The case demonstrates the high importance of the logistics chain optimisation, in which data-based coordination of the vessels and port authorities will improve the operations performance and lead to the decarbonization objectives (Cebir and Akkartal 2024; Su et al. 2024).

At the port level, a decarbonisation strategy implemented by Busan New Port in the form of a digital twin is a demonstration of a path of incremental transition, positioned in prevailing maritime regimes. This method maximises the assets and processes in operation in comparison with wholesale technological changes like fuel switching resulting in less emission due to increased system efficiency (Eom et al. 2023). The case highlights the importance of integrating data from shipping operations, port infrastructure, and hinterland freight transport to achieve end-to-end logistics optimization (Wang et al. 2024). This holistic integration ensures smoother cargo flows and minimises bottlenecks, thus amplifying the environmental benefits beyond the port boundaries. It demonstrates how cyber-physical systems like digital twins can advance sustainability without disrupting established value chains.

Furthermore, the Busan New Port case underscores the interconnectedness of shipping, port operations, and hinterland freight corridors, showcasing how improved data infrastructure can enhance the entire maritime logistics ecosystem. The digital twin's capability to simulate complex operational scenarios and respond dynamically to changing conditions assists in aligning port efficiency with environmental objectives (Zerbino et al. 2019; Lee et al. 2009). This integration facilitates a more resilient and flexible freight corridor, essential for handling global trade disruptions and increasing demand variability. In sum, the Busan digital twin embodies an effective innovation within the prevailing regime, enabling smarter, greener maritime logistics and reinforcing the value of digitalisation for port sustainability and competitiveness.

6 Discussion

The results of this chapter support the sustainability transitions literature that decarbonization of hard-to-abate industries is determined by how the niche innovations, regimes, interrelate with the broader governance environments. Following Geels (2018), the formation of green shipping routes and port electrification programs can be viewed as niche experiments that oppose the mainstream fossil-powered shipping regime. The fact that the study focuses on coordinated policy mixes is also very similar to the focus on regime segmentation by Bergek et al. (2021), who also note that this aspect allows facilitating sustainability transitions in coastal shipping. Moreover, the significance of the international regulatory pressure is representative of the ideas of Bach and Hansen (2023) who hold that the lack of uniformity in policy signals of global institutions may impede innovation adoption. The findings also uphold the opinion of Li and Hu (2024) that the key to successful shipping governance is the ability to strike a balance between conflicting economic and environmental interests. Altogether, the chapter confirms that the change in technology cannot materialize without the support of the multi-level institutional alignment.

The chapter also concurs with the scholars that focus on the economic and operational aspects of decarbonisation of the maritime. The importance of alternative fuels and carbon pricing schemes can be explained by the results of Lagouvardou et al. (2023), who confirm that effective market-driven initiatives are necessary to eliminate cost differences between the fossil and renewable marine fuels. In a similar manner, Serra and Fancello (2020) claim that efficiency metrics, though useful, cannot bring about the profound emission reductions necessary under the IMO long-term goals bearing in mind that this argument is also reflected in this research. The combination of digital technologies like port digital twins assists Eom et al. (2023), who demonstrate that logistics optimisation can result in a substantial decrease of emissions without direct replacement of fuel. Also, Mallouppas and Yfantis (2021) emphasize that the paths to innovation should go together with the fuel switching, operational efficiency, and regulatory reform. Thus, it is established in the discussion that a resilient and competitive maritime transition requires an integrated approach to technology and governance.

The chapter also is similar to the literature that emphasizes the larger systemic and trade-associated consequences of maritime decarbonization. Urban et al. (2023) highlight that the connection between shipping and energy systems is a key sector that is critical to ensuring deep reduction of emissions, especially in cases where renewable electricity is needed to generate green hydrogen and ammonia. This contributes to the theme of the chapter of infrastructure integration and energy system alignment. Similarly, Clora and Yu (2022) show that unilateral carbon emission restrictions may disrupt the balance of trade and pose the risk of carbon leakage, which supports the necessity of global and international regulation as discussed in this study. Another reason why Luderer et al. (2019) warn of decarbonization pathways is that they can create environmental trade-offs across sectors, and lifecycle and systems-based assessment can be crucial. With the context of maritime transition established in the context of global trade and global energy, the chapter conforms to the arguments of scholars who do not see decarbonization as a specific sectoral change but as a multi-layered, interconnected global change.

7 Further Research

The empirical knowledge of integrated policy mix effects on long-term decarbonisation in maritime and other hard-to-abate sectors should be further developed. Although the literature investigates the dynamics of alternative fuels, green corridors, and digital optimization individually, there is a scarcity of literature that captures the synergistic impact of these three systems in global supply chain systems. The comparative case analysis across regions applying various regulatory models including the EU, Asia–Pacific and emerging economies could be used in future research to evaluate the policy coherence and effectiveness. The lifecycle emissions, trade impacts and energy system coupling would also be useful in form of quantitative modelling to explain the unintended environmental or economic trade-offs. Moreover, the finance mechanisms such as green bonds, carbon contracts for difference and blended finance structures require further research in order to facilitate large-scale infrastructure transitions. Certain behavioural and organisational aspects including industry resistance, stakeholder coordination, and innovation diffusion also need to be looked at more closely. Longitudinal research of green shipping routes development and port electrification project evolution projects might shed some light on the dynamics of the transition over time. Lastly, the evidence base on scalable and equitable decarbonisation trajectories within international maritime transport systems could be reinforced by interdisciplinary research that incorporates governance theory, engineering analysis and economic modelling.

8 Managerial/Policy Implications

The results of the present research have significant implications to managers and policymakers who aim at speeding up the process of decarbonisation in maritime. The findings point to the necessity of consistent and rational policy combinations to unite carbon pricing, fuel standards, and infrastructure incentives, to policymakers. The regulatory transparency lowers uncertainty and fosters the long-term growth of the business of the alternative fuels, port electrification, and digital optimization system. The coordinated international frameworks should be given priority by governments to eliminate their policy fragmentation and leakage of carbon, as well as provide developing economies with financial and technical help. In the case of industry managers, strategic planning needs to shift away under efficiency gains to multiplex transition strategies which encompass fuel diversification, fleet retrofit and coordination within green shipping corridors. Digital technologies like port digital twins, smart logistics platforms can create short-term emission cuts in the meantime training firms to take larger structural action. The collaborations with the populace will be mandatory to distribute the risks of new fuels such as ammonia and hydrogen. Managers are also expected to use lifecycle assessment in making decisions as a way of ensuring that the environmental burdens are not transferred across sectors. Comprehensively, the active involvement in the changing regulatory environment and joint innovation networks will play a key role in being competitive in the shift towards climate-neutral maritime transport.

9 Conclusion

To sum up, the decarbonisation of hard-to-abate maritime transport necessitates a multidimensional change that incorporates technological innovation, policy change, infrastructure development, and digital transformation. The discussion indicates that other alternative fuels like ammonia, hydrogen, and methanol have a great potential but require favourable regulation systems and carbon pricing systems to gain commercial sustainability. The examples of green shipping routes and port electrification represent how small-scale innovations may disrupt the existing fossil-fuel regimes in case they are backed by coordinated governance. Online technologies such as port digital twins also demonstrate that operational optimization can provide significant emission cuts in the current systems. Nevertheless, global development is still lagging behind disjointed international policies and conflicting economic of interests. The paper highlights the key role of consistent policy bundles, multilevel governance, and cross-sectoral integration in breaking technology lock-ins and market obstacles. There is a need to be able to balance environmental ambition and trade competitiveness in order to be able to guarantee climate mitigation and economic sustainability. To accelerate the adoption of the maritime industry by net-zero emissions, long-term cooperation between the governments, the stakeholders of the industry, financial institutions, and international organizations will be essential in ensuring that the systemic change is permanent.

References

Akhavan M (2025) Decarbonising maritime transport: the role of green shipping corridors in making sustainable port-city ecosystems. Ocean Society 2. https://doi.org/10.17645/oas.9411

Ayvalı T, Edman Tsang SC, Van Vrijaldenhoven T (2020) The position of ammonia in decarbonising maritime industry: an overview and perspectives: Part I. Johnson Matthey Technology Review 65(2):275–290. https://doi.org/10.1595/205651321x16043240667033

Bach H, Hansen T (2023) Flickering guiding light from the international maritime organisation's policy mix. Environ Innov Soc Trans 47:100720. https://doi.org/10.1016/j.eist.2023.100720

Baumgärtner N, Deutz S, Reinert C, Nolzen N, Kuepper LE, Hennen M, Hollermann DE, Bardow A (2021) Life-cycle assessment of sector-coupled national energy systems: environmental impacts of electricity, heat, and transportation in Germany Till 2050. Front Energy Res 9. https://doi.org/10.3389/fenrg.2021.621502

Beckmann J, Klöckner K, Letmathe P (2024) Scenario-based multi-criteria evaluation of sector coupling-based technology pathways for decarbonization with varying degrees of disruption. Energy 297:131248. https://doi.org/10.1016/j.energy.2024.131248

Bei Z, Wang J, Li Y, Wang H, Li M, Qian F, Xu W (2024) Challenges and solutions of ship power system electrification. Energies 17(13):3311. https://doi.org/10.3390/en17133311

Bengue AA, Alavi-Borazjani SA, Chkoniya V, Cacho JL, Fiore M (2024) Prioritizing criteria for establishing a green shipping corridor between the ports of sines and Luanda using Fuzzy AHP. Sustainability 16(21):9563. https://doi.org/10.3390/su16219563

Bergek A, Bjørgum Ø, Hansen T, Hanson J, Steen M (2021) Sustainability transitions in coastal shipping: the role of regime segmentation. Transp Res Interdiscip Perspect 12:100497. https://doi.org/10.1016/j.trip.2021.100497

Bistline JET (2021) The importance of temporal resolution in modeling deep decarbonization of the electric power sector. Environ Res Lett 16(8):084005. https://doi.org/10.1088/1748-9326/ac10df

Bows-Larkin A (2014) All adrift: aviation, shipping, and climate change policy. Climate Policy 15(6):681–702. https://doi.org/10.1080/14693062.2014.965125

Čampara L, Hasanspahić N, Vujičić S (2018) Overview of MARPOL ANNEX VI regulations for prevention of air pollution from marine diesel engines. SHS Web of Conferences 58:01004. https://doi.org/10.1051/shsconf/20185801004

Cebir Y, Akkartal E (2024) Digital Twin in Logistics and Supply Chain Management. IGI Global, pp 170–192. https://doi.org/10.4018/979-8-3693-3575-8.ch008

Clora F, Yu W (2022) GHG emissions, trade balance, and carbon leakage: Insights from modeling thirty-one European decarbonization pathways towards 2050. Energy Economics 113:106240. https://doi.org/10.1016/j.eneco.2022.106240

Damman S, Steen M (2021) A socio-technical perspective on the scope for ports to enable energy transition. Transp Res Part d: Transp Environ 91:102691. https://doi.org/10.1016/j.trd.2020.102691

Dong J, Cai B, Zhang S, Wang J, Yue H, Wang C, Mao X, Cong J, Guo F (2023) Closing the gap between carbon neutrality targets and action: technology solutions for china's key energy-intensive sectors. Environ Sci Technol 57(11):4396–4405. https://doi.org/10.1021/acs.est.2c08171

Dong J, Zeng J, Yang Y, Wang H (2022) A review of law and policy on decarbonization of shipping. Front Marine Sci 9. https://doi.org/10.3389/fmars.2022.1076352

Dumortier J, Elobeid A, Carriquiry M (2024) Impacts of geopolitics and policy on Latin American biodiversity and water resources. Agribusiness. https://doi.org/10.1002/agr.21965

Eom J-O, Yoon J-H, Yeon J-H, Kim S-W (2023) Port digital twin development for decarbonization: a case study using the pusan newport international terminal. J Marine Sci Eng 11(9):1777. https://doi.org/10.3390/jmse11091777

Geels FW (2018). Socio-Technical Transitions to Sustainability. https://doi.org/10.1093/acrefore/9780199389414.013.587

Geels F (2024) The multi-level perspective on sustainability transitions: background, overview, and current research topics. Cambridge University. https://doi.org/10.33774/coe-2024-c15gb

Harahap F, Nurdiawati A, Conti D, Leduc S, Urban F (2023) Renewable marine fuel production for decarbonised maritime shipping: Pathways, policy measures and transition dynamics. J Clean Prod 415:137906. https://doi.org/10.1016/j.jclepro.2023.137906

Huang J, Fan H, Xu X, Liu Z (2022) Life cycle greenhouse gas emission assessment for using alternative marine fuels: a very large crude carrier (VLCC) case study. J Mar Sci Eng 10(12):1969. https://doi.org/10.3390/jmse10121969

Huang J, Duan X (2023) A comprehensive review of emission reduction technologies for marine transportation. J Renew Sustain Energy 15(3). https://doi.org/10.1063/5.0150010

Issa-Zadeh SB, Garay-Rondero CL (2025) Decarbonizing seaport maritime traffic: finding hope. World 6(2):47. https://doi.org/10.3390/world6020047

Jančić D (2016) TTIP and legislative-executive relations in EU trade policy. West Eur Polit 40(1):202–221. https://doi.org/10.1080/01402382.2016.1240407

Karountzos O, Kagkelis G, Kepaptsoglou K (2023) A decision support GIS framework for establishing zero-emission maritime networks: the case of the greek coastal shipping network. J Geovisualization Spat Anal 7(2). https://doi.org/10.1007/s41651-023-00145-1

Karvounis P, Theotokatos G, Boulougouris E (2024) Environmental-economic sustainability of hydrogen and ammonia fuels for short sea shipping operations. Int J Hydrogen Energy 57:1070–1080. https://doi.org/10.1016/j.ijhydene.2024.01.058

Kattner L, Mathieu-Üffing B, Burrows JP, Richter A, Schmolke S, Seyler A, Wittrock F (2015) Monitoring compliance with sulfur content regulations of shipping fuel by in situ measurements of ship emissions. Atmos Chem Phys 15(17):10087–10092. https://doi.org/10.5194/acp-15-10087-2015

Kim H, Koo KY, Joung T-H (2020) A study on the necessity of integrated evaluation of alternative marine fuels. J Int Marit Saf, Environ Aff, Shipp 4(2):26–31. https://doi.org/10.1080/25725084.2020.1779426

Kopela S (2017) Making ships cleaner: Reducing air pollution from international shipping. Rev Eur, Comp & Int Environ Law 26(3):231–242. https://doi.org/10.1111/reel.12220

Lagouvardou S, Lagemann B, Psaraftis HN, Lindstad E, Erikstad SO (2023) Marginal abatement cost of alternative marine fuels and the role of market-based measures. Nat Energy 8(11):1209–1220. https://doi.org/10.1038/s41560-023-01334-4

Lee H, Lee J, Roh G, Lee S, Choung C, Kang H (2024) Comparative life cycle assessments and economic analyses of alternative marine fuels: insights for practical strategies. Sustainability 16(5):2114. https://doi.org/10.3390/su16052114

Lee G, You S (Iris), Ritchie SG, Saphores JD, Sangkapichai M, Jayakrishnan R (2009) Environmental impacts of a major freight corridor. Transp Res Rec: J Transp Res Board 2123(1):119–128. https://doi.org/10.3141/2123-13

Li W, Hu Z (2024) Pathways in the governance of shipping decarbonization from perspective of balancing the conflicting interests. Front Mar Sci 11. https://doi.org/10.3389/fmars.2024.1479528

Lindstad E, Lagemann B, Rialland A, Gamlem GM, Valland A (2021) Reduction of maritime GHG emissions and the potential role of E-fuels. Transp Res Part d: Transp Environ 101:103075. https://doi.org/10.1016/j.trd.2021.103075

Lister J (2014) Green shipping: governing sustainable maritime transport. Global Pol 6(2):118–129. https://doi.org/10.1111/1758-5899.12180

Luderer G, Pehl M, Arvesen A, Gibon T, Bodirsky BL, De Boer HS, Fricko O, Hejazi M, Humpenöder F, Iyer G, Mima S, Mouratiadou I, Pietzcker RC, Popp A, Van Den Berg M, Van Vuuren D, Hertwich EG (2019) Environmental co-benefits and adverse side-effects of alternative power sector decarbonization strategies. Nat Commun 10(1). https://doi.org/10.1038/s41467-019-13067-8

Mallouppas G, Yfantis EA (2021) Decarbonization in shipping industry: a review of research, technology development, and innovation proposals. J Mar Sci Eng 9(4):415. https://doi.org/10.3390/jmse9040415
Meckling J (2025) The geoeconomic turn in decarbonization. Nature 645(8082). https://doi.org/10.1038/s41586-025-09416-x
Metzger D (2021) Market-based measures and their impact on green shipping technologies. WMU J Marit Aff 21(1):3–23. https://doi.org/10.1007/s13437-021-00258-8
Nguyen HP, Nguyen PQP, Nguyen TP (2022) Green port strategies in developed coastal countries as useful lessons for the path of sustainable development: a case study in Vietnam. Int J Renew Energy Dev 11(4):950–962. https://doi.org/10.14710/ijred.2022.46539
Reddy VJ, Hariram NP, Maity R, Ghazali MF, Kumarasamy S (2023) Sustainable E-fuels: green hydrogen, methanol and ammonia for carbon-neutral transportation. World Electr Veh J 14(12):349. https://doi.org/10.3390/wevj14120349
Sahin AU (2024) IMO 2023 revised emission reduction strategy and regulatory challenges of maritime decarbonization. Eur J Commer Contract Law 16(1):25–33. https://doi.org/10.7590/187714624x17132716463900
Schwarzkopf DA, Petrik R, Hahn J, Ntziachristos L, Matthias V, Quante M (2023) Future ship emission scenarios with a focus on ammonia fuel. Atmosphere 14(5):879. https://doi.org/10.3390/atmos14050879
Serra P, Fancello G (2020) Towards the IMO's GHG Goals: a critical overview of the perspectives and challenges of the main options for decarbonizing international shipping. Sustainability 12(8):3220. https://doi.org/10.3390/su12083220
Shen B, Kahrl F, Satchwell AJ (2021) Facilitating power grid decarbonization with distributed energy resources: lessons from the United States. Annu Rev Environ Resour 46(1):349–375. https://doi.org/10.1146/annurev-environ-111320-071618
Shi J, Zhu Y, Feng Y, Yang J, Xia C (2023) A prompt decarbonization pathway for shipping: green hydrogen, ammonia, and methanol production and utilization in marine engines. Atmosphere 14(3):584. https://doi.org/10.3390/atmos14030584
Song Z-Y, Chhetri P, Ye G, Tae-Woo Lee P (2023) Green maritime logistics coalition by green shipping corridors: a new paradigm for the decarbonisation of the maritime industry. Int J Log Res Appl 28(4):363–379. https://doi.org/10.1080/13675567.2023.2256243
Spinelli F, Mancini S, Vitiello L, Bilandi RN, De Carlini M (2022) Shipping decarbonization: an overview of the different stern hydrodynamic energy saving devices. J Mar Sci Eng 10(5):574. https://doi.org/10.3390/jmse10050574
Su Z, Liu Y, Gao Y, Park K-S, Su M (2024) Critical success factors for green port transformation using digital technology. J Mar Sci Eng 12(12):2128. https://doi.org/10.3390/jmse12122128
Tu H, Liu Z, Zhang Y (2024) Study on cost-effective performance of alternative fuels and energy efficiency measures for shipping decarbonization. J Mar Sci Eng 12(5):743. https://doi.org/10.3390/jmse12050743
Urban F, Nurdiawati A, Harahap F (2023) Sector coupling for decarbonization and sustainable energy transitions in maritime shipping in Sweden. Energy Res Soc Sci 107:103366. https://doi.org/10.1016/j.erss.2023.103366
Vogler J, Stephan HR (2007) The European Union in global environmental governance: Leadership in the making? Int Environ AgreemS: Polit, Law Econ 7(4):389–413. https://doi.org/10.1007/s10784-007-9051-5
Von Malmborg F (2024) At the controls: Politics and policy entrepreneurs in <scp>EU</scp> policy to decarbonize maritime transport. Rev Policy Res. https://doi.org/10.1111/ropr.12609
Wang H, Daoutidis P, Zhang Q (2022) Ammonia-based green corridors for sustainable maritime transportation. Digit Chem Eng 6:100082. https://doi.org/10.1016/j.dche.2022.100082
Wang, K, Xu H, Wang H, Qiu R, Hu Q, Liu X (2024) Digital twin-driven safety management and decision support approach for port operations and logistics. Front Mar Sci 11. https://doi.org/10.3389/fmars.2024.1455522

Watanabe MDB, Cherubini F, Tisserant A, Cavalett O (2022) Drop-in and hydrogen-based biofuels for maritime transport: Country-based assessment of climate change impacts in Europe up to 2050. Energy Convers Manage 273:116403. https://doi.org/10.1016/j.enconman.2022.116403
Wu Y, Huang Y, Wang H, Zhen L, Shao W (2022) Green technology adoption and fleet deployment for new and aged ships considering maritime decarbonization. J Mar Sci Eng 11(1):36. https://doi.org/10.3390/jmse11010036
Xiao G, Pan L, Lai F (2025) Application, opportunities, and challenges of digital technologies in the decarbonizing shipping industry: a bibliometric analysis. Front Mar Sci 12. https://doi.org/10.3389/fmars.2025.1523267
Zerbino P, Aloini D, Dulmin R, Mininno V (2019) Towards analytics-enabled efficiency improvements in maritime transportation: a case study in a Mediterranean port. Sustainability 11(16):4473. https://doi.org/10.3390/su11164473
Zhang J, Zhang Z, Liu D (2024) Comparative study of different alternative fuel options for shipowners based on carbon intensity index model under the background of green shipping development. J Mar Sci Eng 12(11):2044. https://doi.org/10.3390/jmse12112044

Green Transport Systems and Adaptive Strategies for Managing Barriers to Transition

Munyaradzi Chibaro, David Chikwere, Moeti Lysson S. Josephine, Tookhona Tefelo Kacceth, and Justine Mbudaya

Abstract The process of transitioning to green transportation systems in developing countries is an important way for developing countries to achieve their sustainable development goals, reduce their greenhouse gas emissions and increase their ability to be resilient in cities. However, there are many barriers which make it difficult for developing countries to move towards green transportation. These barriers include structural barriers for example lack of infrastructure, lack of finance, limitations of technology, institutional barriers for example fragmentation within government departments, lack of policy implementation/enforcement, lack of coordination and behavioral barriers for example citizen opposition to change, citizen's lack of environmental knowledge/awareness, citizen's preference for traditional modes of transportation. This chapter will evaluate these barriers from both a multidimensional perspective and identify alternative ways for developing countries to successfully implement their plans for green transportation. This chapter will emphasize the necessity of using flexible, country-specific methods which can adapt to changing circumstances and socioeconomic conditions. Several examples of alternative strategies which have been identified to support a successful green transportation system include improving policy implementation through the development of integrated

M. Chibaro (✉) · T. T. Kacceth
Department of Logistics and Supply Chain Management, University of Botswana, Gaborone, Botswana
e-mail: munyachibaro@gmail.com

T. T. Kacceth
e-mail: tefelotookhona@gmail.com

D. Chikwere
Leeds Beckett University, Leeds, UK
e-mail: D.chikwere@leedsbeckett.ac.uk

M. L. S. Josephine
Department of Management, University of Botswana, Gaborone, Botswana
e-mail: moeti-lysson@ub.ac.bw

J. Mbudaya
BA ISAGO University, Gaborone, Botswana
e-mail: jussymbudaya@gmail.com

J. Kanyepe and I. Chirisa (eds.), *Advancing the Green Agenda in the Transport and Logistics Sector in Global South Countries*, World Sustainability Series,
https://doi.org/10.1007/978-3-032-31374-4_11

policies; investing in sustainable infrastructure; increasing capacities among stakeholders; engaging with stakeholders; utilizing new technologies such as digital technology to improve the efficiency of transportation systems. This chapter will also highlight the significance of using tools such as scenario planning and systems thinking in order to manage complexities associated with transitions and assist decision makers with making decisions over the long term. Through the synthesis of theoretical concepts and practical examples, this chapter provides additional insight into how developing countries may address the challenges associated with green transportation transitions.

Keywords Green transport systems · Adaptive strategies · Developing countries · Transport barriers · Sustainable mobility

1 Introduction

The increasing trend of moving toward sustainable transportation has created an urgent need to develop alternative, environmentally friendly and resource-efficient transportation methods. Transportation continues to be one of the largest contributors to greenhouse gases (GHG), air pollution and energy usage globally (Ghose et al. 2024). Moreover, as many developing countries experience rapid urbanization, population increases and expanding economies continue to create increased demands for mobility; this will make it even more difficult to adopt alternative forms of transportation such as low emission vehicles, non-motorized transportation and combined public transit. As a result, transforming transportation systems from green transportation systems is now a high-priority objective of policy makers and other stakeholders working to achieve a balance between economic growth and environmental sustainability (Achillopoulou et al. 2024).

Although the necessity of adopting green transportation systems in developing countries cannot be overstated, the transformation to green transportation systems is fraught with numerous complex obstacles stemming from structural, institutional, and social behaviours. Most cities have poor infrastructure, limited financing capabilities and technological restrictions that limit their ability to implement sustainable transportation options (Kozlov et al 2024). Additionally, inconsistent institutional arrangements, conflicting policies and fractured government structures can inhibit coordination among stakeholders. At the community level, widespread resistance to behaviour modification, low awareness of environmental issues and dependency on traditional transportation modes also hinder the transition process. As a result, there is increasingly a consensus that traditional linear policies are no longer adequate to deal with the dynamic and uncertain aspects of transitioning to new forms of transportation (Chibaro et al. 2024).

Therefore, flexible strategies that involve participation by various stakeholders and utilize systems thinking are necessary. In order to identify and mitigate the multi-dimensional barriers to green transportation system adoption, this chapter outlines multiple adaptive strategies which can help manage uncertainty while promoting a more robust and equitable transition to sustainable transportation systems in developing nations (Khurshid et al. 2023).

2 Background to the Study

The transport sector has become a key component of the overall economy of all nations (enabling the exchange of goods, services and people) in an increasing number of interdependent national economies; however, it is also the most polluting sector globally accounting for large percentages of global greenhouse gas emissions & urban air quality problems (Ardini, and Fahlevi 2024). The UNFCCC and IPCC have continuously stressed that "the transport system" needs to be rapidly decarbonised as part of the global effort to reduce the negative impacts of climate change. As a result, several developed nation-states have increased investment in green transport options, which include electric vehicles, public transportation, pedestrian/bicycle-friendly streetscapes. These projects are generally supported by institutions with the capacity to implement new technology; well-financed; and technologically advanced (Ardini, and Fahlevi 2024).

In addition to the aforementioned obstacles (policy/infrastructure), a growing number of factors are influencing the way in which the world's transport sectors will transition from fossil fuel-based systems to green/sustainable systems through technological innovation and digitalization (Kozlov et al. 2024). Digital technologies (such as smart transportation systems; intelligent transportation systems; data-based traffic management systems; Mobility-As-A-Service Platforms) are changing how transport systems are planned and implemented (Popkova et al. 2026). In industrialized regions, these technologies enable citizens to live within increasingly efficient, environmentally-sound mobility ecosystems (Chibaro et al. 2026a). Unfortunately, there are barriers preventing many developing country governments from taking advantage of new innovative digital solutions. These barriers include lack of adequate digital infrastructure; high costs to deploy the technologies; and a scarcity of skills needed to implement them (Banister 2008; Geels 2012; ITF 2021).

Developing nations face an even greater challenge with regard to their own transitional landscapes due to conflicting priorities in the pursuit of their respective development needs, and resource restrictions. Urbanization and population increases are placing a higher burden upon transportation demands that is far surpassing the ability of current infrastructures to meet these demands. As a result, many cities within Africa, Asia and Latin America are experiencing significant levels of congestion, inefficient public transportation systems and excessive dependence upon fossil fuel-based methods of transportation. The need for sustainable transportation systems to foster both inclusive growth and regional coordination has been recognized at a continental level (African Union), but the degree to which this is achieved continues to be inconsistent. Constraints associated with financial resources, inadequate regulatory frameworks and limited access to environmentally friendly technologies remain barriers to continued progress throughout much of the world (Chibaro et al. 2026a).

An additional key element concerns financing models and investment priorities. A large amount of funding is available worldwide to support green transitions via climate finance programs sponsored by institutions such as the World Bank and the Green Climate Fund. These organizations also perform important roles in providing capacity-building assistance and enabling policy reforms in developing countries regarding infrastructure development (Ruiz et al. 2023). However, developing country's access to such funding is typically restricted by complexity associated with applying for funding, limited institutional capability to process funding applications and competing national priorities. Many African nations, including some in southern Africa, continue to invest primarily in expanding roadway infrastructure versus investing in alternative forms of sustainable transportation because of immediate economic interests rather than long term environmental concerns (World Bank 2022; GCF 2020; AfDB 2019).

The focus on Southern Africa will further emphasize the challenges facing transportation planning due to current spatial plans which generally favour private vehicles and invest very little in coordinated public transportation systems (Chikwere et al. 2024). The cities of Harare, Gaborone and others have become increasingly congested; emissions have risen, and their infrastructure has reached its maximum carrying capacity as the population continues to grow. Although national policy recognizes the importance of sustainability, there remain significant limitations to the implementation of green transportation (Ogola 2021). Therefore, in order to create context-sensitive, adaptable strategies for transitioning from unsustainable to sustainable transportation systems within the region, it is necessary to develop structural efficiencies (and address them), improve institutional capacity (in order to implement) and encourage behaviour modification (so that individuals will be willing to make changes).

In addition to infrastructural and policy-based barriers to the development of green transportation systems around the world, advancements in digital technology and innovative uses of technology are creating new avenues for shaping how green transportation systems are planned and implemented (Brears 2023). "Smart" mobility solutions based upon intelligent transportation systems, data-driven traffic management systems and mobility-as-a-service platforms are changing the way that transportation systems are planned and delivered. Developed regions are using these innovations to develop more efficient, lower carbon footprint mobility systems (Yin et al. 2022). However, in developing regions, the inability to implement these same types of systems is caused by insufficient digital infrastructure, high cost of implementing the required digital infrastructure, and/or the absence of skilled workers to operate them. This creates a "digital divide" between developed and developing regions that makes it more difficult for developing regions to effectively utilize newer emerging mobility solutions when they transition to sustainable transportation systems (Chibaro et al. 2026b).

A final important aspect is related to funding mechanisms and investment priority. A large amount of money is being raised globally to fund transitions to green transportation through climate finance initiatives promoted by organizations such as the World Bank and the Green Climate Fund. These organizations provide significant assistance in developing country's infrastructure development, institution-building, and policy reform. However, access to this type of funding is typically restricted by the complexity of applying for funding, limited ability of institutions in developing countries to apply for funding, and other national priorities. In several African countries including those in southern Africa even though there are a number of climate change funds available to assist developing countries transition to cleaner energy sources, transportation investments continue to give priority to expanding roads rather than developing alternative environmentally friendly transportation options. As is typical of many developing countries, this reflects a preference for short-term economic benefits over long-term environmental benefits (Chibaro et al. 2026b).

Regional inequities in terms of income and geography further complicate the transition to green transportation. In many urban and peri-urban areas, there is no dependable or affordable public transportation option (Chikwere et al. 2025). Thus, people living in those areas are forced to rely on informal/unorganized transportation options that are often inefficient and polluting. In addition to contributing to increased environmental problems, informal/ unorganized transportation options contribute to social inequality by limiting opportunities for work and economic advancement (Li et al. 2026). In many developing countries, the prevalence of informal/unorganized transportation options coupled with weak regulatory oversight present both an opportunity and a challenge for reform. To achieve greater equity and inclusion in green transportation transitions, it is essential to employ integrated planning approaches that link transportation planning with broader socio-economic objectives (Cervero 2013; Schalekamp and Behrens 2010; SLOCAT 2021).

Research Objectives

a. To examine the structural barriers that hinder the adoption of green transport systems in developing countries.
b. To analyze institutional barriers affecting the transition to sustainable transport systems.
c. To investigate behavioral barriers, focusing on public attitudes, mobility preferences, and levels of environmental awareness influencing the uptake of green transport solutions.
d. To develop and evaluate adaptive strategies that can effectively address identified barriers and manage uncertainties in transitioning toward green transport systems in developing country contexts.

3 Research Methodology

This chapter utilized a qualitative desk review research methodology to conduct an examination of the barriers to green transportation systems transition and adaptive strategies to manage complexity associated with green transportation system transition in developing countries. The research methodology utilized secondary data analysis (data drawn from peer-reviewed journal articles, policy documents, institutional reports and development publications) and examines the literature from global organizations (World Bank, United Nations, International Transport Forum), regional sources (African and Southern Africa specific literature) and other relevant sources. The research methodology is suited for this study due to the exploratory/interpretative nature of the study; therefore, no primary empirical data will be generated through the study.

Following the research design, the literature review utilized a structured approach (literature identification, selection and critical evaluation) to identify and select relevant literature. Based upon the thematic relevance, methodological rigor and recency criteria the selected literature was included within the analysis to reflect contemporary sustainable transportation debates and practices. Using the thematic analysis technique, the findings were categorized into structural, institutional and behavioral barriers to green transportation systems transition and corresponding adaptive strategies. Through the use of this analytical technique, patterns, gaps and emerging insights within diverse contexts were identified.

Additionally, the research methodology employs a system thinking perspective to understand interdependencies between different types of barriers and policy responses. Through the integration of insights derived from multiple sources, the study generated a holistic understanding of the transition process. Although the reliance upon existing data may limit the desk review methodology, it provides a cost-effective means to generate comprehensive and policy-relevant insights into green transportation transitions in developing countries.

4 Literature Review

Green transportation systems and adaptive strategies for transitioning over green barriers have grown as concern about climate change, environmental sustainability, urban traffic jams and energy security continue to increase (Nikolova 2025). The study examined the role of green transport systems (electric vehicles, modernization public transportation, non-motorized transportation and renewable-energy powered mobility) in reducing carbon emissions and supporting sustainable development. Furthermore, the study identified the major obstacles that prevent the use of green transportation systems including lack of infrastructure; high cost of investment; limitations in technology; policy inconsistency; institutional resistance; and low levels of acceptance by the public. Furthermore, this chapter also discussed the adaptative approaches that governments, agencies responsible for transportation and private stakeholders can take to overcome these challenges such as policy reform, financial incentives, public-public partnership, innovation assistance, stakeholder engagement and smart transport technologies.

5 Green Transport Systems

Green transportation systems are networks for moving people and goods using environmentally friendly methods. Green transportation systems aim to reduce pollution, ghg emissions from fuel use, traffic congestion and reliance on fossil fuels while providing economic growth and efficient travel. Examples of green transportation systems include electric vehicles, renewable energy powered public transportation, bike lanes, rails, pedestrian paths and smart transportation technologies that provide increased efficiency through reduced energy consumption and damage to the environment. Green transportation systems are becoming more important because of rising concerns over climate change, urbanization and air quality impacts on human health and ecosystems (Rehman et al. 2023a). As a result, governments around the world are implementing policies and plans to promote low carbon modes of travel, invest in clean energy infrastructure and develop plans for sustainable urban transportation. However, despite the many positive attributes of green transportation systems there remain significant obstacles to transitioning to green transportation systems, such as high costs for implementation, lack of supporting infrastructure, technical limitations, institutional opposition and low levels of awareness by the general public. In order to overcome these barriers, additional supportive strategies will need to be employed. The key supportive strategies needed to support successful and sustainable transformations to green transportation systems include policy reform, public–private partnerships, financial incentives, stakeholder involvement and support for innovation (Bešinović 2020).

6 Transport Systems

Transport systems represent an interconnected network of transportation infrastructure, technologies, institutions and services used to move individuals, goods and/or information from point A to B. The components of transport systems consist of railway, air transport, marine, pipeline, and land based public transit systems. All elements of these systems support the goals of economic development, trade, tourism, industrialization, and social interactions (Mezhokh and Smirnova 2023). The betterment of our transportation system through technology will be able to provide greater access to locations, less time spent traveling, thus improving the connection between markets within a country or region and enhancing economic development at both the national and international level. In addition to being a key component in facilitating economic development, modern transportation systems are increasingly using advanced technologies (such as intelligent transportation systems (ITS), digital logistics platforms (DLP) and smart mobility solutions (SMS)) to increase efficiency, safety and sustainability of the transport system (Song and Lu 2026). On the other hand, there are many challenges facing transportation systems today. Some examples of these challenges include traffic congestion, environmental degradation due to pollution caused by vehicles and infrastructure deterioration. Additionally, increased cost of fuel for vehicles and rapid urban population growth have created additional burdens on transportation systems. Consequently, government agencies and policy makers are working towards developing a long-term strategy for sustainable and resilient transportation planning with the goal of increasing mobility and reducing the negative environmental impact associated with transportation (Song and Lu 2026). Therefore, the importance of developing effective transportation systems cannot be overstated in order to achieve social inclusion and economic competitiveness while meeting sustainable development objectives globally (Rodrigue 2024).

7 Adaptive Strategies for Transport Systems

Transportation systems' adaptive responses (strategies) refer to intentional plans or policies and/or new ways that transportation systems will be able to respond appropriately to environmental, technological, social, and economic changes. Adaptive strategies become increasingly important as our transportation systems experience increasing challenges due to a rapidly changing environment (climate change), rapid urbanization, increasing levels of traffic congestion, increased fuel prices, continued environmental degradation and an increase in various forms of mobility (Bauer, and Titz 2025). The ultimate goal of these strategies is to increase the ability of transportation systems to be resilient, sustainable, efficient and flexible with regard to providing reliable mobility for both people and freight. Strategies have included investing in "green" infrastructure; adopting Intelligent Transportation Systems (ITS); promoting public transportation options; integrating renewable energy into the transportation system; developing alternative modes of transportation to motor vehicles; and implementing supportive policies and regulatory environments. Collaboration among stakeholders, data driven decision making and on-going innovation are all key elements necessary for successfully adapting to change throughout the transportation industry. Governments and transportation planners who adopt adaptive strategies will be able to address uncertainty and remove transition barriers associated with creating a sustainable transportation system which will provide a foundation for long-term economic growth, environmental protection and societal welfare (Lomotko et al. 2026).

8 Theoretical Perspectives

8.1 Sustainability Theory

The theory of sustainability is a way of thinking about how we can responsibly use resources today so that our children and grandchildren will be able to meet their own needs (Brundtland 1987). This concept of sustainability is based on the idea of sustainable development. Sustainable Development means balancing three key areas (environmental protection, economic growth and social equity) so that there is a fair distribution of the benefits of development. Sustainability Theory believes that if we want to have long term sustainable development then we need to find ways to manage all three areas equally well. For example, with regard to sustainable transport systems this means developing green forms of mobility that create less greenhouse gases through reduced carbon emissions, reduced air pollutants, lower levels of energy consumption and less traffic congestion. Examples include electric cars, public transport, cycle paths, or using renewable energy sources to power your vehicle. The theory also encourages Governments, Organizations and Communities to develop policies and practices that promote resource efficiency, protect the environment and encourage economic development in an inclusive manner.

Sustainability Theory is applicable to transport planning and policy development because green transportation systems seek to mitigate the environmental and social impact of conventional transport systems. Traditional transport systems heavily use fossil fuels, contributing to greenhouse gas emissions, climate change and environment degradation, and need implement adaptive strategies to achieve a low-carbon mobility and energy usage, and more broadly sustainable urban development. Such strategies include investments in mass public transport, smart transport technologies, promoting non-motorized transport and including environmental regulation to help reduce pollution levels from transport. The sustainability theory is more recent, highlighting the importance of stakeholder participation, institution support and long-term policy commitment to moving along the sustainable transport pathway. Moreover, the more sustainable transport systems are potentially healthier for public health purposes, better for business economic productivity, and contributing to social inclusion by providing safer and cleaner transport options (Rehman et al. 2023b). Thus, sustainability theory plays a front-line role in most discussions about green transport systems that support the planet.

Sustainability Theory is relevant to this study because it demonstrates the rationale for green transport systems and the need to develop them, and the principles of what constitutes a sustainable transport system. The basis is shown in term to be the movement of transport systems away from fossil fuel use and to systems like more electric vehicle transport, public transport and non-motorized transport. The high-cost barrier is also one identified for analysis in transport. Specifically, it highlights the potential transport systems that support to analyse (green policies, investments in clean transport technology) and works toward a more sustainable transport reform.

9 Transition Theory (Socio-Technical Transition Theory)

Transition Theory, or Socio-Technical Transition Theory, describes how societies transition from old-fashioned systems and technologies to new systems and technologies that promote sustainability (Schot 1998). Transition theory examines the relationships between technological advancements, institutions, governmental policies, market systems, infrastructure, and human behaviours to describe how mass scale societal transitions take place. Most often, Transition Theory is applied to areas like transportation; Transition Theory will explain how our transportation systems go from being based solely on fossil fuels to the use of alternative methods of transportation like electric cars, buses powered by solar energy, smart traffic systems, and improved public transportation. According to transition theory, the process of transitioning does not happen overnight. Rather than occurring quickly, transitional processes evolve slowly due to the introduction of new ideas and experiments, policy changes to provide support for the new ideas, and people accepting the new ideas (Schot 1998).

An important aspect of transition theory is the "Multi-Level Perspective" or MLP (Geels 2002). This perspective is made up of three main levels. They consist of; niche innovations, socio-technical regimes, and landscape pressures (Geels 2002). Each level influences each other and their relationship affects the way in which sustainable transportation innovations come into existence, replace current unsustainable transportation systems, and eventually become part of the mainstream transportation systems. Transition theory states that many of these sustainable transportation transitions are hindered by various obstacles. Some of those obstacles include technological uncertainty, insufficient infrastructure, opposition to change from current institutions and organizations, limited finances available to make the transition, and public unwillingness to adapt to new forms of transportation. Transportation systems have been built around many decades of established ways of doing things. Those ways include many aspects of economics, government regulations, business interests, and consumer preferences. As a result, change takes time and can be difficult. For that reason, transition theory emphasizes the need for flexible approaches that allow for innovation and the ability to adapt while changing the entire system. Flexible approaches may include offering government incentives to encourage innovation, investing in R&D, conducting pilot projects to test new transportation concepts, forming public private partnerships to develop new transportation systems, creating favourable laws and regulations that permit the creation of new sustainable transportation systems, and working together with stakeholders who have an interest in developing new transportation systems (Geels 2002).

In addition to recognizing the importance of using flexible approaches when trying to create a new transportation system, Transition theory also acknowledges that there are external pressures that exist outside of any organization or institution that can help accelerate the transition to using environmentally friendly modes of transportation. Examples of external pressures include climate change, increased cost of fossil fuels due to higher fuel costs, environmental regulations requiring greater reductions in greenhouse gas emissions, and increased congestion in urban areas (Bauer and Titz 2025). By identifying the interplay between technological innovations and the social and institutional environments in which they operate, Transition theory offers a theoretical lens through which one can view how new environmentally friendly transportation systems are developed and implemented within contemporary societies.

Transition Theory is applicable to this chapter for two reasons. First, Transition Theory identifies how transportation systems transition from being conventional fossil-fuel-based systems of transportation to being green/sustainable transportation systems utilizing technologies that are less damaging to the environment. Transition theory demonstrates the challenges to transitioning from conventional fossil-fuel-based systems of transportation to green/sustainable systems. Challenges include institutional resistance; lack of adequate supporting infrastructure; lack of sufficient funds; and lower public acceptance of new green technologies for transportation. Transition theory provides a conceptual framework for evaluating flexible strategies that will facilitate successful transitions in transportation systems. Strategies that could be evaluated include government incentives for innovation; collaborative efforts among stakeholders interested in developing green/transportation technologies; investments in sustainable infrastructure; and support for innovation in developing sustainable green/transportation systems. Transition theory will assist the chapter in understanding how technology interacts with policy; institutions; and society during transitions in transportation systems.

10 Innovation Diffusion Theory

The key concept of the innovation diffusion theory is that all new ideas, products and services have to be diffused in a certain way before they can be utilized widely. This process of diffusion takes place through interaction (communication) among persons and social units with regard to their perception of the utility or value of the innovation; its compatibility or fit with existing values and past experiences; the degree to which the innovation is perceived as complex; the degree to which the results of adopting the innovation are observable; and the extent to which users can test the innovation before making a decision regarding its use. According to Rogers (2014), there are five categories of adopter types: innovators, early adopters, early majority, late majority and laggards. Innovators are the first group of adopters. They tend to be venturesome, open-minded, flexible and willing to take risks. Early adopters follow shortly after innovators. These individuals are opinion leaders who help other members of their community make decisions about whether or not to try something new. The early majority represent the next segment of the population. As they begin to see widespread support for an innovation, they become more confident in trying it out. The late majority are somewhat sceptical and wait until most of their friends have already tried something. Laggards are the last group to accept an innovation. They often prefer traditional ways of doing things and may be hesitant to change.

The innovation diffusion theory has been applied to many fields. One example of applying this theory is in the area of transportation systems. In recent years, transportation agencies around the world have implemented various "green" technologies into their systems. Green technology includes alternative fuel vehicles such as hybrid buses and light rail, bicycles, scooters, car-sharing programs, and ride sharing apps. These green technologies were initially met with resistance from citizens and businesses alike. However, once these technologies became available in larger numbers and were seen being used successfully by other cities and states, more people began to consider using them. For instance, in California, where traffic congestion is extremely high, cities like Los Angeles and San Francisco began implementing bike-share programs and other alternatives to driving. Cities like Chicago, Illinois also started investing heavily in green transportation technologies.

Transportation agencies recognize that a major factor in the success of any innovation is the willingness of individuals to try out something new. According to Rogers (2014), social learning is one of the main influences on whether or not someone will accept a new idea. Social learning involves observing what others do and thinking that if everyone else does it then I should too. Therefore, when a city or state begins to implement a new form of transportation, the agency responsible for overseeing it must also focus on creating awareness about the existence of this service. Public awareness campaigns are essential to helping create demand for a new form of transportation. If people don't know that options exist, then they will continue to rely on driving alone.

11 Findings

11.1 Structural Barriers to Green Transport Adoption

Structural impediments to the adoption of green transportation are a subset of the physical, technological, and infrastructure-based impediments that inhibit the widespread deployment and scale-up of green transportation systems in developing nations. A key barrier to the widespread adoption of sustainable transportation is an inadequately developed transportation infrastructure base. This includes poorly constructed or underdeveloped roadways; underdeveloped rail systems; and inadequate numbers of public transportation services. For example, in both urban and rural settings, many communities have not implemented sufficient dedicated infrastructure for example bike paths, pedestrian friendly walkways to accommodate green transportation alternatives for example electric vehicle charging stations (Wang et al. 2023). Therefore, it is very challenging for individuals to implement alternative forms of sustainable transportation within their community due to the existing structure of the community's transportation system. Also, an unreliable source of electrical power limits the ability of communities to deploy and maintain the necessary renewable energy required to support the deployment of electric and other low emission transportation alternatives. High levels of capital investment required to construct new transportation infrastructure represent another significant barrier to the transition toward more environmentally sustainable forms of transportation for governments with constrained financial resources. As a result, these structural constraints limit the pace at which developing nations can transition away from traditional forms of transportation toward green, more sustainable alternatives (Rodrigue 2024).

The second major component of structural impediments is inadequate land-use planning practices. Developing countries have experienced rapid rates of urbanization; however, they have failed to invest in well-coordinated, multi-modal transportation systems (Karam et al 2023). As a result, cities are plagued by traffic congestion; excessive land consumption through urban sprawl; and an over-reliance on private automobiles and unregulated forms of informal transportation. Multi-mode transportation systems do not exist in most developing country cities, making it difficult for travellers to easily move among different types of green transportation for example bus, train, cycling. Moreover, older transportation technology; older vehicle stock; and continued reliance on non-sustainable sources of energy for transportation purposes continue to make travel less efficient and generate more air pollutants than is desirable (Karam et al 2023). An additional constraint on the ability to establish sustainable transportation systems is a lack of technical capacity needed to design, build, and maintain sophisticated green transportation infrastructure. Finally, inadequate data collection capabilities and/or inability to apply appropriate computer modelling tools and other analytical techniques to inform decisions regarding how best to plan and implement sustainable transportation strategies further complicate efforts to develop sustainable transportation systems (Golnar and Beškovnik 2020). Collectively, all these structural impediments present numerous challenges to deploying sustainable transportation systems in developing countries and will likely require considerable investments of money, innovation, and time to successfully address (Bogadi 2022).

12 Institutional Barriers Affecting Transition

Institutional obstacles hindering the transition to green transportation systems in developing nations are largely caused by poor governance, institutional fragmentation, and a lack of coordination between key stakeholders involved in the transportation sector. In many developing nations, responsibilities for both transportation planning and implementation are distributed amongst several ministries, municipal governments, and other agencies resulting in duplicated effort, inefficiency, and inconsistencies in policy. Lack of coordination hinders the development and execution of cohesive sustainable transportation strategies (Tazzie et al. 2025). Weak or absent regulatory frameworks along with ineffective means of enforcing environmental regulations limit the extent to which environmental regulations can be followed; therefore, limiting the rate at which green mobility initiatives are adopted. Institutional resistance to change ("institutional inertia") slows the pace at which innovations occur in transportation systems. Additionally, corruption and/or mismanagement of funds diverts resources from infrastructure and sustainability projects. All these institutional issues prevent an effective transition from traditional transportation systems to more sustainable and environmentally conscious alternative transportation options.

Another obstacle to green transportation transitions is a lack of long-term commitments to green transportation initiatives. Frequent governmental priority changes result in inconsistent transportation policies, which make it difficult to sustainably develop and implement long term infrastructure and innovation projects. Developing country institutions limited institutional capacity for example lacking skilled employees and/or technological capabilities limits their ability to plan, implement, and monitor sustainable transportation systems. A lack of sufficient funding for green transportation projects is indicative of weak prioritization of this type of initiative in developing nation development agendas. Also, developing countries frequently have ineffective monitoring and evaluation processes to measure progress and hold accountable individuals involved in the reforming process of transportation sectors. Inadequate incentives for public private partnerships hinder the opportunity for developing country institutions to invest in innovative sustainable mobility solutions. In many developing countries, outdated institutional frameworks fail to include new technology applications for example electric vehicles and intelligent transportation systems. Therefore, all these institutional issues combine to create a significant barrier to transitioning to sustainable transportation systems and require comprehensive reform of governance systems to address them adequately.

13 Behavioural Barriers Influencing the Uptake of Green Transport Systems

Behavioral constraints driving the acceptance of green transportation systems in developing countries stem from consumer behaviour based on public attitudes, beliefs and established practices concerning personal mobility. Consumers favour traditional modes of transportation including private automobiles, minibus services and motorbikes because they perceive them to be convenient, comfortable, and flexible. These preferences are frequently supported by years of habituated travel behaviour that can only be changed with great difficulty despite availability of alternative sustainable transportation options. Additionally, consumers have little understanding of the benefits of green transportation systems which include reducing air pollution and improving public health through reduction of emissions while providing long term savings (Postorino et al. 2021). Users who choose to use green transportation believe these modes are unreliable and/or take longer than using their own vehicle (Musau 2023). Consumer perceptions and attitudes towards sustainability will inhibit changes in behaviour to support sustainable mobility. Attitudes and beliefs related to social norms and culture continue to prevent consumers from adopting new ways of transporting themselves.

A significant behavioral constraint is the consumer's fear of the unknown or uncertainty related to new technology in the form of electric vehicles and other forms of green transportation. For example, many consumers do not want to purchase an electric vehicle because they are concerned with having a sufficient supply of electricity for charging, longevity of the batteries, the potential for increased maintenance costs and whether the technology is reliable. Lacking first hand experience with alternative sustainable transportation modes creates scepticism and reduces willingness to make changes. Financial behaviour is also relevant since some consumers value short term affordability over long term environmental and economic advantages. As well as the lack of familiarity with educational programs and lack of education regarding environmental issues results in decreased motivation to alter behaviours. Consumers' psychological opposition to change along with continued reliance upon traditional transportation modes continues to reinforce this dependency. Finally, social inequality affects accessibility to green transportation as well as excluding lower income populations from utilizing new technology. Together these behavioral factors create a significant obstacle to increasing the rate at which green transportation systems are being accepted and therefore necessitate the development of specific strategies focused on raising awareness and encouraging incentives for education.

14 Adaptive Strategies for Managing Uncertainties in Transitioning Toward Green Transport Systems

Adaptive approaches to address barriers to transitioning to green transportation in developing countries include a combination of policy, institutional, financial, and technological initiatives. These initiatives aim to overcome structural, institutional, and behavioral barriers. Several different investments in sustainable transportation infrastructure provide one example of how governments may implement their green transportation policies (Poli and Imbesi 2022). Examples of this type of infrastructure include installing electric vehicle charging stations; investing in improved public transportation systems (e.g. buses); creating bike-lane networks; and designing cities to be pedestrian friendly. Governments may also implement policies to encourage the purchase of green vehicles by providing subsidies or tax incentives. Reducing import duties on electric vehicles and renewable energy technologies may also be used to lower costs associated with purchasing these types of vehicles. The success of these types of initiatives relies heavily upon successful public private partnership arrangements which facilitate access to both the necessary financial and technical resources required to implement large scale changes to a nation's transportation network. Finally, implementing integrated urban and transportation plans will ensure that the way land is developed and utilized to develop efficient and low carbon transportation networks (Zhao and Kan 2024).

An additional area of focus includes institutional and behavioral transformations to establish long term sustainability within a nation's transportation system. Public education campaigns to raise awareness regarding the impact of green transportation options may aid in altering users' perceptions and increase acceptance of alternative modes of transportation. Community involvement and stakeholder engagement aids in building trust among stakeholders and provides assurance that users' needs and preferences are being reflected in all levels of government decisions regarding transportation. Enhancing capacity building within the agencies responsible for implementing transportation plans increases technical capability, thereby enhancing agencies' ability to plan, monitor and evaluate sustainable transportation projects. Utilization of digital technology for example intelligent transportation systems, data-driven decision-making tools enhance operational efficiency, improve response time and increase utilization rates. Furthermore, establishing a culture that promotes innovative research and development facilitates the creation of localized green transportation solutions that meet the unique requirements of developing nations. Implementing organizational restructuring efforts, improving interagency coordination, and ensuring consistent application of policies across various agencies assist in ensuring that there is no lack of commitment towards achieving long term sustainability in all aspects of a nation's transportation system.

15 Discussion

The results show that structural barriers in developing countries are due to inefficient systems in transportation and urban development (not just lacking infrastructure) and so they can't be resolved by simply building new roads or upgrading old ones. Additionally, past investments in transportation have led to a history of dependence on fossil fuels and have limited options for alternative modes of transportation. As a result, there exists "path dependency" in which current decisions are made based upon what already exists and will continue to limit the ability to rapidly move away from traditional fuel-based transportation methods. Planning policies related to land use have further contributed to an increased need to travel; residential areas, commercial/industrial districts, etc., are planned separately. This contributes to an increasing reliance on private and informal modes of transportation. Therefore, it appears that these structural barriers represent much more than simply a limitation of technology. Instead, they appear to be directly influenced by long-range planning philosophies and how money is invested. This supports the broader body of literature concerning sustainability transitions that states that "lock-in," specifically to the physical components of a system's infrastructure, acts as a significant constraint to the speed at which systemic changes occur in mobility systems.

Findings regarding institutions illustrate that fragmentation in governance systems play a large role in impeding the movement toward sustainable modes of transportation. In many cases multiple organizations participate in making plans for transportation. With this comes overlapping responsibilities among those organizations and inconsistencies in implementing policies and guidelines. Also, since there exist no clear lines of authority, the accountability processes for effective implementation of environmental regulations and scaling up green transportation programs are very weak. The study shows that policy uncertainty and lack of stability also greatly diminishes investor confidence especially when investing in high-cost projects like electric buses and smart mobility solutions (Popkova et al. 2026). When institutional coordination is poor, policy signalling becomes ambiguous causing investors to delay adopting environmentally conscious technologies. Finally, the study found that institutional capacity also affects not only the ability to implement but also the ability to assess and adapt their efforts over time. Thus, creating a cycle where weaknesses in planning lead to failure in implementation.

The behavioral aspects of the data show that users' choice of transportation is primarily influenced by what they perceive as useful (reliable, timely, and cost predictable) rather than by environmental factors. Therefore, users do not consider environmental concerns when making choices related to their daily commute. Furthermore, even if there are "green" transportation options available, users will generally be reluctant to adopt them due to uncertainty regarding service quality and performance. A second aspect of the study shows that users' trust in transportation systems is an important factor for widespread use; thus, new technologies are often seen as unproven or high-risk. Such perceptions create a gap between when these technologies become available and when people begin using them. Finally, although there may be some degree of social learning among people who have adopted green transportation, this process appears to occur slowly at best because there is relatively little peer pressure experienced by those who choose traditional forms of transportation. Thus, it appears that the primary barrier to adopting green transportation is not the idea of being sustainable but whether one can manage the risks associated with changing their usual transportation habits.

Another way to interpret the data indicates that both financing arrangements and the structure of markets play an important role in determining how quickly green transportation transitions can occur. In many developing countries, lack of accessible affordable long-term capital restricts government and private investments in sustainable transportation infrastructure. Commercial banks and development finance organizations also view investment opportunities in green transportation as too great a risk, resulting in very few loans to support green transportation projects. Lack of local manufacturing capacity and supply chain capability for key components such as batteries, charging equipment, and intelligent transportation technology makes it necessary for most countries to rely on imported products and raises project costs (Chinomona and Mahlatsi 2020). Consequently, green transportation initiatives are frequently funded through grants from donors and pilots and are rarely scalable.

16 Implications

Below are the study's theoretical, policy, practical and economic implications.

17 Theoretical Implications

The results of this research have significant theory-building implications for both Transition Theory and Sustainability Theory by supporting the claim that sustainable transportation transitions are always multi-dimensional and never linear. This data demonstrates that an objective related to environment will only produce systemic change when it is coupled with structural and institutional alignment. This extends Sustainability Theory by demonstrating that environmental, economic and social dimensions can only be successfully integrated through institutional and infrastructural coherency. In addition, the data supports Transition Theory by illustrating how path dependency (and the associated regime stability) influences the development of mobility systems. That is, the presence of 'green' alternatives has little impact on conventional forms of mobility; therefore, socio-technical regimes appear to be much more stable than theorists assume. Therefore, these results support the notion that transition processes take longer time-horizon and/or greater disrupting force than many theorists have suggested; particularly in developing country contexts where the problem of infrastructure lock-in is most pronounced.

Furthermore, the results provide additional contributions to Innovation Diffusion Theory and Behavioral Theory by identifying two of the primary limitations to awareness driven and informational models of innovation adoption. Specifically, while prior diffusion theories emphasize knowledge, persuasion and social influence as the main driving forces behind innovation adoption; however, the current research highlights that there exist other types of structural and institutional impediments to innovation adoption that can serve to constrain or inhibit behavioral response to innovations. These impediments include physical access to alternative technologies, reliability of services offered by alternative technologies, and levels of trust in institutions that offer alternative technologies. Overall, this indicates that innovation adoption is influenced by more than just the individual's decision-making process, but also by other factors outside of their control, i.e., infrastructure availability, etc. Thus, this research provides evidence suggesting that Innovation Diffusion Theory should consider other contextually specific impediments that affect the relationship between awareness/acceptance and actual behaviour. Finally, the results from this research also provide further empirical support for Behavioral Theory. Specifically, it was found that perceived risks, functional utility, and environmental motivation were all ranked high in terms of influencing respondents' decisions regarding their mode(s) of transportation. However, the relative ranking of these factors indicated that perceived risks and functional utility far exceeded respondents' environmental motivations. This provided further evidence for a more complete understanding of respondent's travel behaviour within developing countries. It appears that developing countries have different priorities (practical vs. normative values) compared to developed countries in terms of what influences respondent's travel choices.

18 Policy Implications

The results of the study have considerable impacts on government policy regarding developing a transition to green transportation systems in developing nations. Firstly, it shows that there is a need for long term and holistic transportation policy to develop from infrastructure projects. The long-term transportation policy needs to be developed by all departments of the government and include developing an overall national transportation plan which develops an integration of local city planning, energy planning, and environmental planning. It would include establishing reliable power generation and distribution systems throughout the country, increase public transit services (buses, trains), and creating specific infrastructure for electric vehicles and bicycles/non motorized vehicles. Fiscal policy could provide incentives to purchase green vehicle technology via subsidies/tax credits/reduced tariffs for importing green vehicle technology to make these vehicles affordable to consumers and generate interest in them. Government departmental cooperation to ensure consistent pursuit of sustainability objectives is needed to prevent disjointedness within government policies. Creating a strong regulatory framework along with implementing enforcement of regulations will help ensure environmental regulations are complied with and assist in the gradual reduction/elimination of polluting transportation modes.

The study indicates that behavioral and institutional changes in transportation policy are required in conjunction with physical transportation investments. Public outreach programs, environmental education, and programs promoting behaviour change to alter consumer preference towards use of environmentally friendly transportation modes are essential. Building consensus through inclusive decision-making processes amongst stakeholders, private industry, and community groups is essential in order to build confidence and demonstrate that policies are relevant. Development of new governance structures to facilitate communication amongst various transportation related departments/agencies and increase their ability to plan, implement, monitor, and evaluate alternative forms of transportation is essential. Creation of innovation clusters through grants/funding for R&D, pilot projects/public–private partnerships to test/emphasize emerging environmentally friendly transportation technologies is important to promote and advance knowledge creation for environmentally friendly transportation systems. Developing data driven methods for analysing transportation patterns/intelligent transportation systems will allow for better decision making and improved service delivery.

19 Managerial Implications

At the managerial level, the results show that managers of transportation systems have a greater need to transition away from traditional operational methods toward a more strategic and systemic approach to management. Long-term planning of transportation assets and life cycle management of transportation infrastructure appear to be the primary focus of the managerial efforts due to ongoing and persistent structural and institutional limitations. To provide reliable and consistent service to users, it is important for managers to develop an integrated plan for maintaining fleets, charging equipment and transportation systems. Managers may want to develop processes that foster cooperation and coordination between different functional areas within their organizations. Additionally, performance measurement systems need to be enhanced so they include metrics related to sustainability (i.e., emission reductions, use of renewable energy, service access). The goal would be to enable decisions made at the managerial level to better align with the organization's broader sustainability and developmental objectives. In addition, the data indicate that successful operation of green transportation systems will require managers to engage in more anticipatory planning and to be able to effectively manage uncertainty during periods of technological and infrastructural transition.

Additionally, the results indicate that transportation managers need to make significant improvements in stakeholder outreach and communications to increase both the rate of system deployment and confidence in using these types of systems. Users' behavioural resistance and uncertainty regarding new technology significantly impact the degree to which users will adopt a new type of transportation. Therefore, managers need to make improvements in how well services are communicated to users, including the reliability of services and the quality of the users' experiences. Improved accuracy and timeliness in communicating schedules, availability, and performance data about green transportation systems can help decrease perceived risks associated with using these systems. Providing customers with opportunities to communicate complaints or concerns through formal feedback channels will also facilitate customer participation in identifying problems with services provided by green transportation systems; this will assist with enhancing the quality of those services over time. Finally, the human resources practices used by many transportation agencies need to evolve to accommodate developing skills in employees who work with new and evolving forms of transportation for example electric vehicles, intelligent transportation platforms. Employees working with newer forms of transportation require training programs and other continuing education opportunities to acquire the specialized knowledge needed to operate increasingly sophisticated green transportation systems.

The research shows that at a broad strategic level, managers of transportation organizations need to use an innovative type of leadership to help transition their organization toward a sustainable mobility future. One way this could happen is by encouraging a "culture of testing" - supporting pilot projects and small-scale innovations in the hopes they will be successful enough to be implemented on a larger scale.

Organizations may also want to create internal departments (or task teams) focused on finding new and developing current green transportation ideas. In addition, how financials are managed needs to be changed to provide decision makers with options for conducting cost–benefit analyses that include environmental and social impacts. This would allow them to make more educated choices about investing in sustainable transportation related infrastructure and technologies. The study also found that it is necessary for transportation organizations to collaborate with outside partners like technology companies, universities, and development agencies to have access to funding and technical knowledge needed to implement these types of programs. If transportation organizations embed innovative thinking, collaboration, and sustainability into all aspects of managing their organizations they will be better equipped to manage the complexity involved in moving from carbon-based and vulnerable transportation systems to lower-carbon and more resilient ones (Zhao and Kun 2024).

20 Conclusion

In conclusion, this study has demonstrated that the transition to green transport systems in developing countries is a complex and multi-dimensional process shaped by interacting structural, institutional, behavioral, and financial factors. The findings indicate that structural constraints such as inadequate infrastructure, weak multimodal integration, and unreliable energy systems significantly limit the scalability of sustainable mobility solutions. Institutional weaknesses, including fragmented governance, inconsistent policy frameworks, and limited technical capacity, further impede coordinated implementation and long-term planning. Behavioral factors such as risk perception, habitual travel preferences, and limited environmental awareness also reduce the uptake of green transport options, even where such systems are available. Despite these challenges, the study shows that adaptive strategies involving infrastructure investment, policy reform, stakeholder engagement, innovation support, and capacity building can significantly enhance the prospects for successful transition. Theoretically, the study reinforces the relevance of transition, sustainability, diffusion, and behavioral perspectives in explaining transport system change. Practically, it highlights the need for integrated policy and managerial approaches that address both demand-side and supply-side constraints. The study concludes that achieving sustainable transport systems requires coordinated, long-term, and context-sensitive interventions that align technological innovation with institutional reform and behavioral transformation to ensure resilient and low-carbon mobility futures (Zhao and Kun 2024).

21 Future Directions

Future studies could continue with this work by studying the context-dependent and time-variant elements of the transition from traditional to green transportation in developing countries. An important part of future studies will include understanding the impact of structural, institutional, and behavior-related barriers as they evolve through time as new transportation technologies are developed and changes occur within existing regulatory environments. Longitudinal studies will allow researchers to understand if the strategies and policies designed to adapt to these evolving barriers have lasting positive impacts. Comparative studies across different developing countries/regions will assist researchers in identifying those contextual variables which positively influence successful implementation of transitional models and where best practices may be applied/scaled. It is anticipated that future studies will utilize advanced quantitative modelling techniques for example system dynamics, structural equation modelling to enhance our ability to model the relationships between enabling and barrier factors in transitioning to green transportation. In addition, it is expected that future studies will evaluate the roles of emerging technologies like artificial intelligence, big data analytics, smart mobility platforms in accelerating the use of sustainable transportation. Further, future studies are anticipated to place greater emphasis on equity issues related to transportation transitions, particularly on low-income populations and rural communities (Popkova et al. 2026). Lastly, future studies utilizing an interdisciplinary approach (transportation planning/environmental science/economics/behavioral studies) will be necessary to develop comprehensive and practical solutions to creating sustainable and resilient global transportation systems.

References

Achillopoulou DV, Stamataki NK, Makhoul N, Kromanis R (2024) Adaptive transition of critical transport infrastructures. Bridg Maint, Saf, Manag, Digit Sustain 3267–3275. https://doi.org/10.1201/9781003483755-387

African Development Bank (AfDB) (2019) Transport Infrastructure and Regional Integration in Africa

Ardini L, Fahlevi M (2024) Circular economy from an environmental accounting perspective: Strengthening firm performance through green supply chain management and import regulation in Indonesia's plastic recycling industry. Uncertain Supply Chain Manag 12(3):1633–1646. ISSN 2291-6822. https://doi.org/10.5267/j.uscm.2024.3.017

Banister D (2008) The sustainable mobility paradigm. Transp Policy 15(2):73–80

Bauer W, Titz A (2025) Manoeuvring barriers: assessing adaptive strategies for and persistent barriers to urban Nature-based Solutions in Lilongwe, Malawi. Nat-Based Solut 7:100224. ISSN 2772-4115. https://doi.org/10.1016/j.nbsj.2025.100224

Bešinović N (2020) Resilience in railway transport systems: a literature review and research agenda. Transp Rev 40(4):457–478. ISSN 0144-1647. https://doi.org/10.1080/01441647.2020.1728419

Bogadi A (2022) Overcoming barriers in green infrastructure implementation. In: The Palgrave encyclopedia of urban and regional futures. Springer International Publishing, pp 1187–1195. https://doi.org/10.1007/978-3-030-87745-3_117
Brears RC (2023) Adaptive management and blue-green infrastructure. Blue and Green Cities 61–98. https://doi.org/10.1007/978-3-031-41393-3_4
Brundtland GH (1987) Our common future world commission on environment and development
Cervero R (2013) Transport infrastructure and the environment in developing countries. Transp Rev 33(1):1–19
Chibaro M, Mupfiga C, Mahwine ML, Chikwere D, Mbudaya J, Kanyepe J (2024) Green supply chain management and agricultural performance. Adv Mark, Cust RelatShip Manag, E-Serv 285–316. ISSN 2327-5502. https://doi.org/10.4018/979-8-3693-6468-0.ch010
Chibaro M, Chikwere D, Chisungo C, Manyanga W, Munuhwa S (2026a) Transport and logistics in Zimbabwe: policy and future directions. In: The changing terrain of transport and logistics in Zimbabwe, Springer Nature Switzerland, pp 239–277. https://doi.org/10.1007/978-3-032-11384-9_10
Chibaro M, Mahwine M, Manyanga W, Maringosi U, Chikwere D, Mahwine MV (2026b) A literature review on supply chain analytics in Africa. Strat Supply Chain Anal 126–147. https://doi.org/10.4324/9781003674962-7
Chikwere D, Mbudaya J, Chibaro M (2024) Exploring green energy alternatives for a sustainable future. In: Advances in finance, accounting, and economics. IGI Global, pp 183–206. ISSN 2327-5677. https://doi.org/10.4018/979-8-3693-7570-9.ch012
Chikwere D, Munuhwa S, Chibaro M (2025) Advancing sustainable green marketing agenda in Africa through green supply chain management. In: Palgrave studies of marketing in emerging economies. Springer Nature Switzerland, pp 161–182. ISSN 2730-5554. https://doi.org/10.1007/978-3-031-81627-7_9
Chinomona E, Mahlatsi MK (2020) Green supply chain practices and supply chain performance in the beverage industry of South Africa. J Contemp Manag 17(2):295–325. ISSN 1815-7440. https://doi.org/10.35683/jcm20053.77
Geels FW (2002) Technological transitions as evolutionary reconfiguration processes: a multi-level perspective and a case-study. Res Policy 31(8–9):1257–1274
Geels FW (2012) A socio-technical analysis of low-carbon transitions. Res Policy 41(6):991–1002
Ghose A, Ali SM, Aamir DA, Inamdar P (2024) Assessing global impacts of green transition. In: Advances in finance, accounting, and economics, pp 267–282. ISSN 2327-5677. https://doi.org/10.4018/979-8-3693-7570-9.ch016
Golnar M, Beškovnik B, (2020) Green maritime transport as a part of global green intermodal chains. J Marit Amp; Transp Sci 3(3):21–31. ISSN 1848-9052. https://doi.org/10.18048/2020.00.01
Green Climate Fund (GCF) (2020) Programming Manual
International Transport Forum (ITF) (2021) Transport Innovation for Sustainable Development
Karam A, Jensen AJK, Hussein M (2023) Analysis of the barriers to multimodal freight transport and their mitigation strategies. Eur Transp Res Rev 15(1). ISSN 1866-8887. https://doi.org/10.1186/s12544-023-00614-0
Khurshid A, Khan K, Chen Y, Cifuentes-Faura J (2023) Do green transport and mitigation technologies drive OECD countries to sustainable path? Transp Res Part D: Transp Environ 118:103669. ISSN 1361-9209. https://doi.org/10.1016/j.trd.2023.103669
Kozlov PA, Vakulenko SP., Osokin Oleg V, Yakushev Nikolay V (2024) On the sustainability of transport systems. Transport Urals (2):11–15. ISSN 1815-9400. https://doi.org/10.20291/1815-9400-2024-2-11-15
Li H, Ma J, Tao S (2026) Analyzing the effects of the transportation environment and green spaces on residents' satisfaction. J Transp Geogr 131:104536. ISSN 0966-6923. https://doi.org/10.1016/j.jtrangeo.2025.104536

Lomotko D, Arsenenko D, Lomotko M, Vlasenko K (2026) Green logistics in railway transport: modeling environmental aspects and sustainable solutions. Transp Syst Transp Technol (31):16–27. ISSN 2313-8688. https://doi.org/10.15802/tstt2026/352110

Mezhokh Z, Smirnova N (2023) managing the risk control system of a transport company in a green economy. Transp Bus Russ (2):60–62. ISSN 2072-8689. https://doi.org/10.52375/20728689_2023_2_60

Musau EG (2023) Green logistics and transport processes. Adv Environ Eng Green Technol 252–274. ISSN 2326-9162. https://doi.org/10.4018/978-1-6684-6366-6.ch013

Nikolova C (2025) Improving sustainable mobility through intelligent transport systems Deployment. https://doi.org/10.2139/ssrn.5186593

Ogola JS (2021) Environment, climate change and the green economy. Green Econ Transp Sect 9–20. https://doi.org/10.1007/978-3-030-86178-0_2

Poli I, Imbesi PN (2022) Green infrastructures and water management. Urban regeneration strategies to face global change. In: Lecture notes in networks and systems. Springer International Publishing, pp 2790–2799. ISSN 2367-3370. https://doi.org/10.1007/978-3-031-06825-6_266

Popkova EG, Shadybekov KB, Orozbekov AA, Studenikin NV (2026) Green transition of digital economy infrastructure to support decarbonization through smart transport systems. In: Lecture notes in intelligent transportation and infrastructure. Springer Nature Switzerland, pp 347–355. ISSN 2523-3440. https://doi.org/10.1007/978-3-032-12181-3_35

Postorino MN, Mantecchini L, Paganelli F (2021) Green airport investments to mitigate externalities. In: Research anthology on reliability and safety in aviation systems, spacecraft, and air transport. IGI Global, pp 1180–1200. https://doi.org/10.4018/978-1-7998-5357-2.ch048

Rehman FU, Islam MM, Miao Q (2023) Environmental sustainability via green transportation: a case of the top 10 energy transition nations. Transport Policy 137:32–44. ISSN 0967-070X. https://doi.org/10.1016/j.tranpol.2023.04.013

Rehman FU, Islam MM, Miao Q (2023) Environmental sustainability via green transportation: a case of the top 10 energy transition nations. Transport Policy 137:32–44. ISSN 0967-070X. https://doi.org/10.1016/j.tranpol.2023.04.013

Rodrigue JP (2024) The Geography of Transport Systems. Routledge. https://doi.org/10.4324/978 1003343196

Rogers EM, Singhal A, Quinlan MM (2014) Diffusion of innovations. In: An integrated approach to communication theory and research. Routledge, pp 432–448

Ruiz A, Flores A, García E, Olea F (2023) A pathway for a green transition of the transport sector in Mexico. World Resources Institute. https://doi.org/10.46830/wriib.20.00121

Schalekamp H, Behrens R (2010) Engaging paratransit on public transport reform. Res Transp Econ 29(1):185–194

Schot J (1998) The usefulness of evolutionary models for explaining innovation. The case of the Netherlands in the nineteenth century. Hist Technol, Int J 14(3):173–200

SLOCAT (2021) Transport and Climate Change Global Status Report

Song, M, Lu X (2026) Digitalization and green transition in modern transport systems. Resour, Clim Sustain Dev 323–338. ISSN 3005-0502. https://doi.org/10.1007/978-981-95-5289-4_17

Tazzie YD, Adugna, D, Woldetensae B (2025) Sustainable transport strategies and their implementation barriers in Addis Ababa, Ethiopia. Front Sustain Cities 7. ISSN 2624-9634. https://doi.org/10.3389/frsc.2025.1524020

Wang D, Li J, Liu Y (2023) Evaluating barriers and strategies to green energy innovations for sustainable development: developing resilient energy systems. Front Energy Res 11. ISSN 2296-598X. https://doi.org/10.3389/fenrg.2023.1201692

World Bank (2022) Sustainable Mobility for Development Report

Yin Y, Lam J, Siu L (2022) Impacts of energy transition on liquefied natural gas shipping: a case study of China and its strategies. Transport Policy 115:262–274. ISSN 0967-070X. https://doi.org/10.1016/j.tranpol.2021.11.008

Zhao C, Kun W (2024) Is green transportation efficiency the panacea for alleviating carbon lock-in? The case of China. Transp Econ Manag 2:348–358. ISSN 2949-8996. https://doi.org/10.1016/j.team.2024.10.001

Trade Competitiveness and Green Logistics Standards in Global South Economies: A Desktop Review

Kanyepe James and Manyanga Wilbert

Abstract Green logistics standards and environmental regulations in global trade are changing how companies conduct business across international borders. Export-dependent countries in the Global South face two challenges: they need to be competitive in international trade while also complying with the growing list of environmental requirements that affect their companies' operations. This study examines the relationship between green logistics and trade competitiveness within the economies of the global South. Using a qualitative desk review methodology, we investigated this relationship by analyzing peer-reviewed journals, governmental and institutional policy documents, and reports. Institutional theory and Porter's diamond model were employed as conceptual frameworks to analyze how institutional, logistical infrastructure, and enterprise-level factors influence the adoption of green logistics. The review revealed that compliance with green logistics standards provides potential benefits for both market access and supply chain strength. However, many structural and institutional impediments remain to implementing green logistics in global southern economies. These impediments include underdeveloped logistics infrastructure, a lack of supply chain integration, poor regulatory enforcement, and the limited absorptive capacities of SMEs. Additionally, new environmental regulations, including carbon border adjustments and mandated greenhouse gas emission disclosures, further increase the costs associated with compliance with environmental regulations and widen the existing disparities in trade participation across global borders. This study integrates previously isolated research on electronics and environmental regulations to illustrate how such regulations affect export competitiveness in developing economies. Specifically, the study demonstrates that the positive effects of adopting green logistics practices in developing economies depend on the quality of

K. James (✉)
Department of Logistics and Supply Chain Management, Faculty of Business, Economics and Law, University of Botswana, Gaborone, Botswana
e-mail: jameskanyepe@gmail.com

M. Wilbert
Department of Marketing, Logistics and Sport Management, Faculty of Commerce, Human Sciences and Education, Namibia University of Science and Technology, Windhoek, Namibia

J. Kanyepe and I. Chirisa (eds.), *Advancing the Green Agenda in the Transport and Logistics Sector in Global South Countries*, World Sustainability Series,
https://doi.org/10.1007/978-3-032-31374-4_12

institutions, the extent of developed logistics infrastructure, and government policies that support exporting companies.

Keywords Carbon border mechanisms · Green logistics · Green logistics systems · Sustainability compliance · Emissions reporting

1 Introduction

The world's increased focus on reducing carbon emissions, creating cleaner supply chains, etc., all of which we call environmental sustainability, is becoming a greater force in shaping the rules of global trade (Lumempouw 2024). There are increasing calls for the global logistics sector, the main infrastructure behind international trade, to be held accountable for its role in contributing to greenhouse gas emissions, energy use, and environmental degradation (Stojanovic et al. 2021). With many countries imposing stricter regulations on environmental practices and carbon-related trading issues, exporting countries are facing even greater pressure to remain internationally competitive (Brandi et al. 2020). Many of these countries are found in the Global South, where they rely heavily on resource-intensive goods. Therefore, these countries face two competing demands: meeting internationally agreed-upon sustainable practices and competing in terms of pricing and market share. Therefore, practicing green logistics is not only an environmental necessity but also a critical component of maintaining trade competitiveness (Banswal et al. 2025).

Green logistics involves integrating environmental sustainability into the planning and execution of movement, storage, and distribution functions with respect to logistics operations and balancing environmental impacts against economic/operational objectives (Nagy and Szentesi 2024a, b). Examples include using alternative route options to reduce vehicle distance traveled, investing in newer generation vehicles, and utilizing technological innovations. Regulatory compliance requirements play a major role in influencing the implementation of such practices. Companies are expected to comply with government-mandated sustainability and environmental reporting standards (Kanyepe et al. 2023). More broadly, the increasing emphasis on environmental protection due to globalization and the increasing requirement for member countries of international trade agreements to adopt green logistics standards (Zhao 2024) reinforce the regulatory influences described above.

Some countries use green logistics practices by requiring importers to pay a carbon fee for products that contain carbon-based materials that do not comply with the emission levels defined by their country of origin. For example, the European Union (EU) has implemented a carbon border adjustment mechanism and requires importers to report the total amount of greenhouse gas emissions created during the manufacturing process for each product sold in the EU (Leal-Arcas et al. 2022). The goal of the CBAM and GHG emissions reports is to charge imports for their emissions at a rate similar to that of domestic production. This method helps avoid "carbon leakage" and encourages production processes with lower emission levels (Sun et al. 2023).

However, several countries within the Global South have expressed concerns about the limitations placed on them because of their resource constraints, poorly developed infrastructure, and institutional capacity to adopt and implement new environmental standards (Maimele 2025). Of all the limitations mentioned previously, the obsolescence of logistics systems is likely the largest barrier to implementing new standards. Specifically, there is no ability to track their supply chain digitally. This creates inefficiencies in logistics operations, reduces transparency in logistics operations, and reduces compliance with green logistics requirements (Kuteyi and Winkler 2022a, 2022b). Additionally, many developing countries do not possess the modern roads, ports, and railroads necessary to facilitate logistical operations efficiently and economically. This makes it very difficult for developing countries to establish green logistics systems and provides another barrier to exporting firms (Sinkovics et al. 2021). Most export-dependent firms in the Global South do not have sufficient resources to comply with all the strict standards for green logistics practices.

The institutional structure and policies surrounding logistics directly affect a company's ability to adapt to changing environmental regulations (Khan et al. 2019). This position is supported by Venter et al. (2019), who argued that companies located in countries without robust governance structures, with fractured policy environments, or with limited technological sophistication find it challenging to change transportation practices in accordance with international standards. Conversely, companies located in countries with well-coordinated policies, supportive policies designed to assist in the transition to green logistics, and long-term plans for investment in logistics facilities can effectively become competitive participants in international markets (Hsu et al. 2016). Both public officials and business executives need to fully appreciate the relationships among trade competitiveness, regulatory structures, and logistics infrastructure if they wish to increase successful exports while abiding by global environmental standards.

This study was prompted by the recognition that environmentally sustainable practices and trade competitiveness are no longer mutually exclusive. For manufacturers in Global South countries, "green" logistics present a double burden. First, meeting environmental regulations is likely to result in higher short-term costs associated with logistics operations. Second, compliance can result in improved market quality, supply chain resilience, and corporate image. Unfortunately, the existing academic literature does not provide a complete understanding of how global standards affect environmental systems and trade performance in developing countries.

Owing to these identified gaps in the literature, this study examines the effect(s) of global green logistics standards on trade competitiveness among countries within the Global South. To achieve this objective, this study pursues three primary objectives: (1) evaluate compliance and emissions reporting requirements associated with sustainable logistics systems; (2) analyze the impact of carbon-related trade policies on export performance; and (3) assess institutional or infrastructural impediments to implementing green logistics standards. By synthesizing knowledge across multiple disciplines, including trade, sustainability, and logistics, this research offers value-added contributions to the extant literature by providing a unified perspective on both the opportunities and challenges presented by exporting firms when implementing Green Logistics Standards (GLSs) and Carbon-Related Trade Measures (C-RTMs).

2 Literature Review

2.1 Trade Competitiveness

Trade competitiveness refers to the ability of a country, region, or enterprise to efficiently produce and distribute goods and services in international markets (Bilali 2022). It reflects the extent to which a country's exports exceed those of another country on the basis of price (cost) and/or nonprice factors (such as product quality, design, and features) (Novinc and Škuflić 2025). Several variables contribute to trade competitiveness, including cost, quality, innovation, market access, and compliance with international standards (Fabrizi et al. 2023; Fabrizi et al. 2024). According to Baah et al. (2024), the quality of transportation networks, logistics systems, regulatory frameworks, and institutional strength significantly influences a firm's ability to produce and transport goods efficiently. Similarly, Ur-Rashid and Khan (2024) argue that external market forces, including tariffs, nontariff barriers, and sustainability-related policies, play a decisive role in determining countries' international trade performance.

García Juárez et al. (2025), however, state that with an increasing focus on the use of nonprice factors (such as sustainability practices and certification from international bodies) and a demand for greater transparency and accountability within supply chains, low-cost production, although still important to achieving competitive pricing and being able to export goods cheaply in the global market place, will not be the only factor in determining trade competitiveness. Consequently, exporters are increasingly required to demonstrate adherence to environmental standards to access higher value markets. Failure to meet these emerging requirements may restrict market access, reduce participation in global value chains, and increase the costs associated with certification and compliance.

2.2 Green Logistics Standards

The concept of "green logistics" encompasses a comprehensive array of internationally recognized standards, principles, and practices designed to mitigate the environmental impact associated with the physical movement, storage, and handling of goods (Nicoletti and Appolloni 2024). According to Nagy and Szentesi (2024a, b), the primary aim of green logistics is to develop strategies for reducing CO_2 emissions, energy consumption, and the utilization of natural resources throughout logistics operations. This multifaceted concept includes vehicle efficiency, routing and packing strategies, return-based logistics, the development of low-emission technologies, and green warehousing practices. In response to increasing pressure to comply with green logistics standards, businesses engaged in international trade will be required to meet various reporting obligations related to their carbon footprint, disclose greenhouse gas emissions, and obtain certifications on the basis of adherence to globally established environmental standards (Ahmad et al. 2025a, b).

Green logistics activities are intricately associated with various environmental initiatives at both the global and regional levels. They specifically align with the objectives established by the United Nations' Sustainable Development Goals (SDGs) and the targets delineated in the Paris Agreement (Messaoudi et al. 2026). In addition, adhering to green logistics standards is essential for entering the markets of numerous developed countries (Afum et al. 2022). However, the implementation of green logistics in developing countries faces significant challenges, including inadequate infrastructure, deficient transportation systems, and limited institutional capacity (Kanyepe et al. 2023). Green logistics links environmental sustainability to competitive advantages in international trade. Consequently, compliance with these standards ensures that products from developing countries can access markets where environmental considerations increasingly influence buyer decisions (Tetteh et al. 2025a, b). Conversely, failure to adhere to these standards can limit trade opportunities, increase transaction costs, and reduce developing countries' participation in global value chains.

2.3 Theoretical Perspectives

This study relies on the institutional and Porter's Diamond models. Institutional theory examines how the formal and informal "rules" of an organization (e.g., laws, policies, customs) influence its decision-making processes and behaviors (Berthod, 2023). Specifically, within the realm of logistics and trade, institutional theory recognizes that regulatory environments, policies, and institutional support can significantly affect an organization's willingness to adopt green logistics practices. Organizations and governments choose to engage in sustainability initiatives because they want to be efficient while simultaneously attaining social legitimacy and complying with international standards (Bhuiyan et al. 2023). Furthermore, institutional theory illustrates why there have been varying levels of success in adopting green logistics practices in countries located in the Global South; these variations are dependent on a country's institutional capacity (i.e., the quality of legal frameworks, the effectiveness of law enforcement, and the level of policy coordination).

Porter's diamond model of competitiveness theory identifies four key factors that make a nation competitive: factor conditions (available resources and infrastructure); demand conditions; related and supporting industries; and firm strategy, structure, and rivalry (Butt et al. 2023). In terms of environmental regulation and green logistics practices, Porter's Diamond Model helps illustrate the relationship between the implementation of green logistics practices and improved export performance. Nations and businesses can create a competitive advantage in foreign markets by implementing green logistics practices in their operations, especially where consumers demand carbon-free/low-carbon products. Porter's model shows that adhering to environmental regulations increases business efficiency and reduces the costs associated with logistics optimization, thus creating opportunities for entry into higher-value market segments (Ambec and Barla 2006). Porter's model allows exporters to understand that the implementation of green logistics practices should be viewed not only to meet present regulatory requirements but also to strategically differentiate themselves globally and improve their market position relative to competitors.

3 Methodology

This chapter applies a qualitative desk review method to evaluate the impact (GLS) effect on the competitive position of trade within the countries of the Global South. A desk review involves a systematic evaluation of the relevant literature, policies, and institutional reports to determine emerging trends, challenges, and good practices. This methodology was appropriate for this study because it utilized secondary sources (i.e., previously collected data) rather than collecting new data through surveys or interviews. As such, a desk review allowed for an extensive examination of both local and international developments regarding transportation, logistics, and environmental compliance. The sources included academic journals and books; conference proceedings; reports published by international organizations such as the WTO, UNCTAD, World Bank, and OECD; and policies related to transportation, trade, and environmental regulation. Thematic synthesis and comparative policy evaluation were used in conjunction to evaluate common themes among the studies, including barriers to compliance, inadequate infrastructure, and institutional capability. Concurrently, the comparative analysis of policy frameworks enabled researchers to examine variations in policy development across different countries and identify "best practices" for Global South economies in developing logistics systems that adhere to international standards.

4 Findings

The results of the qualitative desk review are presented in this section. As part of a systematic process, a qualitative and quantitative review of peer-reviewed academic journals, government and institutional reports, and other related documents was systematically performed to determine the themes and trends that exist, the obstacles to implementing sustainable logistics practices, and growth opportunities. Three major interrelated factors were found to directly impact exporter performance, namely, sustainability compliance and emissions reporting, carbon border mechanisms (CBMs), and institutional and infrastructure constraints. These findings demonstrate how the ability of firms to participate in international trade is positively or negatively affected by regulatory requirements (environmental) and their operational capabilities (infrastructure). Empirical evidence, along with conceptual studies and policy reviews, provides an overall assessment of the ways in which green logistics systems (GLSs) affect the competitive advantage of countries participating in international trade.

5 Sustainability Compliance Requirements and Emissions Disclosure

Recent studies have increasingly underscored sustainability compliance and emission disclosures as competitive factors for companies engaged in international logistics and trade (Herold and Lee 2017). In European and North American export markets, exporters are compelled, either by law or regulation, to implement environmentally conscious logistics practices and provide measurable and reportable data on greenhouse gas (GHG) emissions across their entire supply chains. As trade regulations and purchasing requirements evolve into an integral component of sustainability reporting, both regulatory and market influences shape the extent to which companies engage in sustainability compliance. For example, Pallis and Rodrigue (2022) reported that trade agreements that require GHG reporting for companies operating in specific markets, such as the EU's Emissions Trading Scheme (ETS), create a direct link between company carbon performance and market access. Therefore, companies seeking to enter markets that enforce strict environmental regulations must demonstrate compliance with rigorous reporting and monitoring standards.

Moreover, research indicates a large variance in compliance costs across industry segments. Industries with the highest levels of carbon emissions, such as steel, cement, aluminum, and petrochemicals, experience higher compliance costs, as they must invest in emission-measuring devices, energy-efficient equipment, and low-emission logistics operations. However, because all industries must track and report emissions regardless of their level of emissions (e.g., agro-processing, textiles, and electronics), even those industries with significantly lower carbon emissions incur much lower compliance costs than "high" emitting industries do (Kumar and Khan 2024). Companies that begin implementing emission-disclosing systems early in their development process may also gain operational efficiency via several approaches, such as improving logistics efficiency, reducing fuel consumption, and increasing visibility along the supply chain (Onukwelu et al. 2021).

Companies located in developing economies face significant barriers to complying with these requirements (Sulemana et al. 2025). Developing economies generally lack modern logistics infrastructure, consisting of inefficient road networks and congested ports, limiting companies' ability to design and deploy green logistics solutions. Moreover, companies based in these regions often do not possess the digital systems needed to monitor and measure emissions, preventing them from providing accurate data on their carbon footprint (Aderibigbe et al. 2024; Kuteyi and Winkler 2022a, 2022b). Thus, the exports of SMEs in developing economies are disadvantageous. While SMEs account for most exports in most developing economic sectors, they generally lack the financial and technological means to comply with sustainability reporting requirements (Tetteh et al. 2025a, b).

However, despite these limitations, there is overwhelming evidence in the academic literature that companies that proactively implement comprehensive sustainable compliance systems achieve multiple strategic advantages. Studies have empirically shown that when companies use route optimization technology in conjunction with emissions monitoring technology and energy-efficient logistics operations, their overall carbon footprint is reduced, and their operational performance is increased (Briand et al. 2024). Compliance with sustainability standards also enables companies to build trust among consumers and strengthen their relationships with international buyers. Additionally, companies that comply with sustainability standards have greater ease of entry into environmentally conscious, regulated markets (Nagiah and Mohd Suki 2024).

Additionally, regional examples illustrate the varying degrees of compliance capability among exporting companies in different regions of the Global South. Compared with other developing economies in Asia, Southeast Asian countries (such as Thailand and Vietnam) have been active in developing programs to help SMEs establish emission reporting systems and sustainable logistics practices (Hsu 2025; Le 2024; Setyaningsiha et al. 2024). Similarly, comparable programs exist in Latin America, where the Brazilian government has established digital platforms for emissions tracking, and Chile has initiated training programs for SMEs on how to follow environmental regulations (Bastidas 2026). However, similar initiatives to support SMEs in developing sustainable logistics practices are lacking in most African countries. Although initial pilot studies conducted in Kenya and South Africa suggest a growing interest among some African countries in combining logistics efficiency with emissions tracking (Mafini 2018; Okeke 2024), the academic literature supports the notion that sustainability compliance and emissions reporting are evolving from voluntary environmental practices to mandatory business strategies for participating in increasingly stringent global trade markets (Le 2024; Bastidas 2026).

6 Carbon Border Mechanisms and Export Performance

The use of the carbon border mechanism (CBM), which was adopted by governments worldwide for the first time, presents a fundamental shift in how an "environmental" border will be defined internationally. A prime example is the EU carbon border adjustment mechanism (CBAM). One of the primary issues that CBAM addresses is "carbon leakage," which occurs when a company relocates production from one country to another because of either a high tax rate or a fee on greenhouse gas emissions in its original home country.

The literature has demonstrated that the implementation of the carbon border adjustment mechanism (CBAM) has significantly influenced the export competitiveness of numerous developing economies, particularly within carbon-intensive sectors (Morone and Alfino, 2025a). Compared with developed economies, developing economies typically exhibit higher levels of embodied emissions and have limited opportunities to adopt cleaner technologies. Consequently, these economies are likely to experience negative impacts on their export competitiveness (Okoro 2025; Magacho et al. 2023). Exporters from the Global South will require the ability to measure and report embedded carbon emissions, acquire carbon-tracking systems, and enhance their production capabilities. These capabilities are unevenly distributed. As a result, exporters from developing nations may incur increased administrative and production costs associated with CBAM compliance (Lin and Zhao 2024; Okoro 2025). Furthermore, in the absence of targeted support and capacity-building measures, CBAM could prevent exporters from developing countries from competing on price with other exporting nations, thereby limiting their access to regulated markets (Lin and Zhao 2024).

Moreover, the impact of CBAM is anticipated to vary considerably across different sectors and regions, particularly in the Global South. Research on the EU-CBAM indicates that certain energy-intensive industries, such as steel, cement, aluminum, and fertilizers, are especially vulnerable to the increased costs associated with CBAM. These industries possess high levels of embodied carbon and consequently face elevated costs related to compliance with CBAM regulations, in addition to expenses for carbon emissions in regions where carbon pricing is implemented (Morone and Alfino 2025a, b). Further studies focusing on developing countries suggest that diminished export competitiveness under CBAM disproportionately affects major exporting nations integrated into global value chains that rely heavily on carbon-intensive activities. This situation could result in a decline in exports and/or overall well-being as carbon prices increase (Li et al. 2026).

Conversely, although industries with low-emission profiles in the Global South do not directly incur substantial costs owing to the CBAM, they face indirect pressures from expanding supply chain reporting and emissions accounting requirements. These requirements impose administrative burdens on firms engaged in low-carbon product lines and increase compliance costs (Li et al. 2026). Ultimately, CBAM is expected to reshape global trade patterns by providing economic incentives for producers in affected countries to invest in cleaner production technologies and develop robust carbon governance structures to maintain market access (Zhang and Sovacool 2025). Early responses to externally generated carbon-pricing pressures can also promote institutional learning and reform carbon governance in Global South countries. With adequate technical and financial support, low-carbon transitions and enhanced competitiveness can be achieved (Zhang and Sovacool 2025).

Institutional capacity specific to each country significantly influences exporters' responses to CBMs. Countries with strong regulatory frameworks, technical assistance programs, and financial incentives provide optimal opportunities for companies to report greenhouse gas emissions properly. In contrast, exporting industries from developing countries, whose governments are characterized by a lack of coordination and structural fragmentation, experience higher costs and greater barriers to complying with carbon budget management requirements (Bah et al. 2021). Regional comparisons also revealed differences in adaptation capacity. Exporters from developing countries, primarily in Africa and South Asia, which are largely concentrated in heavy industrial manufacturing sectors, are likely to be the most negatively affected by limited access to low-carbon technologies and supporting infrastructure (Nwokolo et al. 2023; Srivastava et al. 2018). However, many Latin American countries and some Southeast Asian countries have implemented policies aimed at assisting exporters in adapting to carbon-based trade regulations, that is, establishing technical assistance programs for employees and investing in energy-efficient transportation infrastructure (Lobo et al. 2025; Morone and Alfino 2025a, b).

Capacity Constraints in Implementing Green Logistics Standards. Many researchers have reported that institutional capacity and logistics infrastructure are key determinants of the rate of green logistics adoption by developing-economy companies in the Global South. Companies can pursue several different types of green logistics strategies, such as low-carbon transportation systems, energy-efficient warehouses, emission monitoring technologies, and sustainable packaging (Kanyepe et al. 2023). The evidence also suggests that an organization's potential to adopt green logistics practices is highly dependent on its surrounding environment, specifically the institutions and logistics infrastructure where it operates (Lawhon et al. 2023).

As demonstrated by the studies examined in this dissertation, inadequately developed institutional environments limit an organization's opportunity to gain the capabilities needed to comply with existing green logistics standards. Although many countries in the Global South currently do not have well-developed regulatory frameworks that clearly outline the expectations for international environmental standards (He 2025), when environmental regulations exist, they typically lack sufficient technical expertise to assist in their implementation. Consequently, companies operating in these countries may find it difficult to identify the reporting requirements they must adhere to, the sustainability standards they should achieve, and the processes needed to ensure adherence to national and international environmental laws (Amoah and Eweje 2022).

To enable the effective functioning of green logistics networks, high-quality networks, roadways, modern port facilities, efficient rail services, and advanced digital platforms for optimizing logistics operations and measuring carbon dioxide emissions are needed (Ahmad et al. 2025a, b). Ouni and Mraihi (2025) reported that in countries with substandard infrastructure, efficiency decreases, resulting in increased fuel consumption and CO_2 emissions. Therefore, attempts to implement environmentally friendly logistics practices will be unsuccessful. Additionally, owing to the limitations of digital logistics platforms and carbon-monitoring technologies available to companies, accurate carbon accounting information cannot be reliably generated (Nosike 2024). From a regional perspective, the literature demonstrates how unique combinations of institutional arrangements and logistics infrastructure affect adoption rates at both the national and regional levels. For example, nearly all countries in sub-Saharan Africa continue to struggle with issues of logistics infrastructure availability and institutional disarray that impede the widespread diffusion of best practices (Kuteyi and Winkler 2022a, 2022b). In contrast, many countries in other parts of the Global South, such as Southeast Asia, have achieved higher rates of green logistics adoption through regulatory reforms and investments in logistics infrastructure and employee education/training (Phinijphara and Lakkhongkha 2024). Similarly, public–private partnerships (PPPs) have allowed small– and medium–sized enterprises in Latin America, especially those located in Brazil and Chile, to obtain energy-efficient vehicles and carbon monitoring equipment (Lobo et al. 2025).

Moreover, the literature has consistently emphasized a direct relationship between institutions and the development of physical–logistics infrastructure. While regulatory requirements alone will not facilitate the adoption of green logistics if no accessible physical-logistics infrastructure is available or if appropriate technical resources exist (Khayyat et al. 2024), merely investing in new physical-logistics infrastructure will not automatically lead to the implementation of sustainable-logistics practices without an effective institutional framework governing the use of the said infrastructure (Cervero 2014a, (b). Thus, cooperative relationships between relevant stakeholders (including government agencies, private logistics companies, and industry associations) are necessary for exchanging knowledge, providing capacity-building activities, and obtaining funding for green logistics projects (Vazquez-Brust et al. 2014).

7 Discussion

The findings of the desk review show that environmental regulation has become a major factor affecting international trade in recent years. A firm's capacity to meet its obligations regarding sustainable development and the disclosure of greenhouse gas emissions are now major factors determining its comparative advantage in the global marketplace. In regions such as Europe and North America, where governments have implemented stringent carbon-related reporting mandates (Herold and Lee 2017; Solakivi et al. 2026), firms that use systematic approaches to measure emissions establish energy-efficient routes in their logistics processes and create sustainable business models can satisfy strictly defined governmental obligations related to environmental protection, increase their operational efficiency, and ensure better transparency in their supply chain management (Briand et al. 2024; Nagiah and Mohd Suki 2024). This shows that the implementation of environmental laws and regulations may lead to a strategic advantage for a firm through increased effectiveness in conducting business operations and improving its reputation.

Additionally, carbon border mechanisms (CBMs), exemplified by the European Union's carbon border adjustment mechanism (CBAM), used by developed countries, illustrate the link between a country and its access to international markets. Studies based on empirical data show that the application of CBMs by developed countries negatively affects carbon-intensive exports from less-developed countries through significant increases in their production costs and possible reductions in their ability to obtain or retain market share (Magacho et al. 2023; Okoro, 2025). Nevertheless, substantial evidence shows that CBMs provide an important stimulus for nations to undergo decarbonization and institutional reforms as part of numerous initiatives, such as technical assistance programs, capacity-building initiatives, and coherent policy frameworks (Zhang and Sovacool 2025). This dual effect illustrates the dependency of CBMs on context. On the one hand, producers/exporters experience difficulties connected to both the operational and economic aspects of fulfilling CBMs' obligations; however, these difficulties can stimulate long-term structural changes toward more sustainable activities.

Infrastructural capabilities are essential prerequisites for adopting environmentally friendly logistics practices. Low-quality regulatory enforcement, disorganized policy frameworks, and limited technical support hinder businesses from realizing environmentally friendly standards. Additionally, inadequate transport networks and the limited availability of digital monitoring systems limit the opportunities for green logistics solutions (Kanyepe et al. 2023; Ahmad et al. 2025a, b). Comparative studies conducted among countries have shown that countries that implement coordinated measures to improve the quality of the regulatory environment and develop both physical and technological infrastructure realize higher levels of adoption than those without such coordination (Phinijphara and Lakkhongkha 2024). Therefore, there are close connections between governance, infrastructure, and the degree of green logistics adoption.

Moreover, there is a dependent relationship between environmental regulations/institutions, infrastructure investment/development, and the competitiveness of export-oriented countries. Firms that successfully manage their compliance obligations related to greenhouse gas emissions associated with environmental regulations and adapt to carbon border mechanisms while implementing sustainable logistics strategies will have optimal conditions to preserve access to international markets and achieve a competitive advantage owing to the growing environmental orientation of the global market. Nonetheless, since most of the literature focuses on the short-term reactions of companies to environmental regulations, additional research is needed to investigate how firms can weigh conflicting interests related to compliance with regulatory requirements, operational efficiency, and investments in green logistics in different institutional and infrastructural settings.

8 Implications

8.1 Theoretical Implications

This study adds to the theoretical knowledge of sustainable international trade by demonstrating the interconnected nature of institutional regulatory compliance, carbon pricing mechanisms (s), and green logistics adoption. One of the key findings is that firm-specific (institutional) capacity and infrastructure serve as intermediate variables. Compliance with environmental regulations alone does not have an independent positive effect on organizational performance. Therefore, this study suggests an integrated framework for describing competitive advantage when organizations can combine their internal capabilities and competences appropriately with effective governance structures, appropriate infrastructure, and necessary outside resources. Additionally, this study expands the body of literature related to global value chains, primarily by demonstrating how regulatory and infrastructural contexts influence firms' strategic decisions.

8.2 Policy Implications

In addition to providing a strong assessment, this review emphasizes the importance of establishing reliable institutional frameworks (i.e., policies/standards, compliance mechanisms, and technical support) to ensure that export-oriented businesses receive policy support from the government. Further empirical studies show that investments in physical transport infrastructure (e.g., port facilities), railway links, and digital solutions are needed to create viable low-carbon logistics systems. Governments in developing countries can accelerate the implementation of sustainable practices among exporting companies via targeted training programs, financial incentives for reducing emissions, and structured government–private sector collaborations to help SMEs meet the requirements of new environmental regulations.

8.3 Managerial Implications

As sustainability becomes an increasing concern for many managers, proactively monitoring greenhouse gas emissions, utilizing sustainable logistics strategies, and improving energy use efficiency are essential. Early compliance with environmental regulations can help reduce the costs associated with the "compliance burden model" while improving operational efficiency and protecting a company's reputation. Companies within international supply networks have a unique opportunity to improve their competitive position through technological advancement, knowledge transfer, and collaboration on environmental protection efforts. Thus, investments in "green logistics" not only support companies' long-term financial obligations related to reducing environmental impact and regulatory requirements but also may build stronger relationships with environmentally responsible customers and enhance a company's ability to compete in a regulated environment.

9 Limitations

The results of this study are limited by the methodological constraints of desk reviews. For example, in a desk review, the research relied entirely on secondary data from various academic disciplines (i.e., peer-reviewed articles), government agencies, institutions, and policy makers. However, it does not include empirical data collected directly from firms. Therefore, the detailed day-to-day operational decisions made by companies at the firm level related to adapting their operations to address climate change mitigation requirements, complying with regulations, and making logistics decisions on the basis of emissions information may not be accurately captured. In addition, the review was restricted to English-language and publicly available documents. As a result of this limitation, useful gray literature, non-English scholarly work, and local studies may provide insight into how firms operate within different regional markets to respond to unique problems and emerging trends.

Owing to the rapid evolution of carbon pricing mechanisms, carbon border taxes, and environmental regulations, many of the older studies reviewed in this paper may no longer represent contemporary environmental policies or recent changes in global sustainability regulations. Another constraint of a desk review is that while the use of thematic synthesis and comparative policy analysis allows researchers to identify common themes and best practices among countries, these approaches also limit the ability to establish causality between variables and measure the size of the effects on trade competitiveness. Further empirical research is needed to support conclusions about causal relationships, determine the impact of carbon pricing on individual company competitiveness, and develop more effective management and policy initiatives.

10 Future Research Agenda

Future research should critically assess the degree to which the implementation of sustainability compliance frameworks and the adoption of green logistics practices impact export performance within specific industries and countries. Therefore, cross-sectional analyses are needed. Researchers should rely on longitudinal rather than cross-sectional methods. Longitudinal methods are best suited for measuring changes in a company's behavior over time, that is, organizational learning processes, a company's response to changing regulatory and market-based sustainability pressures, and an organization's performance trajectory over time. Additionally, comparative cross-national research among African, Latin American, and Southeast Asian regions can offer significant insights into the existence of institutional differences. This type of research will allow scholars to understand how different forms of regulation, infrastructure, and governance create both opportunities and obstacles for companies' adoption of sustainable business practices and further reduce the likelihood that the results of previous research are generalized beyond the original research setting without considering context-specific factors.

Moreover, small- and medium–sized enterprises (SMEs) have traditionally been difficult for researchers to study because of their constrained structures and limited ability to absorb new information and technology related to sustainability compliance. Therefore, future research should focus on understanding the types of institutional, financial, and capability constraints that hinder or facilitate SME engagement in the green logistics transition and the strategies that SMEs use to overcome these challenges. Finally, future research should place increased emphasis on examining the roles of emerging digital technologies (such as the IoT and blockchain systems) as facilitating infrastructures for achieving sustainability. Specifically, researchers should explore how these digital technologies enable greater transparency throughout the supply chain, enhance compliance mechanisms, increase visibility throughout the supply chain, and support strategic decision-making. Researchers should also consider examining the interactions among digitalization, institutions, and capabilities at the firm level when assessing the effectiveness of green logistics adoption.

11 Conclusions

This study shows that in developing countries, sustainable compliance with regulations, adapting to carbon budget management (CBM), and building capacity in institutional infrastructure are major factors determining export competitiveness. Companies that strategically develop low-carbon operational capabilities, use "green" logistics, monitor their greenhouse gas (GHG) emissions, and build other low-carbon operational systems will have an improved opportunity to continue competing for market share and improve their performance as global markets become increasingly governed by sustainability-based rules. Policymakers must identify and address regulatory, infrastructure, and technical gaps to help companies adapt to low-carbon logistics systems. Companies should proactively adopt sustainable practices to ensure their competitiveness in increasingly governed and sustainability-driven markets. The results show that environmental compliance is not just a requirement from regulatory bodies; it is also a strategic necessity for export-oriented companies competing in diverse and increasingly governed global markets.

References

Acquaye AA, Yamoah FA, Ibn-Mohammed T, Quaye E, Yawson DE (2023) Equitable global value chain and production network as a driver of enhanced sustainability in developing economies. Sustainability 15(19):14550

Aderibigbe OO, Fadare SO, Gumbo T (2024) Transport Situation in the Global South: insights from Nigeria, South Africa and India. In: Emerging Technologies for Smart Cities. Springer, Cham. https://doi.org/10.1007/978-3-031-66943-9_3

Afum E, Agyabeng-Mensah Y, Baah C, Asamoah G, Kusi LY (2022) Eco-market orientation in the logistics industry: a conveyor belt for achieving organizational outcomes via green logistics practices. Int J Logist Manag 33(2):712–734

Ahmad AU, Jeevan J, Ruslan SMM (2025) Effect of green logistics performance on sustainability in African countries: AU Ahmad et al. WMU J Maritime Affairs 24(2):249–292

Ahmad AU, Jeevan J, Ruslan SMM (2025) Effect of green logistics performance on sustainability in african countries: AU Ahmad et al. WMU J Marit Aff 24(2):249–292. https://doi.org/10.1007/s13437-025-00373-w

Ambec S, Barla P (2006) Can environmental regulations be good for business? An assessment of the Porter hypothesis. Energy Stud Rev 14(2)

Amoah P, Eweje G (2022) Barriers to environmental sustainability practices of multinational mining companies in Ghana: an institutional complexity perspective. Corp GovAnce: Int J Bus Soc 22(2):364–384

Baah C, Agyabeng-Mensah Y, Afum E, Acquah ISK, Essel D (2024) Government influence on logistics and supply chain innovations: assessing implications for firm performance and societal impact in an emerging economy. Int J Emerg Mark 19(11):3646–3666

Bah M, Ondoa HA, Kpognon KD (2021) Effects of governance quality on exports in Sub-Saharan Africa. Int Econ 167:1–14

Banswal B, Sharma S, Arora R (2025) Exploring green provisions in international trade agreements for economic sustainability through promoting trade flows. Discov Sustain 6(1):871

Bastidas CR (2026) Disclosure information on environmental, social, and governance practices in Latin America. Sustain Dev. https://doi.org/10.1002/sd.70324

Berthod O (2023) Institutional theory of organizations. Global encyclopedia of public administration, public policy, and governance. Springer International Publishing, Cham, pp 6789–6793

Bhuiyan F, Rana T, Baird K, Munir R (2023) Strategic outcome of competitive advantage from corporate sustainability practices: Institutional theory perspective from an emerging economy. Bus Strateg Environ 32(7):4217–4243

Bilali JB (2022) Small/medium sized enterprises (SMEs) competitiveness: a global perspective competitiveness of small and medium-sized enterprises. J Enterp Bus Intell 2(2):109–119

Brandi C, Schwab J, Berger A, Morin JF (2020) Do environmental provisions in trade agreements make exports from developing countries greener? World Dev 129:104899

Briand Y, Gupta D, D'Agosto M, de A, Goes GV, Gonçalves DNS, Garg A, Koning M (2024) A pathway design framework for national freight transport decarbonization strategies. Climate Policy 1–17. https://doi.org/10.1080/14693062.2024.2412709

Butt MA, Katuse P, Namada JM (2023) Role of government as moderator: Porter's diamond firm strategy, structure, and rivalry attribute and industry's performance. Int J Res Bus Soc Sci 12(7)

Cervero R (2014a) Transport infrastructure and the environment in the Global South: Sustainable mobility and urbanism. J Reg City Plan 25(3):174–191

Chirambo, D. (2021). Corporate sector policy innovations for sustainable development goals (SDGs) implementation in the Global South: the case of sub-Saharan Africa. J Sustain Res 3(2)

Fabrizi A, Gentile M, Guarini G, Meliciani V (2024) The impact of environmental regulation on innovation and international competitiveness. J Evol Econ 34(1):169–204

García Juárez HD, Montes Ninaquispe JC, León Luyo SL, Marquez Yauri HY, Mendoza Ocaña CE, De La Cruz Ruiz NV, Apaza Miranda SJ, Corrales Otazú CD, Rodríguez Abraham AR, Villanueva Butrón GV (2025) Competitiveness and diversification in grape exports: keys to their

sustainability in global markets. Agriculture 15(17):1894. https://doi.org/10.3390/agriculture15171894
He Y (2025) Navigating political uncertainty and mineral policy: pathways to global South's environmental sustainability. Glob Environ Chang 92:103002
Herold DM, Lee KH (2017) The influence of the sustainability logic on carbon disclosure in the global logistics industry: The case of DHL. FDX and UPS. Sustainability 9(4):601
Hsu CC, Tan KC, Mohamad Zailani SH (2016) Strategic orientations, sustainable supply chain initiatives, and reverse logistics: Empirical evidence from an emerging market. Int J Oper Prod Manag 36(1):86–110
Hsu E (2025) Global green logistics: challenges and opportunities in Vietnam's sustainable development. Supply Chain Sustain Res 3(3):49–62. https://doi.org/10.14456/scsr.2024.22
Kanyepe J, Alfaneta M, Zizhou B (2023) Green logistics adoption and operational performance within the road freight sector: the moderating role of management commitment. Logist Res 16(1):1–14
Khan SAR, Sharif A, Golpîra H, Kumar A (2019) A green ideology in Asian emerging economies: from environmental policy and sustainable development. Sustain Dev 27(6):1063–1075
Khayyat M, Balfaqih M, Balfaqih H, Ismail M (2024) Challenges and factors influencing the implementation of green logistics: a case study of Saudi Arabia. Sustainability 16(13):5617. https://doi.org/10.3390/su16135617
Kumar R, Khan AK (2024) Supply chain sustainability and SDG compliance: challenges and opportunities for agro-based industries. Sustain Dev Goals Bus Sustain 442
Kumari R, Dwivedi N, Meehnian H (2025) Industrial framework for net zero emission transition in emerging economies. In: Sustainable biotechnological remedial frameworks for the rejuvenation of heavily polluted environments. CRC Press, pp 387–412
Kuteyi D, Winkler H (2022a) Logistics challenges in Sub-Saharan Africa and opportunities for digitalization. Sustainability 14(4):2399. https://doi.org/10.3390/su14042399
Kuteyi D, Winkler H (2022b) Logistics challenges in sub-Saharan Africa and opportunities for digitalization. Sustainability 14(4):2399
Lawhon M, Follmann A, Braun B, Cornea N, Greiner C, Guma P, Dannenberg P (2023) Making heterogeneous infrastructure futures in and beyond the global south. Futures 154:103270
Le LT (2024) Impact of environmental, social and governance practices on financial performance: Evidence from listed companies in Southeast Asia. Cogent Bus Manag, 11(1): Article 2379568. https://doi.org/10.1080/23311975.2024.2379568
Leal-Arcas R, Faktaufon M, Kyprianou A (2022) A legal exploration of the European Union's carbon border adjustment mechanism. Eur Energy Environ Law Rev 31(4)
Li M, Sun Y, Xia Y, Yang Z, Chen C, Yuan Y (2026) The impacts of carbon border adjustment mechanism (CBAM) on international trade and policy responses: from an economic and environmental equity perspective. Energy Policy 210:115014. https://doi.org/10.1016/j.enpol.2025.115014
Lin B, Zhao H (2024) Threatening the Poor? The economic impacts of carbon border adjustment mechanism on developing countries. Struct Chang Econ Dyn 71:582–593
Lobo RCG et al (2025) Electric bus adoption for sustainable urban mobility: Public policies, technological maturity of cities, and small business innovation. Res Transp Bus Manag **62**:101463. https://www.sciencedirect.com/science/article/pii/S2210539525001786
Lumempouw GJ (2024) Sustainability in Global Trade: Integrating principles for inclusive and resilient systems. Asian J Logist Manag 3(1):20–28
Mafini C (2018) Extending green supply chain management activities to developing country SMEs: Evidence from South Africa. South African J Econ Manag Sci. https://sajems.org/index.php/sajems/article/view/1996/1301
Magacho G, Espagne E, Godin A (2023) Impacts of the CBAM on EU trade partners: consequences for developing countries. Climate Policy 24(2):243–259. https://doi.org/10.1080/14693062.2023.2200758

Maimele S (2025) Unpacking and responding to Border Carbon Adjustments (BCAs): a Global South perspective. In: Environmental tax reforms for a just energy transition. Edward Elgar Publishing, pp. 95–113

Messaoudi O, Samet K, Derbel A (2026) AI-enabled strategies for sustainable cross-border logistics: sustainable development goals (SDGs) and decarbonized logistics. In: AI-Enabled strategies for sustainable cross-border logistics. IGI Global Scientific Publishing, pp 355–420

Morone P, Alfino A (2025b) Impact of the EU carbon border adjustment mechanism on the Global South. Discov Sustain 6(1):315

Morone P, Alfino A (2025a) Impact of the EU carbon border adjustment mechanism on the Global South. Discov Sustain 6: Article 315. https://doi.org/10.1007/s43621-025-01131-x

Nagiah GR, Mohd Suki N (2024) Linking environmental sustainability, social sustainability, corporate reputation and the business performance of energy companies: insights from an emerging market. Int J Energy Sect Manage 18(6):1905–1922

Nagy G, Szentesi S (2024) Green logistics: transforming supply chains for a sustainable future. Adv Logist Syst Theory Pract 18(3):29–42. https://doi.org/10.32971/als.2024.026

Nagy G, Szentesi S (2024b) Green logistics: transforming supply chains for a sustainable future. Adv Logist Syst-Theory Pract 18(3):29–42

Nicoletti B, Appolloni A (2024) Green Logistics 5.0: A review of sustainability-oriented innovation with foundation models in logistics. Eur J Innov Manag 27(9):542–561

Nosike RDCJ (2024) Digitalization in developing countries: opportunities and challenges. Niger J Arts HumIties (NJAH) 4(1)

Notteboom T, Pallis A, Rodrigue JP (2022) Port economics, management and policy (1st ed.). Routledge. https://doi.org/10.4324/9780429318184

Novinc F, Škuflić L (2025) Is cost competitiveness a sufficient driving force for croatian exports? South East Eur J Econ Bus 20(2):16–33

Nwokolo SC, Singh R, Khan S, Kumar A, Luthra S (2023) Decarbonizing hard-to-abate sectors in Africa. In: Africa's path to net-zero: Exploring scenarios for a sustainable energy transition. Cham: Springer Nature Switzerland, pp 211–236

Okeke A (2024) Evaluating sustainable practices and supply chain management effectiveness in African SMEs. J Sustain Res. https://sustainability.hapres.com/htmls/JSR_1605_Detail.html

Okoro C (2025) Impact of carbon border adjustment mechanisms (CBAMs) on the export competitiveness of developing nations. Int J Econ Policy 5(4):53–61. https://doi.org/10.47941/ijecop.3252

Onukwulu EC, Agho MO, Eyo-Udo NL (2021) Framework for sustainable supply chain practices to reduce carbon footprint in energy. Open Access Res J Sci Technol 1(2):012–034

Ouni M, Mraihi R (2025) Impact of transport infrastructure on poverty, income inequality, and unemployment in developing countries: an assessment of the sustainable development goals. Afr Transp Stud 3:100050

Phinijphara S, Lakkhongkha K (2024) Revolutionizing regional trade: digital logistics in Thailand's eastern economic corridor. J ASEAN PLUS Stud 5(1):22–37. https://so06.tci-thaijo.org/index.php/aseanplus/article/view/270407

Setyaningsiha S, Widjojo R, Stephenson PK (2024) Challenges and opportunities in sustainability reporting: a focus on small and medium enterprises (SMEs). Cogent Bus Manag 11(1):Article 2298215. https://doi.org/10.1080/23311975.2023.2298215

Sinkovics N, Sinkovics RR, Archie-Acheampong J (2021) Small and medium-sized enterprises and sustainable development: In the shadows of large lead firms in global value chains. J Int Bus Policy 4(1):80–101

Solakivi T, Saurama A, Ojala LM (2026) Contradictory emission regulations across transport chains: challenges for international supply chains. Int J Phys Distrib Logist Manag 56(11):1–33

Srivastav S, Fankhauser S, Kazaglis A (2018) Low-Carbon Competitiveness in Asia. Economies 6(1):5

Stojanović Đ, Ivetić J, Veličković M (2021) Assessment of international trade-related transport CO2 emissions a logistics responsibility perspective. Sustainability 13(3):1138

Sulemana I, Cheng L, Agyemang AO, Osei A, Nagriwum TM (2025) Stakeholders and sustainability disclosure: Evidence from an emerging market. Sustainable Futures 9:100445

Sun X, Mi Z, Cheng L, Coffman D, Liu Y (2023) The carbon border adjustment mechanism is inefficient in addressing carbon leakage and results in unfair welfare losses. Fundam Res 4(3):660–670. https://doi.org/10.1016/j.fmre.2023.02.026

Tetteh FK, Kwateng KO, Tukue T, Mensah J (2025b) Green supply chain management practices: review, framework and future research directions. J Responsible Prod Consum 2(1):112–150

Tetteh FK, Kwateng KO, Tukue T, Mensah J, Atiase DD (2025) Effect of logistics performance on transport environmental quality: comparative insight from Sub-Saharan Africa and OECD markets. Asian J Shipp Logist

Ur-Rashid HH, Khan MA (2024) Standards and nontariff barriers in trade–A case study of South Asia. Struct Chang Econ Dyn 69:427–437

Vazquez-Brust DA, Sarkis J, Cordeiro JJ (2014) Collaboration for sustainability and innovation in the Global South: a cross-border, multistakeholder perspective. In: Vazquez-Brust D, Sarkis J, Cordeiro J (eds) Collaboration for sustainability and innovation: a role for sustainability driven by the Global South? Greening of industry networks studies. Springer, Dordrecht, vol 3. https://doi.org/10.1007/978-94-007-7633-3_1

Venter C, Mahendra A, Hidalgo D (2019) From mobility to access for all: Expanding urban transportation choices in the global south. World Resources Institute, Washington, DC, pp 1–48

Villena VH, Dhanorkar S (2020) How institutional pressures and managerial incentives elicit carbon transparency in global supply chains. J Oper Manag 66(6):697–734

Vîrjan D, Manole AM, Stanef-Puică MR, Chenic AS, Papuc CM, Huru D, Bănacu CS (2023) Competitiveness—the engine that boosts economic growth and revives the economy. Front Environ Sci 11:1130173

Wang L, Lin S, Li C (2026) The synergistic effect of china's dual track carbon reduction policies: a joint test based on carbon trading and green certificate trading. SAGE Open 16(1):21582440251410300

Zhang W, Sovacool BK (2025) The "triple burden" effect and "pressure-opportunity paradox" of net-zero transitions: Exploring the political economy of Carbon Border Adjustment Mechanism (CBAM) implementation in the Global South. Energ Strat Rev 62:101913

Zhao R (2024) Green trade barriers in global play: legal challenges and strategic opportunities for developing countries. J Educ, HumIties Soc Sci 39:0048–55. https://doi.org/10.54097/kz926p47

Regional Integration and Cross-Border Green Transport Corridors

Munyaradzi Mahwine, Munyaradzi Chibaro, and David Chikwere

Abstract Cross-border green transport corridors have emerged as a strategically significant response to the dual imperatives of regional economic integration and environmental sustainability. This study examines the intersection of green corridor development and regional integration dynamics, drawing on recent empirical and conceptual scholarship to interrogate the mechanisms through which cross-border transport infrastructure facilitates economic cohesion, reduces carbon emissions, and strengthens inter-state governance frameworks. The study adopts a qualitative, interpretive methodology grounded in systematic literature synthesis and conceptual modelling. Findings reveal that while green transport corridors hold substantial promise for advancing sustainable regional development, their effective implementation is contingent upon coordinated policy frameworks, adequate financing architectures, and the resolution of institutional asymmetries across participating states. The study advances a conceptual model linking transport connectivity, green corridor governance, and regional integration outcomes, and concludes with implications for policy and managerial practice, as well as directions for future empirical inquiry.

Keywords Green transport corridors · Regional integration · Cross-border cooperation · Sustainable logistics · Transport policy · Africa · Europe

M. Mahwine
Department of Risk and Supply Chain Management, Midlands State University, Gweru, Zimbabwe

M. Chibaro (✉)
Department of Logistics and Supply Chain Management, University of Botswana, Gaborone, Botswana
e-mail: munyachibaro@gmail.com

D. Chikwere
Leeds Business School, Leeds Beckett University, Leeds, UK
e-mail: D.chikwere@leedsbeckett.ac.uk

J. Kanyepe and I. Chirisa (eds.), *Advancing the Green Agenda in the Transport and Logistics Sector in Global South Countries*, World Sustainability Series,
https://doi.org/10.1007/978-3-032-31374-4_13

1 Introduction

Transport has long occupied a central position in the discourse on regional economic integration. The capacity of nations and sub-national territories to exchange goods, people, and services efficiently across shared borders is not merely a logistical matter; it is fundamentally political, economic, and, increasingly, ecological. Over the past two decades, the framing of cross-border transport infrastructure has undergone a discernible shift, moving from a predominantly connectivity-centred paradigm towards one that foregrounds environmental sustainability alongside economic functionality (Panagakos 2016; Aditjandra et al. 2016). The concept of the green transport corridor has emerged from this reorientation, offering a framework in which freight and passenger movements are not only optimised for speed and cost but are evaluated against criteria of carbon intensity, modal efficiency, and ecological impact.

This study is situated at the intersection of two bodies of scholarship that have developed largely in parallel but have rarely been brought into direct, sustained dialogue: the literature on cross-border regional integration and the emerging literature on green logistics and sustainable transport corridors. The tendency to treat these as separate domains is understandable given their distinct disciplinary origins the former rooted in political economy, international relations, and regional planning, the latter in logistics science, environmental economics, and transport engineering yet it has produced a conceptual gap that this study seeks to address (Castanho et al. 2017; Medeiros et al. 2021). Regional integration processes, whether in Europe, Africa, or Asia, are inescapably infrastructural: their success depends in no small measure on the quality, capacity, and sustainability of the transport networks that connect their member states.

The argument advanced here is that green transport corridors are not merely an environmental add-on to conventional cross-border infrastructure development but constitute a qualitatively distinct mode of corridor governance that carries profound implications for how regional integration is imagined and practised. A green corridor does not simply move freight more cleanly; it restructures the incentive landscape for cross-border cooperation, creates new institutional interdependencies, and generates distributional effects that cut across national interests in ways that conventional grey infrastructure development does not (Basche and Spera 2023; Bertram et al. 2024). Understanding these dynamics is essential for scholars and practitioners engaged with the governance of regional integration in the twenty-first century.

2 Background to the Study

The study of cross-border transport connectivity has a relatively long pedigree, stretching from early work on European economic integration through to more recent scholarship on inter-regional corridors in the Global South. What has changed markedly in recent years is the normative frame within which this scholarship operates. The classic work of Castanho et al. (2017) on accessibility and connectivity in European borderlands, for instance, established the importance of transport linkages as enabling conditions for cross-border cooperation projects, framing connectivity as a critical prerequisite for integration success. Environmental sustainability, where it appeared at all, tended to be treated as a secondary consideration or as a separate domain of analysis.

The past five years have witnessed a substantive shift in this regard. Driven partly by the mainstreaming of climate commitments within international development frameworks and partly by the recognition that transport accounts for a disproportionate share of global greenhouse gas emissions, scholars and policymakers have increasingly engaged with the question of how cross-border transport infrastructure can be redesigned, regulated, and financed to deliver environmental as well as economic dividends (Aditjandra et al. 2016; Panagakos 2016). The green corridor concept which Panagakos (2016) situates within a broader project of decoupling freight movement from carbon output has gained traction as an analytical and operational framework in this context.

In the European context, cross-border cooperation in transport has been institutionalised through structures such as the European Territorial Cooperation framework, Euroregions, and trans-European transport networks (Medeiros et al. 2021; Basche and Spera 2023). These arrangements have generated a rich empirical record of both the achievements and the limitations of cross-border transport planning, with scholars identifying persistent obstacles including the mismatch of national regulatory frameworks, asymmetric investment capacities between neighbouring states, and the political salience of border regions as sites of identity contestation (Bertram et al. 2024; Cappellano et al. 2021). The Euregio Meuse-Rhine case, analysed by Basche and Spera (2023), illustrates both the integrative potential of cross-border public transport and the complex web of institutional interactions that must be navigated to realise that potential.

In Africa, the context is substantively different. Here, transport infrastructure remains chronically underdeveloped relative to the continent's integration aspirations, and the relationship between transport corridors and regional economic integration is both more consequential and more contested (Masuku 2024; Mwesigye 2021). The East African Community, the Southern African Development Community, and the African Continental Free Trade Area have all identified transport connectivity as a prerequisite for meaningful economic integration, yet progress in building seamless, interoperable cross-border networks has been uneven (Flory and Nyaronga 2025; Walumweya and Kawimbe 2024). The additional dimension of environmental sustainability adds complexity to an already demanding policy agenda, particularly given the financing constraints facing African states and the imperative to avoid locking in carbon-intensive infrastructure in the name of development.

It is against this dual backdrop the maturation of green corridor thinking in European contexts and the urgent, if more challenging, transport-integration nexus in African settings that this study develops its analysis. The central contention is that green cross-border transport corridors represent a promising but insufficiently theorised instrument of regional integration, one whose potential can only be realised through purposive institutional design, sustained political commitment, and a genuinely comparative scholarly conversation that crosses the Europe-Africa divide.

3 Research Objectives and Hypotheses

The study is guided by the following overarching research objective: to critically examine the relationship between green cross-border transport corridor development and regional integration, with particular attention to the governance, policy, and institutional dimensions of this relationship. From this, three specific objectives are derived:

RO1: To assess the extent to which green transport corridor frameworks advance regional integration objectives in cross-border contexts.

RO2: To identify the institutional, policy, and financing conditions that facilitate or obstruct the development and sustainability of green cross-border transport corridors.

RO3: To develop a conceptual model that captures the relationship between green corridor governance, transport connectivity, and regional integration outcomes.

From these objectives, three research hypotheses are formulated for analytical orientation. First (H1), it is hypothesised that green transport corridors that are embedded within formal regional integration frameworks demonstrate higher levels of institutional sustainability and modal efficiency than those developed outside such frameworks. Second (H2), it is hypothesised that the effectiveness of green cross-border transport corridors in promoting regional integration is moderated by the degree of regulatory harmonisation and governance alignment between participating states. Third (H3), it is hypothesised that financing mechanisms that combine public investment with green finance instruments produce more durable corridor outcomes than either pure public or pure private financing arrangements. These hypotheses are treated as analytical propositions structuring the literature review and discussion rather than as claims to be tested through statistical inference.

4 Conceptual Model

The conceptual model developed in this study draws on three interrelated constructs: transport connectivity, green corridor governance, and regional integration outcomes. These are linked by a series of mediating and moderating relationships informed by the existing literature, as elaborated below.

Transport connectivity is understood here in a broad sense, encompassing both the physical infrastructure of cross-border movement roads, railways, ports, and border crossing facilities and the functional accessibility that this infrastructure generates for people, goods, and services (Castanho et al. 2017; Vulevic et al. 2020). Connectivity is not simply a technical attribute of infrastructure but is shaped by the regulatory environment, the quality of logistics services, and the institutional arrangements that govern the use of border facilities (Furuichi 2021; Aslam and Khan 2022). Yuan and Yang (2022) propose a systems dynamics approach to evaluating cross-border transport infrastructure interconnection, highlighting the non-linear and feedback-driven nature of connectivity outcomes an insight that this model incorporates by treating connectivity as a dynamic rather than static variable.

Green corridor governance refers to the ensemble of institutional arrangements, policy instruments, regulatory frameworks, and financing mechanisms through which the environmental performance of cross-border transport corridors is managed and sustained. Panagakos (2016) identifies green corridor governance as a multi-actor process spanning public authorities, transport operators, logistics firms, and civil society, with each actor carrying distinct interests and capacities. Aditjandra et al. (2016) extend this by emphasising the importance of technology deployment particularly in rail freight as a governance lever for reducing the carbon intensity of corridor operations. The governance dimension is particularly salient in cross-border contexts, where the absence of a single regulatory authority means that green performance standards must be negotiated, aligned, and enforced across multiple jurisdictions.

Regional integration outcomes are understood here as encompassing economic, political, and social dimensions. Economic outcomes include trade facilitation, reduced transaction costs, enhanced market access, and the stimulation of cross-border investment (Mwesigye 2021; Flory and Nyaronga 2025). Political outcomes relate to the strengthening of inter-state relationships, the institutionalisation of cooperative governance arrangements, and the reduction of political border effects (Medeiros et al. 2021; Nshimbi et al. 2018). Social outcomes include the mobility gains of border populations, the development of cross-border labour markets, and the emergence of shared cultural or regional identities facilitated by improved connectivity (Stoffelen 2018; Cavallaro and Dianin 2020).

The model proposes that green corridor governance mediates the relationship between transport connectivity and regional integration outcomes: improved connectivity alone does not guarantee integration dividends unless it is governed in ways that align incentives, coordinate investment, and manage the distributional effects of corridor development. This mediation is further moderated by two contextual factors: the degree of institutional maturity of the regional integration framework in question and the financing architecture available for corridor development. Where regional institutions are well developed and green finance instruments are accessible, the positive relationship between governance and integration outcomes is expected to be stronger. Conversely, where institutional capacity is weak and financing is constrained conditions characteristic of many African cross-border corridors the model predicts that even well-designed green corridor frameworks will struggle to generate durable integration outcomes without targeted capacity support.

5 Research Methodology

This study employs a qualitative, interpretive methodology centred on systematic literature synthesis and conceptual analysis. Given that the primary objective is to develop theoretical understanding of the relationship between green corridor development and regional integration rather than to generate statistical generalisations from primary field data a qualitative approach is both appropriate and well aligned with established practice in transport planning and regional development scholarship (Cappellano et al. 2021; Basche and Spera 2023). The study does not seek to test hypotheses in the inferential statistical sense but uses them as orienting propositions that structure the reading and synthesis of existing scholarship.

The analytical approach draws on what Tranfield et al. (2003) term structured narrative synthesis, in which themes emerging from the literature are systematically organised and interrogated in relation to the study's research objectives, rather than subjected to quantitative meta-analytic aggregation. This approach is appropriate given the heterogeneity of methodological traditions represented in the literature which spans engineering system analysis (Yuan and Yang 2022), political geography (Bertram et al. 2024), logistics science (Furuichi 2021), and development economics (Flory and Nyaronga 2025) and the consequent incomparability of outcome metrics across studies. The study's interpretive framework is further informed by the conceptual model articulated in the preceding section, which guided the thematic organisation of the literature review and the construction of the discussion.

Recognising the inherent limitations of a desk-based, literature-synthesis approach, the study draws explicitly on comparative case evidence where available, particularly the empirical studies of European cross-border transport arrangements (Medeiros et al. 2021; Basche and Spera 2023; Bertram et al. 2024) and African corridor development initiatives (Flory and Nyaronga 2025; Masuku 2024; Walumweya and Kawimbe 2024), to ground its conceptual claims in observable practice. The study acknowledges the limitations of generalising from these cases to the full diversity of global cross-border transport contexts, and this acknowledgement is reflected in the directions for further research outlined in a later section.

6 Literature Review

Cross-Border Transport Connectivity and Regional Integration (RO1)

The relationship between transport connectivity and regional economic integration is one of the most extensively studied topics in regional development scholarship. At its most fundamental, the argument is straightforward: reducing the costs and time associated with cross-border movement of goods, services, and people unlocks trade flows, encourages factor mobility, and creates the conditions for deeper economic interdependence between neighbouring territories (Castanho et al. 2017; Aslam and

Khan 2022). What makes this relationship analytically interesting is the degree to which it is conditioned by factors beyond the physical availability of infrastructure.

Medeiros et al. (2021) examined cross-border regions across Europe and found that improvements in cross-border transport services consistently boosted regional economic performance, but that the magnitude of integration gains was strongly differentiated by the institutional quality of cross-border cooperation arrangements. Regions embedded in formalised cooperation structures such as European Groupings of Territorial Cooperation—extracted substantially larger integration dividends from transport investment than regions relying on informal or ad hoc coordination mechanisms. This finding is significant for the purposes of the present study because it establishes empirically what the conceptual model proposes: that infrastructure investment alone is insufficient to generate integration outcomes and that governance architecture matters fundamentally.

Vulevic et al. (2020) analysed accessibility dynamics in the Portuguese-Spanish borderland and demonstrated that cross-border cooperation projects that addressed both physical connectivity and accessibility governance generated more durable integration effects than those focused exclusively on infrastructure construction. This distinction between connectivity the technical capacity of infrastructure to facilitate movement and accessibility the actual ease with which people and goods can move across borders is an important one, picked up also by Castanho et al. (2017), who argued that connectivity and accessibility function as complementary rather than interchangeable drivers of cross-border integration success. From a green corridor perspective, this distinction is particularly salient because environmental regulations and green performance standards can, if poorly designed, reduce accessibility even while enhancing the environmental profile of corridor infrastructure.

In the Asian context, Menon (2024) examines the Asian Development Bank's role in fostering regional integration through cross-border transport connectivity and identifies a persistent tension between the bank's infrastructure-first development model and the growing recognition that connectivity investments need to be embedded within broader regional governance frameworks to generate lasting integration effects. Dai et al. (2024) extend this to the specific case of Xinjiang's cross-border transport network optimisation, arguing that route-level efficiency improvements must be accompanied by corridor-level governance reforms to translate local connectivity gains into regional integration outcomes. Together, these contributions reinforce the central thesis of this study: that the relationship between transport connectivity and regional integration is fundamentally mediated by governance.

Hanaoka et al. (2021) introduce a further dimension by examining the factors that influence modal choice in inter-regional cross-border transport, finding that the decision between land and maritime routes is shaped not only by cost and time considerations but also by the regulatory environment and the availability of multimodal transfer facilities at border points. This is directly relevant to the green corridor agenda, since modal shifts particularly from road to rail or maritime are among the most powerful instruments for reducing the carbon intensity of cross-border freight movement. The finding that modal choice is institutionally conditioned suggests that green corridor policies must engage with regulatory and institutional barriers to modal shift rather than assuming that infrastructure provision alone will drive behaviour change.

Green Corridor Governance: Frameworks, Policies, and Financing (RO2)

The green corridor concept, in its contemporary form, emerged from the European logistics and transport policy community in the first decade of this century, crystallising around the idea that specific high-volume freight corridors could serve as test-beds for the integrated deployment of clean technologies, optimised logistics practices, and supportive regulatory frameworks (Panagakos 2016; Aditjandra et al. 2016). Panagakos (2016) provides the most comprehensive conceptual elaboration of this framework, defining green corridors as transport networks in which the environmental performance of freight operations is systematically measured, reported, and improved through a combination of technology deployment, operational optimisation, and governance intervention. The breadth of this definition is both its strength and its analytical challenge: it encompasses a wide range of corridor configurations and governance models, making comparative assessment difficult.

Aditjandra et al. (2016) situate green rail transportation within this broader corridor framework, arguing that rail is uniquely positioned as a green corridor backbone technology given its substantially lower carbon intensity relative to road freight. However, they note that the structural shift from road to rail in cross-border freight contexts faces significant institutional barriers, including the incompatibility of railway gauge standards across national boundaries, differences in electrification standards, and the fragmentation of rail operating markets along national lines. These barriers are not primarily technological but institutional and political, reinforcing the governance-centred analytical frame adopted in this study.

Basche and Spera (2023) provide a particularly instructive analysis of green corridor governance in the Euregio Meuse-Rhine context, mapping the interactions between public transport authorities, regional governments, and national ministries in the management of cross-border public transport services. Their study identifies a characteristic tension in cross-border green transport governance between the drive for integrated, corridor-level performance management and the persistence of nationally bounded administrative structures and financing arrangements. This tension which they describe as a structural feature of cross-border governance rather than a resolvable problem has direct implications for green corridor sustainability: governance arrangements that cannot bridge national administrative silos will struggle to maintain consistent environmental performance standards across corridor segments.

Bertram et al. (2024) adopt a space–time analytical framework to examine modal choice between rail and road in cross-border mobility contexts, demonstrating that travel time competitiveness rather than environmental performance is the primary driver of modal decisions for most cross-border travellers. Their findings have sobering implications for green corridor governance: if travellers choose modes on the basis of time rather than carbon footprint, then green corridor policies that do not address the time-competitiveness of low-carbon modes risk generating corridors that are green in design but not in practice. This insight calls for governance frameworks that combine environmental standards with targeted investments in the speed and reliability of sustainable transport modes.

The financing of green cross-border transport corridors presents its own distinctive governance challenges. Bolgova et al. (2024) examine financing of Eurasian Economic Union international transport corridors and highlight the complexity of multi-state investment coordination in corridor development, noting that the misalignment of national financing cycles, investment priorities, and risk appetites frequently delays or distorts corridor development plans. Yuan and Yang (2022) demonstrate through a system dynamics approach that the interconnection performance of cross-border transport infrastructure is highly sensitive to financing consistency: corridors that experience investment gaps or financing discontinuities at individual segments show disproportionate system-level performance degradation. These findings point to the importance of corridor-level financing architectures such as dedicated corridor funds or multilateral green bonds that can smooth over the national financing inconsistencies that characterise multi-state corridor development.

Desiderio (2025) advances a particularly ambitious vision for Africa, proposing the combination of green and digital corridor approaches to revolutionise continental logistics. This dual-corridor concept in which environmental sustainability and digital transformation are pursued simultaneously as integrated corridor governance objectives represents a qualitative step beyond conventional green corridor frameworks and carries significant implications for how regional integration is conceptualised on the continent. The digital dimension, Desiderio (2025) argues, is not merely an operational enhancement but a governance enabler: digital platforms for cargo tracking, customs pre-clearance, and corridor performance monitoring can dramatically reduce the transaction costs associated with cross-border trade while simultaneously providing the data infrastructure needed to manage and verify environmental performance.

In Africa, the intersection of corridor governance and regional integration takes on additional complexity given the diversity of institutional frameworks, the varying degrees of state capacity among corridor member states, and the particular financing constraints of the continent's infrastructure development challenge. Masuku (2024) surveys the landscape of transport infrastructure and regional integration in Africa, identifying the African Union's Programme for Infrastructure Development in Africa (PIDA) as the continent's primary framework for aligning corridor development with integration objectives, but noting persistent weaknesses in PIDA's implementation architecture. Eke (2025) argues that the African Continental Free Trade Area creates a qualitatively new imperative for cross-border logistics optimisation, one that will require not only infrastructure investment but a fundamental rethinking of the governance of corridor operations across the continent.

7 Tourism, Mobility, and the Social Dimensions of Green Corridors

The integration effects of green cross-border transport corridors are not confined to freight and economic dimensions. Cavallaro and Dianin (2020) demonstrate that cross-border public transport functions as a driver of tourism development in Alpine border regions, creating new visitor flows and stimulating the development of cross-border tourism products. Stoffelen (2018) extends this observation to the specific case of cycling tourism trails as tools for cross-border integration, showing that the Vennbahn cycling route along the Belgian-German border has become a vehicle for cultural exchange, cross-border community building, and the development of shared regional identity. These findings suggest that green transport corridors which, by definition, include sustainable and active mobility modes carry integration potential that extends well beyond the freight and economic efficiency dimensions that typically dominate corridor policy discourse.

This social and cultural dimension of green corridor integration is particularly important in border regions that have historically been defined by division or conflict, where the development of shared mobility infrastructure can serve as a confidence-building and normalisation mechanism. Cappellano et al. (2021) document the role of cross-border regional planning in the Cascadia region of North America in building shared institutional capacity and regional identity across the US-Canadian border, noting that transport infrastructure planning has functioned as a practical entry point for deeper cross-border cooperation that extends into social policy, environmental management, and economic development. The lesson for green corridor development is that corridors should be conceived not merely as freight arteries but as multi-functional integration instruments with significant social and cultural dimensions that deserve explicit attention in corridor governance frameworks.

8 Sub-Saharan Africa: Specific Challenges and Emerging Opportunities

The African dimension of cross-border green transport corridor development merits separate treatment given the scale of the infrastructure deficit, the specificity of institutional challenges, and the stakes of getting corridor development right in a context where path dependencies are still being established. Flory and Nyaronga (2025) examine the Central Corridor Transit Transport Facilitation Agency and its contribution to economic growth through transport corridor development, finding that institutionalised corridor management bodies generate significantly better logistics performance outcomes than informal or bilateral corridor arrangements. Their study provides direct empirical support for the first hypothesis advanced in this study: that green corridors embedded in formal regional frameworks perform better than those developed outside such structures.

Mwesigye (2021) examines the role of transport and logistics in East African Community trade promotion and regional integration, identifying non-physical barriers—including inconsistent customs procedures, duplicative documentation requirements, and inadequate border crossing facilities as more significant impediments to cross-border trade than physical infrastructure gaps in many EAC corridors. This finding has important implications for green corridor governance: it suggests that the governance of corridor operations including the harmonisation of customs and regulatory procedures may deliver greater integration dividends than additional infrastructure investment, particularly in corridors where the physical connectivity already meets basic functionality thresholds.

Walumweya and Kawimbe (2024) analyse cross-border trade optimisation in Eastern and Southern Africa, providing policy insights that complement the institutional analysis of Flory and Nyaronga (2025) and Mwesigye (2021). Their work highlights the particular importance of digital trade facilitation tools in reducing the transaction costs of cross-border commerce in the region, echoing Desiderio's (2025) argument about the combined potential of green and digital corridor approaches. Onyekachi and Nwosu (2021) situate these trade corridor dynamics within the broader conceptual frame of regional integration, arguing that effective trade corridors are a necessary but not sufficient condition for the forms of deep integration including factor market integration and regulatory convergence that generate the most substantial economic dividends. This distinction between shallow and deep integration is important for green corridor policy: corridors that merely reduce physical transport costs may facilitate shallow integration without generating the institutional convergence and governance alignment that deep integration requires.

Baranzelli et al. (2022) contribute a distinct analytical dimension by examining the intersection of EU-Africa strategic corridors and critical raw materials supply chains, arguing that the development of green transport infrastructure connecting African resource-producing regions with European markets serves both regional development and strategic supply security objectives. This dual-purpose framing of green corridor development simultaneously serving African integration and European strategic interests complicates the governance of such corridors and raises questions about the equity of the arrangements through which they are financed and managed. Nshimbi et al. (2018) provide essential background on the role of informal cross-border economies in African regional integration, a dimension that formal corridor governance frameworks frequently overlook but that represents a substantial share of actual cross-border economic activity in many African corridors.

9 Results and Discussion

The synthesis of the literature across the preceding review sections yields a set of findings that, taken together, constitute a substantive response to the study's research objectives and allow the research hypotheses to be assessed in analytical terms. The discussion is organised around the three hypotheses, with each serving as a focal point for integrating findings from the literature and developing their implications.

10 Hypothesis One: Formal Integration Frameworks and Corridor Sustainability

The evidence supporting H1 that green transport corridors embedded in formal regional integration frameworks demonstrate higher institutional sustainability and modal efficiency is substantial and consistent across geographical contexts. Medeiros et al. (2021) provide direct European evidence that institutionalised cross-border transport cooperation generates larger integration dividends than informal arrangements. Flory and Nyaronga (2025) provide analogous African evidence in the context of the Central Corridor Transit Transport Facilitation Agency. Menon (2024) draws the same conclusion from the Asian Development Bank's extensive experience with cross-border transport investment across Asia.

What the literature also reveals, however, is that the relationship between formal frameworks and corridor performance is not automatically positive: formal frameworks that are poorly designed, inadequately resourced, or lacking genuine political commitment from member states can become obstacles rather than enablers of corridor development. Basche and Spera (2023) document precisely this dynamic in the Euregio Meuse-Rhine, where the multiplication of formal cooperation structures has in some cases complicated rather than simplified the governance of cross-border transport. This qualification is important for policy: the implication is not simply that more formalisation is always better, but that the design and political underpinning of formal frameworks matters as much as their existence.

The modal efficiency dimension of H1 finds support in the work of Hanaoka et al. (2021) and Bertram et al. (2024), both of which demonstrate that modal shifts towards greener transport modes require not only physical infrastructure investment but sustained governance support. Green corridor frameworks that are anchored in formal regional integration institutions are better positioned to sustain the long-term governance commitment—including regulatory harmonisation, investment in intermodal transfer facilities, and the maintenance of time-competitiveness for sustainable modes—that modal shift requires. The implication is that green corridor governance cannot be an ad hoc or project-bounded endeavour but must be institutionalised within durable regional governance structures to generate lasting modal efficiency gains.

11 Hypothesis Two: Regulatory Harmonisation and Governance Alignment

H2 that regulatory harmonisation and governance alignment moderate the effectiveness of green corridors in promoting regional integration receives particularly strong support from the African literature. Mwesigye (2021) and Walumweya and Kawimbe (2024) both identify non-physical regulatory barriers as more significant impediments to cross-border trade integration than physical infrastructure gaps in East and Southern African corridors. Eke (2025) argues that the African Continental Free Trade Area's transformative potential will only be realised if accompanied by a comprehensive programme of logistics and customs regulatory harmonisation across member states.

The European literature provides important lessons about both the potential and the limits of regulatory harmonisation in cross-border green transport governance. Aditjandra et al. (2016) highlight the persistence of technical incompatibilities between national rail systems as a structural barrier to the development of seamless green rail corridors across European borders, noting that decades of integration effort have not fully resolved the gauge, electrification, and signalling incompatibilities that fragment European rail. Bertram et al. (2024) add that even where physical incompatibilities have been resolved, regulatory differences in operating licences, safety certifications, and passenger rights create friction that reduces the competitiveness of cross-border rail relative to road. These findings suggest that regulatory harmonisation is a necessary but slow-moving and politically demanding process that green corridor governance must actively manage rather than assume will occur organically.

Dai et al. (2024) offer a contrasting perspective from Central Asia, where cross-border transport network optimisation is conditioned by the overlay of multiple competing regional integration frameworks including Chinese Belt and Road Initiative structures, Eurasian Economic Union arrangements, and bilateral state agreements each with its own regulatory logic. In contexts of regulatory pluralism, the governance challenge for green corridors is not harmonisation in the classical sense but the navigation of a complex institutional landscape in which different regulatory frameworks apply to different corridor segments. This is an analytically distinctive challenge that the European integration literature, which assumes a single overarching regulatory framework, is poorly equipped to address and that represents a significant direction for further theoretical development.

12 Hypothesis Three: Financing Architecture and Corridor Durability

The evidence on H3 that mixed public–private financing with green finance instruments produces more durable corridor outcomes is suggestive but less conclusive than the evidence on H1 and H2, partly because empirical studies specifically examining the financing architecture of green cross-border corridors are still relatively sparse. Bolgova et al. (2024) and Yuan and Yang (2022) both establish the importance of financing consistency in cross-border corridor performance, but neither specifically examines the differential effects of financing mechanism composition. Masuku (2024) and Eke (2025) identify financing as a critical constraint on African corridor development without providing detailed analysis of which financing architectures perform best in that context.

The most substantive contribution to H3 comes from Desiderio (2025), whose analysis of combined green and digital corridor approaches for Africa implicitly addresses the financing dimension by proposing a combination of development finance, green bonds, and public–private partnership structures as the appropriate architecture for transformative corridor development. Baranzelli et al. (2022) add a strategic dimension by noting that EU-Africa corridor financing arrangements that incorporate critical raw materials supply security objectives tend to attract more stable and long-term financing commitments than those framed purely in development or environmental terms. This finding suggests that the political economy of green corridor financing specifically, the degree to which corridor development aligns with the strategic interests of major financing actors may be as important as the technical design of financing instruments.

13 The Africa-Europe Comparative Dimension

One of the more striking findings to emerge from the literature synthesis is the degree of structural similarity between the governance challenges facing green cross-border transport corridors in European and African contexts, despite the profound differences in infrastructure maturity, institutional capacity, and economic development between these settings. In both contexts, the literature identifies the mismatch between nationally bounded administrative structures and the corridor-level governance requirements of integrated green transport as a fundamental challenge. In both contexts, regulatory harmonisation is identified as necessary but politically demanding. In both contexts, the time-competitiveness of sustainable transport modes relative to carbon-intensive alternatives is highlighted as a critical governance concern.

What differs is the scale and urgency of the challenge. In Europe, the green corridor agenda is a refinement of a relatively mature cross-border transport system; in Africa, it must be built from much more fragile foundations, in a context where basic infrastructure connectivity remains incomplete and institutional capacity is severely constrained. This difference has important implications for the transferability of European green corridor governance models to African settings. Direct transplantation of European governance frameworks is unlikely to be appropriate; what is needed instead is a process of contextual adaptation that retains the core governance principles corridor-level performance management, integrated financing, regulatory harmonisation while designing institutional arrangements that are calibrated to African institutional realities.

14 Directions for Further Research

The analysis presented in this study generates several directions for further empirical and theoretical inquiry. First, the conceptual model developed here requires empirical testing across a range of cross-border corridor contexts. The model's core propositions that green corridor governance mediates the connectivity-integration relationship, and that this mediation is moderated by institutional maturity and financing architecture are theoretically coherent and consistent with the existing literature, but they have not been directly tested through primary field research. Future studies employing mixed methods designs combining corridor-level performance data with qualitative governance analysis would be well positioned to provide such tests.

Second, the African context is substantially under-researched relative to its analytical and policy importance. The studies by Flory and Nyaronga (2025), Masuku (2024), Walumweya and Kawimbe (2024), and Eke (2025) represent important contributions, but they barely scratch the surface of the governance, financing, and institutional diversity of African cross-border corridor development. Longitudinal case studies of specific African corridor management bodies examining how they develop institutional capacity over time and how this development relates to corridor environmental and economic performance would be particularly valuable.

Third, the social and cultural dimensions of green corridor integration, touched upon in the work of Stoffelen (2018) and Cavallaro and Dianin (2020), deserve more systematic theoretical and empirical attention. The existing literature is heavily weighted towards economic and technical dimensions of corridor governance, with social impacts, border community experiences, and cultural integration effects receiving relatively limited scholarly attention. Research that integrates social geography and cultural studies perspectives into the green corridor framework could substantially enrich understanding of what integration actually means for the people who live in border regions.

Fourth, the intersection of digital transformation and green corridor governance highlighted by Desiderio (2025) represents an emerging research frontier that merits dedicated scholarly attention. The potential for digital platforms to simultaneously reduce transaction costs, enhance supply chain transparency, and provide the data infrastructure for green performance management is substantial, but the governance implications of digital corridor platforms including questions of data sovereignty, regulatory jurisdiction, and equitable access are poorly understood and require urgent scholarly attention.

Fifth, comparative research examining the differential experiences of green corridor development in Europe, Africa, and Asia would contribute substantially to the development of a genuinely global theory of green corridor governance. Much of the existing theoretical work is implicitly or explicitly Eurocentric, and the insights from African and Asian corridor development including Dai et al. (2024) on Xinjiang and Menon (2024) on the Asian Development Bank suggest that these contexts carry distinctive theoretical insights that a more globally attentive scholarship could exploit.

15 Managerial and Policy Implications

The findings of this study carry a number of implications for policymakers, corridor managers, and development finance practitioners engaged with green cross-border transport development. These are organised below around three principal themes.

16 Institutional Design

The evidence that formal regional integration frameworks enhance the sustainability and modal efficiency of green corridors carries a clear institutional design implication: green corridor governance should be deliberately anchored within existing or purpose-designed regional integration institutions rather than being managed through ad hoc or project-bounded arrangements. For African policymakers, this means ensuring that the continent's emerging corridor management bodies including those operating within the frameworks of the African Continental Free Trade Area and the Programme for Infrastructure Development in Africa are equipped with the mandate, resources, and institutional authority to pursue integrated green performance management across national boundaries. For European policymakers, it reinforces the case for consolidating rather than fragmenting the institutional architecture of cross-border transport governance, taking to heart the lessons of the Euregio Meuse-Rhine about the costs of institutional multiplication without corresponding integration (Basche and Spera, 2023).

Corridor managers need to develop governance capacities that span both technical and political domains. Managing a green cross-border corridor requires not only expertise in transport engineering and logistics optimisation but also the political and diplomatic skills to navigate the multi-jurisdictional governance landscapes that are inherent in cross-border settings. The development of specialised corridor management professional competencies combining technical transport knowledge with inter-institutional governance and cross-border negotiation skills is a training and capacity-building agenda that has received insufficient attention from both academic and professional institutions.

17 Regulatory and Policy Harmonisation

Policymakers in regional bodies should treat regulatory harmonisation as a strategic priority rather than a residual policy task. The evidence that non-physical regulatory barriers are more significant obstacles to integration than physical infrastructure gaps in many African corridors (Mwesigye 2021) suggests that the regulatory reform agenda deserves at least as much political and financial investment as infrastructure construction. Specific priorities for harmonisation include customs procedures, documentation requirements, transport operator licensing, vehicle safety standards, and environmental performance reporting frameworks. The development of shared digital customs and corridor management platforms as advocated by Desiderio (2025) and Walumweya and Kawimbe (2024) offers a practical pathway to accelerating harmonisation whilst building the data infrastructure needed for evidence-based corridor governance.

18 Financing Architecture

Development finance institutions and national governments should design corridor financing architectures that combine the long-term stability of public investment with the discipline and innovation that market-based green finance instruments introduce. Specifically, the development of dedicated corridor green bonds in which financing is tied to measurable environmental performance outcomes across the full corridor could provide both the financing stability and the performance incentive that current corridor financing arrangements frequently lack. Menon's (2024) analysis of the Asian Development Bank's corridor financing approach offers useful lessons in this regard, particularly the bank's use of performance-linked disbursement mechanisms that tie financing flows to the achievement of specific connectivity and environmental performance milestones. For African corridors in particular, financing architectures need to incorporate capacity-building components that address the institutional absorptive capacity constraints that frequently limit the effectiveness of externally financed infrastructure investment (Masuku 2024).

19 Conclusion

This study has examined the relationship between green cross-border transport corridors and regional integration, arguing that this relationship is fundamentally mediated by governance architecture and moderated by institutional maturity and financing design. The analysis has drawn on a broad range of recent scholarship spanning European, African, and Asian contexts to develop a conceptual model, test three analytical hypotheses, and generate substantive findings with both theoretical and practical implications.

The central argument of the study that green transport corridors are not merely environmental enhancements to conventional cross-border infrastructure but constitute a qualitatively distinct governance challenge with profound implications for regional integration has been borne out by the literature synthesis. The evidence supports the view that corridors embedded in formal regional frameworks outperform those managed informally (H1), that regulatory harmonisation and governance alignment are critical moderators of corridor effectiveness (H2), and that mixed financing architectures incorporating green finance instruments produce more durable outcomes than single-mechanism approaches (H3), although the evidence on H3 is less fully developed in the existing literature than on H1 and H2.

From a comparative perspective, the analysis reveals striking parallels between the governance challenges facing green corridors in European and African contexts, despite the substantial differences in infrastructure maturity and institutional capacity between these settings. In both contexts, the fundamental challenge is to bridge the gap between nationally bounded administrative structures and the corridor-level governance requirements of integrated green transport management. The solutions must, however, be context-sensitive: European governance models cannot be transplanted wholesale to African settings, and the distinctive institutional realities, financing constraints, and political economies of African corridor development require adapted approaches rather than borrowed templates.

The study has also identified significant directions for further research, including the empirical testing of the conceptual model, longitudinal case studies of African corridor management bodies, deeper engagement with social and cultural integration dimensions, examination of the governance implications of digital corridor platforms, and the development of a more genuinely global comparative theory of green corridor governance. These research directions represent a substantial agenda, one that is commensurate with the scale and complexity of the challenge that green cross-border transport corridors pose for regional integration scholarship and practice in the decades ahead.

In sum, realising the integrative and environmental potential of green cross-border transport corridors requires not only the construction of sustainable infrastructure but the patient, politically savvy work of building the governance institutions, regulatory harmonies, and financing architectures within which that infrastructure can perform. Scholars, policymakers, and practitioners all have essential roles to play in this endeavour, and the sustained dialogue between these communities is itself a precondition for the kind of evidence-based corridor governance that the green transport agenda demands.

References

Aditjandra PT, Zunder TH, Islam DMZ, Palacin R (2016) Green rail transportation: Improving rail freight to support green corridors. In Green Transp Logist: Quest Win-Win Solut (pp. 413–454). Springer International Publishing.

Aslam K, Khan M (2022) Cross-border connectivity and economic integration in regional contexts. J RegNal Connect Dev 1(1):22–32

Baranzelli C, Blengini GA, Josa SO, Lavalle C (2022) EU–Africa strategic corridors and critical raw materials: Two-way approach to regional development and security of supply. Int J Min Reclam Environ 36(9):607–623

Basche H, Spera F (2023) Interactions between key factors that influence cross-border cooperation in public transport: The case of the Euregio Meuse-Rhine. J BordLands Stud 38(5):681–698

Bertram D, Chilla T, Hippe S (2024) Cross-border mobility: Rail or road? Space-time-lines as an evidence base for policy debates. J BordLands Stud 39(5):913–930

Bolgova E, Bolgov S, Kurnikova M (2024) EAEU international transport corridors in the sustainable development of a regional economy. In E3S Web of conferences 471, p. 06005. EDP Sciences

Cappellano F, Richardson K, Trautman L (2021) Cross border regional planning: Insights from Cascadia. Int Plan Stud 26(2):182–197

Castanho RA, Vulevic A, Fernández JC, Fernández-Pozo L, Gómez JMN, Loures LC (2017) Accessibility and connectivity—movement between cities, as a critical factor to achieve success on cross-border cooperation (CBC) projects: A European analysis. Sustain Cities Soc 32:181–190

Cavallaro F, Dianin A (2020) Cross-border public transport as a driver for tourism in the Alps. Transp Res Procedia 48:2446–2461

Dai X, Liu M, Lin Q (2024) Research on optimisation strategies of regional cross-border transportation networks implications for the construction of cross-border transport corridors in Xinjiang. Sustain 16(13):5337

Desiderio D (2025) Unlocking Africa's potential: Combining the green and digital corridor approaches to revolutionise logistics. Available at SSRN 5253143

Eke EI (2025) Optimising cross-border logistics in Africa: Unlocking the potential of the African Continental Free Trade Area. Int J LeadShip, Qual Educ Afr Dev 2(2):192–213

Flory OO, Nyaronga GJ (2025) Enhancing economic growth through transport corridors: a study of the central corridor transit transport facilitation agency. Afr Dev Financ J 9(1):18–39

Furuichi M (2021) Cross-border logistics practices, policies, and its impact. In Glob Logist Netw Model Policy (pp 47–69). Elsevier

Hanaoka S, Matsuda T, Saito W, Kawasaki T, Hiraide T (2021) Identifying factors for selecting land over maritime in inter-regional cross-border transport. Sustain 13(3):1471

Masuku S (2024) Transport infrastructure and regional integration in Africa. In The Quest for Unity (pp 373–401). Routledge

Medeiros E, Ferreira R, Boijmans P, Verschelde N, Spisiak R, Skoniezki P, Berzi M (2021) Boosting cross-border regions through better cross-border transport services: The European case. Case Stud Transp Policy 9(1):291–301

Menon J (2024) The role of the Asian Development Bank in fostering regional integration through cross-border transport connectivity. In Elgar Companion Asian Dev Bank (pp 101–115). Edward Elgar Publishing

Mwesigye EB (2021) The role of transport and logistics in promoting trade and regional integration within the East African Community (EAC). J Public Policy & GovAnce 5(2):1–20

Nshimbi CC, Moyo I, Oloruntoba SO (2018) Borders, informal cross-border economies and regional integration in Africa: An introduction. Africa Insight 48(1):3–11

Onyekachi AB, Nwosu C (2021) Understanding the concept of trade corridor and regional integration: A key factor towards the development of Sub-Saharan Africa. Icheke Journal of the Faculty of Humanities, 19(3)

Panagakos G (2016) Green corridors basics. In Green Transportation Logistics: The Quest for Win-Win Solutions (pp 81–121). Springer International Publishing

Stoffelen A (2018) Tourism trails as tools for cross-border integration: A best practice case study of the Vennbahn cycling route. Ann Tour Res 73:91–102

Tranfield D, Denyer D, Smart P (2003) Towards a methodology for developing evidence-informed management knowledge by means of systematic review. Br J Manag 14(3):207–222

Vulevic A, Castanho RA, Naranjo Gomez JM, Loures L, Cabezas J, Fernandez-Pozo L, Martin Gallardo J (2020) Accessibility dynamics and regional cross-border cooperation (CBC) perspectives in the Portuguese-Spanish borderland. Sustain 12(5):1978

Walumweya M, Kawimbe S (2024) Enhancing economic integration through cross-border trade optimisation in Eastern and Southern Africa: Insights for policy and practice. GSJ, 12(8)

Yuan H, Yang B (2022) System dynamics approach for evaluating the interconnection performance of cross-border transport infrastructure. J Manag Eng 38(3):04022008

Scenario Planning in Managing Supply Chain Risk and Uncertainty: A Supply Chain Dashboard

Munyaradzi Chibaro, Unity Maringosi, David Chikwere, Munyaradzi Mahwine, and Wilbert Manyanga

Abstract Transport transition dynamics are developing in an environment of high levels of uncertainty relative to technological evolution; climate shocks; economic turmoil; and political upheaval. A number of tools and frameworks will be developed for Global South countries to support adaptive policy-making through this period of transition. The chapter focuses on methods for assessing the potential vulnerability of investments in transport infrastructure using stress tests; mechanisms for adjusting policies dynamically; and ways of modeling longer term impacts. The focus on resilience and adaptability provides a basis for planning and implementing transport-related green initiatives. The research draws upon resilience theory; adaptive policy pathways; dynamic capabilities theory; and systems thinking. Together these theories provide the basis for proposing a structured methodological approach for evaluating the impact of investments in transportation and logistics across multiple possible future scenarios. In addition to employing exploratory scenario construction techniques; incorporating infrastructure vulnerability models; conducting carbon-transition sensitivity analyses; and establishing real time dashboard monitors, the chapter presents a scenario-based framework that has particular utility in the context of Global South economies which face fiscal challenges; existing infrastructure gaps;

M. Chibaro (✉)
Department of Logistics and Supply Chain Management, University of Botswana, Gaborone, Botswana
e-mail: munyachibaro@gmail.com

U. Maringosi
Department of Education, Great Zimbabwe University, Masvingo, Zimbabwe

D. Chikwere
Leeds Beckett University, Leeds, UK
e-mail: D.chikwere@leedsbeckett.ac.uk

M. Mahwine
Department of Risk and Supply Chain Management, Midlands State University, Gweru, Zimbabwe

W. Manyanga
Department of Management Sciences, Workwell Research Unit, North–West University, Potchefstroom, South Africa

J. Kanyepe and I. Chirisa (eds.), *Advancing the Green Agenda in the Transport and Logistics Sector in Global South Countries*, World Sustainability Series,
https://doi.org/10.1007/978-3-032-31374-4_14

climate vulnerabilities; and institutional weakness. The chapter also illustrates how scenario-based dashboards may integrate early warning indicators; risk heat maps; adaptive threshold triggers; and carbon performance metrics to develop a comprehensive decision-making tool. The findings indicate that integrating analytical capability (operational analytics) with strategic planning (long-term foresight) can significantly reduce the risk associated with stranded assets; enhance policy adaptability; and improve sustainability outcomes. Therefore, resilience and adaptability represent two fundamental design principles in promoting sustainable transformation within global supply chains. Finally, the chapter highlights some practical implications for government agencies; logistics organizations; development finance institutions; and regulatory bodies seeking to manage uncertainty while facilitating accelerated low-carbon transport transitions.

Keywords Supply chain risk · Uncertainty · Resilience · Supply chain dashboard

1 Introduction

The structural changes occurring in today's global supply chains are taking place with an unprecedented velocity. Decarbonizing our supply chains, new technologies that include AI and Automation, new architectures of global trade, and increasing frequency and intensity of climate-related disruptions are all transforming transportation systems globally. Developing countries including those located in Africa, Latin America, and parts of Asia face significant challenges in addressing these transformations due to fragile institutional and limited physical infrastructures. Developing country transport system transitions will occur under extreme uncertainty from rapid technology development, climate variability and extremes, political realignments at the global level, demographic changes, and increased economic volatility. Long term investment decisions on both supply chain investments and transport infrastructure policies require frameworks beyond conventional linear planning models and static risk assessment methodologies. An integrated scenario-planning model which includes a dynamic supply chain dashboard has been developed to support enhanced resilience; increased adaptive capacity; and sustainable development/transition governance of transportation systems (Sazvar et al. 2021)

Green logistics is creating a lot of uncertainty. The main way for companies to get their products from suppliers to customers is through transportation systems. Transportation determines how quickly they can move goods; what it costs them to move those goods; how much carbon dioxide they produce per mile travelled; and how resilient they are when there is a problem (Syahchari et al. 2022). As governments plan for an electric truck and hydrogen fuel corridor future, with increased use of rail and inland waterway services, this creates great uncertainty. There will need to be many billions invested into new transportation systems that may have limited value based on rapidly changing technologies and policies. Logisticians will also need to create supply chains that operate in the face of changing fuel prices; changing regulations; and other changes due to climate conditions.

Traditional forecasting methods use the assumption that current trends will continue in the future to forecast what may happen next. Due to increasing rates of transformational change, as well as "deep uncertainties" related to model systems, the relative likelihoods of events, or the desirability of potential future outcomes, these forecasting methodologies are no longer reliable. The term deep uncertainty describes a state where an organization does not have knowledge about some aspect of its environment, or there is substantial disagreement among decision makers about how to model their system, assign probabilities to possible outcomes, or evaluate which desired outcomes would be best. Given this type of uncertainty, organizations can use structured processes to envision numerous alternative plausible futures based on different assumptions regarding the evolution of key factors and trends, referred to as scenarios. In addition to providing a process for envisioning multiple alternatives, the structured nature of the methodology also enables organizations to integrate new information from emerging sources into their thinking regarding the future at any time during the process.

This chapter suggests that integrating scenario planning with digital supply chain dashboards offers an effective way for managing the transition in transportation services associated with uncertain environmental challenges. A digital supply chain dashboard functions as the primary user interface that links long-range strategic foresight regarding the characteristics of likely future environments with ongoing short- to medium-range assessments of current performance against planned targets, including continuous assessment of risk exposure and opportunities to respond proactively to changing circumstances through policy adjustments. Furthermore, by incorporating flexibility and resilience into the design and management of the infrastructure that supports supply chains, it is possible to reduce the negative impacts of uncertainty on long-term sustainability goals while continuing to advance a green agenda (Mahajan 2025).

2 Background to the Study

Over the last few decades the complexity of the supply chain has increased, with an increasing amount of interconnection between countries on a global scale, through globalization, technological advances, and increasing international trade networks. Although these changes have led to improvements in efficiency and market integration, they have also increased the likelihood of risk and uncertainty within global supply network systems. Examples of disruptions to supply chains include geopolitical issues, pandemics, climate related disasters, and economic volatility. Global supply networks were significantly impacted by the COVID-19 pandemic including major disruption to logistics and transportation networks world-wide. This resulted in significant weaknesses being revealed in the areas of supply chain planning and risk management (Ivanov and Dolgui 2020). Additionally, increasing climate variability and geopolitical tensions are continuing to challenge the stability and predictability of global supply chain operations. As a result, organizations will be required to utilize proactive risk management strategies that promote supply chain resiliency (Wieland 2020).

Supply chain planning is typically based upon traditional, deterministic forecasting models that presuppose generally stable markets. However, these types of methodologies are becoming less effective in today's environment of high levels of uncertainty and the increasing speed of environmental change. Therefore, it is being recommended that organizations begin using scenario planning as an approach to making strategic decisions with respect to managing complex and uncertain environments. The process of developing future scenarios is called scenario planning. This methodology allows organizations to develop a number of different possible future states of affairs so they can identify possible disruptions and create plans to respond to those disruptions. When decision-makers explore several possible future states of affairs, they can better appreciate how uncertainty will affect their organization and be able to design flexible strategies that provide greater ability for their organization to adapt and recover from any type of disruption or failure (Chermack 2011). Some examples of the use of scenario planning include; assessing the impact of various types of disruptions including, but not limited to; disruptions caused by fluctuations in customer orders; disruptions associated with the availability and reliability of transportation services; disruptions associated with failures of suppliers; and disruptions caused by regulatory changes affecting supply chains.

The rapid growth of digital technology and data analysis also offers organizations the opportunity to improve supply chain management decisions by using advanced techniques such as the use of data analytics in order to monitor supply chains (Ghanbari et al. 2025). Most modern supply chains produce large volumes of "operational" data from sources including enterprise systems; sensor networks; logistics tracking applications; and market intelligence tools. To support the integration of these data streams into a single analytical environment, there have been developments in the area of supply chain dashboards. These are essentially graphical decision-support tools which combine the outputs of real-time data and analytics into an integrated single platform for viewing and analyzing performance. Supply chain dashboards provide managers with metrics related to key performance indicators; enable managers to identify early warning signs of disruptions; and assess the potential implications of multiple different strategic alternatives (Waller and Fawcett 2013). Therefore, organizations may be able to further enhance their capability to predict risk and proactively address evolving supply chain issues when they utilize both scenario planning and dashboard-based analytics (Zsidisin et al. 2024).

Although numerous developments have occurred within the field of integrated risk management – such as the use of scenario planning, digital analytics and real-time monitoring systems by companies across various industries; and many companies continue to struggle to implement these types of combined approaches to managing risk in developing economies. The limited availability of digital infrastructure, lack of analytical capability and poor data governance are common obstacles faced by many organizations which prevent them from being able to take full advantage of the opportunities available to them for making strategic decisions based on these tools (Christopher and Peck 2004). As such, the increasing need for research into the ways in which supply chain dashboards may be used to operationalise scenario planning has been identified as an important area for further investigation. Therefore, this paper aims to investigate whether/and if so how, the integration of scenario planning frameworks and decision-support systems based on dashboard technology may provide enhanced supply chain resilience and better prepare organisations for uncertainty and volatility.

To take advantage of the increasing sophistication of supply chain systems, and to provide additional emphasis for effective systemic risk management methodologies, supply chain disruptions continue to be both more common and more severe. A number of factors contribute to these risks, as they can come from many different areas. These include failures by suppliers; disruptions to transportation; government regulations; fluctuations in customer orders/demand; and/or technological breakdowns. Because of the increasingly integrated nature of global supply networks, according to Christopher and Holweg (2011) when there is a disruption in one area of the network it may rapidly disrupt other tiers of suppliers and distributors. Examples of how this occurs are port congestion/shortages of critical raw materials/unexpected geopolitical sanctions which can greatly affect the timing of deliveries and production schedules. Therefore, companies need sophisticated planning tools that allow them to assess their vulnerability to disruptions; identify what could happen if a disaster were to occur; and enable managers to make decisions promptly in response to potential disasters.

Scenario planning has become an important strategic foresight tool for businesses to navigate through the uncertainty and volatility of business environments. It originated in strategic management and energy planning but has recently been used in Supply Chain Management to evaluate the impact of possible future uncertainties on operations. Forecasting traditionally predicts one probable future; whereas scenario planning evaluates many possibilities and considers all the interactions among various factors based on different scenarios (Schoemake 1995a). As a result, it provides the ability for Supply Chain Managers to prepare alternative strategic plans, i.e., sourcing from alternative suppliers, redesigning their distribution systems, or changing their inventory levels to mitigate risks. As a practical application, scenario planning allows companies to proactively manage their supply chain risk rather than reactively respond to uncertainty (Wube et al. 2025).

The advent of digital technology, big data analysis, and the Internet of Things (IoT) has also dramatically changed how companies operate their supply chains. Because these technologies allow companies to collect and use huge amounts of information about their operations in real time, they are able to gather much more accurate information than before about the movement of goods through their logistics networks, including tracking shipments, assessing the performance of their suppliers, and evaluating customer demand (Saran and Kesswani 2025). With this technology, supply chain members can make decisions based on current data from a wide variety of sources. Digital supply chains offer increased transparency of logistics movements which allows companies to quickly respond to problems that may occur at any point within the global or regional supply network using advanced data-driven predictive modelling methods as described by Ivanov and Dolgui (2021a, b, c).

One such tool is the supply chain dashboard. The supply chain dashboard serves as an interactive portal to visualize and monitor Key Performance Indicators (KPI) throughout supply chains. Dashboards take input from multiple locations within the supply chain and organize it into an organized and easy-to-use format to support manager decisions. Real time monitoring and visual analysis tools embedded in dashboards allow managers to identify when their KPIs deviate from expected values, anticipate future problems/risks, and analyze the possible outcomes of various strategic options (Few 2013). For example, KPI's displayed on a supply chain dashboard might include inventory levels, transportation delays, supplier reliability and demand fluctuations. Using a combination of indicator and simulated based scenario planning, you can better determine how an issue will spread through your supply network and develop effective mitigations (Yassine 2022).

Although both using supply chain dashboards with scenario planning has several potential advantages; however, numerous organizations have difficulty utilizing these tools successfully. Most commonly, this occurs due to an inability to integrate data across various supply chain partners, lack of technology to utilize as needed, and limited analytical ability of decision makers (Ji and Liu 2022). To further compound difficulties associated with implementing scenario planning requires coordination among all parties involved, such as, suppliers, logistics providers and regulatory agencies. This coordination effort can become even more complicated within developing economies when they continue to develop their digital infrastructure and data governance structures. Therefore, there exists a great need for research on how to use scenario planning and dashboard technologies together to provide better supply chain risk management (Ghanbari et al. 2025).

Against that background, the purpose of this research is to analyze whether or not using scenario planning to manage supply chain risk and uncertainty will utilize a supply chain dashboard. The goal of this research is to examine how data visualization tools as well as analysis based on scenarios can assist in supporting strategic decision-making so as to aid in the development of more resilient and adaptable supply chain systems. As such, it is crucial for companies to understand the mechanisms by which scenario planning may be operationalized with digital dashboards if they are going to increase their preparedness and responsiveness within an increasing number of global supply chain uncertainties.

3 Research Objectives

i. To examine the role of scenario planning in managing supply chain risk and uncertainty **by** identifying potential disruptions and evaluating strategic response mechanisms within supply chain systems

ii. To analyse how supply chain dashboards can enhance decision-making and operational visibility through the integration of real-time data, performance indicators, and analytical tools
iii. To evaluate the effectiveness of integrating scenario planning with dashboard-based decision-support systems in improving supply chain resilience, risk mitigation, and strategic preparedness in uncertain environments

4 Theoretical Developments

4.1 Scenario Planning in Managing Supply Chains

The growing complexity and interdependence of today's global supply networks have increased organisations' exposure to a variety of types of risk, including but limited to demand variability, political instability, supplier disruption, and climate-induced shocks. However, traditional planning methodologies rely heavily upon static, linear forecast models that assume market stability. These models are ineffective under highly uncertain conditions. Scenario planning offers organisations the ability to identify, develop and analyze multiple possible futures. Thus, organisations may utilize scenario planning methodology to provide decision makers with a means to proactively identify possible disruptive forces and/or evaluate alternative strategic options.

Through developing and analyzing a range of possible future scenarios, organizations are able to create more flexible supply chain designs that will enable them to react to changing environmental factors. In terms of Supply Chain Management (SCM) practices, scenario planning supports managers who seek to identify potential risks and be strategically prepared. SCM managers utilize scenario-based analysis techniques to predict how various sources of uncertainty (including but limited to transportation disruptions, supplier failure, regulatory change, and changes in demand from consumers) could potentially impact their organization. The ability to analyze how these sources of uncertainty might affect an organisation provides the opportunity to evaluate a range of operational strategies (such as creating diverse sources of supply, increasing inventory levels (safety stocks), designing new distribution networks or investing in alternative forms of transportation). By systematically developing and evaluating the feasibility of each possible disruption, as well as assessing the potential consequences associated with each identified disruption, organizations are then able to develop contingency plans designed to mitigate vulnerabilities to unknown/unpredictable events (Ivanov 2021).

A key feature of scenario planning is that it enables businesses to develop supply chains which are more robust and adaptable. Scenario analysis can be used by organisations to continue to review their supply chain performance through data-based decision support tools. The use of such tools will enable the organisation to react to emerging risks (Keskin 2022). Therefore, this forward-looking method enhances an organisation learning capability, enabling it to respond more rapidly to any disruptions. Consequently, there has been growing recognition of the value of using scenario planning as an important mechanism for enhancing supply chain resilience and for achieving sustainable operation of supply networks in the face of increasing uncertainty worldwide (Prompatanapak and Lopetcharat 2020).

5 Supply Chain Risk and Uncertainty

Globalization and increased complexity in today's global production/distribution systems has made supply chain risk/uncertainty an important area of concern for many firms engaged in supply chain management. Supply chain risk is defined as the possibility of a disruption occurring that will hinder or eliminate the movement of products, financial transactions, and information through one or more processes within the supply chain. The causes of these disruptions are numerous; some include supplier failure, transportation delay, fluctuations in consumer demand, changes in regulations, acts of nature (such as hurricanes), and political instability. Uncertainty occurs when firms cannot predict with certainty their future operational environment. Many firms operate in very volatile markets where they lack complete knowledge about current events, and therefore, find it difficult to make accurate forecasts regarding demand, logistics coordination and/or maintaining consistent and stable supplier relationships (Shahzad and Irfan 2024).

The integration of supply chains into a single system allows issues with one section of the supply chain to spread to many other levels throughout the entire system (Ahmadi et al. 2024). For example, shortages of critical raw materials, port congestion, and global health crises such as pandemics can significantly disrupt production schedules and delivery timelines can severely impact an organization's ability to meet its production schedule or to get products delivered on time. This is typically followed by significant increases in the cost of doing business, lower levels of customer satisfaction and a loss of revenue (Shahzad and Irfan 2024).

To successfully address supply chain risk and uncertainty in today's environment, companies need to implement a strategic approach for managing risks proactively by increasing visibility and resiliency throughout their entire supply network; increase supply chain partner collaboration; and use digital technology to provide real time monitoring and predictive analytics as well as develop contingency plans for all possible disruption scenarios. Companies will be able to reduce the effects of disruptions on supply chain operations and maintain business continuity when they have identified the possible vulnerability points and developed contingency plans.

6 Supply Chain Resilience

Supply chain resilience is about how well an organisation's supply chain system will be able to predict, prepare for, react to, and recover from disruptions to its overall operational effectiveness (Ghosh and Sar 2022). Due to increasing levels of uncertainty related to climate change, technological disruption, geopolitical events and global pandemics, resilience in supply chains has become even more significant than it was previously. modern global supply networks are geographically dispersed and have many different stakeholder participants, for example suppliers, manufacturers, logistics providers and distributors etc. The geographic dispersion of these stakeholders creates increased vulnerabilities in the event of unplanned interruptions in the flow of goods and/or services; for example failure of one of your key suppliers, transportation delays, unanticipated changes in demand. To address this increased risk exposure to the potential occurrence of unplanned events (shocks) to their supply chains, Organisations need to design and implement supply chain systems that are capable of absorbing and rapidly recovering from unplanned events that create disruptions (Ivanov 2021).

Resilience in supply chains can be created using four approaches; namely, by increasing flexibility, implementing redundancy, enhancing visibility and facilitating collaboration. By creating flexibility within their organisation's production schedule, procurement strategy and distribution channel(s), organisations are able to respond and adjust as needed during times of disruption. For instance, organisations can maintain diversity with respect to their supplier base (having multiple sources) or create alternate routes for product delivery via transportation (for example trucking companies). Additionally, redundant systems (for example inventory levels and/or suppliers) provide organisations with additional protection against supply chain shocks (both positively and negatively). In addition to increased redundancy, utilising enhanced digital technology to improve organisational visibility (via tools such as real-time tracking and big data/analytical techniques) enable organisational decision makers to quickly identify potential disruptions to the supply chain and act accordingly (Sharma et al. 2025).

Furthermore, collaboration among supply chain partners plays a critical role in strengthening resilience. Effective information sharing and coordinated planning allow organisations to respond collectively to disruptions and maintain the stability of supply networks. By integrating risk management strategies, digital technologies, and collaborative practices, organisations can build more resilient supply chains capable of adapting to uncertainty and sustaining long-term operational performance (Alkhudary et al. 2022).

7 Evolution of Supply Chain Dashboards

Supply chain dashboards have evolved with other digital transformation efforts within supply chain management and have moved from static reporting tools into intelligent real-time decision-support systems. Early-stage organizations used spreadsheet-based tools (e.g., Microsoft Excel), enterprise systems (e.g., SAP ERP) to produce retrospective reporting that provided historical information for monitoring rather than providing information to support proactive decisions. Enterprise resource planning systems and data warehouses continued to evolve during the 2000's and enabled the integration of dashboards and thus cross-functional visibility and standardized key performance indicator (KPI) tracking (Gunasekaran and Ngai 2004). Business intelligence platforms such as Tableau and Power BI emerged and were able to enhance dashboards through the use of real-time data visualization, interactivity and self-service analytics; therefore, enhancing managerial insight and responsiveness (Chen et al. 2012). More recently, artificial intelligence and machine learning have been integrated with IoT technologies to transform dashboards that are now predictive and prescriptive. They can predict demand, recognize risk factors, and recommend optimal action (Wang et al. 2016). Today's advanced dashboards help to enable planners for future scenarios as well as provide resiliency and agility through more complex and global supply chain networks. This transformation represents a transition from descriptive systems to predictive autonomous systems which will allow organizations to move away from reactive operations toward data driven strategic decision making (Birkel and Hartmann 2020).

Supply chain dashboards create many benefits for companies by allowing greater visibility into their supply chain operations; improved communication & coordination among departments involved in those operations; better decision making on a timely basis. Companies can use the software that is used to build supply chain dashboards (for example, Tableau, Power BI) to view metrics about their supply chain operations in near real-time. This allows them to respond quickly if there are disruptions. Supply chain dashboards combine data from various business units within an organization. As a result, they help eliminate information barriers between functional areas. Supply chain dashboards provide data driven decision making. For example, AI & ML can be applied to historical data to improve forecast accuracy and manage risks associated with your supply chain (Eichenseer et al. 2025). Supply chain dashboards have been shown to reduce costs; improve organizational agility and allow planners to proactively plan supply chains in today's increasing complexity and dynamic supply chains (Zheng and Brintrup 2025).

8 Theoretical Framework

8.1 Resilience Theory for Adaptive Policy Pathways

Hollings' (1973) ecological study is where resilience theory first originated; then in organizational and systems research it was further developed. According to Hollings (1973) and other researchers in this area, resilience theory defines an organization's/supply chain's capacity for absorbing Disturbances (Disruptions), adapting to change or changing circumstances, and continuing to operate without collapse. When discussing organizations and supply chains in terms of resilience theory, its focus is on the organization/supply chain's capacity to anticipate disruptions, respond effectively when those disruptions occur, and recover quickly after the disruption(s) has/have occurred while maintaining operational stability. Unlike other theories such as efficiency theory, which focuses on operating efficiently, resilience theory focuses on the organization's/supply chain's capability to be adaptable, flexible, and to learn so they can better handle uncertainty and complexity.

Resilience theory within supply chain management will provide a conceptual framework to help understand how companies develop their supply chains so they can be able to withstand disruptions, which include but are limited to supplier failure, transportation delay, volatile demand and outside shocks. Resilience of supply chain is typically developed by the development of diverse supplier bases, flexible logistics networks, collaborative information sharing and the utilization of technology for real time monitoring. From the standpoint of this research into using scenario planning to manage the risks and uncertainties associated with supply chains utilizing a supply chain dashboard; it will serve as the base for understanding how organizations can predict and react to disruption (Tarei et al. 2022). The ability to explore future possible disruptions via scenario planning, coupled with an increased level of transparency from dashboards, and enhanced data driven decision making capabilities, will increase the overall resilience and adaptability of supply chain systems.

9 Dynamic Capabilities Theory

Teece, Pisano, and Shuen (1997) established Dynamic Capabilities Theory to show how organizations can create and connect their internal/external resources to respond to the rapid changes in the environment. According to the authors, when a business is in an unpredictable or unstable environment; it has to do something more than just have useful assets, i.e., build its own abilities to detect new opportunities/threats before they are too late, use these opportunities strategically, and modify existing organizational processes. As such, dynamic capabilities emphasize learning from experience, organizational agility, and the ability of organizations to be flexible with strategies to maintain competitive advantage in increasingly turbulent environments.

The Dynamic Capabilities Theory provides an appropriate framework for organizations to develop adaptable strategies (for example adaptive policy pathways) in response to changing environmental conditions. An example of such an adaptive pathway is the development of a series of sequential strategic options; these are developed as new information becomes available about evolving environmental trends or uncertainties. The development of alternative strategic options based upon dynamic assessment of a firm's internal/external environment is reflective of the dynamic capability theory's call to continuously assess its current state and adapt its operational strategy in order to mitigate against potential emerging threats. As part of a larger study on using scenario-based planning to manage supply chain risk and uncertainty through a supply chain dashboard, the dynamic capabilities theory offers insight into how firms may utilize scenario-planning and data analysis to identify potential interruptions to the supply chain, determine viable alternatives for mitigating those interruptions and adjust policies regarding supply chain management in order to maintain greater levels of resilience and longer term sustainability (Manhart et al. 2020).

10 Systems Thinking Theory for Adaptive Policy Pathways

Systems thinking theory originated with Ludwig von Bertalanffy's (1968) development of General Systems Theory. This theoretical framework identifies all complex phenomenon to be viewed as an interconnected system as opposed to individualized components. The focus for Systems Thinking is on the interconnectivity and interactional relationship or feedback loop between the various components within a system. When viewing organizational and/or policy-based systems, Systems Thinking helps demonstrate how a decision made in one area of a system has a cascading effect upon another component of the same system. As such, Systems Thinking supports a holistic approach to problem solving, long term strategic planning, and integration of decision-making processes when addressing problems which are complex and/or contain uncertainty.

Systems thinking is an important tool for understanding how policies and strategies develop in interconnected settings and as part of adaptive policy pathways. The primary goal of adaptive policy pathways is to address uncertainty in systems, such as, supply chains by developing flexible policies which are adaptable in response to changes in the operating environment. Systems thinking allows policy makers and managers to examine the relationships among actors in the supply chain and their supporting structures including technology and infrastructure at different points in time. As stated by Walters (2021), systems thinking also enables the creation of holistic decision-making support systems that support the development of an integrated approach to decision making when considering scenario planning as a means of identifying potential disruption risks at a systemic level through a supply chain dashboard that displays real-time information about the dynamics associated with the interaction of all components of the supply chain (Walters 2021).

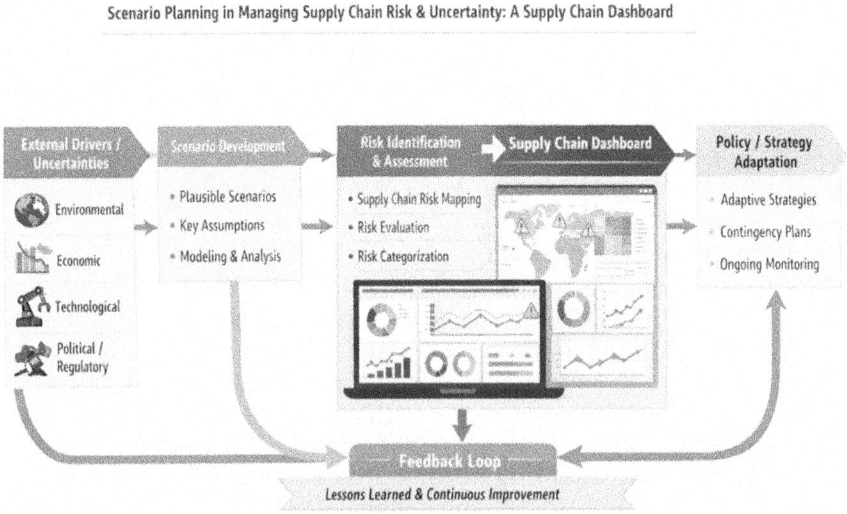

Source Developed by the Authors

The conceptual framework for managing uncertainty in a global supply chain is shown in the conceptual framework. It describes how to utilize a dashboard to proactively address risks and uncertainties in the supply chain using scenario planning. Scenario planning starts with understanding the various drivers outside of the organization and the possible uncertainties such as environmental, economic, technological and political changes that will affect the entire supply chain. Each driver will help develop and refine possible future scenarios. Organizations use these future scenarios to determine critical assumptions and then conduct modelling and analysis to identify possible disruptions. Once all of the possible disruptions have been identified, an evaluation of each disruption is done by identifying its likelihood and impact on the supply chain. Additionally, based upon its' likelihood and/or impact, each disruption is classified as either an operational, financial or strategic risk. Lessons learned from this process are incorporated into a supply chain dashboard. The dashboard includes real time data; displays KPI's for current performance; uses colour coding to display a "risk heatmap" showing current risk levels associated with each node in the supply chain; provides ability to simulate future possible disruptions. The information provided via the dashboard assists management in making informed decisions regarding what policies and strategies should be implemented to adapt to changing circumstances; developing and implementing contingent plans; and continuing to monitor current events. Finally, there is a continuous loop of providing lessons learned from past policy/strategy implementations back into the development of new possible future scenarios. Thus, creating opportunities for continued learning and improving the resiliency of the organization (Manyanga et al. 2026).

11 Methodology

To analyze the potential impact of scenario planning and supply chain dashboard applications on reducing uncertainty in global logistics, this research adopted an analytical methodology called a desk-based literature review. As a desk review method allows for the systematic compilation of previously collected information that has been documented in peer-reviewed journals, academic books, industry studies and governmental documents; and can be used to create a theoretical framework for an area of interest, this type of methodology provides the best opportunity to achieve a theoretically based overview of the topic. In addition, the methodology adopted for this project is a systematic literature review, which followed a transparent process with clear inclusion/exclusion criteria. Therefore, academic data bases (Scopus, Web of Science, Google Scholar) are being utilized to locate relevant documentation that reflects the development of Digital Supply Chain strategies since 2000 and through to 2025. Keywords associated with "Scenario Planning", "Supply Chain Risk Management", "Uncertainty Reduction", "Supply Chain Dashboards" and "Decision-Support Systems" have been developed to assist with identifying literature related to these topics. Additionally, grey literature from reputable sources (World Bank, World Economic Forum) is being incorporated into the analysis to provide additional insight regarding the perspective of industry leaders and emerging trends within the field of supply chain management.

A thematic content analysis methodology was employed to evaluate selected source material. These sources were analyzed, coded and categorized according to study-specific objectives; namely, the use of scenario planning in identifying potential disruptions and the evaluation of appropriate strategic responses, (ii) the role of dashboards in providing timely, real-time visibility and information necessary to make decisions, and (iii), how integrating the use of scenario planning into dashboard-enabled systems can increase resiliency and enhance overall preparedness. Through this methodological approach, the researcher was able to identify commonalities among the various studies reviewed including conceptual linkages and areas of disparity in terms of research on this topic. In addition, by emphasizing credibility, dependability and confirmability through the use of multiple triangulated sources and standardized analytical processes, the study also demonstrated its rigorous methodology. Furthermore, due to the complexity of digital supply chain phenomena and the fact that empirical evidence may be limited or difficult to obtain while rich conceptual understanding exists; this methodology provided the most viable approach to developing an integrated model linking scenario planning and dashboard-enabled decision support systems that facilitate effective management of supply chain risk and uncertainty.

12 Findings

This research identifies that for organizations dealing with an unpredictable environment, both scenario planning (to be able to predict) and supply chain dashboards (to be able to respond), will complement each other when it comes to mitigating risks associated with the unpredictability of complex supply chains. The review further states that through the use of scenario planning, organizations can increase their strategic foresight capabilities and develop plans based on what may occur, while using dashboard systems, that are enhanced with tools like Power BI, increases the ability of the organization to react quickly and effectively in response to issues occurring in real time. When used together, they enable better decision making, increased resilience, and the ability for an organization to adapt more easily to changing conditions, which ultimately allows them to transition from reactive to proactive risk management in today's ever-changing and volatile global supply chain environment.

13 The Role of Scenario Planning in Managing Supply Chain Risk and Uncertainty

The findings indicate that the use of scenario planning provides value as an essential strategic tool for companies to manage the uncertainty and risks within their supply chains through anticipating a wide variety of possible disruption paths including disruptions due to volatility in consumer demand and supply shortages to disruptions caused by changing global political climates and changes in weather patterns. The World Economic Forum has identified that the use of a scenario-based approach enables firms to improve their ability to respond to rapidly changing environments through increased organizational agility, improved strategic flexibility and enhanced preparedness however, there are significant limitations to using this approach (Ivanov 2020). Due to the subjectivity involved in developing scenarios they are typically influenced by manager's perception and biases therefore limiting the completeness of representing potential future uncertainties (Schoemaker 1995a). In addition, due to the qualitative nature of the approach it lacks precision therefore providing little to no basis for quantifying the likelihoods associated with different types of risk or prioritizing them accordingly (Tang 2006). The development of a scenario plan is generally a time-consuming process and may occur at irregular intervals and/or at best on a periodic basis, thereby failing to provide timely updates relative to the dynamic nature of many business environments. Additionally, despite the creation of a high-level strategy via scenario planning the translation of that strategy into meaningful action/operations across all departments within an organization can be limited, especially in those organizations that lack strong analytical capabilities. Therefore, while the use of scenario planning can significantly enhance an organization's strategic foresight and risk management capabilities, the effectiveness of that capability is heavily dependent on integrating the outputs from this approach with outputs generated from data driven analytics and real-time monitoring systems so that the information developed from these multiple inputs can be used as a foundation for making informed decisions regarding how to address a wide range of potential risks.

The study shows that the use of scenario planning is a useful strategic tool for companies to manage and help mitigate the potential impacts (costs) in terms of their supply chain(s), due to several types of uncertainty which are, volatility in demand, shock in supplies, geopolitical disasters and environmental disasters. As indicated by data collected by organizations like the World Economic Forum, using a scenario planning based approach will assist with increasing organizational agility; increase strategic flexibility; improve preparedness by making the transition from predictive certainty paradigm to an adaptive resilience paradigm (Schoemaker 1995a; Ivanov 2020).

However, an evaluation of the literature identifies significant shortcomings. The subjectivity inherent within the process of developing scenarios results in a reliance upon managerial cognition and bounded rationality. As such, the resultant representation of future potential uncertainty will likely contain biases and may be incomplete (Tang 2006). Furthermore, the qualitative nature of the process used to develop scenarios makes quantifying probability and prioritizing risk challenging (Tang 2006). In addition, the time consuming and episodic nature of the process may make it less than ideal for use in dynamic and rapidly changing environments. Finally, there exists a substantial gap between converting scenario insight into practical action. This gap is most pronounced in those organizations with lower levels of analytical maturity. Therefore, although using scenario planning provides improved strategic foresight and raises awareness of risks associated with uncertainty; its usefulness is dependent upon integrating it with data driven analytic tools and real time monitoring systems to increase robustness and provide actionable decisions (Chikwere et al. 2026).

This supports evidence that demonstrates that while a key function of scenario planning is identifying possible risks such as pandemic disruption, supplier failure, or regulatory shock; it also provides an opportunity for companies to stress test their current and potential future supply chain configuration designs and assess the effectiveness of different strategic alternatives. Furthermore, research conducted by the World Economic Forum and other studies support the premise that companies utilizing scenario planning will be more resilient than those who do not utilize this methodology in order to achieve redundancy, flexible response to changing circumstances, and diversification of sourcing options (Ivanov and Dolgui 2021a, b, c). Finally, scenario planning encourages organizational learning with cross-functional communication and long-term thinking; two components essential in complex and unpredictable supply chains.

Yet, an even more profound scholarly criticism of the methodology for conducting a thorough assessment of the structure of the practice indicates that there are continued structural limitations. One such limitation is the lack of standardized methods used for developing scenarios, this lack of standardization results in inconsistent approaches to both scenario design and analysis by different companies. Additionally, because the methodologies used rely upon subjective, qualitative judgments rather than objective, quantifiable measures, it limits their ability to be integrated with quantifiable risk assessments, limiting their predictive strength (Schoemaker 1995a). A number of companies have difficulty integrating scenario plans' output into their company's operational decision-making process. As such, the gap between the use of strategic foresight to anticipate future business conditions and the actual execution of those strategies continues to grow. Additionally, as global supply chains continue to become increasingly complex, and therefore susceptible to disruptions from unexpected "low probability/high impact" or "black swans," this type of event is likely underrepresented in the outputs developed through traditional scenario planning (Taleb 2007). In addition, if a firm does not continuously update its scenario planning outputs to reflect changes occurring in real time, then the outputs will quickly become irrelevant in rapidly changing environments. Therefore, although scenario planning has made significant contributions to identifying potential risks and preparing companies strategically, the most effective way to maximize the value derived from using scenario planning is to integrate it with dynamic analytical tools and digital applications along with company-wide capabilities that provide for continuous adaptation and implementation (Chibaro et al. 2026a, b).

14 Supply Chain Dashboards Enhancing Decision-Making and Operational Visibility

Research has shown that supply chains using dashboards have improved their ability to make decisions and see how operations are running because they use current, integrated information from all parts of the supply chain. Dashboards today can be run on a variety of software platforms like Microsoft's Power BI or Tableau. Organizations now have the capability to track and identify issues in the entire end-to-end process of the supply chain earlier and react faster than before. Studies support the idea that dashboards provide situational awareness; encourage data-based decision making; and help cross-functional teams work together better through one platform for truth (Chen et al. 2012; Wamba et al. 2017). Furthermore, the integration of Artificial Intelligence and predictive analytics capabilities allows dashboards to move beyond descriptive reporting toward forecasting demand variability, identifying risks, and recommending optimal interventions, thereby strengthening operational agility and resilience (Wang et al. 2016).

Although an extensive review of literature presents several potential disadvantages in using dashboards as part of the decision-making process, there are also at least three major constraints that limit the use of dashboards. These constraints include data quality issues such as poor integration, poor governance and incomplete or inaccurate data which limits the ability to make informed decisions. Dashboards provide greater visibility into the data however this increased visibility results in too much information (information overload) with too many metric and visualization options for decision makers to clearly understand the underlying trends. Finally, although dashboards can be implemented to provide both descriptive and analytical insight into the data; most current implementations remain largely descriptive in nature and do little to assist organizations with the type of complex, strategic decision making required (Wamba et al. 2017). There are potential risks associated with the development of an excessive reliance by managers on information from the dashboard to make decisions (i.e., "dashboard paralysis") which could negatively impact the manager's ability to use his/her own experience and knowledge to understand the context surrounding a particular event or issue. The high cost and technical complexity associated with developing an integrated, real-time dashboard will likely prevent many small organizations from using such technology. In addition, while supply chain dashboards have several positive impacts including enhanced visibility into operations, the utilization of these technologies is dependent on the availability of sufficient data infrastructure, the capability of users to properly utilize them, and that the organization uses them strategically as part of its internal business process (Mehravaran and Nemati 2025).

As well as increasing real time visibility for supply chains, dashboards can assist in both predictive and collaborative decision making within ever increasing complex supply networks (Lucchetta 2023). Dashboards provide end to end transparency when integrated with multiple data sources including IoT enabled tracking systems, an organization's internal platforms and external market intelligence, allowing for quicker response to disruption based on empirical evidence (Saran and Kesswani 2025). Advanced analytics embedded through tools like Power BI or Tableau enables organizations to embed simulation models and scenario-based insights into operational workflow processes, therefore enhancing both responsiveness and alignment of procurement, production, and distribution functions (Wamba et al. 2017; Wang et al. 2016). Furthermore, dashboards increase cross functional integration via the creation of a common information sharing platform, which creates improved communication, responsibility, and performance measurement. In addition to these benefits AI adds to this capability by providing additional capabilities such as anomaly detection, auto alerting and adaptive learning which collectively lead to increased resiliency and agility of supply chains (Chibaro et al. 2026a, b).

Although they represent an improvement in operational visibility and analytics-based decision making, these dashboards are still hindered by numerous challenges which prevent them from realizing their full transformational potential. While many dashboards provide sophisticated analysis, there is no mechanism for integrating non-quantitative, outside-in data sources (i.e., political risks, behaviour of suppliers), and therefore limit their ability to be comprehensive. The issues of latency in accessing data, along with the inability to interoperate across legacy systems also hamper the promise of using this type of technology for "real time" decision making. In addition, the use of dashboards can promote a techno centrism approach to managing (for example focus solely on quantifiable measures and ignore qualitative judgments and contextual information) that can lead to reduced quality decision making (Chen et al.

2012). There is significant evidence that organizations continue to resist change and have limited digital capability to effectively adopt and utilize these types of dashboards. Finally, new risks associated with data security/privacy/system vulnerabilities exist when using dashboards to enable digitally based supply chains. Therefore, although supply chain dashboards improve operational visibility and the ability to make better decisions through advanced analytics, their ultimate success will depend on having a robust ecosystem of data, organization's preparedness for technology adoption and utilizing a balanced approach to providing both technical knowledge and human experience.

15 The Effectiveness of Integrating Scenario Planning with Dashboard-Based Decision-Support Systems

The study's results confirm that using a combination of scenarios for planning as well as the use of a decision support system based on a dashboard, will increase an organization's ability to be resilient in its supply chain, reduce risk and prepare strategically better than either method alone. According to literature, scenario planning allows organizations to anticipate what could potentially disrupt their supply chain and evaluate how best to react to those disruptions. Dashboards collect data from various locations, monitor KPIs and provide management with usable information for making decisions (Ivanov and Dolgui 2021a, b, c; Schoemaker 1995a; Chen et al. 2012). Together they allow organizations to simulate different possible future states of affairs, estimate the impact that uncertainty may have upon them and take proactive steps to mitigate it throughout all parts of the procurement, production and logistics network (Haque et al. 2023). Together, these methods also support agility in decision-making so that when an organization senses a developing risk or changes in market conditions they can quickly respond accordingly. Furthermore, the utilization of advanced analytical techniques such as AI and predictive analytics strengthen the capability of the integrated system to recognize early warnings of potential disruptions, predict fluctuation in demand and allocate resources effectively (Wanga et al. 2016; Wamba et al. 2017). Studies that empirically examine the effects of utilizing this type of integrated approach show that organizations employing it see increased efficiencies, decreased downtime, and increased resiliency to known and unknown events that disrupt supply chains.

However, a detailed examination of the shortcomings which limit the efficiency of this type of combination are evident. The first limitation to successfully combine the process of scenario planning and the use of dashboards for decision-making will be dependent upon the quality, timeliness, and compatibility of data from numerous systems. If the data collected within an organization is fragmented, inaccurate, or simply out-of-date it will hinder the ability to predict future events effectively, thereby leading to decisions based on incomplete information. In fact, Tang (2006) states that "inaccurate and/or old data can lead to poor/incorrect decision making."

The second limitation lies in the methodological challenges associated with developing qualitative narrative scenarios related to the quantitative analytical outputs generated by dashboards. These two types of analyses require complex modelling techniques and experience that many organizations do not possess. The third limitation to the effective application of these technologies relates to the aforementioned organizational and cultural limitations including resistance to change, limitations in analytical abilities, and limited cross-functional communication among employees. Ultimately, relying too heavily on technology to generate decision-making support through integrated systems may also decrease the reliance of humans on their own judgments, understandings of context, and intuitive strategic thinking. Therefore, although utilizing both scenario planning and dashboards in combination provides substantial increases in supply chain preparedness and resiliency; ultimately, the greatest benefit is achieved when combined with good data governance practices, ongoing employee training and education, and organizational capabilities that foster flexible decision-making processes that are aware of specific contexts (Safari et al. 2023).

The findings further reveal that the integration of scenario planning with dashboard-based decision-support systems foster continuous strategic monitoring, enabling firms to evaluate the effectiveness of implemented risk mitigation measures in near real-time (Khan et al. 2020). Dashboards provide dynamic visualizations of scenario outcomes, allowing organizations to track performance against predefined KPIs, simulate "what-if" scenarios, and adjust contingency plans accordingly. This capability enhances the anticipatory and adaptive dimensions of supply chain resilience, supporting not only operational continuity but also strategic learning over time (Ivanov 2020). By linking scenario-derived insights with live operational data, organizations can identify emerging vulnerabilities, such as supplier bottlenecks or logistics delays, before they escalate into critical disruptions.

In addition to being able to assess the effectiveness of integration (by means of its usage), the critical evaluation also considers how the same effective utilization varies depending on the specific type of business. As an example, although an integrated system is likely to enhance a company's ability to respond, it has generally been implemented at very different levels across various industries. The reasons for this variation in implementation depend largely upon differences in technological preparedness within each respective industry and differences in the analytical capability of each industry. The challenges presented to small and medium-sized enterprises (SMEs) in implementing these systems are largely a function of their limitations in terms of available resources. However, SMEs also face other limitations beyond just those related to resources such as limited availability of high-quality data sources providing meaningful input into a system and limited infrastructural capacity to implement real-time monitoring processes (Babu et al. 2023). Finally, there are methodological complexities involved with linking qualitative, narrative-based scenarios to quantitative analytics-based dashboards. Linking qualitative, narrative-based scenarios to quantitative analytics-based dashboards typically require sophisticated modelling approaches to ensure consistent and accurate representation of the potential insights derived from the analysis. Furthermore, a possible disadvantage of linking qualitative, narrative-based scenarios to quantitative analytics-based dashboards is the likelihood of cognitive overload. Decision-makers may find it difficult to interpret large volumes of outputs from complex scenario analyses combined with large quantities of real-time metrics. Thus, delays in response times from management to critical issues affecting the organization are possible. The results of the study suggest that organizational culture plays a major role in determining the success or failure of an integrated planning process. Therefore, successful integration is dependent upon more than simply deploying the correct technology; successful integration depends upon creating a collaborative work environment that encourages cross-functional cooperation, supports data-driven decision making, and continuously offers training opportunities for managers so they can effectively utilize their own professional judgment and/or utilize the insights generated through automation.

In summary, combining a dashboard-based decision support system with a scenario-based approach to create strategic preparedness for your company increases your overall supply chain resiliency. The realization of the full potential of this combined approach is contingent upon having adequate data management; strong methodological capability to connect scenario-based approaches with analytical tools; and the organization's ability to combine technical solutions with their own collective experience and knowledge. By doing so, the combination of these two approaches will convert what are now typically static and rigidly defined supply chains into dynamic and adaptive systems capable of successfully navigating very high uncertainty and complexity.

16 Discussion

The use of supply chain dashboards along with scenario planning indicates that organizations are able to manage and adapt to changes in supply chains and risks due to their reliance on a systematic, data driven approach. In addition to being proactive and anticipating potential disruptions or interruptions in business operations, the process of planning allows companies to explore alternative strategies for responding to potential disruptions as well as develop flexible contingency plans based upon these alternatives. In addition, previous research has demonstrated that scenario planning encourages organizational learning and resiliency particularly when dealing with complex dynamic environments and therefore it provides companies with a better means of preparing strategically to react to environmental disruption (Tang 2006; Schoemaker 1995a; Ivanov 2020). However, while scenario planning does aid companies in achieving strategic readiness to react to disruptions there still remains a disconnect between recognizing an issue and translating this knowledge into tangible and actionable operational decisions (Ivanov 2020; Schoemaker 1995a). As such, the development of additional decision-support tools has become increasingly important.

Supply chain dashboards are a way to close this knowledge gap. Supply chain dashboards provide real time visibility into operations through integrating and monitoring supply chain data. Supply chain dashboards can assist with making better decisions quicker based on the real time information being provided (Chen et al. 2012; Wamba et al. 2017). Supply chain dashboards integrate data from various sources such as internet of things (IOT) Sensors, enterprise resource planning systems (ERP), market intelligence and present key performance indicators (KPI) and predictive analytics in an easy-to-understand format. Supply chain dashboards can enhance situational awareness, improve operational coordination and help identify risks. However, the impact that supply chain dashboards have will be limited if there is poor data quality, organizational capability, or the ability for managers to understand/interpret large amounts of data. Managers who rely too heavily on dashboards may make less strategic judgments than managers who have experience using both quantitative and qualitative analysis (Susitha et al. 2024).

Integrating scenario planning with dashboard-based decision-support systems offer the most robust approach to enhancing resilience and strategic agility. This integration enables organizations to link qualitative foresight with quantitative analytics, simulate potential disruption impacts, and monitor responses in real time (Ivanov and Dolgui 2021a, b, c). The literature suggests that such integration strengthens risk mitigation, enables proactive adaptation, and facilitates continuous learning within supply chains. Yet, challenges remain, including aligning qualitative scenarios with quantitative models, ensuring data interoperability, mitigating cognitive overload, and embedding a culture of cross-functional collaboration. Additionally, resource limitations and technological maturity, especially in SMEs, can hinder full utilization (Babu and Yadav 2023). Thus, while the combined approach offers significant advantages in navigating uncertainty, its success depends on robust data governance, methodological rigor, and the development of organizational capabilities that balance technological intelligence with human decision-making. Ultimately, this discussion underscores that achieving resilient, responsive, and adaptive supply chains requires a deliberate integration of foresight, analytical tools, and strategic implementation frameworks.

The combination of scenario planning and the use of a dashboard-based decision support system can help an organization develop proactively resilient approaches to managing disruption risks within its supply chains. Scenario planning provides strategic vision for anticipating future trends while a dashboard provides operational visibility into current performance. The two tools provide a synergistic ability to allow an organization to continue to monitor and adapt to changing conditions in the supply chain (Ivanov and Dolgui 2021a, b, c). In addition, the continuous monitoring of real time data will enable an organization to make changes in its short-term operations based upon long term strategic goals identified through the development of scenarios. A positive feed back loop is created by using scenario derived insight to define dashboard metrics while at the same time using real time data to refine the assumptions used in developing those scenarios. This type of dynamic interaction will enhance both an organization's ability to anticipate and respond to disruption risks within its supply chain (high uncertainty) whether it be due to a global pandemic or other types of disruptions.

Additionally, the integration enables risk assessment and allocation of resources by enabling decision makers to link a myriad of different operational indicators in real time with each possible scenario, which will enable them to determine their most vulnerable areas and deploy resources that are best suited to support contingency operations. Furthermore, this will allow for the creation of an adaptive supply chain network that is capable of being quickly modified. As such, while contributing to resilience of organizations through ability to provide continuous service during disruption, it will also contribute to competitive advantage (Wang et al. 2016; Wamba et al. 2017).

The assessment also demonstrates that success depends upon both technology and organization to be mature. A solid quality data architecture, interoperable it-systems and high analytical skills are essential for running valid simulations, predictions and monitoring. In addition, organizations have to provide the appropriate environment; this includes cross-functional teams, data-literate employees, and an organizational culture that supports decisions based on empirical evidence and at the same time allows managers to use their own judgement (Chen et al. 2012). As a result, if these elements do not exist then the integration of scenario dashboards into management practice could lead to a low utilization of the tools or even to false conclusions. Furthermore, a problem exists with respect to cognitive overload. Complex dashboard combined with different scenarios analysis can easily overwhelm decision-maker, possibly leading to delays in taking actions which can significantly harm business operations. Lastly, SMEs may experience limitations with respect to resources which will limit how scalable and adoptable such integrated systems would be within the organization. This has led to significant inequality in terms of resilience capability among supply chain partners. (Helgeson et al. 2022). While each approach individually strengthens supply chain risk management, their integration provides a holistic, dynamic, and adaptive framework for navigating uncertainty. Realizing its full potential requires careful attention to data quality, organizational capability, and methodological rigor, ensuring that technology augments not replace human strategic insight.

17 Implications

17.1 Theoretical Implications

This study provides an important insight into the theoretical aspects of how strategic foresight and decision-support are converging in supply chain management. Through the examination of three key elements (Scenario Planning, Supply Chain Dashboards and their combination), this research provides evidence for both Complexity Theory and Dynamic Capabilities Theory as ways to understand how organisations operate through uncertain environments. Scenario Planning is based upon the tenets of Complexity Theory that describe nonlinear interactions, uncertainty, and emergence of risk in global supply chains. Dashboards represent the operationalization of Dynamic Capabilities Theory where the capability to sense, seize and reconfigure organisational resources in response to real time information enhances adaptability and resilience (Teece 2007). The combination of these two approaches creates a new theoretical framework that demonstrates how qualitative foresight, and quantitative analytics can be used together to develop anticipatory adaptive supply chain systems, closing the gap between strategic planning and operation. These findings will form the basis for future research using data-driven theory-informed Frameworks to manage supply chain uncertainty and complexity.

18 Policy Implications

The findings of this study suggest many potential policy issues (implications) indicating that the government agencies, regulatory bodies and private sector companies should develop collaborative supply chains that are agile, responsive and data driven. The results of the scenario planning and dashboard integration studies also identified the types of policies that could effectively support the development of digital infrastructure through investment in digital infrastructure; standardization of data-sharing protocols; and in the deployment of real-time monitoring systems to improve the transparency and responsiveness of supply chain networks. Policies may also assist in regulating the use of predictive analytics and AI-enabled decision-support tools by encouraging their adoption, as well as ensuring that appropriate safeguards exist regarding data security; privacy; ethical use. In addition, cross-industry collaboration and knowledge-sharing policies will contribute to improved collective readiness for large-scale systemic disruptions such as pandemics, climate events, geopolitical shocks. Through enabling organizational capacity and technological readiness, policy interventions can help firms especially Small to Medium Sized Enterprises to implement scenario-based planning and dashboard driven decision-making to improve supply chain resilience, risk mitigation and continuity in global environments where uncertainty has increased.

Government agencies, regulatory organizations, and businesses must develop and support a robust and dynamic supply chain through the use of technology. Scenario analysis and dashboard integration indicate that government can create an environment to invest in digital infrastructure; establish common standards for sharing data; and monitor their supply networks in real time to be responsive to disruptions and improve overall visibility (Ivanov and Dolgui 2021a, b, c; Wang et al. 2016). Regulatory frameworks can support both the use of predictive analytics and AI-enabled decision-making tools, by facilitating secure collection, storage, transmission, and usage of data (security), and by providing legal recourse for individuals whose personal information has been mishandled (privacy) and for misuse of data by organizations (ethics). Furthermore, policies that facilitate intersectoral cooperation and the sharing of knowledge will increase the likelihood of collective preparedness against systemic threats like pandemic outbreaks, extreme weather/climate disasters, and geopolitical shock waves. Therefore, regulatory frameworks can enhance both the technology readiness and organizational capacity of companies (especially small- and medium-sized enterprises), which will enable them to develop and implement scenario-based planning systems and dashboard driven decision making systems. In doing so these types of programs will improve their supply chains' ability to be resilient to disruptions, mitigate risks associated with those disruptions, and ensure business continuity in a growing number of unpredictable global markets (Helgeson et al. 2022).

19 Managerial Implications

The findings of this research have a number of implications for management. Specifically, they highlight the need for firms to combine strategic foresight in order to improve their ability to make operational decisions during times of disruption by using data-based tools that are used to monitor their supply chains. To achieve better preparedness for possible future disruptions, managers can use scenario planning as an anticipatory tool in order to explore different strategies for mitigating risk and creating contingency plans (Chen et al. 2012; Ivanov 2020). Dashboards enable managers to track key performance indicators and receive real time data regarding operational performance on a regular basis and/or predict what may happen in the near term. By utilizing both of these types of tools together, managers will be able to integrate their long-term strategic goals into their daily activities which will support proactively managing risks, allocating resources dynamically, and improving communication among functionally distinct departments (Wamba et al. 2017; Wang et al. 2016). Practically, managers can make effective use of dashboards for making decisions through scenario analysis by investing in robust data governance, good analytical capability and training employees; therefore, developing an environment within their organization where they encourage collaborative problem solving (cross-functionally) and evidence based management will allow managers to translate technological insight into strategic actions that enhance both operational effectiveness and the ability of organizations to adapt in fast changing supply chain environments.

20 Conclusion

This research has demonstrated that when combined with supply chain dashboards (SCDs), scenario planning is a robust and holistic method to manage Risk & Uncertainty within Complex SCs. Strategic Foresight is provided through scenario planning by outlining possible disruptions as well as assessing viable alternatives for response; operational visibility and real-time monitoring are enhanced through SCDs which provide data-driven decision making (Ivanov and Dolgui 2021a, b, c; Chen et al. 2012). When combined, these tools enable organizations to align long-term strategies with daily operations, improve resilience, and foster adaptive capabilities in the face of dynamic and unpredictable events (Wang et al. 2016; Wamba et al. 2017). However, the effectiveness of this integration is contingent upon robust data governance, organizational capability, and the ability to translate analytical insights into actionable strategies. The findings highlight the importance of combining technological intelligence with human judgment to create agile, anticipatory, and resilient supply chains capable of sustaining performance under uncertainty.

References

Ahmadi T, Hesaraki AF, Morsch JPM (2024) Exploring IT-driven supply chain capabilities and resilience: the roles of supply chain risk management and complexity. Supply Chain Manag: Int J 30(1):50–66. ISSN 1359-8546. https://doi.org/10.1108/scm-11-2023-0561

Alkhudary R, Queiroz MM, Féniès P (2022) Mitigating the risk of specific supply chain disruptions through blockchain technology. Supply Chain Forum: Int J 25(1):1–11. ISSN 1625-8312. https://doi.org/10.1080/16258312.2022.2090273

Babu H, Yadav S (2023) A supply chain risk assessment index for small and medium enterprises in post COVID-19 era. Supply Chain Anal 3:100023. ISSN 2949-8635. https://doi.org/10.1016/j.sca.2023.100023

Birkel HS, Hartmann E (2020) Internet of things–the future of managing supply chain risks. Supply Chain Manag: Int J 25(5):535–548. ISSN 1359-8546. https://doi.org/10.1108/scm-09-2019-0356

Bryan R, Kai N (2006) 1 Dynamics of global supply chain supernetworks in a new era of risk and uncertainty. Towar Better Perform Transp Netw 225–249. https://doi.org/10.4324/9780203965573-18

Chen H, Chiang RHL, Storey VC (2012) Business intelligence and analytics. MIS Q 36(4):1165–1188

Chermack T (2011) Scenario planning in organizations. Berrett-Koehler

Chibaro M, Muchemwa M, Mahwine M, Mpfiga C (2024) Supply chain risks and organisational competitiveness in corporate timber manufacturing companies in Mutare metropolitan Region, Zimbabwe. Logistics Research 17:8. https://doi.org/10.23773/2024_8

Chibaro M, Kanyepe J, Manyanga W, Lysson JM, Musasa T (2026b) Blockchain and big data integration for transparent and traceable supply chains. Enhancing Sustain Glob Supply Chain Big Data Anal 147–174. https://doi.org/10.4018/979-8-3373-6896-2.ch005

Chibaro M, Mahwine M, Manyanga W, Maringosi U, Chikwere D, Mahwine MV (2026a) A literature review on supply chain analytics in Africa. Strat Supply Chain Anal 126–147. https://doi.org/10.4324/9781003674962-7

Chikwere D, Chibaro M, Govere E (2026) Assessing the interplay between critical success factors for supply chain analytics and operational performance in a data-driven Era. Strat Supply Chain Anal 35–58. https://doi.org/10.4324/9781003674962-3

Christopher M, Holweg M (2011) Supply chain 2.0: managing supply chains in the era of turbulence. Int J Phys Distrib Logist Manag 41(1):63–82

Christopher M, Peck H (2004) Building the resilient supply chain. Int J Logist Manag 15(2):1–14

Dani S, Rajesh R (2008) Agility and supply chain uncertainty: a scenario planning perspective. Int J Agil Syst Manag 3(3):178. ISSN 1741-9174. https://doi.org/10.1504/ijasm.2008.021208

Eichenseer P, Hans L, Winkler H (2025) A data-driven machine learning model for forecasting delivery positions in logistics for workforce planning. Supply Chain Anal 9:100099. ISSN 2949-8635. https://doi.org/10.1016/j.sca.2024.100099
Few S (2013) Information dashboard design: displaying data for at-a-glance monitoring. Analytics Press
Ghanbari H, Shabani M, Mohammadi E, Seiti H (2025) A comprehensive bibliometric review of digitalization impacts on supply chain risk management. Supply Chain Anal 12:100165. ISSN 2949-8635. https://doi.org/10.1016/j.sca.2025.100165
Ghosh SK, Sar AK (2022) Impact of effective supply chain management and supply chain risk management capabilities on construction project performance. Indian J Sci Technol 15(12):505–517. ISSN 0974-6846. https://doi.org/10.17485/ijst/v15i12.2194
Gunasekaran A, Ngai EWT (2004) Information systems in supply chain integration. Eur J Oper Res 159(2):269–295
Haque M, Kumar PS, Ruhul S, Daryl E (2023) A review on uncertainty modeling for decentralized supply chain systems. Flex Syst Manag 23–50. ISSN 2199-8493. https://doi.org/10.1007/978-981-99-2629-9_2
Helgeson JF, Roa-Henriquez A (2022) A theoretical framework for supply chain resilience planning. Int Ser Oper Res Manag Sci 159–189. ISSN 0884-8289. https://doi.org/10.1007/978-3-031-09183-4_8
Ivanov D (2020) Predicting the impacts of epidemic outbreaks on global supply chains. Transp Res Part E 136:101922
Ivanov D, Dolgui A (2020) Viability of intertwined supply networks. Int J Prod Res 58(10):2904–2915
Ivanov D, Dolgui A (2021a) A digital supply chain twin for managing disruptions and resilience. Int J Prod Res 59(18):1–18
Ivanov D, Dolgui A (2021b) A digital supply chain twin for managing disruptions. Int J Prod Res 59(14):1–17
Ivanov D (2021) Managing supply chain resilience. ClRoom Companion: Bus 29–61. ISSN 2662-2866. https://doi.org/10.1007/978-3-030-70490-2_2
Ji C, Liu X (2022) Design of risk sharing and coordination mechanism in supply chain under demand and supply uncertainty. RAIRO–Operat Res 56(1):123–143. ISSN 0399-0559. https://doi.org/10.1051/ro/2021186
Keskin B (2022) Digital supply chain transformation at Jabil: integrated supply and demand planning via concurrent planning. Case Women 1–15. ISSN 2732-4443. https://doi.org/10.1108/cfw.2022.000005
Khan S, Haleem A, Khan MI (2020) Assessment of risk in the management of Halal supply chain using fuzzy BWM method. Supply Chain Forum: Int J 22(1):57–73. ISSN 1625-8312. https://doi.org/10.1080/16258312.2020.1788905
Lucchetta F (2023) Predicting and managing supply chain risks. J Supply Chain Manag, Logist Procure 5(2):131. ISSN 2516-1822. https://doi.org/10.69554/mtro5006
Mahajan R (2025) Collaborative planning and design of digital supply chain. Flex Syst Manag 41–61. ISSN 2199-8493. Springer Nature Singapore. https://doi.org/10.1007/978-981-96-3556-6_3
Manhart P, Summers JK, Blackhurst J (2020) A meta-analytic review of supply chain risk management: assessing buffering and bridging strategies and firm performance. J Supply Chain Manag 56(3):66–87. ISSN 1523-2409. https://doi.org/10.1111/jscm.12219
Manyanga W, Kanyepe J, Chibaro M, Manyanga T (2026) Ethical considerations in supply chain analytics. Strat Supply Chain Anal 77–105. https://doi.org/10.4324/9781003674962-5
Mehravaran M, Nemati A (2025) A resilient poultry vaccine supply chain network design considering perishability and stress test. Supply Chain Anal 9:100098. ISSN 2949-8635. https://doi.org/10.1016/j.sca.2024.100098

Prompatanapak A, Lopetcharat K (2020) Managing changes and risk in seafood supply chain: a case study from Thailand. Aquaculture 525:735318. ISSN 0044-8486. https://doi.org/10.1016/j.aquaculture.2020.735318

Safari L, Sadjadi SJ, Sobhani FM (2023) Resilient and sustainable supply chain design and planning under supply disruption risk using a multi-objective scenario-based robust optimization model. Environ, Dev Sustain, *26*(11), 27485–27527, ISSN 1573–2975, Springer Science and Business Media LLC, https://doi.org/10.1007/s10668-023-03769-x

Saran N, Kesswani N (2025) An analytics-driven framework for securing industrial IoT-enabled supply chain management systems. Supply Chain Anal 11:100128. ISSN 2949-8635. https://doi.org/10.1016/j.sca.2025.100128

Sazvar Z, Tafakkori K, Oladzad N, Nayeri S (2021) A capacity planning approach for sustainable-resilient supply chain network design under uncertainty: a case study of vaccine supply chain. Comput Ind Eng 159:107406. ISSN 0360-8352. https://doi.org/10.1016/j.cie.2021.107406

Schoemaker P (1995a) Scenario planning: a tool for strategic thinking. Sloan Manag Rev 36(2):25–40

Shahzad A, Irfan M (2024) Enhancing supply chain resilience through supply chain finance: moderating role of supply chain disruption. SSRN Electron J. ISSN 1556-5068. https://doi.org/10.2139/ssrn.4915925

Sharma P, Pandey D, Manuj HK (2025) Impact of supply chain risk and supply chain finance on firm performance: Evidences from Indian SMEs. Risk, Reliab Resil Oper Manag 87–102. https://doi.org/10.1016/b978-0-443-29812-7.00005-x

Susitha E, Jayarathna A, Herath HMRP (2024) Supply chain competitiveness through agility and digital technology: a bibliometric analysis. Supply Chain Anal 7:100073. ISSN 2949-8635. https://doi.org/10.1016/j.sca.2024.100073

Syahchari DH, Sudrajat D, Lasmy L, Herlina MG, Estefania F, Zanten EV (2022) Achieving supply chain resilience through supply chain risk management and supply chain partnership. In: 2022 5th International conference on computers in management and business (ICCMB), pp 209–212, ACM. https://doi.org/10.1145/3512676.3512712

Taleb NN (2007) The Black Swan. Random House

Tang CS (2006) Perspectives in supply chain risk management. Int J Prod Econ 103(2):451–488

Tarei PK, Kumar G, Ramkumar M (2022) A mean-variance robust model to minimize operational risk and supply chain cost under aleatory uncertainty: a real-life case application in petroleum supply chain. Comput Ind Eng 166:107949. ISSN 0360-8352. https://doi.org/10.1016/j.cie.2022.107949

Waller M, Fawcett S (2013) Data science, predictive analytics, and big data. J Bus Logist 34(2):77–84

Walters J (2021) A renewed focus on supply chain risk management. J Transp Supply Chain Manag 15. ISSN 2310-8789. https://doi.org/10.4102/jtscm.v15i0.678

Wamba SF, Gunasekaran A, Akter S et al (2017) Big data analytics and firm performance. J Bus Res 70:356–365

Wang G, Gunasekaran A, Ngai EWT, Papadopoulos T (2016) Big data analytics in logistics and supply chain management. Transp Res Part E 95:98–115

Wieland A (2020) Dancing the supply chain: toward transformative supply chain management. J Supply Chain Manag 57(1):58–73. ISSN 1523-2409. https://doi.org/10.1111/jscm.12248

Wube W, Berhan E, Tesfaye G (2025) Decision support systems for a resilient and sustainable closed loop supply chain under risk: A systematic review and future research directions. Clean Logist Supply Chain 15:100217. ISSN 2772-3909. https://doi.org/10.1016/j.clscn.2025.100217

Yassine N (2022) Inventory planning under supplier uncertainty in a two-level supply chain. Int J Logist Manag 34(2):497–516. ISSN 0957-4093. https://doi.org/10.1108/ijlm-02-2021-0104

Zheng G, Brintrup A (2025) A machine learning approach for enhancing supply chain visibility with graph-based learning. Supply Chain Anal 11:100135. ISSN 2949-8635. https://doi.org/10.1016/j.sca.2025.100135

Zsidisin GA, Gaudenzi B, Pellegrino R (2024) Decision analysis techniques in supply risk assessment. Strat Sourc 37–62. https://doi.org/10.1007/978-3-031-52592-6_3

Policy, Regulation, and Multilevel Governance for Sustainable Transport Transitions

Ernest Mugoni and Pardon J. Muzondo

Abstract The transition to sustainable transport systems requires coherent policy frameworks, effective regulatory instruments, and coordinated governance across multiple levels of decision-making. This chapter examines the policy and governance dimensions of transport transitions, reviewing regulatory frameworks and policy tools currently deployed to promote sustainable mobility. Through a systematic analysis of multi-level governance literature (2019–2026), the chapter identifies persistent challenges including vertical fragmentation between national and local priorities, horizontal fragmentation across policy domains, and implementation deficits arising from weak institutional capacity. The analysis reveals that sustainable transport outcomes depend critically on three interconnected factors: policy coherence across governance levels and sectors, institutional capacity for policy design and enforcement, and effective compliance mechanisms that bridge the gap between regulatory intent and on-the-ground implementation. The chapter develops a conceptual framework integrating policy design theory with multi-level governance perspectives, illustrated through comparative analysis of European and developing country cases. Findings demonstrate that successful transport transitions require moving beyond mono-level, mono-modal policy approaches toward integrated governance arrangements that align funding, planning, and regulatory functions. The chapter concludes with recommendations for policymakers seeking to enhance policy coherence, build institutional capacity, and strengthen regulatory enforcement for sustainable transport.

Keywords Transport policy · Multi-level governance · Regulatory frameworks · Policy coherence · Institutional capacity · Sustainable mobility · Implementation

E. Mugoni (✉) · P. J. Muzondo
Department of Supply Chain Management, Faculty of Agribusiness and Entrepreurship, Marondera University of Agricultural Sciences and Technology (MUAST), Marondera, Zimbabwe
e-mail: ernmugoni@gmail.com

P. J. Muzondo
e-mail: pmuzondo@muast.ac.zw

J. Kanyepe and I. Chirisa (eds.), *Advancing the Green Agenda in the Transport and Logistics Sector in Global South Countries*, World Sustainability Series,
https://doi.org/10.1007/978-3-032-31374-4_15

1 Introduction

Transport systems globally stand at a crossroad. The sector accounts for about 25 percent of carbon dioxide emissions attributable to energy-related emissions globally, emissions are still escalating across most regions, and decarbonisation is vital in order to meet climate commitments set by nations under the Paris Agreement (Liu, Dijk and Colombo 2023). Even decades of policy initiatives promoting sustainable mobility are no substitute for action. The proportion of overall greenhouse gas emissions caused by transport increased from 16.5% in 1990 to 28.9% in 2022, as passenger cars continued to be the most common type of transportation in Europe (Budnitz et al. 2025). This continued disconnect between a policy aspiration and reality points to deeper problems across the manner in which transport is governed, regulated and financed.

The multi-layered nature of transport governance complicates matters even further. Transport system decisions are taken at several levels of government which are supranational, national, regional and local with different competences, resources and accountabilities (Marsden 2025). Transport policy intersects with a host of other related areas of governance such as land use planning, environmental protection, energy, economic development and social welfare, leading to horizontal coordination challenges as well as vertical ones (Hansson 2023). This institutional complexity indicates that to deliver sustainable transport outcomes, good policies are essential but governance systems must enable coherent action across levels and sectors. The multi-level perspective (MLP) from socio-technical transitions theory offers valuable insights here, as it conceptualises transport systems as consisting of interconnected niches, regimes, and landscapes, highlighting how governance interventions must address each of these levels to achieve lasting change (Geels 2012; Whitmarsh 2012).

This chapter examines the policy, regulatory and governance aspects of transport transitions. It responds to three related questions: What are the existing regulatory frameworks and policy levers to support sustainable transport? How does governance at a multi-level structure mediate the design, development, and delivery of policies? And what factors affect whether policies achieve their objectives? Despite a growing body of literature on transport governance, three critical gaps remain underexplored. First, existing studies tend to treat policy coherence, institutional capacity, and regulatory enforcement as separate analytical concerns rather than as an interdependent system; the mechanisms through which deficits in one dimension undermine the others have not been systematically theorised. Second, comparative analysis of governance configurations across the Global North and Global South remains limited, with most studies defaulting to European cases and neglecting the distinct institutional challenges facing rapidly urbanising, lower-income economies. Third, the integration of transition management theory (particularly the multi-level perspective) with multi-level governance frameworks has been largely conceptual, with insufficient empirical grounding in how governance interventions at niche, regime, and landscape levels interact in practice. This chapter directly addresses these gaps by developing an integrative framework and applying it through systematic comparative analysis. Indeed, the chapter contends, sustainable transport transitions rely crucially on policy coherence at all levels of governance, institutional capacity for policy implementation and enforcement, and capable vehicles for bridging the gap between regulatory intent and the ground realities.

The attention to governance is timely and necessary. Policy developments such as around 90% reduction in transport emissions by 2050 in the European Green Deal, the EU Urban Mobility Framework, the national strategies for net-zero, both in Paris and in other parts of the world create unprecedented pressure for transport transformation (European Commission, 2019, 2020). Yet as Marsden et al. (2014) note, climate change targets in the transport sector have often been seen as 'muddling through', rather than strategically delivered, in response to governance failures that fragment accountabilities and water down accountability. To understand how governance designs can facilitate rather than impede transport transitions is therefore critical for academics and practitioners. As Poyraz and Szalmáné Csete (2023) observe in their analysis of EU sustainability transitions, multi-level governance arrangements are fundamental to whether ambitious policy goals translate into tangible outcomes on the ground.

The chapter is structured as follows. Contextualising the evolution of transport policy and governance thinking is given in Sect. 2. Section 3 sets forth the aims of the research. The fourth section introduces a conceptual model integrating policy design theory with multi-level governance perspectives. Methodological framework is described in Sect. 5. Section 6 is a systematic literature review structured around three analytical themes: regulatory frameworks and policy tools; multi-level governance arrangements; and policy coherence and implementation. Section 7 presents results and their impact. Sections 8–10 provide research directions, policy implications, and conclusions.

2 Background to the Study

2.1 The Evolution of Transport Policy Paradigms

Transport policy has undergone major paradigm shifts in recent decades. The time-honoured 'predict and provide' approach, the dominant paradigm during most of the twentieth century, aimed to align transport supply with the rise in travel demand through ongoing infrastructure development (Liu et al. 2023). This assumption held that transport demand was exogenous and that the task of policy was to build capacity to fulfil it. The result was self-reinforcing car dependence: more roads meant more traffic, which also caused more demand for roads.

Subsequently, from the 1990s, this approach came under increasing criticism. The European Transport White Paper of 1992 showed a departure toward sustainable mobility, and that it was necessary to manage demand rather than to simply cope with it (Todella et al. 2024). An alternative organizing principle emerged as the 'avoid-shift–improve' framework: avoid unnecessary travel through integrated land use and transport planning; shift trips to more sustainable modes; and improve the environmental performance of all modes through technological innovation (United Nations, 2016 cited in Liu et al. 2023).

Sustainable mobility rhetoric has been widespread, though implementation has been difficult. This has led researchers to identify a number of the ways that the political economy of car dependence, cultural preferences for car usage, technological lock-ins and institutional path dependence underpin planning practices, funding streams and organisational structures (Budnitz et al. 2025). These elements establish what Budnitz et al. (2025) refer to as 'modal thinking' the assumption present in policy processes that sustainable and just urban mobility necessarily involves modes other than the private car, despite a parallel national focus on the electrification of private vehicles. Hoffmann et al. (2017) apply the multi-level perspective to analyse the entrenched automobility regime, demonstrating how governance interventions at niche, regime, and landscape levels interact to either reinforce or challenge car dependence.

2.2 *The Emergence of Multi-Level Governance Perspectives*

The notion of multi-level system of governance derived from discussions of European integration and emphasised the distribution of power among government tiers and the participation of non-state actors in decision-making (Hooghe and Marks 2010). In the transport sector, multi-level governance is particularly pertinent as responsibilities in transport systems are divided into supranational, national, regional, and local levels (Marsden 2025). The multi-level perspective on socio-technical transitions, pioneered by Geels (2012), complements this governance lens by explaining how regime-level stability interacts with niche innovations and landscape pressures, providing a framework for understanding the dynamics of low-carbon transitions in transport.

There are two general types of multi-level governance generally described (Hooghe and Marks 2010). Type I governance is similar to federalism where there are fewer general-purpose jurisdictions with no overlapping memberships and stable, system-wide architectures. Type II governance comprises task-specific jurisdictions which are transport authorities, port authorities, metropolitan planning organisations in which memberships flexibly overlap and are tailored to certain policy problems. Most transport governance reflects aspects of both types which results in complex institutional landscapes (Roe 2009). Ehnert et al. (2018) demonstrate in their comparative study of four European states that these governance configurations significantly shape the capacity for urban sustainability transitions, with federal systems exhibiting different coordination dynamics than centralised ones.

The multi-level governance lens has several analytical advantages for examining transport policy. Firstly, it directs attention to vertical relationships i.e. how policies produced at elevated levels transform (or fail to translate) into practice at lower levels (Marsden et al. 2014). Second, horizontal relationships where the coordination across sectors and territorial areas, at the same level of governance (Hansson 2023). Third and perhaps most important, it draws attention to the role of non-state actors which include businesses, civil society organisations, expert networks in influencing policy processes and outcomes (Kin et al. 2017). Research by Guo and Li (2025) on low-carbon tourism transport in rural China illustrates how multi-level governance arrangements mediate the participation of diverse actors across administrative scales, shaping both policy evolution and implementation pathways.

2.3 Policy Coherence and Implementation Challenges

One of the consistent themes in research on transport policy is the gap between policy development and implementation outcomes. National policy guidelines to design strategies for sustainable mobility rarely follow practical local implementation processes due to misalignments between policy design and adoption settings (Jain 2024). It is a manifestation of several reasons for the failure to be realised these include inadequate monitoring structures at various levels of governance, disjointed and incoherent institutional frameworks, and a limited local capacity for implementation of a policy (Rosales and Haarstad 2023). Massuga et al. (2024) provide conceptual clarity on modes of governance for sustainability transitions, distinguishing between hierarchical, network-based, and market-based governance modes and their implications for implementation effectiveness.

Policy coherence involves such a policy harmonisation of activities across sectors and at multiple levels of governance to achieve mutually reinforcing results has become salient in addressing policy failures (Khatoon et al. 2025). Incoherence can manifest along three dimensions which include vertical incoherence between national targets and local strategies, horizontal incoherence between planning of transport and land use, and temporal incoherence where long-term sustainability objectives are in conflict with short-term political and financial cycles (Van Straalen and Witte 2018). Cevheribucak (2024) examines how multilevel governance structures affect the design and implementation of Sustainable Urban Mobility Plans (SUMPs), revealing that coherence deficits often stem from misaligned competencies and resources across governance tiers.

The COVID-19 pandemic drew attention to the potential of fast transport policy change and the governance challenges involved. Hirschhorn (2021) examines the multi-level governance responses in public transport to the pandemic and argues that crisis conditions facilitated rapid decision-making at the same time as they laid bare underlying vulnerabilities in governance arrangements. The experience of the pandemic shows that while governance innovation is feasible, it takes institutional arrangements shaped by flexibility to respond. Kotilainen et al. (2026) extend this analysis to conditions of urgency in the battery industry development for electric mobility, demonstrating how multi-level governance under pressure requires both adaptive capacity and institutional resilience.

2.4 Research Objectives

In this chapter, three main research objectives are covered:

RO1: Analyse and compare the regulatory framework and policy tools employed currently to contribute to sustainable transport transitions. This includes the wide range of available policy instruments, such as regulatory, economic, informational and organisational policy and the way they are applicable to various forms of governance. Particular focus is on how policy mixes combine instruments to achieve more than one aim or overcome barriers to implementation.

RO2: To examine multi-level governance arrangements in transport and what it means for policy design and delivery. This includes analysis of vertical relationships between supranational, national, regional, and local governance levels; horizontal relations across policy sectors and territorial jurisdictions; and the work of other non-state actors in governance.

RO3: To investigate what the determinants of policy coherence, institutional capacity and regulatory enforcement would be determinants of sustainable transport success. This deals with the factors that enable or constrain effective implementation such as institutional design, resource availability, monitoring mechanisms, and accountability arrangements.

These aims are interrelated i.e. regulatory frameworks operate within multi-level governance systems, and their effectiveness depends on coherence across levels and sectors as well as institutional capacity for implementation and enforcement.

3 Conceptual Model

3.1 Theoretical Foundations

The chapter takes a theoretical framework combining the insights of policy design theory with multilevel governance frameworks. Policy design theory is a literature that offers tools to understand policy content and processes of policy change through analysis (Howlett and Rayner 2013; Cashore and Howlett 2007). The policy composition framework organises policy in six elements with three levels of abstraction which include macro-level goals and instrument logics; meso-level objectives and instrument choices; and micro-level targets and instrument calibrations (Capano and Howlett 2024). Such a fine-grained analysis allows the extraction of tensions and inconsistencies in policy mixes.

Multi-level governance theory provides insight into how authority is distributed across governance levels and how the interactions between levels lead to policy processes and their resultant consequences (Hooghe and Marks 2010; Marsden 2025). In doing so, the framework differentiates vertical relationships (across levels) from horizontal relationships (across sectors and jurisdictions), as well as formal institutional arrangements and informal governance practices. Complementing this, the multi-level perspective from transition studies (Geels 2012) offers a dynamic understanding of how socio-technical systems evolve through interactions between niches (sites of innovation), regimes (entrenched practices and institutions), and landscapes (exogenous trends and pressures). This integration of governance and transition perspectives is increasingly recognised as essential for analysing sustainable mobility transformations (Whitmarsh 2012; Hoffmann et al. 2017).

3.2 The Policy Coherence and Implementation Framework

With these theoretical principles as bases, Fig. 1 proposes a conceptual model connecting policy design, governance structures, and the policy implementation results. The framework refers to three interrelated dimensions:

Policy Coherence refers to the alignment of policies across levels of governance (vertical coherence), across policy sectors (horizontal coherence), and over time (temporal coherence). Comprehensively integrated policy mixes utilise different instruments to achieve synergy, rather than working in tension with each other, with

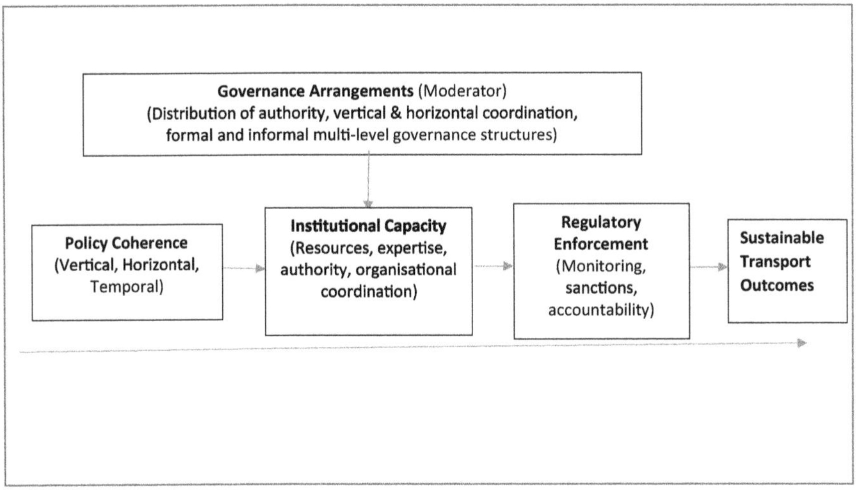

Fig. 1 Conceptual framework linking policy coherence, institutional capacity, regulatory enforcement, and sustainable transport outcomes. *Source* Authors' own compilation

clear articulation of how different layers of governance contribute to shared objectives. Cevheribucak (2021) demonstrates in the Turkish context how energy transition and sustainable road transport policies require coherence across these dimensions to achieve inclusive change.

Institutional Capacity includes the resources, capabilities, and authority needed for effective policy implementation and enforcement. Such a framework should include budgetary provisions for infrastructure investment and service provision; technical and managerial capabilities for policy design and execution; regulatory authority backed by adequate enforcement mechanisms; and organisational structures that support coordination between key players. Pinno (2025) proposes a Regional Sustainability Maturity Model that conceptualises capacity development as a strategic, process-oriented journey, offering insights for how regions can progressively enhance their governance capabilities.

Regulatory Enforcement is the process through which the regulatory body translates policy into compliance and behaviour change. Monitoring systems for tracking the progress of implementation and regulatory compliance; sanctioning mechanisms for addressing non-compliance; accountability arrangements in which implementing actors are held responsible for outcomes. The framework proposes that sustainable transport outcomes rely on all three dimensions. Well-designed policies (policy coherence) will not prevail without sufficient institutional capacity for implementation. Strong institutional capacity can be misdirected without coherent policy guidance Effective enforcement mechanisms need both clear policy targets and capable implementing entities. Hansel (2025) illustrates this interplay in his study of mobility hub governance, showing how the dynamics of stability and change in sustainable mobility transitions depend on the alignment of these three dimensions.

The conceptual framework further specifies that governance arrangements, particularly the distribution of authority across levels and the mechanisms for inter-level coordination, moderate relationships among coherence, capacity, enforcement, and outcomes. Multi-level governance structures can either enable or obstruct policy coherence depending on how they align responsibilities with resources and create incentives for coordination. Fathoni, Lovett, and Rifansha (2025) apply the multilevel perspective to the Indonesian transport sector, demonstrating how governance arrangements at national and subnational levels interact to shape renewable energy transitions, with implications for the coherence-capacity-enforcement nexus.

3.3 Research Methodology

This chapter adopts a systematic literature review methodology using established guidelines for transparent and replicable synthesis (Tranfield et al. 2003). The review follows a structured protocol informed by the PRISMA (Preferred Reporting Items for Systematic Reviews and Meta-Analyses) framework, adapted for policy and governance research (Moher et al. 2009). The protocol specified: (a) a priori research questions aligned with the three objectives; (b) explicit search strings and database coverage; (c) pre-defined inclusion and exclusion criteria applied independently by both authors, with disagreements resolved through discussion; (d) a quality assessment process evaluating methodological rigour, relevance, and contribution to the theoretical framework; and (e) a data extraction template coding each source for governance level, policy instrument type, context (developed/developing country), and key empirical or theoretical contribution. This transparent and replicable approach reduces selection bias and enables readers to assess the scope and limitations of the review's evidential base. The review protocol outlined research questions relevant to the three objectives, search strategies for a variety of databases, inclusion and exclusion criteria, and quality assessment process.

3.4 Search Strategy and Database Selection

The following databases were searched: Scopus, Web of Science, ScienceDirect, SpringerLink, and Google Scholar. The search strings combined terms such as transport policy ("transport policy" OR "transport regulation"), governance ("multi-level governance" OR "multilevel governance" OR "governance arrangements"), and policy implementation ("policy coherence" OR "policy coordination" OR "regulatory enforcement"). The search was restricted to publications from 2019–2026 to ensure the work's currency, other than foundational theoretical works which included seminal contributions from Geels (2012), Whitmarsh (2012), and Hoffmann et al. (2017) that established the multi-level perspective in transport studies.

3.5 Inclusion Criteria and Screening

Studies were eligible if they: (1) covered transport policy, regulation, or governance; (2) were published in peer-reviewed journals, academic books, or high-quality research reports based on empirical or theoretical methods; (3) came from reputable and peer-reviewed journals, books, or high-quality research reports; (4) were written in English. Articles on technical aspects of transport without policy or governance components were excluded. The screening process proceeded in three stages: (i) title and abstract screening against the inclusion criteria; (ii) full-text review of candidates passing stage one; and (iii) backward and forward citation tracking to identify additional relevant sources not captured in the database searches. The initial database searches yielded approximately 420 records after deduplication. Following title and abstract screening, 98 records were retained for full-text review. Of these, 47 were excluded at full-text stage (primarily for lacking policy or governance content, being technically focused, or being inaccessible). An additional 12 sources were identified through citation tracking. The final corpus comprised 30 sources used directly in this chapter, selected to achieve both thematic coverage across the three research objectives and geographic diversity across developed and developing country contexts.

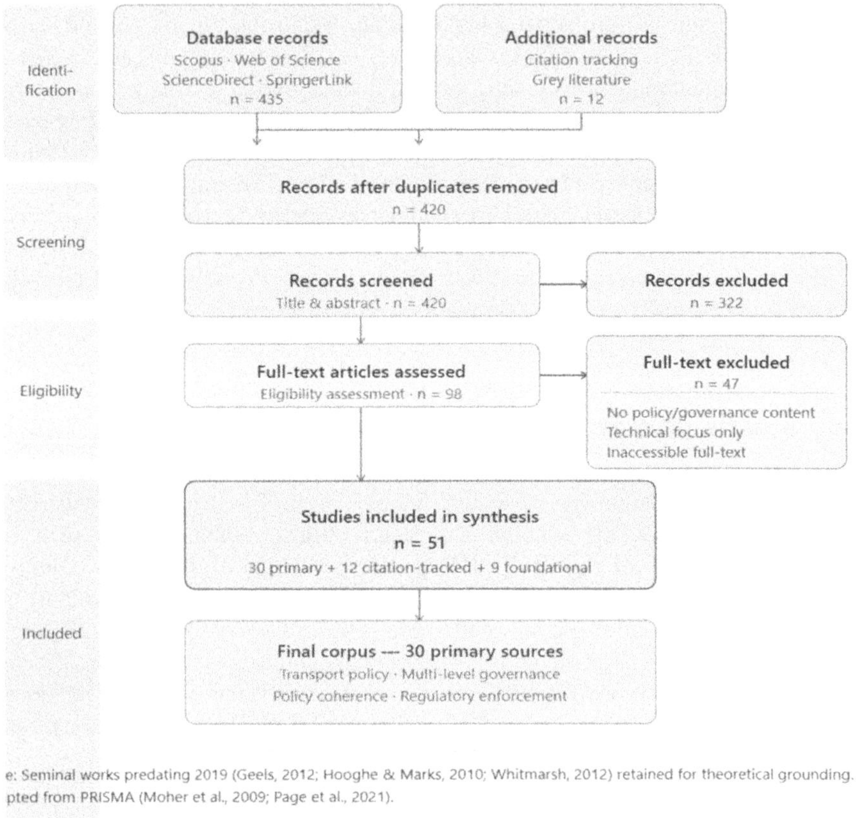

Fig. 2 PRISMA Flow Diagram—Systematic literature review study selection process. *Source* Authors' own compilation, adapted from Moher et al. (2009)

A PRISMA flow diagram documenting this selection process is as shown in Fig. 2 (Table 1).

3.6 Analytical Approach

Analysis was thematic across the three research objectives. Evidence concerning the policy instruments, governance arrangements, methods of implementation, and impacts was synthesised for each theme. Special attention was paid to research of other governance models with different contexts as well as of factors associated with successful policy implementation. The integration of multi-level perspective studies (Geels 2012; Ehnert et al. 2018; Fathoni et al. 2025) enabled analysis of how socio-technical dynamics interact with governance arrangements across niche, regime, and landscape levels.

Table 1 Inclusion and exclusion criteria for the systematic literature review

Criterion	Inclusion	Exclusion
Topic focus	Transport policy, regulation, governance, or multi-level governance	Purely technical transport studies with no policy or governance dimension
Publication type	Peer-reviewed journal articles, academic book chapters, edited volume contributions, high-quality institutional/government research reports	Opinion pieces, blogs, conference abstracts without peer review, popular media
Methodology	Empirical (qualitative, quantitative, mixed methods) or rigorous theoretical/conceptual studies	Studies without clear methodological grounding or without sufficient analytical rigour
Publication period	2019–2026 (contemporary literature); seminal theoretical works pre-2019 retained if foundational to the MLP or multi-level governance frameworks	Studies published before 2019 unless meeting the seminal works exception
Language	English language publications	Publications in languages other than English
Geographic scope	Any country or region; Global North and Global South contexts actively sought for comparative coverage	No geographic exclusion applied; however, studies with no transferable governance insights were excluded
Thematic relevance	Studies addressing at least one of: regulatory frameworks and policy instruments; multi-level governance arrangements; policy coherence, institutional capacity, or regulatory enforcement	Studies addressing transport technology, engineering, or operations without governance or policy analysis
Accessibility	Full text accessible through institutional database subscription or open access	Full text inaccessible despite reasonable retrieval effort

Source Authors' own compilation, adapted from Tranfield et al. (2003) andMoher et al. (2009)

4 Literature Review

4.1 Regulatory Frameworks and Policy Tools for Sustainable Transport

4.1.1 Typologies of Policy Instruments

Transport policy has a range of tools which uses a wide range of tools and means to influence and promote behaviour, to form investments, and to meet sustainability objectives. The conventional typologies point out that there are regulatory instruments (commands and controls), economic instruments (prices and incentives), informational instruments (communication and education), and organisational instruments (planning and coordination) (Syzdykbayeva et al. 2019). There are unique strengths and limitations of each category.

Regulatory instruments establish binding requirements (such as vehicle emissions standards, fuel quality, parking regulations and low-emission zones), with enforcement supported by sanctions. Where they succeed is in giving certainty about minimum standards and creating level playing fields for regulated agents. Nevertheless, regulatory approaches may be stifling and often face opposition from politicians who have no desire to interfere, with the requisite resources for monitoring and enforcement required (Teles and Sousa 2021).

Economic instruments function via price signals instead of orders. Fuel taxes, congestion charges, vehicle purchase taxes and benefits in the form of subsidies for alternative transport or clean vehicles all introduce financial incentives for environmentally friendly behaviour. They can be economic devices that are easy to use; they can yield differential responses depending on the specifics of the situation. Their effectiveness is context dependent on price levels and decision-makers' sensitivity, and they may have fairness concerns concerning regressive effects (Zhang and Witlox 2019). Informational instruments aim to change behaviour with communication, not regulation or price. Eco driving training, travel awareness, custom travel and eco-labelling for cars, for example, and other information-promotion tools, are some attempts to influence consumption. These instruments are ultimately less coercive, more cost effective, and do less damage than regulation or pricing (Conteh 2013), but their effects tend to be low and may diminish over time.

Organisational instruments are planning mechanisms, coordination mechanisms and institutional arrangements. Sustainable Urban Mobility Plans (SUMPs), transport strategies, partnership arrangements, and integrated governance structures govern decisions and who participates in those choices. Often overlooked in policy analysis, such instruments may be indispensable, and key tools in the success of other instruments (Knapskog et al. 2025). Cevheribucak (2024) provides detailed analysis of how SUMP design and implementation are shaped by multilevel governance structures, revealing that organisational instruments are not neutral tools but embed particular governance logics.

4.1.2 Policy Mixes and Instrument Interactions

Today's policy ideas are heavily based on the idea that instruments are not used in isolation. Policy mixes i.e. mixtures of instruments that help achieve common goals are increasingly acknowledged as an approach for tackling complex sustainability issues (Liu et al. 2023). The effect of policy mixes is that the two instruments are not mutually exclusive: one combination produces synergetic (reinforcing effects), some produce conflicting (counteracting effects), and others generate neutral or uncertain interrelations.

A study of policy mixes as a tool for sustainable urban mobility points to a number of design principles. First, mixes need to couple 'push' (disincentives for unsustainable behaviour, such as congestion charging) with 'pull' (incentives for sustainable alternatives, such as better public transport) to prevent political reactions and facilitate changes in behaviour (Marcucci and Stathopoulos 2012). Second, mixes should consider all the barriers to sustainable mobility such as infrastructural, economic, informational, and cultural acknowledging that no single tool can dismantle all barriers. Third, mixes should be coherent at all levels of governance with national frameworks generating local action while local innovations inform national policy (Knapskog et al. 2025).

Liu et al. (2023) describe two conflicts between multi-level policy mixes in urban mobility transitions. The first is that funding mechanisms are still biased toward 'solving bottlenecks through infrastructure', favouring capital-intensive projects over softer measures and reinforcing automobility. Second, national governments, meanwhile, often have considerable influence over local policy while being relatively insulated from influence by other levels. These conflicts lead to systematic distortions in policy mixes, where these institutional mechanisms create systematic distortions in the mix of policy outputs, and therefore favour certain types of instruments and governance levels over others. Geels (2012) explains such distortions through the lens of regime resistance, where entrenched automobility regimes shape policy instrument selection to protect incumbent interests.

4.1.3 The Avoid-Shift-Improve Framework in Practice

The avoid-shift-improve framework offers a heuristic for organising policy tools around complementary strategies. 'Avoid' policies aim to reduce the need for motorised travel through integrated land use planning, teleworking, and local service provision. 'Shift' policies aim to move trips to more sustainable modes through public transport investment, cycling infrastructure, and walking-friendly urban design. 'Improve' policies emphasize reducing the environmental impact of all modes, particularly in vehicle technology, clean fuels, and operational efficiency.

A systematic exploration of how these strategies are implemented reveals systematic biases. Khatoon et al. (2025) study the low-emission transport policy landscapes in Dhaka City, Bangladesh and observe that policy frameworks were constructed around politically prioritised supply-side solutions (i.e. metro rail) without alignment with land use and housing policies and with insufficient consideration of local need-based alternatives (i.e. walking and cycling). This phenomenon emphasising visible, capital-intensive projects above dispersed, behavioural measures is demonstrated under various circumstances.

In the EU, there are explicit adoptions of the avoid-shift-improve model with the European Union's Sustainable and Smart Mobility Strategy (European Commission 2020), with the implementation of the goal of automated mobility, zero-emission vessels and high-speed rail in line with the goal of ensuring everyone can have access to sustainable choices. However, Budnitz et al. (2025) note this core tension; European national governments focus on the low carbon transition to electric vehicles (the 'improve' strategy) while cities target the reduction of car traffic altogether (a mixture of 'avoid' and 'shift' strategies). This tension, coming from different governance responsibilities and political constituencies, is a drag on coherence in policy. Geelhuijzen (2025) analyses similar dynamics in Dutch cities, applying the multi-level perspective to understand how national and local policy tensions shape sustainable mobility transitions.

4.1.4 Emerging Policy Instruments

In recent years, policy instruments have seen innovation in transport policy instruments. Digitalisation facilitates these developments by introducing new methods: real-time information frameworks to support modal choice; Mobility-as-a-Service applications that leverage ticketing and payment; data-driven enforcement of low-emission zones; and AI-assisted policy design and evaluation (European Commission 2020). These tools have potential for a more tailored, more nimble and more responsive delivery of policy interventions, but also raise the issues of data privacy, algorithmic bias and digital exclusion.

Behavioural insights have informed the development of novel instrument designs. 'Nudge' approaches structuring decisions to make sustainable choices the default or most convenient have been used in studies on travel mode choice, eco-driving, and commuting behaviour. Less coercive compared with regulation and less economically burdensome than pricing, nudges usually have modest but sometimes insufficient impacts to meet ambitious targets (Teles and Sousa 2021). Participatory instruments have come to be prominent, reflecting the awareness that when it comes to transport transitions, public acceptance will need public involvement and engagement. Kin et al. (2017) develop a stakeholder-based methodology that addresses barriers and builds support to enhance the success of urban freight transport measures in multi-level governance contexts with stakeholder involvement. The EU's Urban Agenda partnerships provide examples of participatory governance and involve cities, member states, and stakeholders in the co-development of policy (Knapskog et al. 2025). Hansel (2025) demonstrates in his study of mobility hubs how participatory governance can facilitate the dynamics of stability and change in sustainable mobility transitions.

4.2 *Multi-Level Governance Arrangements in Transport*

4.2.1 Vertical Governance Relationships

Transport policy effectiveness requires vertical relationships between levels of governance. Supranational organisations such as the European Union establish frameworks e.g., emissions standards, state aid rules, Trans-European Transport Network (TEN-T) guidelines that determine which national and local policy possibilities are available. National governments usually regulate the country's major infrastructure, rail networks, vehicle regulations, and taxation. Regional authorities are usually involved in planning and funding public transport networks. And local government controls streets, parking, land use, and local mobility services (Marsden, 2025). Poyraz and Szalmáné Csete (2023) provide a comprehensive analysis of how these vertical relationships function within the EU's sustainability transition governance, highlighting both coordination mechanisms and persistent tensions.

The division of responsibilities thus also provides opportunities and challenges for policy coherence. A transparent distribution of competences can promote specialisation and accountability. However, fragmentation may result in gaps (issues falling between levels), overlaps (multiple levels asserting competence), and conflicts (conflicting policies at different levels) (Veeneman and Mulley 2018). Transport policy is especially susceptible to these difficulties since transport systems are characterised by their multi-scalar nature, a journey consisting of local streets, regional public transport, national roads, and international connections. Ehnert et al. (2018) compare four European states to demonstrate how different institutional architectures produce distinctive patterns of vertical coordination and conflict in urban sustainability transitions.

Studies of climate change governance in transport show challenges of vertical coordination. Marsden et al. (2014) investigate approaches to climate change targets across governance levels in the UK transport sector and describe how target-setting occurs primarily at national level while delivery depends heavily on local action, thus creating a 'delivery gap', whereby responsibility without authority compromises implementation. Butterfield and Low (2017) investigate monitoring hurdles for multi-level governance at an important level of transport carbon reduction to show how data availability, methodological consistency, and attribution issues make accountability very hard.

India's move towards net-zero transport goals is reviewed in Jain (2024), where he finds a mismatch between national policy intent and local realization. Transport projects despite constitutional amendments and national schemes to facilitate city development are often seen in isolation without being integrated into city planning or the assessment of impacts on sustainable mobility. The transport department cannot steer sustainable mobility on its own and should collaborate with land development authorities; but the institutional frameworks do not mandate or permit it. Fathoni, Lovett, and Rifansha (2025) identify similar vertical governance challenges in Indonesia's renewable energy transition for transport, where national targets for biofuel adoption confront limited local implementation capacity and coordination mechanisms.

4.2.2 Horizontal Governance Relationships

Inter alia, horizontal coordination among sectors of policy is just as important. Transport policy and its integration with land use planning, environmental protection, energy, economic development, health, and social welfare is also very important. Coherence between these fields is crucial for optimal transport outputs to be sustainable, but sectoral silos between sectors and competing professional cultures affect unity (Hansson 2023).

The relationship between transport and land use planning has particular ramifications. Land use patterns drive travel demand, which sets the conditions for sustainable mobility. On the flip side, transport investment has a direct effect on land development, forming urban form for decades. Thus efficient use of transportation and land use planning should form the essential basis of 'avoid' strategies. However, institutional arrangements frequently separate these functions, with different agencies, professional communities, funding streams, and decision cycles (Van Straalen and Witte 2018). Guo and Li (2025) examine how horizontal coordination between transport, tourism, and environmental agencies shapes low-carbon tourism transport outcomes in rural China, revealing that integration deficits undermine policy effectiveness. Rosales and Haarstad (2023) investigate governance issues for urban logistics in Norwegian cities and reveal that the transport of freight often falls between sectoral responsibilities—too commercial for transport planning, too urban for logistics policy, too fragmented for environmental regulation. This 'governance gap' results in policy that underprioritizes urban freight as a sector to focus on despite its increasing importance with e-commerce practices, as well as the impacts they have on the environment.

Cross-border coordination adds another layer of complexity. Dörry and Decoville (2016) examine governance and policy networks in the cross-border metropolitan region of Luxembourg and uncover how disparate national frameworks, languages, and planning cultures obstruct coordination. Panagakos and Psaraftis (2017, 2018) consider multi-level cooperation for transport among the Baltic Sea Region in terms of best practices and lessons from bottom-up corridor initiatives, which required coordination across multiple national jurisdictions in the region.

4.2.3 The Role of Non-State Actors

Governance at multiple levels is about much more than the government: it has to also include businesses, civil society organizations, experts and citizens. These non-state actors engage with the policy process across a variety of agendas: as advocates influencing agendas, as experts informing policy design, as partners in implementation, as subjects of regulation, and as citizens expressing preferences through consultation and political participation (Kin et al. 2017). Hoffmann et al. (2017) demonstrate how non-state actors, particularly automotive industry incumbents, can exert significant influence on regime stability or transformation depending on their alignment with niche innovations or landscape pressures.

Business actors, including transport operators, logistics firms, vehicle manufacturers, technology providers, play an important role in the transport transitions. Their decisions about investment, mode of operation, and innovation activities, guide what is possible and how quickly change occurs. Involving business in policy-making can enhance instrument design, consolidate support for implementation and channel private resource efforts for public purposes. Business influence can also tilt policy towards the interests of powerful incumbents and away from transformative change (Roe, 2007).

Civil society organisations seek to protect the environment, support social equity, promote active mobility, and promote community interests. They may enable the elevation of voices otherwise marginalised and have the capacity to hold governments to account for commitments and to galvanise public support for ambitious policies. Their involvement may also serve to improve legitimacy and to assure that transitions are 'just' rather than merely low-carbon (Jain, 2024). Expert communities give knowledge that informs policy. Research institutions, consultancies, universities and professional bodies generate evidence on the issue and develop methodologies and training for practitioners. Expert opinion helps maintain the quality of policy analysis and design (Conteh, 2013), but expertise does not necessarily represent neutrality—framing, methodological assumptions, and disciplinary perspectives contribute to production of knowledge and its interpretation.

4.2.4 Multi-Level Governance Challenges in Developing Countries

Multilevel governance challenges in transport among developing countries. Institutional capacity is generally poor i.e. financial resources, technical expertise, and enforcement ability. Transport provision and economic activity are largely guided by informal sectors, which function outside formal regulation. Rapid urbanisation and motorisation create urgent pressures that outpace policy development (Khatoon et al. 2025). Fathoni, Lovett, and Rifansha (2025) apply the multi-level perspective to analyse renewable energy transitions in Indonesian transport, revealing how developing country contexts exhibit distinctive dynamics where international landscape pressures (climate agreements, technology transfer) interact with national regime structures and local niche innovations.

Khatoon et al. (2025) analyse the transport policy coordination in Dhaka City, Bangladesh, and concluded that policy frameworks and institutional arrangements were in place to facilitate coordination between local and global actors (development partners). Development partners were involved in mobilising resources, transferring low-emission transport technologies, and local capacity building. But policies prioritising politically motivated supply-side solutions were not sufficiently cohesive with land use and housing policies, nor sufficiently focused on unintended consequences and boundary-spanning challenges.

The idea of 'codified and reflexive policy coordination' was formulated by Khatoon et al. (2025) for developing country settings. Codified coordination involves formal policy frameworks and institutional arrangements that structure the interactions among actors. Reflexive coordination is evaluating policy interventions in a formalised model, which allows learning and adjustment. Both are necessary, codified frameworks as the source of certainty and direction, reflexive processes for responding to context and feedback. Guo and Li (2025) provide complementary evidence from rural China, demonstrating how codified frameworks at national and provincial levels interact with reflexive, place-based implementation at local levels in low-carbon tourism transport initiatives. International development assistance impacts transport governance in developing countries through conditionalities, technical support, and project financing. Although such assistance encourages capacity and knowledge sharing, it can lead to dependencies, distort priorities, and impose unworkable models on local contexts (Syzdykbayeva et al. 2019). Finding the right balance between international engagement and local control/ownership and accountability continues to be a constant problem.

4.3 Policy Coherence, Institutional Capacity, and Regulatory Enforcement

4.3.1 Policy Consistency: A Concept and Measurement

Policy coherence has become a focal issue in transportation governance research. The concept is a systematic alignment of policies across sectors and levels of governance to work towards mutually reinforcing results. Coherence is separate and distinct from coordination: coordination is the processes of interaction between actors; coherence is the substantive compatibility of policies (Khatoon et al. 2025).

Vertical coherence is about alignment across governance levels. Local strategies should be informed by national targets; national policy should be an outcome of local innovations. Without vertical coherence, policy at various levels will operate contradictory, e.g., national fuel subsidies may undermine local public transport investment, or local parking measures may not align with regional road network designs (Marsden 2025). Cevheribucak (2021) illustrates these vertical coherence challenges in Turkey's energy transition and sustainable road transportation, revealing how national energy priorities sometimes conflict with local mobility needs. Horizontal coherence has to do with alignment of policies of policy sectors. Transport policy must be compatible with land use, environmental, energy, and social policies. There should be multiple forms of incoherence: transport investment (sprawl against land use plans); environmental regulations (car use regulation) that drive up transport costs and provide no access to mobility alternatives; and energy policies (green energy for

EV markets without regard to grid or charging capacity) (Hansson 2023). Massuga et al. (2024) provide conceptual tools for understanding how different governance modes (hierarchical, network, market) produce characteristic patterns of horizontal coherence or fragmentation.

Temporal coherence issues when this alignment occurs over time. "Achieving long-term sustainability goals will require continuity of policy direction that will be sustained over the course of political cycles." Decisions made in the short term that have lasting consequences–infrastructure, land use, technology choices–should also be measured against long-range goals. When policy changes, cessation of funding and priorities change and damage investor confidence, and they undermine planning and the transition (Van Straalen and Witte 2018). Pinno (2025) addresses temporal coherence through the Regional Sustainability Maturity Model, which conceptualises capacity development as a longitudinal process requiring sustained commitment across political and budgetary cycles.

It's methodologically challenging to measure policy coherence. Coherence varies but is not binary. That begs a counterfactual evaluation how would this stack up without alignment? It has to do with normative judgements about what kind of outcomes are better. And it must accommodate dynamic interaction: coherence now may foster or restrict coherence later on (Todella, Abastante and Cotella 2024). Geelhuijzen (2025) addresses these measurement challenges through longitudinal analysis of sustainable mobility transitions in Dutch cities, tracking coherence patterns over time.

4.3.2 Institutional Capacity for the Governance of Transportation

Institutional capacity is the resources, capabilities and authority to implement and enforce policy effectively. Impairment of capacity is clearly cited as one of the impediments to sustainable transport results, it is more so reported at the local government level where the implementation responsibility usually lies. The multi-level perspective highlights how capacity deficits at particular levels can constrain the entire transition process (Geels 2012; Whitmarsh 2012).

Financial capacity consists of capital expenditures for infrastructure and services, and maintenance and business as usual expenditure funds. Transport infrastructure is a capital-intensive process that requires continuous funding streams over extended windows of time. This requires higher-level institutional funding, which does not lend to their local needs as much as it can, as many local governments don't have sufficient fiscal resources or borrowing capacity and these funds can either be inconsistent or conditional or be out of line with the local needs (Jain 2024). Kotilainen et al. (2026) examine how financial capacity for battery industry development in the EU is shaped by multi-level governance arrangements, with supranational funding mechanisms interacting with national industrial policies and local implementation capacities.

Technical capacity means the ability to analyse, design and evaluate a policy. Sustainable transport planning requires competencies across multiple domains such as traffic engineering, urban design, behavioural science, data analytics, stakeholder engagement that may be scarce, particularly in smaller or less wealthy jurisdictions. Developing capacity needs investment into training, recruitment and knowledge transfer, but these investments compete with day—to—day operational aspects of doing business (Rosales and Haarstad, 2023).

Capacity The capacity of an organisation is characterised by the structures and processes that make effective action possible. Organisational capacity is comprised of things such as clear allocation of responsibilities, effective coordination mechanisms, streamlined decision processes, and performance management systems. But transport governance often takes multiple agencies over and is fragmented, with overlapping responsibilities, unclear accountabilities, and poor coordination (Veeneman and Mulley 2018). Ehnert et al. (2018) demonstrate how organisational capacity for urban sustainability transitions varies systematically across different national governance contexts, with federal systems exhibiting different coordination patterns than unitary states.

Regulatory capacity indicates the authority to make an enforce action along with sufficient measures to monitor and sanction it. Without monitoring and sufficient sanctions to deter non-compliance, regulation fails. Many jurisdictions lack resources for monitoring (inspectors and surveillance technology), and many faces political barriers to enforcing these rules. These create implementation gaps regulations exist on paper, but not in practice (Sack 2011).

4.3.3 Regulatory Enforcement and Compliance

Regulatory enforcement is the part that links policy intent and actual outcomes on the ground. For effective enforcement, there should be clarity about what is required, monitoring to detect non-compliance, and sanctions sufficient to deter violations. Each element poses challenges in transport contexts.

Clarity of requirements depends on regulatory design. Vague standards, ambiguous definitions, and complicated compliance processes breed uncertainty that undermines compliance and enforcement. Transport regulations are often technically complex (e.g., emissions standards, vehicle specifications, fuel quality parameters) and beyond the capacity of regulated actors or enforcement agencies to interpret and apply (Syzdykbayeva et al. 2019). Fathoni, Lovett, and Rifansha (2025) identify enforcement challenges in Indonesia's biofuel mandate for transport, where clarity of requirements is undermined by inconsistent fuel quality standards and limited testing capacity across the archipelago.

Monitoring compliance is resource intensive. Vehicle emissions testing usually requires equipment and trained personnel; parking and traffic enforcement requires patrols or automated systems; monitoring fuel quality involves sampling and laboratory analysis. Resource constraints limit monitoring coverage, resulting in enforcement gaps that undermine deterrence. Technological options including automated number plate recognition, remote emissions sensing, and satellite tracking can extend monitoring reach, yet they raise privacy concerns and require investment (Butterfield and Low 2017).

Sanctions for non-compliance must be sufficiently severe to deter violations but proportionate to the offence. Fines that are too low become a simple cost of doing business; enforcement that is too sporadic fails to deter; sanctions that are too severe may provoke resistance or be withheld by enforcement authorities. Consistency in sanctioning matters for perceived legitimacy: arbitrary or discriminatory enforcement erodes both trust and compliance (Sack 2011).

Enforcement is also a matter of political will. Regulatory agencies might feel pressured to go easy on powerful interests; enforcement might slacken during economic downturns; political appointees can override professional enforcement decisions. Sustained enforcement presupposes institutional insulation from short-term political pressures; there must be accountability for performance (Conteh 2013). Kotilainen et al. (2026) examine how urgency in battery industry development creates pressures for accelerated enforcement procedures, sometimes at the expense of environmental and social safeguards.

4.3.4 Learning from Implementation Failures

Implementation failures provide opportunities for learning. An examination of why well-designed policies fails to achieve intended outcomes may reveal systemic weaknesses and contribute to improvements in policy design, governance arrangements, and enforcement mechanisms. Sack (2011) examines governance failures in integrated transport policy in Germany and the mismatches of 'co-opetition' (simultaneous cooperation and competition) in multi-level systems. Policy integration demands cooperation across actors with varied interests and accountabilities, but institutional arrangements commonly incentivise competition over collaboration. The consequence is a 'mismatch' problem: policy ambition exceeds institutional capacity to deliver.

Van Straalen and Witte (2018) examine multi-level governance challenges for regional planning strategies in the Netherlands and report that these scales become 'entangled' when planning issues cross jurisdictional boundaries. Strategies formulated at the regional level ultimately have to be implemented through municipal decisions, but municipalities may have different priorities and face different pressures. Without mechanisms to align incentives and ensure follow-through, regional strategies remain aspirational. Geelhuijzen (2025) provides detailed empirical analysis of such entanglement in Dutch sustainable mobility transitions, tracing how implementation failures at municipal level undermine regional and national policy ambitions.

The COVID-19 pandemic provided a natural experiment in rapid policy change. Hirschhorn (2021) examines multi-level governance responses to the crisis in public transport, observing that crisis conditions facilitated accelerated decision-making and relaxation of normal procedures. But crisis governance also demonstrated fragilities: coordination mechanisms designed for routine operation became inadequate for emergencies; data sharing across levels was impeded by legal and technical barriers; and accountability suffered as decisions were made swiftly with limited scrutiny.

4.4 Results and Discussion

4.4.1 Synthesis of Findings

Several recurring themes were identified of policy, regulation and multi-level governance on sustainable transport as seen in the literature reviewed. First, policy mixes are important but hard to put together well. Sustainable transport transitions require combinations of instruments in order to overcome a wider range of obstacles and interests, but interactions between them may generate unintended consequences, and coherence and consistency across a spectrum of government levels is often lacking (Liu et al. 2023). The avoid-shift-improve framework is a useful heuristic, but is inconsistently applied, with the 'improve' strategies of vehicle technology often prioritised over preventative 'avoid' and 'shift' strategies that contest car dependence more fundamentally (Budnitz et al. 2025). The multi-level perspective helps explain this pattern by revealing how incumbent regimes favour technological fixes that require minimal behavioural change (Geels 2012; Hoffmann et al. 2017). Critically, however, the reviewed evidence also reveals important contradictions and limits to this generalisation. Khatoon et al. (2025) demonstrate that in Dhaka City, the prioritisation of capital-intensive metro infrastructure was not simply regime capture by incumbents but was partly driven by development partner conditionalities and political economy dynamics specific to the Global South. This suggests that the explanatory reach of the MLP regime resistance argument may be context-dependent in ways that the European-centred literature has not adequately theorised. Similarly, Fathoni et al. (2025) find that in Indonesia, niche innovations in biofuel adoption were as likely to be stifled by national regulatory inconsistency as by incumbent fossil fuel interests, pointing to governance failure rather than regime resistance as the primary explanatory factor. These findings do not invalidate the MLP framework but indicate the need for its analytical categories to be applied with greater sensitivity to institutional context, particularly outside OECD settings.

The second is that multiple tier governance structures result in policy effectiveness in systematic ways. Vertical fragmentation at the levels and especially between national governments (national governments prioritise technological solutions and cities pursuing modal shift) generate tensions that erode coherence (Marsden 2025; Budnitz et al. 2025). Horizontal separation in sectors of transportation, land use, environment, energy, etc., hampers coordinated efforts (Hansson 2023). Non-state actors can strengthen policy formulation and support but also may be captured by existing interests (Kin et al. 2017). Comparative research by Ehnert et al. (2018) and Poyraz and Szalmáné Csete (2023) demonstrates that these patterns vary systematically across national contexts, with governance architectures shaping the possibilities for coordination. Yet the literature also reveals important boundary conditions on this generalisation. The assumption that vertical fragmentation is uniformly detrimental to coherence is complicated by cases where decentralisation has enabled more responsive, place-based policy innovation (Guo and Li 2025; Jain 2024). In developing country contexts, weak national institutions may mean that local actors and international development partners effectively fill governance gaps in ways that formal multi-level frameworks do not anticipate. The reviewed evidence therefore suggests a non-linear relationship between governance structure and policy coherence: both excessive centralisation and fragmented decentralisation can impede coordination depending on institutional context, resource endowments, and the nature of the transport challenge being addressed.

Third, coherence of policy, institutional capacity and regulatory enforcement are interdependent factors in the effectiveness of implementation. Coherent policies have clear rules guiding the behaviour of implementing actors. Institutional capacity allows for effective execution. Enforcement mechanisms keep compliance and accountability in place. Any weakness in any dimension is a drag on the others: coherent policies do not scale without capacity; capacity lacks coherence without targeting; enforcement is ineffective without clear targets and capable institutions (Khatoon et al. 2025). Massuga et al. (2024) provide conceptual clarification of how different governance modes (hierarchical, network, market) shape these interdependencies, while Pinno (2025) offers a framework for assessing capacity maturity across regions.

Fourth, developing countries face unique governance challenges associated with fragile institutional capacity, the dynamics of the informal sector, and reliance on international development assistance. At the same time, however, they also hold the promise of leapfrogging to sustainable systems and of learning from the mistakes of industrialised nations (Jain 2024; Khatoon et al. 2025). The idea of 'codified and reflexive coordination' is provided as a way of establishing stable institutional arrangements and flexible learning processes and a means for both of them. Studies by Fathoni et al. (2025) on Indonesia and Guo and Li (2025) on rural China illustrate how these dynamics play out in diverse developing country contexts.

4.5 *Theoretical Implications*

The results are consistent with and extend the theoretical framework in which the theory of policy-makers' design in governance models is embedded and the multi run governance perspective is developed and extended. The policy composition theory (Cashore and Howlett 2007; Capano and Howlett 2024) can be a helpful tool in disentangling policy content and discerning tensions among goals, objectives, instruments and calibrations. Ex-studies regarding layering, drift, conversion and replacement of policy showed that small incremental gains may add up to massive shift or, in turn, generate path dependence (Budnitz et al. 2025).

Multi-level governance theory (Hooghe & Marks 2010; Marsden 2025) demonstrates how the role of authority spread over different levels is a determining factor leading to policy processes and policies' results. So this distinction between Type I and Type II governance can shed light on how some issues of coordination become much less manageable. These vertical and horizontal relationships have an inherent analytical leverage for diagnosing governance failures and indicating potential reform strategies. The results also offer suggestions for theory extensions.

The integration of the multi-level perspective from transition studies (Geels 2012; Whitmarsh 2012) with governance theory proves particularly fruitful. This synthesis enables analysis of how governance arrangements at niche, regime, and landscape levels interact to shape transition pathways. Hoffmann et al. (2017) demonstrate this integration by analysing automobility regime discontinuation through a combined MLP-governance lens. Hansel (2025) extends this approach to mobility hub governance, revealing dynamics of stability and change in sustainable mobility transitions. Fathoni, Lovett, and Rifansha (2025) apply this integrated framework to renewable energy transitions in Indonesian transport, demonstrating its utility in developing country contexts.

The notion of 'modal thinking' (Budnitz et al. 2025, p. 65) describes how assumptions embedded in policy processes shape what is regarded as possible or desirable, creating a feedback loop that reinforces automobility and marginalises sustainable alternatives. This concept connects policy design theory to cultural and cognitive dimensions of governance that are under-theorised in institutional analysis, drawing on earlier work by Hoffmann et al. (2017) on automobility regime discontinuation and echoing Geels' (2012) account of how socio-technical regimes shape the cognitive frames of policy actors. Together, these contributions point toward a cognitive turn in transport governance research: not only institutional arrangements but also the interpretive frames and normative assumptions of actors need to be analysed as governance phenomena.

The 'codified and reflexive coordination' framework (Khatoon et al. 2025) is a way regarding policy learning in multi-level contexts. Codified arrangements grant stability and predictability; reflexive processes allow for adaptation and learning. They both need to be harnessed, but they draw in different directions and so create tensions that need to be addressed, not alleviated. Guo and Li (2025) provide empirical validation of this framework in rural Chinese contexts, showing how codified national frameworks interact with reflexive local implementation in low-carbon tourism transport.

4.6 Practical Implications

The findings imply several requirements for policy makers. First, coherence across levels and sectors of governance should be accounted for in policy design from the start, rather than being considered as a second-stage concern after implementation and change. This requires approaches to capture the involvement of different levels as well as sectors in policy-making and a comparison of the proposed policies with existing ones in terms of compatibility. Second, institutional capacity development should be considered as essential to form the bedrock of effective policy—not something added on later. Financial investments and technical expertise, organisational structures and enforcement capability are crucial if policies are to have their desired impact. Building capacity is a process that takes time and a consistent investment; funding a project in the short term is not a sufficient amount.

Third, monitoring and evaluation should be integrated into policy design, along with precise indicators, data collection mechanisms and accountability mechanisms. The lack of monitoring enables gaps in implementation to go undetected and also hampers the learning about the policy. Without evaluation, you can't tell what works, what doesn't and why. Fourth, there should be governance at multiple levels that is designed so that incentives across levels are aligned. This entails providing lower-level governments with resources that are commensurate with their responsibilities, that higher-level policies create enabling conditions rather than constraints for local action, and that coordination mechanisms facilitate mutual adaptation rather than a top down path.

More implications exist for other developing nations. International development assistance should develop local capacity and not create dependency on them, and can only help if and only if it is tailored to local priorities, not donor priorities. Informal sector actors must become part of the governance landscape and must not be ignored, nor suppressed. Leapfrogging opportunities should be seized, but with an eye to local context rather than blind absorption of models from elsewhere.

4.7 Further Research

4.7.1 Comparative Governance Research

In this era, comparative research on transport governance across countries and regions would demonstrate that contextual factors significantly influence the efficacy of the policy. Marsden (2025) calls for research on examining the different implications of multi-level governance configurations to transport outcomes, and how federal versus unitary states, centralised versus decentralised systems, and countries with different legal and administrative traditions may differ. Such studies would guide institutional design by determining governance arrangements that underpin policies' effective implementation.

4.7.2 Policy Learning and Policy Transfer

The study of policy learning the process by which policymakers learn from experience and foreign sources would provide insights into how governance capacity changes over time. Knapskog et al. (2025) scrutinise the process of such policy learning on parcel lockers in multi-level governance networks, illustrating how city/region/national government knowledge circulation influences policy diffusion across multiple cities that influences cross-party learning. More effort can be made to identify when policy learning becomes authentic and innovation is real rather than merely imitative.

4.7.3 Governance of Emerging Technologies

Emerging technology such as autonomous vehicles, Mobility-as-a-Service, artificial intelligence, digital platforms presents governance challenges that legacy systems may not be able to deal with. How such multi-level governance arrangements can adapt to technological disruption must be studied, keeping innovation from simply serving commercial interests to the public interest (Budnitz et al. 2025).

4.7.4 Just Transitions and Governance Equity

The concept of a 'just transition' to climate policy highlights that decarbonisation is something that needs to be coupled with social protection and social equity. Research is demanded on how multi-level governance arrangements may facilitate just transitions in transportation that benefits and losses are distributed fairly, marginal groups in transport policy-making are able to speak and that transition policies will not exacerbate existing inequity (Jain 2024).

4.7.5 Governance of Urban Freight and Logistics

Urban freight has received relatively little attention in transport governance research relative to passenger transport. Hansson (2023) and Rosales and Haarstad (2023) list governance pitfalls of urban logistics that are, in part, sector-specific, are invisible in planning methods, and contain commercial interests versus public goods. Future studies might consider how multi-level governance can be adopted for freight in a more comprehensive manner.

4.7.6 Governance Capacity and Policy Implementation

Research that makes connections between governance capacity and policy implementation outcomes would be evidence for these investments in capacity development. What patterns of financial, technical, organisational and regulatory capacity are required to deliver effective policy implementation? How do capacity deficits interact with other obstacles to implementation? What approaches to building capacity are most effective in different settings? (Veeneman and Mulley 2018).

4.8 *Managerial and Policy Implications*

4.8.1 Implications for Policymakers

For policymakers at every level of governance, the results indicate some priority actions.

At supranational level (European Union, for instance): Offer enabling frameworks which define clear direction while leaving room for adaptation on the ground. Ensure that funding schemes cohere to sustainability goals rather than fuelling automobility. Foster capacity development in the Member States and the cities through sharing knowledge, technical support and peer learning. Create monitoring systems to establish consistent measurements of progress on a regular basis in all jurisdictions, while protecting principle of subsidiarity (Knapskog et al. 2025).

On the national level: Turning the international commitments into national strategies with specific targets, timelines and resources. Make certain national policies create enabling conditions for local actions instead of limiting them. Ensure sustainable funding for transport investment, with multi-year commitments that provide mechanisms for long-run planning. Train and educate sub-national governments in infrastructure and provide technical support and leadership opportunities. Implement coordination structures that include different sectors (transport, land use, environment, energy, finance) to have a cohesive policy (Jain 2024).

At regional level: Plan and finance public transportation at scales that are aligned with travel patterns, coordinating across local government boundaries. Harmonize regional plans with national frameworks and local programs. Offer technical aid to smaller local authorities without expertise. Ensure the exchanging knowledge and cooperation between local authorities (Van Straalen and Witte 2018).

At local level: Develop sustainable urban mobility plans to integrate sustainable urban transport policies at every level to create integrated urban mobility strategies, as well integrate travel and transport with land use, to promote the idea that transportation must conform to land use. Involve stakeholders, including businesses, civil society, and citizens in policy, to generate support and to bring together different points of view. Create capacity for operationalisation of policy such as the implementation of parking and low-emission zones. Monitor results and make evidence-based policy modifications (Liu et al. 2023).

4.9 *Implications for Transport Managers and Practitioners*

For transport managers and operators, the results indicate the need of governance awareness. Technical knowledge of transport planning itself needs to be supplemented with knowledge of the governance space who has control over the decision-making, the flow of resources, what the systems are for coordinating activities, where congestion is created. Good practice must negotiate these governance dimensions and not just apply technical solutions. Multi-level working is becoming more and more important for those working in the transport field. Projects mean working with stakeholders on more than a single level getting funding from national programmes, coordinating with regional plans, working with neighbouring towns, collaborating with local residents, working with local communities. Technical competency is just as important as skills of stakeholder engagement, partnership working and political communication.

Monitoring and evaluation should be incorporated so embedded in practice, as opposed to imposed as requirements. Gathering outcomes data, reflecting on what does work, and why, then sharing learning with peers helps support individual project performance as well as collective growth. Those practicing should push for evaluation tools and use findings to inform improved practice moving forward.

4.9.1 Implications for Capacity Building

Investment in training for transport governance needs to be sustained from many different points of view. On revenue, financial capacity rests on reliable and sufficient sources of funding, be it from national budgets, local revenues or international funding sources. Investment in education and training, recruitment and retention, knowledge exchange networks and other investment in technical capacity. Organisational capacity calls for a focus on institutional design, definitional roles and roles at the interface level, a streamline in processes and accountability mechanisms. Regulatory capacity also involves resources supporting monitoring and enforcement, while simultaneously demanding political will to apply rules consistently. Building new capacity has to be context-relevant. What works in big cities might not be good for small towns (or even great markets); what works in high-income countries may not necessarily be successful in the less developed areas. Peer learning of a similar nature (cities of a similar size, countries with similar governance traditions) can be more valuable than importing models from disparate locales.

5 Conclusion

This chapter has examined the policy, regulatory, and governance dimensions of sustainable transport transitions through a systematic review of multi-level governance literature published between 2019 and 2026. It has made three substantive contributions. First, it has developed and applied an integrative conceptual framework linking policy coherence, institutional capacity, and regulatory enforcement as interdependent determinants of sustainable transport outcomes a synthesis not previously attempted in this form in the transport governance literature. Second, it has critically examined how these relationships play out differently across governance contexts, generating evidence-grounded propositions about the contextual conditions under which each dimension is most constraining. Third, it has identified concrete governance mechanisms including codified and reflexive coordination, capacity maturity frameworks, and participatory governance instruments through which policymakers can address identified deficits.

Findings indicate a critical relationship between governance and sustainable transport outcomes. Policies with proper design are useless without the right institutional capacity for deployment. Without enforcement, defined regulatory requirements don't work. Ambitious objectives at one level of governance are frustrated by contradictory policies at another. The governance of transport is as important as its technology, infrastructure, and operations. From this several interrelated themes emerge. First, the coherence of policy across levels of governance and across sectors is crucial and yet difficult to attain. Vertical tension between national and local interests, horizontal fissures across policy domains, and time lags between long-range targets and short-range rhythms hinder coherence. Responding to these demands active oversight of governance structure, not just policy content.

Second, institutional capacity is a necessary condition of meaningful policy, not an afterthought. Policy cannot deliver or become effective until resources, talent, capacity, and technology are developed and maintained. Capacity deficits are especially pronounced in developing states and at the local government level both of which are where implementation responsibility often lies. Third, regulatory enforcement closes the gap between the intention of policy and the actual results on the ground. Compliance is only enforced with clarifying requirements, availability of resources to monitor compliance, and sanctions that are enough to dissuade noncompliance. In the absence of the enforcement regime, regulations are written down, but they are not implemented on a practical basis, and policy becomes suspect and unsustainable.

Fourth, all of these dimensions are embedded in multi-level governance arrangements. The interplay between authority at different levels dictates who decides what, who is responsible for financing what, who executes what, and who is empowered to take action. These arrangements can either facilitate or hinder policy coherence, capacity building, and regulatory enforcement depending on the way they align responsibilities with resources and promote incentives for coordination. Transport decarbonisation is essential and urgent, as recognised in the European Green Deal, national net-zero strategies globally, and international climate commitments. The evidence reviewed in this chapter, however, cautions against universal prescriptions. Governance reforms that proved effective in federal European states such as multi-year transport funding compacts and integrated spatial planning requirements may face insurmountable political economy obstacles in unitary or capacity-constrained systems. Similarly, participatory governance innovations developed in high-income city contexts cannot be directly transplanted to rapidly urbanising developing country settings where civil society capacity and data infrastructure are limited. Policymakers must therefore begin with honest diagnosis of their specific governance constraints rather than importing best-practice templates. That said, cross-cutting principles do emerge from the evidence: alignment of responsibilities with resources across governance levels; investment in institutional capacity as a precondition rather than an afterthought to policy ambition; and monitoring systems designed to detect and respond to implementation gaps. Sustainable transport transitions require not only technological advancement and behavioural change but institutional reform building governance systems capable of delivering policy consistency, mobilising resources, and maintaining accountability across political cycles.

For researchers, this review identifies five specific and high-priority directions for future inquiry. First, comparative governance studies should move beyond descriptive typologies to test causal propositions about which governance configurations under which contextual conditions produce coherence outcomes; longitudinal designs tracking governance change alongside transport outcomes are particularly needed. Second, policy learning and transfer research should examine whether and how governance innovations diffuse across jurisdictions of similar institutional type, and what conditions transform imitation into genuine adaptation. Third, governance of emerging technologies autonomous vehicles, MaaS platforms, AI-assisted enforcement requires dedicated analysis, given that existing regulatory frameworks were not designed for platform-mediated mobility. Fourth, research on just transitions should centre on distributive questions: who bears the costs of governance failures, and how can multi-level arrangements be redesigned to ensure equitable outcomes. Fifth, urban freight governance deserves substantially greater scholarly attention given its rapid growth and current marginalisation in transport planning frameworks. For practitioners and policymakers the results highlight the need for governance awareness, multi-layered working, and ongoing capacity building.

Transport systems are also an engineering of governance. The challenge of sustainable transport is not only to design better policies but also to create better governance arrangements which unite levels, sectors, and actors around shared purposes; that mobilise resources for implementation; that hold decision-makers accountable for outcomes; and that learn from experience to improve over time. The challenge of governance is critical in addressing all the challenges facing transport if it is to make its significant contribution to an environmentally sustainable future.

References

Ahlqvist V, Norrman A, Jahre M (2020) Supply chain risk governance: Towards a conceptual multi-level framework. Oper Supply Chain Manag: Int J 13(4):382–395

Bache I, Bartle I, Flinders M, Marsden G (2015) Multilevel governance and climate change: Insights from transport policy. Bloomsbury Publishing

Budnitz H, Marsden G, Anable J, Docherty I (2025) Multi-level governance and modal thinking: Tensions in electric mobility transitions in European cities. Transp Policy 160:63–72

Butterfield BJ, Low NP (2017) Reducing carbon emissions from transport: Multi-level governance and the problem of monitoring. Urban Policy Res 35(3):235–247

Capano G, Howlett M (2024) The taxonomy of policy composition: A framework for analysing policy content. Advance online publication, Policy Sciences

Cashore B, Howlett M (2007) Punctuating which equilibrium? Understanding thermostatic policy dynamics in Pacific Northwest forestry. Am J Polit Sci 51(3):532–551
Cevheribucak G (2021) Energy transition and sustainable road transportation in Turkey: Multiple policy challenges for inclusive change. Front Sustain Cities, 3, Article 631337
Cevheribucak G (2024) Effects of multilevel governance on the design and implementation of sustainable mobility plans. Advance online publication, Gestión y Análisis de Políticas Públicas
Conteh C (2013) Policy governance in multi-level systems: Economic development and policy implementation in Canada. McGill-Queen's University Press
Corfee-Morlot J, Kamal-Chaoui L, Donovan MG, Cochran I, Robert A, Teasdale PJ (2009) Cities, climate change and multilevel governance (OECD Environmental Working Papers No. 14). OECD Publishing
Doern GB, Coleman J, Prentice BE (2019) Canadian multimodal transport policy and governance. McGill-Queen's University Press
Dörry S, Decoville A (2016) Governance and transportation policy networks in the cross-border metropolitan region of Luxembourg: A social network analysis. Eur Urban Reg Nal Stud 23(1):69–85
Ehnert F, Kern F, Borgström S, Gorissen L, Maschmeyer S, Egermann M (2018) Urban sustainability transitions in a context of multi-level governance: A comparison of four European states. Environ Innov Soc Trans 24:101–116
European Commission (2019) The European Green Deal (COM(2019) 640 final). European Commission
European Commission (2020) Sustainable and smart mobility strategy – Putting European transport on track for the future (COM(2020) 789 final). European Commission
Fathoni F, Lovett JC, Rifansha MM (2025) A multi-level perspective on transition to renewable energy in the Indonesian transport sector. Energies, 18(21), Article 5723
Geelhuijzen F (2025) Understanding the process and mechanisms of sustainable mobility transitions in Dutch cities: An analysis using multi-level perspective framework [Unpublished manuscript]
Geels FW (2012) A socio-technical analysis of low-carbon transitions: Introducing the multi-level perspective into transport studies. J Transp Geogr 24:471–482
Gritsenko D (2020) Multi-level governance in the Arctic shipping. Advance online publication, Marine Policy
Guo Y, Li S (2025) Multi-level governance of low-carbon tourism in rural China: Policy evolution, implementation pathways, and socio-ecological impacts. Front Environ Sci, 12, Article 1482713
Hansel J (2025) Governing mobility hubs in the sustainable urban mobility transition: Dynamics of stability and change. Transp Policy 163:323–334
Hansson L (2023) Urban freight transport and multi-level governance. In: Browne M, Behrends S, Woxenius J (eds) Handbook on city logistics and urban freight. Edward Elgar Publishing, pp 365–380
Hirschhorn F (2021) A multi-level governance response to the Covid-19 crisis in public transport. Transp Policy 112:13–21
Hoffmann S, Weyer J, Longen J (2017) Discontinuation of the automobility regime? An integrated approach to multi-level governance. Transp Res Part a: Policy Pract 103:391–408
Hooghe L, Marks G (2010) Types of multi-level governance. In: Enderlein H, Wälti S, Zürn M (eds) Handbook on multi-level governance. Edward Elgar Publishing, pp 17–31
Howlett M, Rayner J (2013) Patching vs packaging in policy formulation: Assessing policy portfolio design. Politics and Governance 1(2):170–182
Jain D (2024) Coordinated governance for equitable, net-zero urban transport. *ORF Expert Speak*. Observer Research Foundation. Retrieved January 15, 2026, from https://www.orfonline.org
Khatoon T, Rahman M, Islam M, Hossain M (2025) Codified and reflexive policy coordination: How successful are Dhaka City's policies for a successful sustainable passenger transport transition? Oxford Open Energy, 3, Article oiae009
Kiese M (2012) Regional cluster policies in Germany: A multi-level governance perspective on policy learning. Eur Rev Ind Econ Policy (5)

Kin B, Verlinde S, Mommens K, Macharis C (2017) A stakeholder-based methodology to enhance the success of urban freight transport measures in a multi-level governance context. Res Transp Econ 65:10–23
Knapskog M, Tennøy A, Hanssen JU (2025) Policies on parcel lockers: Multi-level governance networks for urban logistics. Plan Pract Res 40(4):932–953
Kotilainen JM, Sairinen R, Lehtimäki H, Mielonen N (2026) Multilevel governance under urgency: Developing battery industry in the EU. Resources Policy, 112, Article 105825
Liu X, Dijk M, Colombo C (2023) Improving multi-level policy mixes for sustainable urban mobility transition [SSRN Working Paper]. https://ssrn.com/abstract=4378292
Moher D, Liberati A, Tetzlaff J, Altman DG, The PRISMA Group (2009). Preferred reporting items for systematic reviews and meta-analyses: The PRISMA statement. PLOS Medicine, 6(7), Article e1000097
Marcucci E, Stathopoulos A (2012) Multi-level governance and transport policy: The case of local roads in Italy. Int J Transp Econ 39(1):15–38
Marsden G (2025) Multi-level governance, politics and geography in transport. In: Finger M, Audouin M (eds) Handbook of transportation and public policy. Edward Elgar Publishing, pp 76–88
Marsden G, Ferreira A, Bache I, Flinders M, Bartle I (2014) Muddling through with climate change targets: A multi-level governance perspective on the transport sector. Climate Policy 14(5):617–636
Massuga F, Doliveira SLD, Kuasoski M, Mangoni SS (2024) Modes of governance for sustainability transitions: Conceptual definitions. Environ Policy Gov 34(6):663–678
Panagakos G, Psaraftis HN (2017) Best practice in multi-level cooperation for transport and growth in the BSR [Thematic Study WP5]. Baltic Sea Region Programme
Panagakos G, Psaraftis HN (2018) Lessons on multi-level governance learned from the bottom-up corridor projects in the Baltic Sea Region. In Transport Research Arena 2018 [Conference proceedings]
Pinno S (2025) Regional sustainability maturity model: A strategic and process-oriented approach on how to transform regions in a future-orientated way [Manuscript submitted for publication]. SSRN. https://ssrn.com/abstract=5542858
Poyraz AY, Szalmáné Csete M (2023) Multi-level governance of sustainability transition in the European Union. In SDGs in the European region (pp. 291–313). Springer International Publishing
Roe M (2007) Shipping, policy and multi-level governance. Marit Econ & Logist 9(1):84–103
Roe M (2009) Multi-level and polycentric governance: Effective policymaking for shipping. Marit Policy Manag 36(1):39–56
Rosales R, Haarstad H (2023) Governance challenges for urban logistics: Lessons from three Norwegian cities. Environ Policy Gov 33(3):221–231
Sack D (2011) Governance failures in integrated transport policy: On the mismatch of 'co-opetition' in multi-level systems. Ger Policy Stud 7(2):43–68
Syzdykbayeva BU, Jumadilova S, Rainbekov Z, Baigarin K (2019) Government regulation of logistics: Theoretical and practical analysis. J Adv Res Law Econ 10(1):370–385
Teles MDF, Sousa JFD (2021) Sustainable development and morphological analysis: A multi-level strategic planning for the transport sector. Int J Inf Decis Sci 13(4):350–364
Todella E, Abastante F, Cotella G (2024) Practicing multilevel governance: The revision of the Piedmont regional territorial plan. Land, 13(6), Article 755
Tranfield D, Denyer D, Smart P (2003) Towards a methodology for developing evidence-informed management knowledge by means of systematic review. Br J Manag 14(3):207–222
Van Straalen FM, Witte PA (2018) Entangled in scales: Multilevel governance challenges for regional planning strategies. Reg Stud Reg Sci 5(1):157–163
Veeneman W, Mulley C (2018) Multi-level governance in public transport: Governmental layering and its influence on public transport service solutions. Res Transp Econ 69:430–437

Westman LK, Broto VC, Huang P (2019) Revisiting multi-level governance theory: Politics and innovation in the urban climate transition in Rizhao, China. Polit Geogr 70:14–23
Whitmarsh L (2012) How useful is the multi-level perspective for transport and sustainability research? J Transp Geogr 24:483–487
Zhang S, Witlox F (2019) Analyzing the impact of different transport governance strategies on climate change. Sustain, 12(1), Article 200

Reimagining Global Supply Chains: Innovation Pathways to Decarbonization

James Kanyepe, Munyaradzi Chibaro, Nyarai Kasambuwa, and Michael Alfaneta

Abstract This chapter examines how innovation enables decarbonization within contemporary supply chains through a systematic review and synthesis of the existing academic literature, policy frameworks, and industry reports. This study adopted a qualitative desk-based research design, relying exclusively on secondary data sources to develop an integrated conceptual understanding of supply chain decarbonization. The literature indicates that carbon emissions are deeply embedded within supply chain operations, particularly in transportation, warehousing, production, and sourcing activities. However, emissions are not solely operational in nature; they are also structurally shaped by supply chain design decisions, product development processes and global network configurations. The review further demonstrates that technological innovation, particularly digitalization, artificial intelligence, Internet of Things applications, automation, and electrification, plays a central role in improving emission visibility, operational efficiency, and system-wide optimization. In addition, the literature highlights that sustainable product design, material substitution, and circular economy strategies are critical for reducing embodied carbon before products enter operational supply chains. Despite these advancements, this review identifies persistent challenges, including fragmented emissions data, uneven technological capabilities across supply chain partners, coordination complexity in global networks, and limitations in the practical implementation of sustainability strategies.

Keywords Global supply chains · Innovation · Decarbonization · Sustainability

---

J. Kanyepe (✉) · M. Chibaro
Department of Logistics and Supply Chain Management, University of Botswana, Gaborone, Botswana
e-mail: jameskanyepe@gmail.com

N. Kasambuwa
Academy of Teaching and Learning, Chinhoyi University of Technology, Chinhoyi, Zimbabwe

M. Alfaneta
Department of Supply Chain Management, Chinhoyi University of Technology, Chinhoyi, Zimbabwe

J. Kanyepe and I. Chirisa (eds.), *Advancing the Green Agenda in the Transport and Logistics Sector in Global South Countries*, World Sustainability Series,
https://doi.org/10.1007/978-3-032-31374-4_16

1 Introduction

The global supply chain is one of the most complex and influential systems in contemporary economies. It encompasses the extraction of raw materials, transformation of inputs through manufacturing processes, and distribution of finished goods to consumers across national and continental boundaries (Franke et al. 2024). This capacity underpins the global trade, industrial development, and consumption patterns that characterize modern life (Grytsenko et al. 2023). Nevertheless, throughout much of its evolution, the global supply chain system has been constructed on carbon-intensive foundations, heavily dependent on fossil fuels such as coal, diesel, and natural gas for power production, transportation, and logistics (Gamage et al. 2025). Historically, the primary objective of supply chain design has been to maximize efficiency by increasing output and minimizing cost and time (Abaku et al. 2024). Xin (2024) argues that environmental considerations have largely been externalized and are regarded as indirect consequences of industrial progress rather than integral components of operational decision-making. Consequently, the expansion of global supply networks over the past century has been accompanied by parallel increases in greenhouse gas emissions, resource depletion, and environmental degradation (Dubey et al. 2021).

The increasing urgency of climate change, coupled with growing regulatory demands, investor expectations, and consumer awareness, has elevated sustainability from a peripheral issue to a strategic imperative (Shen et al. 2025). Decarbonization has emerged as a central objective for companies seeking long-term competitiveness and resilience in a rapidly changing global environment (Johnson 2025). This endeavor extends beyond isolated enhancements in energy consumption or transportation optimization and requires a comprehensive transformation of supply chain architectures. This transformation involves redesigning products to minimize material intensity, restructuring logistics networks, transforming supplier relationships, and integrating digital technologies that enable real-time emissions tracking and decision-making. Decarbonization necessitates a profound rethinking of how value is created, delivered, and sustained (Brinken et al. 2022).

According to Mubarik et al. (2025), innovation plays a pivotal role in decarbonization, as it encompasses not only technological advancements but also organizational, procedural, and strategic transformations. This includes the incorporation of digital technologies such as artificial intelligence, machine learning, blockchain, and the Internet of Things (IoT), which provide unprecedented visibility across complex supply networks (Olsen 2026). These technologies enable companies to measure emissions at various stages of production and distribution, identify inefficiencies, and optimize operations based on costs, speed, and carbon intensity. Traditional supply chains typically function in a linear and reactive manner, responding to demand fluctuations and disruptions as they occur (Jha and Mohan 2023). In contrast, digitally enabled supply chains are increasingly proficient at anticipating disruptions, simulating alternative scenarios, and dynamically adjusting operations to minimize their environmental impact. This shift from reactive management to proactive optimization represents a fundamental evolution in supply chain logic (Jha and Mohan 2023).

Decarbonization cannot be accomplished solely through internal operational modifications; it requires extensive collaboration among suppliers, logistics providers, distributors and customers (Miklautsch and Woschank 2022). This interdependence has elevated the importance of supply chain collaboration as a conduit for the diffusion of innovation. Cancela et al. (2026) reported that companies are increasingly participating in joint sustainability initiatives, sharing data on emissions performance, coinvesting in low-carbon technologies, and aligning their procurement strategies with environmental objectives. In numerous cases, competitors collaborate within shared logistics networks to minimize empty transportation capacity and enhance overall system efficiency. These developments signify a transition from competitive isolation to ecosystem-based value creation (Wang et al. 2025).

Traditionally, product design has emphasized functionality, cost, and manufacturability, whereas procurement has prioritized price and reliability. However, both functions are now being reoriented to incorporate carbon performance as a core evaluation criterion (Dautel et al. 2024). Almusaed et al. (2024) noted that materials are selected based on life-cycle emissions, suppliers are evaluated using sustainability metrics, and product architectures are simplified to reduce material use and increase recyclability. This upstream focus is particularly significant because it enables emissions to be prevented at the source rather than being mitigated after production (Alpan et al. 2023). In addition, electric vehicles, automated warehouse systems, and electrified port equipment are gradually replacing fossil fuel-based machinery (Kanyepe et al. 2025). This shift not only reduces direct emissions but also generates compounding environmental benefits as electricity grids are increasingly decarbonized. Unlike combustion-based systems, whose emissions are fixed at the point of fuel consumption, electric systems improve over time as energy systems develop, thereby introducing a dynamic pathway for long-term emission reduction (Beck and Pedraza 2026).

Transitioning to low-carbon supply chains necessitates technological advancements and organizational and behavioral transformations. For example, companies face significant challenges, including investment costs, infrastructure limitations, workforce capability, and operational risks (Mupfiga et al. 2024). Resistance to change is prevalent, particularly when new technologies disrupt established workflows or require the acquisition of new skills (Yordanova 2023). Similarly, Anathole et al. (2026) noted that decarbonization is increasingly influenced by external governance mechanisms such as carbon pricing policies, environmental regulations, and sustainability reporting standards. These frameworks redefine corporate incentives by rendering emissions quantifiable and financially significant, thereby strengthening the business case for innovation-driven, sustainable practices. This prompts firms to integrate environmental considerations into their core strategic planning rather than merely treating them as compliance obligations (Polater 2025).

This chapter is situated within this broader transformation, examining how innovation is reshaping supply chain systems, how decarbonization is implemented across various functional domains, and how organizations manage the tensions between efficiency, resilience, and sustainability. This provides a foundation for understanding the mechanisms through which global supply chains are being redefined in response to one of the most pressing challenges of the twenty-first century (Jones 2025). The central argument presented is that decarbonization is not only an external constraint imposed on supply chains but also an internal design principle that shapes their future evolution. As technological capabilities expand and institutional pressures intensify, integrating sustainability into supply chain strategies becomes essential for competitiveness, resilience, and long-term value creation (Johnson 2025). This chapter primarily focuses on examining the role of innovation in facilitating decarbonization within contemporary supply chain systems. To achieve this primary objective, the study is guided by the following specific aims: (1) to analyze the relationship between supply chain activities and carbon emissions, and (2) to evaluate the influence of technological innovation on enabling decarbonization of supply chains. Furthermore, it seeks to investigate how supply chain design and product development processes contribute to emission reductions and assess the significance of collaboration and stakeholder integration in achieving decarbonization goals.

2 Literature Review

2.1 Theoretical Foundations

This chapter synthesizes theoretical frameworks to elucidate the role of innovation in facilitating decarbonization in supply chains. Systems theory conceptualizes the supply chain as an interconnected system, wherein modifications in one segment affect the entire network (Reeb and Folger 2023). In this context, decarbonization is viewed as a comprehensive transformation that requires coordination among various stakeholders. The resource-based view (RBV) asserts that competitive advantage arises from resources that are valuable, rare, inimitable, and non-substitutable (Zvarimwa and Zimuto 2022). Innovation capabilities and low-carbon competencies function as strategic resources for enhancing environmental and economic performance. In addition, companies that invest in analytics, electrified logistics, and sustainable designs can reduce emissions while maintaining efficiency. In contrast, stakeholder theory emphasizes that decisions are influenced by stakeholders, such as customers, suppliers, and regulators. Stakeholder expectations shape sustainability strategies, with regulatory pressure, investor demand, and consumer preferences driving sustainable practices (Sun et al. 2025). The theories used in this study provide a multilayered explanation of decarbonization by integrating internal capabilities, systemic interdependence, and external pressures.

2.2 Innovation-Driven Decarbonization

Innovation is broadly conceptualized to include technological, process, organizational, and design innovation (Li et al. 2023). These forms of innovation interact to reshape the functioning of supply chains and the generation of carbon emissions and their management. At the input level, innovation influences sourcing decisions through sustainable procurement, material substitution, and supplier engagement strategies. At the operational level, innovation manifests through the digitalization, automation, and electrification of logistics and production systems (Seidelson 2024). At the design level, product redesign and circular economy principles reduce material intensity and emissions embedded in goods before they enter the supply chain. At the coordination level, collaboration platforms and data-sharing systems enable integration across multiple actors, ensuring that emissions are monitored and reduced across the entire value chain. Decarbonization is a dependent outcome of these innovation processes. This is achieved through emission reduction, avoidance, and transparency (Allon 2026).

2.3 Supply Chain Operations and Carbon Emissions

The relationship between supply chain operations and carbon emissions has been extensively examined in the fields of logistics, industrial ecology, and sustainable supply chain management. Across this body of literature, there is a broad consensus that supply chain activities constitute a major source of global greenhouse gas emissions, primarily because of their reliance on energy-intensive production systems, transport networks, and storage infrastructures (Jones 2025). However, contemporary research increasingly emphasizes that emissions are not merely a byproduct of isolated operational activities but rather an embedded outcome of system-wide structural and managerial decisions within supply chains (Retegi et al. 2026). Early conceptualizations of supply chain emissions were largely confined to manufacturing processes and direct organizational activities. Over time, this narrow perspective has been expanded through the development of life cycle assessment approaches and carbon accounting frameworks, particularly the Greenhouse Gas Protocol classification of Scope 1, Scope 2, and Scope 3 emissions. This expansion has been critical in revealing that most emissions associated with supply chains occur outside the boundaries of focal firms. In many sectors, especially retail, manufacturing, and logistics-intensive industries, Scope 3 emissions dominate the total carbon footprint, often exceeding 70% of the overall emissions.

Globalization and outsourcing have led to geographically dispersed production networks, resulting in extended transportation distances and increased logistical complexity (Allon 2026). While these strategies have improved cost efficiency and market responsiveness, they have simultaneously increased carbon emissions owing to the spatial separation of production and consumption. Although just-in-time and lean production systems are effective at reducing inventory holding costs, they are also associated with increased transportation frequency and reduced shipment consolidation, thereby contributing to higher emission intensity under certain conditions. Supply chain emissions are not solely determined by firm-level decisions but are also influenced by the carbon intensity of the national energy grids and transport infrastructure (Weng 2025). Regions with a high reliance on fossil fuel-based electricity generation tend to exhibit higher operational emissions, whereas those with a greater integration of renewable energy sources demonstrate lower carbon intensity in logistics and manufacturing operations (Zhang 2024). Similarly, the availability and efficiency of multimodal transport infrastructure significantly influence emission outcomes, with rail and maritime systems generally offering lower carbon intensity than road and air transport systems (Vishwakarma et al. 2025).

In recent years, an expanding body of evidence has illustrated how digitalization affords companies enhanced visibility and control over their carbon footprint associated with supply chain emissions. Technologies such as IoT sensors, artificial intelligence, and advanced analytics provide companies with highly detailed data on energy consumption, transport efficiency, and operational performance (Olsen 2026). This improved visibility enables companies to identify "emission hot spots" within their operations and make more informed, data-driven decisions regarding optimal routes and load consolidation. For example, route optimization systems employing algorithms can reduce fuel usage by minimizing travel distances and avoiding congestion, whereas predictive analytics can help decrease empty miles in freight transport. It is also crucial to acknowledge that while digitalization can influence the reduction of supply chain carbon footprints, it does not automatically yield environmental benefits for all companies. A key factor influencing its success is an organization's commitment to sustainability and the extent to which sustainability objectives are integrated into decision-making. Additionally, cost savings and increased operational efficiency from digitalization may stimulate demand for transport services. Consequently, overall emissions may remain stable or even increase, despite improvements in emission intensity.

2.4 The Role of Technological Innovation in Enabling Supply Chain Decarbonization

The role of technological innovation in enabling supply chain decarbonization has become a central theme in contemporary sustainability and logistics literature. Across a wide range of studies, there is growing agreement that technological advancement is a critical enabler of emissions reduction, particularly in complex and globally dispersed supply chain systems, where traditional management approaches are insufficient (Chikwere et al. 2026). However, the literature also highlights that technology alone does not automatically lead to decarbonization; rather, its effectiveness depends on how it is embedded within organizational processes, decision-making structures, and broader supply chain ecosystems (Mubarik et al. 2025). One of the most significant developments in this field is the rise of digital transformation technologies, including artificial intelligence, the Internet of Things (IoT), big data analytics, and blockchain systems (Olsen 2026). These technologies have fundamentally changed the manner in which supply chains are monitored, managed, and optimized.

Historically, emissions within supply chains have been difficult to measure because of limited visibility across multiple tiers of suppliers and logistics partners (Acquaye et al. 2014). Digital technologies now enable real-time data collection and analysis, allowing firms to track energy consumption, transport efficiency and emission intensity across different stages of the value chain. This enhanced visibility is widely recognized in the literature as a foundational requirement for effective decarbonization, as it shifts supply chain management from retrospective reporting to proactive emissions control (Vishwakarma et al. 2025). Studies have shown that AI-driven systems can significantly reduce emissions by improving demand forecasting, optimizing inventory levels, and enhancing transportation routing (Onukwulu et al. 2023). AI systems can identify inefficiencies that are not easily detectable through traditional planning methods by analyzing large and complex data sets. For example, dynamic routing algorithms can adjust delivery routes in real time based on traffic conditions, fuel consumption patterns, and delivery priorities, thereby reducing unnecessary mileage and the associated emissions.

Similarly, predictive analytics can improve warehouse efficiency by optimizing stock placement and reducing the energy consumption associated with material handling operations. The Internet of Things has also emerged as a key technological enabler of decarbonization (Vishwakarma et al. 2025). IoT devices embedded in vehicles, warehouses, and manufacturing equipment allow continuous monitoring of operational performance. This includes tracking the fuel consumption, machine efficiency, temperature control, and idle time. The literature emphasizes that IoT-enabled supply chains are better positioned to identify emission hotspots and implement targeted interventions. In cold chain logistics, for example, IoT sensors help optimize refrigeration systems, reducing energy waste while maintaining product. In transport operations, telematic systems enable fleet managers to monitor driver behavior and vehicle efficiency, thereby contributing to reduced fuel consumption and emissions (Chibaro et al. 2024).

Blockchain enables the secure and immutable recording of transactions across supply chain networks, thereby improving accountability and data integrity (Happy et al. 2021a, b). In the context of decarbonization, blockchain facilitates emission tracking across multiple tiers of suppliers, ensuring that carbon data are reliable and verifiable (Chibaro et al. 2026). This is particularly important for addressing the persistent challenge of emissions, where data gaps and inconsistencies often hinder accurate measurements. By improving trust and transparency, blockchain systems support more credible carbon reporting and enable stronger collaboration between supply chain partners (Happy et al. 2021a, b). The literature highlights the growing adoption of electric vehicles in logistics operations, including delivery vans, trucks, forklifts, and port equipment. Electrification reduces direct emissions from transport and material handling activities, while also offering long-term environmental benefits as electricity grids transition toward renewable energy sources (Richardson 2013). Unlike internal combustion engines, electric systems have become progressively cleaner over time, making them a strategic and long-term solution for reducing logistics-related emissions. However, studies have noted that the environmental effectiveness of electrification depends on grid composition, battery production emissions, and infrastructure readiness (Carodenuto 2020).

In addition, automated warehouses, robotic picking systems, and intelligent conveyor networks enable more precise and efficient handling of goods, thereby reducing idle time and unnecessary energy consumption (Yarlagadda 2024). The literature suggests that automation not only improves productivity but also enhances energy efficiency by optimizing workflow design and reducing human error. However, it is also acknowledged that the initial energy and resource intensity required to implement such systems must be carefully considered within lifecycle assessments. Despite the clear potential of technological innovation, the literature consistently emphasizes that its impact on decarbonization is not guaranteed. Many studies have highlighted that firms often adopt advanced technologies without fully integrating them into sustainability strategies, resulting in limited environmental benefits (Ejairu et al. 2025). This highlights the importance of organizational alignment, where technological systems must be embedded within broader environmental objectives and supported by appropriate governance structures.

Another critical insight from the literature is the role of data integration and interoperability in maximizing the impact of technological innovation. Supply chains are often characterized by fragmented systems and incompatible data platforms, which limit the effectiveness of digital technology. Without seamless data sharing across suppliers, logistics providers, and customers, the potential benefits of AI, IoT, and blockchain systems cannot be fully realized (Khawla et al. 2025). Consequently, scholars increasingly argue that technological innovation must be accompanied by standardization efforts and collaborative DGFs initiatives. Furthermore, the literature highlights the importance of innovation diffusion in supply chain networks. Leading firms often invest in advanced technologies; however, the overall effectiveness of decarbonization depends on the ability of these innovations to spread across suppliers and partners. Moreover, SMEs may face barriers related to costs, technical capacity, and infrastructure limitations. This creates uneven levels of technological adoption across supply chains, which can undermine system-wide emissions reduction efforts (Khawla et al. 2025).

2.5 Supply Chain Design and Product Development Processes in Emissions Reduction

Supply chain design and product development processes have been increasingly identified in the literature as critical upstream determinants of carbon emissions (Das and Shaw 2017). In contrast to earlier sustainability scholarship, which predominantly emphasized operational efficiency and end-of-pipe mitigation strategies, contemporary research underscores that a substantial share of emissions is effectively "locked in" during the early stages of product conception and supply chain configuration (Chen et al. 2025). Accordingly, the design phase is now widely conceptualized as a pivotal intervention point for achieving substantial and sustained decarbonization across supply chain systems. Within this framework, product design decisions are understood to exert a structuring influence on downstream environmental outcomes by determining material composition, production complexity, logistics and transportation requirements, and end-of-life treatment pathways. Products designed with high material intensity, multiple components, and limited recyclability tend to generate higher emissions throughout their lifecycle (Medhi 2020). Conversely, simplified product architectures, lightweight materials, and modular designs contribute to reduced emissions by lowering resource consumption and improving manufacturing and logistical efficiency. This has led to the growing prominence of eco-design and sustainable product development frameworks that integrate environmental considerations into early stage design decisions, rather than treating them as downstream constraints (Johnson 2025).

Life cycle thinking is foundational in this regard. Life cycle assessment (LCA) quantifies emissions across a product's life cycle, from raw material extraction to disposal. Studies show that a substantial share of emissions is associated with upstream activities, such as material sourcing and production, not just transportation or consumption. This insight shifts the focus to redesigning products to reduce embodied carbon and emissions before they enter the supply chain. Additionally, supply chain design significantly shapes emission outcomes. It involves the configuration of production facilities, distribution centers, sourcing locations, and transport networks. Although geographically dispersed supply chains are cost-efficient, they generate higher emissions owing to increased transportation and carbon-intensive logistics. In contrast, regional or local supply chains can reduce transportation emissions and improve responsiveness, sometimes sacrificing economies of scale.

Network optimization studies further highlight the importance of facility location decisions in determining emission intensity. The placement of manufacturing plants and distribution centers influences not only cost and service levels but also the overall carbon footprint of logistics operations (Sindhwani et al. 2023). Research indicates that integrating carbon considerations into network design models leads to different optimal configurations than traditional cost-minimization approaches. In carbon-aware models, firms may prioritize proximity to demand centers, multimodal transport access, or renewable energy availability when designing supply chain networks. Circular economy principles have also gained prominence in the literature as key drivers of emissions reduction through supply chains and product redesign (Wu 2025). The circular economy emphasizes resource efficiency, reuse, remanufacturing, recycling, extending product life cycles, and reducing virgin material dependence. Circular design strategies can significantly reduce emissions by minimizing material extraction and production demand. However, circular systems require high levels of coordination and infrastructure development to function effectively at scale.

In addition, packaging contributes significantly to emissions through material production, transportation, and waste generation. Research shows that lightweight, recyclable, and reusable packaging solutions can substantially reduce supply chain emissions (Srivastava et al. 2025). Standardizing packaging formats enhances transport efficiency by improving load consolidation and reducing unused space. This demonstrates how small design changes can yield environmental benefits. The literature emphasizes the role of supplier integration in sustainable design. Product development is a collaborative process involving multiple supply chain actors. Early supplier involvement allows firms to access expertise in sustainable materials, cleaner production techniques and alternative sourcing strategies (Bubicz 2022). This collaborative approach enhances innovation capacity and ensures that sustainability considerations are embedded throughout the product development process rather than being added retrospectively (Ejairu et al. 2025).

Depite these advancements, several challenges have been identified in the literature. One major limitation is the trade-off between cost, performance, and sustainability in the design. Firms often face competing priorities, where low-carbon materials or sustainable production methods may increase costs and/or complexity. Additionally, existing supply chain structures may not always support the implementation of redesigned products, particularly when legacy systems are heavily optimized for existing product specifications. Another challenge is the integration of sustainability metrics into design decision-making tools. While life cycle assessment methodologies provide robust frameworks for evaluating emissions, their application in real-time product development processes remains limited in many organizations (Nakao and Kohsaka 2026). This gap between analytical capability and operational implementation reduces the effectiveness of design-led decarbonization strategies.

2.6 Collaboration and Stakeholder Integration in Achieving Decarbonization Goals

Collaboration and stakeholder integration have emerged as central themes in the literature on supply chain decarbonization, reflecting a shift away from firm-centric approaches toward network-based sustainability strategies (Wu 2025). Contemporary research increasingly recognizes that carbon emissions in supply chains are inherently distributed across multiple organizational boundaries, meaning that no single firm can achieve meaningful decarbonization independently (Mańkowska et al. 2021). Instead, emissions reduction requires coordinated action among suppliers, manufacturers, logistics providers, customers, regulators, and other institutional actors operating within and around the supply chain ecosystem. A foundational argument in the literature is that supply chains function as interdependent networks in which decisions made by one actor have direct and indirect consequences for other actors. This interdependence is particularly evident in emissions generation, where upstream suppliers influence material emission intensity, manufacturers determine production-related emissions, and logistics providers shape transportation-emission outcomes (Bhatti et al. 2021). Downstream actors, including retailers and consumers, further influence emissions through their demand patterns and consumption behaviors. Consequently, decarbonization is increasingly conceptualized as a collective action problem that requires alignment across the entire value chain (Chibaro et al. 2024).

Supplier integration is a crucial aspect of collaboration. Early research on supply chains concentrated on transactional relationships with suppliers, focusing on costs, quality, and delivery. However, recent sustainability studies have highlighted the importance of strategic partnerships with suppliers, which involve joint planning, shared sustainability goals, and coordinated investments in low-carbon technologies. Companies that engage suppliers in efforts to reduce emissions often experience improvements in carbon performance, which is vital for Scope 3 emissions and is largely driven by supplier activities. Closely related to this is the importance of information sharing and transparency in the decision-making process. Effective decarbonization depends heavily on the availability of accurate, timely, and standardized emissions data across supply chain tiers (Mańkowska et al. 2021). The literature highlights that data fragmentation remains a significant barrier to emissions management, particularly in complex global supply chains where visibility of lower-tier suppliers is limited. Therefore, collaborative data-sharing platforms and digital traceability systems are increasingly considered essential tools for enabling coordinated carbon management. By improving transparency, firms can identify emission hotspots, benchmark performance, and support continuous improvement initiatives across their supply networks (Demir and Solingen 2021).

Another important aspect of collaboration discussed in the literature is logistics and infrastructure sharing. In traditional supply chain models, firms often operate independent logistics systems, leading to the duplication of transport routes, underutilized capacity, and increased emissions. Recent studies have highlighted the emergence of collaborative logistics models, in which competing firms share transport resources, warehouses, and distribution networks to improve efficiency and reduce environmental impact. These arrangements, often referred to as collaborative transport networks or shared logistics platforms, enable load consolidation and reduce empty return journeys, thereby lowering overall carbon emissions. These initiatives demonstrate how inter-firm cooperation can generate both environmental and economic benefits. The role of multistakeholder governance structures has also been widely emphasized in the literature. Decarbonization initiatives are increasingly influenced by regulatory frameworks, industry standards, and voluntary sustainability commitments. Governments and international organizations play key roles in establishing carbon reporting requirements, emission targets, and compliance mechanisms that influence corporate behavior. Simultaneously, nongovernmental organisations and civil society groups contribute to accountability by advocating for transparency and environmental responsibility. Investors and financial institutions further reinforce these pressures through ESG-linked investment criteria and sustainability expectations. This multistakeholder environment creates a governance ecosystem in which firms are incentivized to collaborate to achieve shared sustainability objectives.

Customer-driven pressure is another important dimension of stakeholder integration highlighted in the literature. Increasing consumer awareness of environmental issues has led to a growing demand for low-carbon products and sustainable supply chains (Chibaro et al. 2024). This shift in consumer preferences has encouraged firms to adopt greener practices and communicate their sustainability performance transparently. Retailers play a critical role in transmitting sustainability expectations upstream to suppliers, thereby influencing production and sourcing decisions across the supply chain. Despite these developments, several persistent challenges in achieving effective collaboration have been identified. Firms may be reluctant to share sensitive operational or emissions data because of concerns about competitive disadvantages or data misuse. This limits the effectiveness of collaborative initiatives and hinders the development of fully transparent supply chain networks. Additionally, differences in technological capabilities, financial resources, and sustainability maturity across supply chain partners create uneven levels of participation, which can weaken collective decarbonization efforts.

Another challenge is related to the coordination complexity inherent in global supply chains. The larger and more geographically dispersed the supply network, the more difficult it becomes to align sustainability goals and implement consistent emission reduction strategies. Variations in regulatory environments, infrastructure quality, and institutional capacity further complicate collaboration across international supply chains. These structural constraints often result in fragmented sustainability efforts rather than fully integrated decarbonization strategies. The literature also highlights the importance of power dynamics within supply chains, where dominant firms, such as large retailers or multinational manufacturers, exert a significant influence over suppliers. While this can facilitate the diffusion of sustainability standards, it may also create dependency relationships that limit the autonomy of smaller suppliers. In some cases, sustainability requirements imposed by lead firms may place financial or operational burdens on suppliers without adequate support, potentially undermining their long-term collaboration (Demir and Solingen 2021).

3 Methodology

3.1 Research Design

This study adopts a qualitative desk-based research design to examine the role of innovation in enabling decarbonization within contemporary supply chain systems. A desk review is appropriate for this study because it allows for the systematic collection, synthesis, and interpretation of existing knowledge without primary data collection. The focus is on developing a conceptual and analytical understanding of how supply chain decarbonization is shaped by technological, organizational, and systemic innovations. Rather than empirically testing hypotheses, this study integrates insights from academic literature, policy documents, and industry reports to construct a coherent narrative of emerging trends and practices.

3.2 Data Sources

This study relies exclusively on secondary data sources drawn from a wide range of credible and authoritative materials. These include peer-reviewed journal articles in supply chain management, logistics, sustainability, and innovation studies. Scholarly books and book chapters were also used to provide theoretical grounding and historical context. The analysis is further informed by reports and publications from international organizations such as the United Nations, World Bank, Organization for Economic Co-operation and Development, and International Energy Agency, which provide globally relevant policy and emissions data. Industry white papers and sustainability reports from logistics, manufacturing, and technology firms were also included to capture practical and applied perspectives. Together, these sources ensure a balance between academic rigor, policy relevance, and industry practice.

3.3 Search Strategy and Selection of Literature

A structured search strategy was employed to identify the relevant literature for inclusion in the review. Academic databases such as Scopus, Web of Science, and Google Scholar were used to retrieve peer-reviewed publications. The search was guided by key terms and combinations, such as supply chain decarbonization, green logistics, carbon emissions in supply chains, sustainable supply chain management, digital transformation and sustainability, and electrification of logistics systems. The selection of literature was guided by clear inclusion criteria. Only sources written in English and directly related to supply chain, logistics systems, or industrial operations were considered. Preference was given to studies that addressed sustainability, emissions reduction, or innovation in a meaningful way, particularly those published within the last decade, to ensure relevance to current developments. Outdated, duplicated, or loosely related studies on supply chain decarbonization were excluded from the analysis. This ensured that the review remained focused, current, and analytically consistent.

3.4 Data Analysis Approach

The selected literature was analyzed using thematic analysis. This approach involved identifying recurring patterns, concepts, and arguments across the reviewed sources and organizing them into coherent themes. Through iterative reading and coding, key thematic areas were developed, including digital transformation in supply chains, electrification of logistics systems, innovation in product and process design, collaborative supply chain sustainability, regulatory and policy influences, and the relationship between decarbonization and supply chain resilience (Johnson 2025). These themes were then synthesized to develop an integrated understanding of how innovation enables decarbonization across different stages of the supply chain. The emphasis was not only on describing existing knowledge but also on identifying linkages, contradictions, and gaps in the literature to support a more comprehensive conceptual interpretation of the findings.

3.5 *Limitations of the Methodology*

Although the desk-based approach provides valuable insights, it has limitations. This study relies entirely on secondary data, which means that the quality and accuracy of the findings depend on the reliability of the literature and reports. There was no primary empirical validation through fieldwork, interviews, or surveys, which limited the ability to confirm real-time industry practices or organizational behavior. Additionally, differences in the methodological approaches across the reviewed studies may have introduced variations in interpretation and comparability. Despite these limitations, a desk review remains appropriate for this chapter, as it enables a broad and integrative perspective on this rapidly evolving subject. This allows for the consolidation of fragmented knowledge across disciplines and provides a strong conceptual foundation for understanding innovation-driven decarbonization in contemporary supply chains.

4 Discussion

Recent research has provided additional evidence for one of the most important concepts in sustainable supply chain management. Emissions are not simply an accident of a company's operations; they are built into the operation, planning, and administration of all supply chain systems. The fact that Scope 3 emissions represent most emissions in almost every sector provides evidence of the need to move away from optimizing each company individually toward creating accountable value chains (Jones 2025; Vishwakarma et al. 2025). This aligns with the recent literature that portrays decarbonization as an issue related to the structure of the supply chain system rather than simply an issue of operational process improvements.

Road transportation represents one of the largest sources of emissions for supply chain companies, especially in trucking. Trucks rely heavily on fossil fuels and are plagued by inefficiencies associated with backhauling with no products loaded, in addition to the fragmentation associated with multiple carriers transporting goods over long distances (Jones 2025). Although airfreight accounts for a relatively small percentage of the total transportation used in supply chains, it emits large amounts of greenhouse gases per tonne of product transported. Maritime transportation is more efficient than airfreight and trucking but produces significant emissions based on volume alone. However, the issues associated with transportation extend beyond simple modal issues. Logistical architectures (e.g., network designs, procurement strategies, and inventory models) create variability in emissions levels. An example of this is the just-in-time (JIT) system. JIT systems reduce costs for companies; however, they produce higher frequencies of shipments, resulting in increased transportation emissions and fewer opportunities for consolidated shipments (Allon 2026). Thus, a conflict exists between the pursuit of economic efficiency and environmentally sustainable solutions.

Another area that has emerged as having a major effect on the level of emissions produced by supply chain companies is warehouse and inventory management (WIM). Warehouse and inventory management are recognized as contributing factors to the production of emissions, rather than being simply passive logistical components. Inefficient layouts and energy-intensive storage systems combine to increase a company's overall carbon footprint (Weng 2025). Overall, these findings indicate that emissions are generated by operational procedures and structural arrangements. Therefore, companies must develop holistic approaches to address the production of emissions. Digital technology offers a promising mechanism to support decarbonization efforts. Technologies such as artificial intelligence (AI), the Internet of Things (IoT), and blockchain offer companies greater ability to monitor their operations, track their products, and optimize processes, enabling them to better identify areas where emissions are occurring and make informed business decisions (Olsen 2026; Vishwakarma et al. 2025). Examples include using AI-driven route optimization to minimize fuel usage or using IoT-enabled real-time monitoring systems to improve logistics performance. Blockchain technology enables companies to establish transparent and accountable reporting mechanisms for emissions (Happy et al. 2021a, b; Chibaro et al. 2026). However, the study cautions against a simplistic view of technology as a panacea for addressing the challenges of decarbonization. Technology works only if organizations provide appropriate organizational support, facilitate data exchange, and commit resources to ensure successful implementation (Ejairu et al. 2025; Khawla et al. 2025).

In addition to the adoption of new digital technologies, upstream interventions in product design and supply chain configuration have emerged as key factors in achieving reductions in emissions. Studies using life cycle assessments have shown that many emissions are "locked-in" at the design stage of a product, thereby increasing the emphasis on proactively designing products for decarbonization (Johnson 2025; Medhi 2020). The implementation of eco-designed strategies, such as material efficiency, modularity, and recyclability, has been found to significantly reduce lifecycle emissions. Supply chain design choices, such as facility locations and sourcing strategies, have also been identified as key drivers of emission levels (Sindhwani et al. 2023). The integration of circular economy principles is another strategy that supports emission reduction by encouraging reuse, remanufacturing, and resource efficiency (Wu 2025).

Finally, this study concludes that supply chain decarbonization is primarily a collaborative activity involving multiple stakeholders. Greenhouse gas emissions exist across multiple stakeholders in a supply chain network. Therefore, no single stakeholder can effectively reduce emissions (Bhatti et al. 2021; Mańkowska et al. 2021). Companies are required to collaborate with suppliers, share information regarding their emissions-producing activities, and jointly invest in low-carbon technologies (Demir and Solingen 2021). Collaborative logistics strategies, such as shared modes of transportation and warehouses, have been identified as viable methods for reducing emissions through increased load density and fewer empty miles (Wu 2025). Although collaborative strategies present viable options for companies seeking to reduce their greenhouse gas emissions, barriers remain for companies participating in globally dispersed supply chains, such as lack of trust, unequal distribution of power, and lack of technological parity among stakeholders (Chibaro et al. 2024).

5 Implications

The findings have strong theoretical implications, primarily in terms of increasing the understanding of supply chains as integrated systems of interdependence in which emissions can be structurally embedded. Moreover, the predominance of emissions clearly indicates an urgent need for sustainability research at both the network and multistakeholder levels. Additionally, the findings highlight the need for clear differentiation in terms of the distinction between efficiency gains and the reduction in absolute levels of emissions, specifically when potential rebound effects are considered. In addition to the theoretical implications, the findings provide several important recommendations for managers. First, the findings demonstrate that firms must adopt an end-to-end approach to decarbonize their supply chains. This requires various actions, including, but not limited to, developing comprehensive carbon accounts, integrating sustainability considerations into supply chain design and product development processes, and ensuring that new or existing digital technologies developed by these firms meet environmental goals.

On the policy front, the study recommends the need for standardized frameworks for carbon reporting, investment in low-carbon infrastructure, and incentives for companies to implement low-carbon technologies. Both policy interventions are likely to be especially important in the Global South, where structural and physical barriers remain. In addition to their direct economic impacts and ability to improve air quality and mitigate climate change through reduced greenhouse gas emissions, the findings of this study identify the broad societal and environmental benefits associated with supply chain decarbonization efforts.

6 Conclusions

The study concluded that carbon emissions are influenced by the configuration of supply chains, including the allocation of responsibilities, geographical locations, and operational methodologies. These emissions are not solely attributable to suboptimal operational performance; rather, they are shaped by organizational structure, governance, and coordination among various actors and locations globally. The study also concluded that technology is a crucial instrument for emission reduction, with the potential to positively impact emissions. However, the effective utilization of technology is contingent upon its alignment with organizational practices and sustainability objectives.

The findings suggest that the optimal strategy for managing emissions is proactive, focusing on product design, supply chain configuration, and stakeholder integration to prevent emissions rather than merely mitigating them. The persistence of structural barriers, such as system fragmentation, capability disparities among system members, and coordination challenges, underscores the need for systemic rather than incremental changes in supply chain systems. Consequently, the transition to low-carbon supply chains requires a shift from efficiency-based systems to those centered on resilience and sustainability, embedding environmental considerations into all decision-making processes.

References

Abaku EA, Edunjobi TE, Odimarha AC (2024) Theoretical approaches to AI in supply chain optimization: pathways to efficiency and resilience. Int J Sci Technol Res Arch 6(1):092–107

Acquaye A, Genovese A, Barrett J, Lenny Koh SC (2014) Benchmarking carbon emissions performance in supply chains. Supply Chain Manag Int J 19(3):306–321

Allon G (2026). AI in supply chain management: disruptions and challenges. Springer Series in Supply Chain Management, 21–33, ISSN: 2365-6395, Springer Nature Switzerland. https://doi.org/10.1007/978-3-032-07054-8_3

Almusaed A, Yitmen I, Myhren JA, Almssad A (2024) Assessing the impact of recycled building materials on environmental sustainability and energy efficiency: a comprehensive framework for reducing greenhouse gas emissions. Buildings 14(6):1566

Alpan G, Bouzekri H, Sanlaville É (2023) Enhancing global supply chain performance by optimizing port resources. Maritime ports, supply chains and logistics corridors, 151–163, Routledge. https://doi.org/10.4324/9781003365013-15

Anathole U, Limei C, Hamza F (2026) Navigating Corporate greenwashing in the carbon era: synergizing internal carbon pricing, ESG governance, and regulatory stringency to advance environmental integrity. Corp Soc Responsib Environ Manag 33(2):1794–1814

Beck A, Pedraza A (2026) Deep and shallow decarbonization in supply chains. Washington, DC: World Bank. https://doi.org/10.1596/1813-9450-11306

Bhatti S, Ferraris A, Khan J (2021) Big data in supply chains- achieving supply chain innovation through capabilities. Acad Manag Proc (1):14635, ISSN: 0065-0668, Academy of Management. https://doi.org/10.5465/ambpp.2021.14635abstract

Brinken J, Pabsch C, Behrendt F (2022) Decarbonization potential of logistic 4.0 technologies in apple supply chains. Procedia Comput Sci 200:461–470, ISSN: 1877-0509, Elsevier BV. https://doi.org/10.1016/j.procs.2022.01.244

Bubicz M (2022). Apparel supply chain in the context of Brazil: social sustainability management and comparison analysis with global supply chains. Open Engineering Inc. https://doi.org/10.31224/2208

Cancela BL, Santos C, Coelho A (2026) Green strategies for circular economy and sustainable supply chains. Corp Soc Responsib Environ Manag 33(1):675–693

Carodenuto S (2020) Addressing deforestation in global supply chains: the industry approach. Copernicus GmbH. https://doi.org/10.5194/egusphere-egu2020-12305

Chen Q, Su H, Gao W, Lv J (2025) Does environmental regulation drive or hinder the development of new quality productive forces? evidence from Chinese listed firms. Front Environ Sci 13:1582399

Chibaro M, Kanyepe J, Manyanga W, Moeti-Lysson J, Musasa T (2026) Blockchain and big data integration for transparent and traceable supply chains. Enhancing sustainability in global supply chains with big data analytics, 147–174, IGI Global Scientific Publishing. https://doi.org/10.4018/979-8-3373-6896-2.ch005

Chibaro M, Muchemwa M, Mahwine M, Mpfiga C (2024) Supply chain risks and organisational competitiveness in corporate timber manufacturing companies in Mutare metropolitan region, Zimbabwe, Logistics Research, ISSN: 1865-0368, Bundesvereinigung Logistik (BVL), Bremen, Vol 17, Iss 1, pp 1–14. https://doi.org/10.23773/2024_8

Chikwere D, Mubaiwa C, Chibaro M (2026) Greening supply chains in Africa. Sustainable innovation in Africa, 64–84, Routledge. https://doi.org/10.4324/9781003745273-4

Das R, Shaw K (2017) Uncertain supply chain network design considering carbon footprint and social factors using two-stage approach. Clean Technol Environ Policy 19(10):2491–2519

Dautel JL, Thakur J, Elberry AM (2024) Enabling industrial decarbonization: a MILP optimization model for low-carbon hydrogen supply chains. Int J Hydrogen Energy 77:863–891, ISSN: 0360-3199, Elsevier BV. https://doi.org/10.1016/j.ijhydene.2024.06.050

Demir NU, Solingen E (2021) Are global supply chains vital to china's leaders?. Geopolitics, supply chains, and international relations in East Asia, 135–152, Cambridge University Press. https://doi.org/10.1017/9781108985468.011

Dubey R, Foropon C, Venkatesh VG (2021) Climate-resilient supply chains. The Routledge companion to global value chains, 58–85, Routledge. https://doi.org/10.4324/9781315225661-7

Ejairu E, Filani OM (2025) Assessing the sustainability practices of global supply chains: challenges and solutions. Eng Technol J 10, ISSN: 2456-3358, Everant Journals. https://doi.org/10.47191/etj/v10i08.14

Franke H, Chae S, Foerstl K (2024) Toward a configurational understanding of global supply chain complexity. J Bus Logist 45(2):e12371

Gamage N, Perera S, Nanayakkara S (2025) Supply chain and construction practices for low-carbon construction. In: Global net zero carbon practices in construction, pp 101–115. Singapore: Springer Nature Singapore

Grytsenko SI, AA T, Polishchuk AV (2023) Modern trends in global supply chains. Electronic scientific and practical publication in economic sciences

Happy A, Rahman KW, Chowdhury MdMH, Chowdhury M, Scerri M (2021) Blockchain adoption challenges in supply chain. The Routledge companion to global value chains, pp 137–155, Routledge. https://doi.org/10.4324/9781315225661-13

Happy A, Rahman KW, Chowdhury Md, Maruf H, Chowdhury, M, Scerri M (2021) Blockchain adoption challenges in supply chain. The Routledge companion to global value chains, pp 137–155, Routledge. https://doi.org/10.4324/9781315225661-13

Jha H, Mohan U (2023) A multiperiod discrete event simulation model for comparing synchronous and asynchronous facility reopening in global supply chains affected by disruption. Supply Chain Analytics, 2, 100010, ISSN: 2949-8635, Elsevier BV. https://doi.org/10.1016/j.sca.2023.100010

Johnson H (2025) Adapting to disruptions: a qualitative study on strategies for building resilience in global supply chains. MDPI AG. https://doi.org/10.20944/preprints202501.0090.v1

Jones J (2025) A systematic literature review on supplier relationship management in the context of global supply chains. MDPI AG. https://doi.org/10.20944/preprints202501.0460.v1

Kanyepe J, Boy R, Chibaro M, Mphela T, Tlhakanelo K (2025) Revolutionizing supply chain forecasting with generative ai and machine learning. Advances in business strategy and competitive advantage, 435–462, ISSN: 2327-3429, IGI Global. https://doi.org/10.4018/979-8-3693-4433-0.ch016

Khawla AT, Abdessamad Z, Wafaa I (2025) Blockchain for supply chain innovation and transparency: a PRISMA-Based systematic review: enhancing supply chains with blockchain. 2025 16th International conference on logistics and supply chain management (LOGISTIQUA), pp 1–9, IEEE. https://doi.org/10.1109/logistiqua66323.2025.11122716

Li W, Yigitcanlar T, Browne W, Nili A (2023) The making of responsible innovation and technology: an overview and framework. Smart Cities 6(4):1996–2034

Mańkowska M, Pluciński M, Kotowska I (2021) Biomass sea-based supply chains and the secondary ports in the era of decarbonization. Energies 14(7):1796, ISSN: 1996-1073, MDPI AG. https://doi.org/10.3390/en14071796

Medhi PK (2020) Blockchain-enabled supply chain transparency, supply chain structural dynamics, and sustainability of complex global supply chains—a text mining analysis. Information for efficient decision making, 273–312, WORLD SCIENTIFIC. https://doi.org/10.1142/9789811220470_0011

Miklautsch P, Woschank M (2022) Decarbonizing industrial logistics. IEEE Eng Manage Rev 50(3):149–156

Mubarik MS, Gunasekaran A, Khan SA, Mubarak MF (2025) Decarbonization through supply chain innovation: Role of supply chain collaboration and mapping. J Clean Prod 507:145492

Mupfiga C, Chibaro M, Mahwine M, Muchemwa M (2024) Implementation of green supply chain management practices in Zimbabwean organizations. Logistic Oper Manag Res (LOMR) 3(1):66–75, ISSN: 2830-2680, Research Synergy Foundation. https://doi.org/10.31098/lomr.v3i1.1801

Nakao K, Kohsaka R (2026). Applicability of supply shed approach in global supply chains: integrating supply shed and landscape approaches for natural capital assessment. Environ Res Commun 8(2):22501, ISSN: 2515-7620, IOP Publishing. https://doi.org/10.1088/2515-7620/ae3d83

Olsen TL (2026) Artificial intelligence in agricultural supply chains: practice and prospects. Springer Series in supply chain management, 255–266, ISSN: 2365-6395, Springer Nature Switzerland. https://doi.org/10.1007/978-3-032-07054-8_18

Onukwulu EC, Agho MO, Eyo-Udo NL (2023) Developing a framework for AI-driven optimization of supply chains in energy sector. Global J Adv Res Rev 1(2):82–101

Polater A (2025) Building resilient supply chains: unveiling the interplay of reflexivity, supply chain innovation and improvisation capability. Benchmarking Inte J 32(9):3524–3555, ISSN: 1463-5771, Emerald. https://doi.org/10.1108/bij-10-2024-0930

Reeb RN, Folger SF (2023) Community outcomes of service learning: Research and practice from a systems theory perspective. In: Research on service learning, pp 389–418. Routledge

Retegi J, Kamp B, Igartua JI (2026) Industrial value chains and greenhouse gas emissions: an EEIOT-based sustainability analysis for assessing policy options. Gases 6(1):12

Richardson DB (2013) Electric vehicles and the electric grid: A review of modeling approaches, Impacts, and renewable energy integration. Renew Sustain Energy Rev 19:247–254

Seidelson C (2024) Collaboration in agile supply chains. Building agile supply chains. SAGE Publications, Inc.https://doi.org/10.4135/9781071941935

Shen J, Zhang Q, Tian S (2025) Decarbonization pathways analysis and recommendations in the green steel supply chain of a typical steel end user-automotive industry. Appl Energy 377:124711, ISSN: 0306-2619, Elsevier BV. https://doi.org/10.1016/j.apenergy.2024.124711

Sindhwani R, Hasteer N, Behl A, Chatterjee C, Hamzi L (2023) Analysis of sustainable supply chain and industry 4.0 enablers: a step toward decarbonization of supply chains. Ann Oper Res 359(1):429–467, ISSN: 0254-5330, Springer Science and Business Media LLC. https://doi.org/10.1007/s10479-023-05598-7
Srivastava S, Shardeo V, Dwivedi A, Paul SK (2025) Analyzing enablers of artificial intelligence for decarbonization: implications for circular supply chains. Ann Oper Res 359(1):953–980, ISSN: 0254-5330, Springer Science and Business Media LLC. https://doi.org/10.1007/s10479-025-06843-x
Sun G, Sulemana I, Agyemang AO (2025) Examining the impact of stakeholders' pressures on sustainability practices. Manag Decis 63(11):3967–3995
Vishwakarma AK, Patro PK, Acquaye A (2025) Applications of AI to low carbon decision support system for global supply chains. Cleaner Logist Supply Chain 17:100261, ISSN: 2772-3909, Elsevier BV. https://doi.org/10.1016/j.clscn.2025.100261
Wang CH, Chen CH, Chen CC, Hsieh HH (2025) Decarbonization of remanufacturing processes in circular supply chains: case study on rubber recycling industry. Indus Manag Data Syst 125(10):2753–2776, ISSN: 0263-5577, Emerald. https://doi.org/10.1108/imds-04-2024-0397
Weng LS (2025) Achieving net-zero supply chains: strategies, technologies, and financial innovations for corporate Decarbonization., Elsevier BV. https://doi.org/10.2139/ssrn.5146993
Wu J (2025) Automation in production and supply chain layout: balancing efficiency and employment. Springer series in supply chain management, 243–263, ISSN: 2365-6395, Springer Nature Singapore. https://doi.org/10.1007/978-981-96-3228-2_10
Xin L (2024) Green supply chain management influence on the economic and environmental aspects in China. Environ Sci Pollut Res 31(14):21144–21156
Yarlagadda VK (2024) Cutting-edge developments in robotics for smart warehousing and logistics optimization. Robot Xplore: USA Autom Digest 1(1):61–79
Yordanova Z (2023) Barriers to organizations to adopt digital transformation for driving eco-innovation and sustainable performance. In: International conference on optimization, learning algorithms and applications, pp 162–171. Cham: Springer Nature Switzerland
Zhang J (2024) Energy access challenge and the role of fossil fuels in meeting electricity demand: promoting renewable energy capacity for sustainable development. Geosci Front 15(5):101873
Zvarimwa C, Zimuto J (2022) Valuable, rare, inimitable, nonsubstitutable and exploitable (VRINE) resources on competitive advantage. Int J Bus Manag Sci 8(1):9–22

State Capacity and Institutional Reform in Advancing the Green Transition in the Transport Sector

David Chikwere, Munyaradzi Chibaro, Munyaradzi Mahwine, and Lazarus T. Sabungu

Abstract With the growing urgency of climate change, the green transition is among the most dominant policy priorities, and governance capability has been increasingly recognized as a fundamental factor for delivering on such transitions. While there has been an ever-increasing investment in green technologies and supportive policy frameworks, less has been written on institutions and their role in implementation, specifically with respect to green transport regulation. This chapter helps fill that gap by looking at how differences in state capacity and institutional design shape the effectiveness of policies for a green transition. The chapter discusses the impact of state capacity and institutional change in driving sustainable transitions. Specifically, it aims to (i) understand the role of state capacity in effective green transition policies, (ii) analyse institutional constraints that prevent implementing green transport regulations, and (iii) explore ways to strengthen institutional capacity for sustainable governance. The study adopts a qualitative research approach grounded in institutional theory, drawing on a systematic review of literature from Scopus and Google Scholar academic databases alongside comparative policy analysis of green transport governance frameworks between 2019 and 2026. The findings reveal that weak inter-agency coordination, inadequate enforcement mechanisms, and institutional fragmentation significantly hinder the translation of green policies into practice. Furthermore, governance deficits are exacerbated by corruption vulnerabilities and insufficient technical expertise, particularly in the developing world. However, evidence also indicates that targeted capacity-building programmes, adaptive policy learning,

D. Chikwere (✉)
Leeds Business School, Leeds Beckett University, Leeds, UK
e-mail: D.chikwere@leedsbeckett.ac.uk

M. Chibaro
University of Botswana, Gaborone, Botswana

M. Mahwine
Supply Chain and Risk Management, Midlands State University, Gweru, Zimbabwe

L. T. Sabungu
Midlands State University, Gweru, Zimbabwe
e-mail: sabungul@staff.msu.ac.zw

J. Kanyepe and I. Chirisa (eds.), *Advancing the Green Agenda in the Transport and Logistics Sector in Global South Countries*, World Sustainability Series,
https://doi.org/10.1007/978-3-032-31374-4_17

and international collaboration can enhance institutional resilience and regulatory performance. Building on these insights, the chapter outlines far-reaching institutional reforms to reinforce coherent regulation, transparency and accountability mechanisms, technical capacity, cross-sectoral linkages and international cooperation. The findings underscore the conclusion that successful green transition is a process which increasingly relies not only on enforcing technological innovation but equally depends on strong state capacity and long-term modernization of institutions, making governance reform a core component of sustainable transformation in the environmental field.

Keywords State capacity · Institutional Reform · Green transition · Environmental Governance · Green transport regulation

1 Introduction

The urgency of the climate crisis worldwide has placed the green transition at the heart of public policy, economic planning and environmental governance (Chikwere et al. 2024). Countries worldwide are under increasing pressure to decarbonise key sectors of their economy, such as transport and energy, so that national approaches to mitigation align with global frameworks like the United Nations Sustainable Development Goals (UNSDGs) and Paris Agreement (UNEP 2023; IPCC 2022). Even if there has been an unprecedented scale-up of low-carbon technologies and sustainability-oriented policies, the success or failure of these efforts is increasingly contingent on how well state institutions can construct, coordinate and implement regulatory regimes (Meckling and Nahm 2019; Rodrik 2021). In this regard, state capacity has been a core driver of successful green transitions that are both about the type of environmental policies developed as well as their translation into real-world outcomes.

State capacity is a set of administrative, technical and regulatory capabilities that a government needs to provide public goods, enforce laws and coordinate complex policy processes (Fukuyama 2013). In the case of environmental governance, this means the capacity to design coherent regulations, ensure inter-agency coordination, monitor compliance and effectively enforce environmental standards (Dubash 2021). Policy incoherence and limited impact of regulation are frequent results because bureaucratic fragmentation, overlapping mandates within the agencies, weak enforcement mechanisms, and corruption risks create challenges (Andrews et al. 2017; Khan et al. 2019). In addition, low technical capacity and limited resources are serious impediments to the adoption of complex environmental technologies and regulatory innovations, which help achieve sustainability goals.

The transport sector is one of the biggest contributors to global greenhouse gas emissions and reflects the governance challenge of the green transition (Munuhwa 2024; Marzouk 2025). Chikwere et al. (2025) contend that integrated policy responses across institutions, including transport authorities, environmental agencies, urban planners and energy regulators, are necessary for effective decarbonisation in this sector. However, insufficient inter-agency coordination and institutional silos often obstruct aligning policy objectives with comprehensive regulatory frameworks (Barrett et al. 2022; Sovacool et al. 2022). These governance gaps underscore the urgent need for institutional reforms not only to strengthen administrative capacity but also in terms of regulatory coherence and accountability mechanisms.

Despite increasing scholarly interest in climate policy and technological innovation, the role of institutional capacity and governance reform in mandating the green transition is still underexplored. The literature at hand tends to concentrate on technological solutions, market-based instruments and financial mechanisms (Stern et al. 2022; Chausson et al. 2023) but pays very limited attention to the institutional context in which they are implemented. In addition, studies investigating governance sensibilities often take broad, cross-national approaches and provide little insight into the specific institutional dynamics undermining policy implementation in different domains of regulation (such as transport versus environmental). This forms an important research gap in understanding how state capacity constraints intersect with institutional design, which leads to the success or failure of green transition policy.

In addressing this gap, the chapter investigates how state capacity and institutional reform drive green transition in the transport sector. Using an institutional theory lens, it analyses how governance structures, administrative capabilities and regulatory frameworks affect outcomes on environmental policies. The chapter examines the limits set by bureaucratic fragmentation, corruption risks, and limited technical expertise to implementation processes in the Global South. It also examines how policy learning, capacity-building initiatives and international cooperation can contribute to institutional effectiveness and promote sustainable development goals.

The need to strengthen state capacity is further compounded in a globalised world where environmental governance has become more complex as states are faced with environmental challenges that transcend their boundaries (Chasek and Downie 2020). Issues such as international agreements, transnational regulatory networks and cross-border environmental challenges compel governments to promote collaborative governance arrangements that necessitate high degrees of institutional coordination and adaptability (Michaelowa and Michaelowa 2019; Hale 2020). In this context, international cooperation is vital in enhancing knowledge transfer (including technical assistance and capacity-building programmes), especially for countries that have considerably limited institutional resources like the Global South. However, the success of such initiatives relies on high domestic institutional readiness and capabilities to embed external learning within local governance systems.

Similarly, the dangers of fragile governance structures must not be underestimated. Corruption and regulatory capture, on the other hand, may lead to ineffective environmental policies because of their transparency failure, which leads to insufficient enforcement, which causes public mistrust as well (Transparency International 2022). These challenges are particularly pressing in areas where there are long-gestation capital investments and complex regulatory systems, such as transport infrastructure and energy systems. Accordingly, these challenges cannot be resolved by merely mainstreaming institutions but need stringent accountability systems and participatory governance modalities to institutionalise stakeholder engagement and oversight.

On this basis, the chapter is structured around these Research Questions:

1. What is the role of state capacity in green transition policy effectiveness?
2. What institutional constraints hinder the implementation of green transport regulations?
3. What strategies can be used to strengthen institutional capacity for sustainable governance?

It thus both engages with and contributes to a broader conversation in environmental governance around the value of getting institutional design right for achieving sustainability outcomes. It provides an antidote to technology-centric approaches and requires attention to governance capacity, regulatory change and administrative capability for pursuing a green agenda. Ultimately, the chapter contends that, for a successful green transition, we must look beyond innovation and investment to consider how states can support the calibre of institutions necessary for implementing, enforcing and sustaining transformative policy change.

2 Literature Review

This section explains in the first instance the Institutional theoretical framework which forms the foundation of the related literature. This is then followed by a review of the literature related to this chapter.

3 Theoretical Framework

3.1 Institutional Theory

This chapter is informed by institutional theory, which provides a robust analytical lens for examining how formal structures, governance arrangements, and normative systems influence policy outcomes in the green transition. Institutional theory proposes that the behaviour of organisations and the effectiveness of policies are not necessarily driven by technical efficiency, but rather embedded in broader regulatory environments, cultural norms and cognitive structures that inform decision-making processes of political actors (North 2025; Jepperson and Meyer 2021). This is critical, especially for the transport sector, where the enactment of low-carbon policies is dependent on interaction between many institutions, governance actors and regulatory systems.

The Institutional framework uses the three pillars of institutions (regulative, normative and cognitive) to explore the dynamics of state capacity (Scott 2014). The regulative pillar focuses on formalised rules, enforcement mechanisms and compliance systems that are particularly important in transport decarbonisation policies like emissions standards and infrastructure regulations. Whereas the normative pillar shows how values, professional norms and inter-organisation coordination play a role in achieving policy coherence, the cognitive pillar helps understand why a knowledge-based approach can produce shared meanings and technical expertise that makes complex environmental reforms possible to implement.

Drawing on the institutional theory, this study provides insights into more persistent, credible governance challenges, including bureaucratic fragmentation, regulatory overlap and corruption that threaten to undermine transport policy implementation. It also allows for the analysis of institutional change through mechanisms like policy learning, capacity-building and international diffusion of best practices (DiMaggio and Powell 1983). By connecting these dimensions, this framework allows for an understanding of their interaction to provide a comprehensive perspective on how institutional constraints and reforms have influenced the effectiveness of such policies. As such, institutional theory serves as a consistent framework that carries over these arguments: State capacity and reforming governance structures are required prerequisites at the national level for sustainable and coordinated transformation of transport sectors.

4 The Green Transition in the Transport Sector

Transport accounts for a significant proportion of greenhouse gas emissions globally, making the green transition in the transport sector a key pillar of global decarbonisation efforts. Thus, the transition challenges carbon-intensive mobility systems and attempts to establish low-carbon, energy-efficient and sustainable transport modes such as electrification or modal shifts alongside holistic urban mobility planning (Marzouk 2025). This transition is not just technological or socio-technical, as it necessitates alignment between infrastructure development, behavioural change and policy frameworks.

Existing literature highlights that effective transport decarbonisation is likely to depend on coherent policy mixes consisting of regulatory standards, market-based instruments, and public investment (Sovacool et al. 2022). For example, emissions norms, fuel pricing and vehicle subsidies (EVs) can go a long way when combined with infrastructure investment in public transport and non-motorised mobility systems. The effectiveness of these interventions, however, strongly relies on governance frameworks able to coordinate across energy, urban development and environmental disciplines.

Additionally, the transition to green transport is also constrained by path dependencies entailed (often fossil fuel-based) systems embedded within existing infrastructure and institutional arrangements (Kotilainen et al. 2019). Then lock-in effects make changes resistant, and the thicket of policy implementation is dense. Consequently, Abram et al. (2022) argue that valid systems change like constant decarbonisation will require not only technological invention and innovation but the institutional determination to demolish lock-in systems so consistent sustainability transitions can be achieved.

5 The Role of State Capacity in Environmental Governance

It is widely held that state capacity is critical to effective governance and implementation of environmental policy, which at least partially explains the tendency to overlook this factor. It encompasses the administrative, technical and regulatory capabilities that enable governments to formulate, promulgate and implement policies (Fukuyama 2013). The complex, multidimensional and long-horizon nature of response to environmental challenges makes state capacity remarkably relevant for the green transition.

As theoretical discussions mention, high administrative capacity would allow governments to efficiently devise coherent policy frameworks, harnessing resources for productive allocation and ensuring continuity of policies (Andrews et al. 2017). However, it is also about improved access to technical knowledge and data systems that will be vital not only for the governance of complex environmental regulation but for new enabling technologies like electric mobility and smart transport systems. Poorly designed, ineffective enforcement is a risk where there is insufficient expertise in regulatory frameworks.

Another critical dimension is regulatory capacity, specifically an organization's ability to monitor compliance and enforce rules. The other vital theme for understanding policy failure in such contexts where the integrity of institutions is threatened through corruption and political interference relates to weak enforcement mechanisms (Khan et al. 2019). Moreover, state capacity is never fixed as it evolves through mechanisms of institutional learning and reform or through external intervention. Changing nature that highlights the need for ongoing investment in governance systems to support the green transition.

6 The Institutional Fragmentation, Coordination, and Regulatory Design

Institutional fragmentation continues to be one of the greatest obstacles to effective environmental governance in the transport sector. Fragmentation occurs when the institutions responsible for policy design and implementation are split across multiple agencies with overlapping or poorly defined mandates, resulting in coordination failures and policy incoherence (Bouckaert et al. 2022). Transport governance often faces such challenges when multiple, very different actors get involved in shaping transport systems, such as the transport ministries, environmental agencies, local governments and energy regulators (Fathoni et al. 2025).

Fragmented governance structures have also been associated with the provision of inconsistent signals from policy, duplication of effort and coverage gaps in regulation, all of which detract from the effectiveness of green transition efforts (Giwah et al. 2020). For instance, if the development of charging infrastructure and energy supply systems is not synchronised, then there will be a gap between the introduction of electric vehicles on the road and their consumption and usage in practice.

Well-functioning regulatory design is more important to counter negative fragmentation effects. This suggests the establishment of integrated policy frameworks that harmonize objectives across sectors and levels of governance. Scholars also argue that there is a need for policy coherence, where different points of entry that act as regulatory mechanisms should not be contradictory or counter to one another but rather mutually reinforcing (Sovacool et al. 2022). A mechanism that facilitates inter-agency coordination, such as the establishment of task forces for joint action or an integrated planning body whose members include a variety of units working on different stages of a policy process (Kamrul Ahsan 2023), can further encourage collaboration. Nonetheless, cooperation is often hindered by institutional indifference, political competition and a shortage of resources. Institutional reforms are thus critical to overcome fragmentation for effective governance, which can be achieved by substantially clarifying roles and responsibilities, strengthening coordination mechanisms and enhancing accountability at multiple levels of government.

7 Institutional Reform, Policy Learning, and Capacity Building

Institutional reform is key to informing the governance challenges around green transition in the transport sector. It is a process that contributes to the restructuring of organisational forms, regulatory regimes and administrative practices aimed at improving efficiency, accountability and adaptability (North 2025; Scott 2014). Reform efforts are particularly pressing in these contexts, where existing institutions struggle to cope with the complexities of environmental governance.

Policy learning is critical to enable institutional reform (Leeuw et al. 2020). This entails the ways in which policymakers and institutions learn from previous experiences, new evidence, and foreign knowledge to adjust their strategies. Policy learning contributes to advances of the green transition by helping governments learn from one another where practices work, or do not work, so policy failures can be avoided and regulatory approaches refined (Leong and Howlett 2022). This is particularly important in the field of transport, where policy frameworks need to adapt continuously to technological advancements at an unprecedented pace.

Institutional reform would also require a mammoth effort in capacity building. It includes efforts to build the capacity of public institutions through knowledge, skills and resources that support policy design and implementation (Williams 2021). This encompasses, among others, training programmes, technical assistance and investments in data and digital infrastructure. For instance, a review highlights the contribution of international cooperation, which, by facilitating knowledge transfer and fostering systemic transfer of financial and technical resources, promotes skills retention mechanisms in developing countries (Pandey et al. 2022). Significantly, institutional reform is not just a process but an ongoing one that necessitates systematic political will and stakeholder involvement. Radtke (2025) argues that adaptive and flexible governance systems are seen as more fit to manage the uncertainties and complexities posing challenges to the green transition. As a result, policy learning and capacity-building mechanisms need to be integrated into institutional frameworks to achieve long-term sustainability targets in the transport sector.

8 Research Methodology

To fulfil the objectives of this study about state capacity and institutional reform to facilitate the green transition in the transport sector, a systematic literature review (SLR) was conducted. In terms of methodological proximity, literature reviews are most like content analysis, as they allow for a systematic assessment of both thematic content and structural elements present in the existing library of scholarship through qualitative (and where appropriate) quantifiable insights (Fink 2019). Such an approach is well-suited for studying complex and multidimensional phenomena such as environmental governance, where much empirical evidence exists but tends to be scattered across studies focused on policy, institutions or specific sectors.

There are two main purposes of literature reviews. First, they comprehensively synthesize existing knowledge by capturing key themes, debates and emerging trends across a field. Second, they contribute toward the development of concepts through elucidation of theoretical constructs and by aiding theory-building activities (Meredith 1993). The review was intended to draw together academic and policy analyses of state capacity, institutional fragmentation, regulatory design and governance mechanisms in the transport sector with respect to implications for the green transition, specifically as they relate to this chapter.

To ensure methodological rigour, the review was conducted following a clear and transparent process. Research questions were well-defined, and relevant studies were identified and included, leading to a clear search strategy across primary repositories, including Scopus and Google Scholar. Clearly defined research questions improve the reliability and replicability of literature reviews by focusing on only the most relevant academic contributions (Nguyen et al. 2018). Relevant publications were identified using keywords including “state capacity,” “institutional reform,” “green transition”, “environmental governance,” and “transport decarbonisation.”

Based on this, a series of inclusion and exclusion criteria was employed to determine the relevance of the selected literature. Peer-reviewed publications were generally prioritised along with policy reports and institutional publications from multinational organisations, favouring studies published more recently. This approach was taken to better reflect current developments around governance and sustainability transitions. It also synthesized foundational literature to develop a robust theoretical foundation that was used to construct ideas of institutional theory and state capacity.

This enabled an identification of research gaps, such as limited institutional consideration in some decarbonising transport study designs, and aspects related to governance capacity to deliver policy. It also helped build a comprehensive analytical framework connecting the detailed empirical review of institutional constraints with policy effectiveness and sustainability outcomes across sectors (aspects that are commonly overlooked in existing work on the subject). This not only covers a cumulative relationship with prior knowledge but also shares clear research agenda and direction for action to increase institutional capacity in support of the green transition of the transport system.

8.1 The Search Strategy

This study incorporated a search strategy that aimed to increase the transparency and rigour with which aligned matched goals of research compared against salient state capacity and institutional reform in accordance with green transition elements within their respective transport domain. Time frames and search boundaries were well-defined by means of detailed inclusion/exclusion criteria aligned with best practices in systematic reviews of the literature (Wang and Chugh 2014; Del Vecchio et al. 2022). These criteria were aimed at ensuring the rigorous relevance, quality and thematic appropriateness of the literature chosen.

The first phase involved the identification of appropriate databases and search terms. Literature was based on the years between 2019 and 2026 to capture contemporary developments in environmental governance, transport decarbonisation and institutional reform. We chose Scopus and Google Scholar as databases commonly utilized to capture peer-reviewed academic publications and policy-relevant research. Keywords to scaffold this search process included, inter alia: "state capacity"; "institutional reform"; "green transition" or "environmental governance"; and "transport decarbonisation". These search terms were used in the titles, abstracts and author-defined keywords of articles to ensure that all relevant studies were included. The scoping search identified a broad and diverse set of publications spanning the fields of governance, policy and sustainability studies.

Exclusion criteria regarding publication type, language, period and thematic relevance were consistently applied following the methodological criteria (Kauppi et al. 2018; Nguyen et al. 2018). Scoping review included only peer-reviewed publications, reports from reputable international organisations and high-quality conference papers, excluding non-academic sources and related studies. Furthermore, only articles in English were included in the analysis to ensure consistency. We excluded studies without constructs specifically framed for their study on governance, institutional capacity or transport-related environmental policy to maintain conceptual consistency with this study.

Moreover, Boolean operators (e.g., AND, OR) and database-specific filters were also utilised to narrow down the dataset further in terms of specificity and duplicate studies. This two-step cleaning process enabled the discarding and management of duplicates or studies without enough relevance in a final sample thematically homogenous, coherent and methodologically sound. Through an iterative screening of articles, this process ensured we retained only studies specifically focused on the institutional dynamics, regulatory frameworks and governance challenges pertaining to the green transition within our first set of articles.

Therefore, a stringent search strategy was followed to develop an exhaustive and concise materials base together with an in-depth literature synthesis on state capacity and transport sector reform. Such a category-based structure, therefore, sufficiently carved out the theoretical developments and analytical insights, as well as the practical lessons learned from policy implementation, for an empirically justified and analytically compact discussion of governance challenges to and pathways for reforming the green transition (Fig. 1).

9 Results

9.1 Geographical Distribution of Authorship

The geographical distribution of the chosen studies shown in Fig. 2 provides highly pertinent insights regarding the state of research on global state capacity and green transition within transport/logistics and related sectors. With the evidence from the UK (6) and Italy (5) as the most contributing countries, as well as the United States (3), the dataset exhibits a strong geographical authorship from developed economies. Other developed countries with many studies are Australia, China, Finland, France, Germany and Sweden (two studies each), while there is one publication from Turkey. This distribution shows a clear Global North bias in academic debate on institutional capability and green transition governance. The top-line performance of European countries in particular reflects the strong policy response to sustainability that has driven ambitious decarbonization goals and targeted regulatory frameworks, which provide increasingly tight price signals through large-scale investments into green infrastructure. Strong institutional competencies likely fuel the capacity for both policy experimentation and research that locks in administration visibility within bodies of literature.

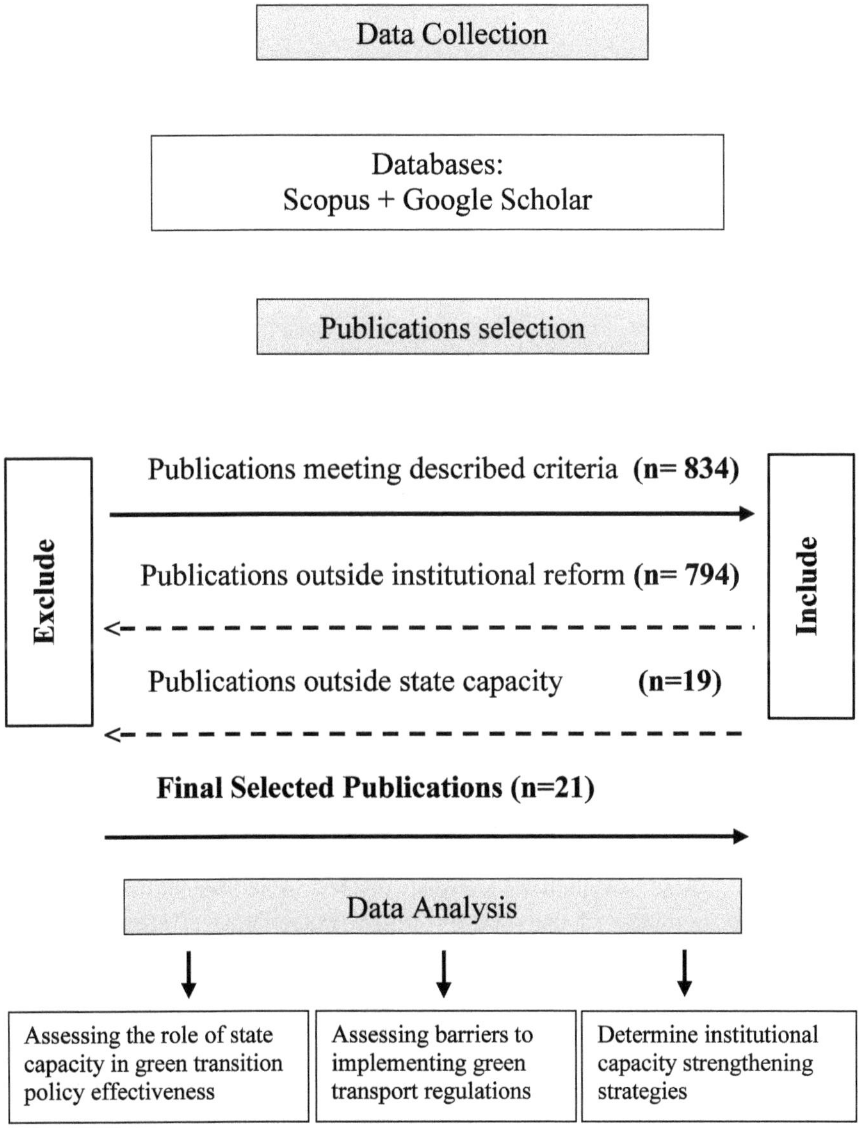

Fig. 1 Research design. *Source* Literature Analysis

In contrast, the relative under-representation from developing and emerging economies (e.g. Ghana, Nigeria, Morocco, Pakistan, Uzbekistan) indicates a significant knowledge gap. Although they encounter specific barriers associated with transport emission, infrastructure shortfalls and institutional capacity constraints, these regions are under-researched in the empirical literature as well as policy. Such an imbalance may relate to less research funding, weaker academic–policy links and

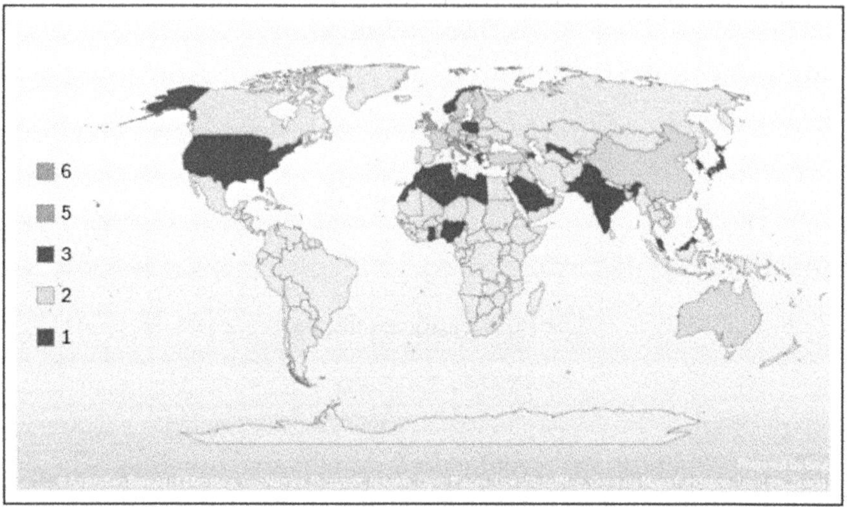

Fig. 2 Geographical distribution of authorship. *Source* Literature Analysis

limited data availability in these contexts. It also highlights its distribution, showing that context-specific institutional challenges in developing countries, such as governance gaps, corruption risks and shortfalls to public outcomes, are under-researched in the literature. This prompts questions of whether findings are largely concluded from developed societies, where institutional frameworks are comparatively stable and resource endowed. Significantly, the emerging diversity of research contexts was also an observation highlighted with contributions from regions such as the Middle East (Saudi Arabia), North Africa (Algeria, Libya) and Asia (India, Japan, South Korea). These contributions are relatively few in number, but they indicate an increasing awareness of the need for critical examination of green transition governance in ways that go beyond normative Western contexts.

9.2 Number and Year of Publications

Figure 3 shows the distribution over time of publications related to state capacity and green transition, especially in the transport and related sectors, since 2019 (up until 2026). A distinct pattern emerges showing a steady increase, followed by some fluctuations and finally an upturn in research output. During the early phase (2019–2020), the number of publications was still very low and stable, with only one (1) publication per year. This indicates that around this time, research on the institutional aspects of the green transition in the transport sector was still growing and had not yet received significant attention in academia. The sparse output might also reflect a

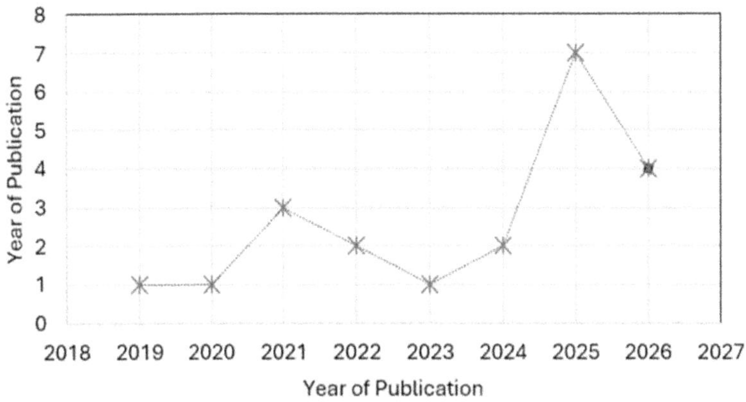

Fig. 3 Number and year of publications. *Source* Literature Analysis

stronger initial interest in understanding the technological and economic dimensions of decarbonisation than in understanding governance and state capacity.

2021 stands out as a year with three (3) publications, pointing towards an increasing academic interest in the governance dimensions of sustainability transitions. This momentum, however, is not maintained consistently as the output decreases to two (2) publications in 2022 and one (1) in 2023. This could reflect shifting research priorities or delays in the academic publishing process. They are on an upward trend for 2024 and beyond. This number increases to two (2) and follows with a steep peak of seven (7) publications in 2025, before subsequently going down to four (4) across the years 2025–2026. The jump in 2025 is more pronounced and likely connected to heightened attention among policymakers worldwide on the topics of climate governance, net-zero commitments and institutional reform. The discounting of city-levelled climate actions as potential solutions for the massive traffic accident has also occurred in parallel with increasing international talk around stricter regulatory frameworks, more green financing and sectoral programs on sustainability, for example, through the transport sector.

Overall, suggestions are emerging for a more focused and collective treatment of state capacity and institutional reform as one important dimension to inclusion across each part of the green transition. The recent uptick in publications reflects a larger shift among scholars away from technical fixes to governance, implementation challenges, and the efficiency of institutions. However, these differences in publication output suggest that this area of study has not delivered quite the same kind of sustained productivity equivalent to more established regions. It highlights the value of continued investment in academic study that aims to build both theory and evidence, especially where less attention has been paid, such as developing economies or topic-specific governance issues within transport.

10 Summary of Key Findings

See Table 1.

11 Thematic Synthesis of Results

The selected literature review illustrates the multiple overlapping channels through which the motives of state capacity and institutional reform produce a green transition in transport. They highlight the strong role of governance capability, persistence in structural barriers and dynamic institutional approaches.

12 The Role of the State in Policy Effectiveness

One common thread in the literature is the importance of state capacity for successful policies for the green transition. Table 1 shows evidence both cross-nationally and across regions of the value of sound governance systems, such as comprehensive regulatory frameworks, administrative efficiencies and technical capacity, so that policy objectives become deliverable results. For instance, Andrinos et al. (2026) emphasize that port cities' sustainable transition is contingent on integrated governance and skills systems, while Mazzocchi et al. (2025) stress that clearly defined policy frameworks are essential to promote ports' function as energy hubs.

Zhang et al. (2020) and Khaleel and Yusupov (2025) also point out that transitions in energy and transport at scale require highly coordinated policies, financial systems, and infrastructural arrangements. These findings complement the assertions that state capacity is not just a residual state, as some researchers argue, but governs policy success connected to the regulatory design, implementation, and enforcement process. Weak institutional capacity is a serious constraint on effective policy in the transport sector, where decarbonisation depends on cross-sectoral integration.

13 Barriers to Implementation at the Institutional and Structural Level

The literature highlights several entrenched barriers that limit the potential uptake of green transport mandates. Collaboration by improving inter-agency communication is a persistent challenge, with overlapping mandates of agencies and inconsistent signals for policy. Siddika et al. (2025) and Nilsson et al. (2020) found that policies have low coherence and general transitional efficiency; regulatory approaches were found to be extremely disjointed.

Table 1 Summary of findings

Author(s)	Role of state capacity in green transition policy effectiveness	Barriers to implementing green transport regulations	Institutional capacity strengthening strategies
Andrinos et al. (2026)	Governance systems and skills ecosystems are central to enabling sustainable transitions in port city regions	Skills gaps, weak training systems, and socio-institutional misalignment	Integrated skills systems, VET reform, and multi-stakeholder (quadruple helix) collaboration
Mazzocchi et al. (2025)	Strong regulatory and policy frameworks enhance ports' role as energy hubs	Fragmented policy integration and uneven territorial development	Development of sustainability indices and targeted policy reforms
Khaleel and Yusupov (2025)	Coherent policy, finance, and infrastructure systems are essential for transition effectiveness	Policy uncertainty, infrastructure gaps, and financing constraints	Public–private partnerships, green finance mechanisms, and integrated policy frameworks
Pode and Diouf (2026)	Government incentives and regulations drive EV adoption and transport decarbonisation	Gap between policy design and actual adoption, technological uncertainty	Subsidies, regulatory incentives, and technological advancement support
Siddika et al. (2025)	Integrated policy strategies enhance innovation and renewable energy transitions	Lack of coordination across policy domains and fragmented strategies	Comprehensive regulatory integration and strategic planning
Sackitey et al. (2025)	Policy instruments (e.g., environmental taxation) influence clean energy adoption	Short-term economic costs, uneven policy impacts, and trade uncertainties	Complementary subsidies, investment in innovation, and infrastructure development
El Hafdaoui et al. (2025)	Strong institutional frameworks enable renewable energy expansion and transport electrification	Weak regulatory effectiveness, grid inefficiencies, and investment barriers	Regulatory reform, infrastructure modernisation, and public–private partnerships
Di Vaio and Ali (2025)	Governance, accountability, and institutional frameworks shape decarbonisation outcomes	Weak ESG integration and limited institutional coordination	Strengthening accountability systems and aligning regulatory frameworks
Aslam et al. (2026)	Effective governance supports net-zero transitions across sectors, including transport	Behavioural, financial, and geopolitical constraints	Integration of digital tools, circular economy practices, and stakeholder engagement

(continued)

Table 1 (continued)

Author(s)	Role of state capacity in green transition policy effectiveness	Barriers to implementing green transport regulations	Institutional capacity strengthening strategies
Oyenuga et al. (2025)	Technological governance enhances environmental monitoring and policy execution	Infrastructure limitations and technological adoption gaps	Use of AI, IoT, and data-driven governance systems
Gordon et al. (2023)	Policy alignment and governance coordination are critical for hydrogen-based transport transitions	Socio-technical barriers, public resistance, and policy misalignment	Integrated socio-technical frameworks and stakeholder engagement
Nilsson et al. (2020)	Strong governance capacity is essential for industrial and transport decarbonisation	Weak transition governance and a lack of capacity-building focus	Policy coherence, capacity building, and creation of green markets
Cheshmehzangi and Chen (2021)	Governance structures and policy planning drive national decarbonisation strategies	Institutional inefficiencies and regional disparities	Long-term planning, regional coordination, and policy alignment
Newell (2021)	Political economy and power structures shape transition effectiveness	Power imbalances and institutional resistance	Institutional restructuring and inclusive governance approaches

Source Literature Analysis

Additionally, financial and infrastructure limitations are other hurdles. Khaleel and Yusupov (2025) cite access to green finance and infrastructure as barriers, whereas El Hafdaoui et al. (2025) point to grid system inefficiencies and regulatory frameworks. Social resistance, behavioural barriers, and uncertainty about newly developed technology are some additional factors that contribute to the complexity of implementation processes (Gordon et al. 2023; Aslam et al. 2026). These obstacles are interdependent and grounded in progressively deeper systemic governance structure failures. Weak institutional capacity makes technical and financial obstacles more challenging to surmount, while policy uncertainty stifles investment. This further supports the requirement for additional adversarial and all-encompassing approaches that consider structural and institutional impediments simultaneously.

14 Disparities Between Policy Design and Execution

The other is the disjunction between policy aspiration and reality. While many governments have developed progressive policies for transport decarbonisation, from subsidies for integration of electric vehicles and renewables to carbon credit schemes, the actual implementation has fallen short in most cases. This gap is well documented in the literature relating to electric vehicle adoption and clearly supportive policies (Pode and Diouf 2026). This implementation gap is primarily attributable to weak enforcement mechanisms, inadequate administrative capacity and a shortage of technical expertise. Pode and Diouf (2026) show that a public policy impact not only importantly rely on a well-designed regulatory setup, but also on the institutional context in which it is rendered functional. Well-designed policies can achieve their objectives only if they are supplemented with sufficient monitoring, enforcement and coordination mechanisms.

15 Institutional Reform and Capacity-Building as Enablers

Literature shows that when it comes to overcoming governance challenges and facilitating the green transition, institutional reform and capacity-building are crucial. Intensive capacity-building programmes (skills development, training and knowledge transfer) have been identified in policy dialogue as one of the key elements to improve institutional performance. Andrinos et al. (2026) stress the crucial role of vocational education and training systems in enabling sustainable transitions, while Oyenuga et al. (2025) argue that digital technologies are promising in terms of enhancing governance and monitoring capacities.

Regulatory reform is another key focus area where research calls for greater coherence in policy, less bureaucratic complexity and clearer accountability of methods. El Hafdaoui et al. (2025) and Nilsson et al. (2021) point to the need for modernised regulatory frameworks and stronger coordination among government institutions in support of transitions to renewable energy and transport. In addition, international collaboration and multi-sectoral ventures are emphasised as key modalities to leverage resources, exchange knowledge and strengthen institutional capacity (Khaleel and Yusupov 2025).

16 The Role of Governance Innovation and Multi-Stakeholder Collaboration

A final transversal theme is the increasing importance of governance innovation and collaborative approaches in managing the complexities associated with the green transition. Such developments emphasize the importance of resilient governance frameworks that can adapt to change, such is technological, economic, and ecological. This involves incorporating digital facilities, such as artificial intelligence and data analytics mechanisms, to enhance policy tracking and decision-making (Oyenuga et al. 2025).

In addition, multi-stakeholder cooperation among governmental, private, academic and non-governmental actors has become universally accepted as key to the effective design and implementation of policies. Andrinos et al. (2026) call for quadruple helix developments to facilitate inclusive and sustainable transitions, while Newell (2021) highlights how power dynamics need to be addressed if equality is to be reflected in the governance process. Overall, the thematic analysis suggests that the frame for the success of the transport sector green transition is ultimately established through interaction among patterns of state capability, institutions and governance. Technological innovation and policy ambition are clearly critical, but any approach will need a stronger institutional underpinning to succeed. Therefore, successfully overcoming these barriers through institutional reform, capacity building, and collaborative governance approaches is necessary to see effective and sustainable transport decarbonisation.

17 Further Research

Although this chapter contributes to existing knowledge on the link between state capacity and institutional reform in advancing the green transition in the transport sector, there are still several avenues for future research. First, it remains critical to run longitudinal and comparative studies that examine how institutional reforms diverge over time and in various governance contexts (preferably across developed/developing economies). These studies would shed more light on the causal linkage between institutional capacity and policy success. Second, future research should be informed by empirical case studies and mixed method approaches that can evaluate the real-world effectiveness of specific modes of transport decarbonisation, such as regulatory interventions, enforcement mechanisms or activities intended to build capacity. Third, there needs to be more attention on the role of emerging technologies, such as digital monitoring systems and data-driven regulatory tools in improving performance by institutions, as well as transparency. For instance, research must examine the political economy of institutional reform, including power relations, interests of stakeholders and resistance to change, which often makes or breaks governance transformations during the green transition.

18 Study Implications

18.1 Theoretical Implications

This chapter advances theory at the connection of institutional analysis, environmental governance and sustainability transitions by foregrounding state capacity as a determinant of green transition outcomes in transport sectors. First, it contributes to institutional theory by illustrating that the effectiveness of regulative, normative and cognitive pillars is contingent on underlying conditions in state capacity. Although the prevalence of isomorphic pressures and legitimacy-seeking behaviour are traditional staples in explaining stability and change in institutional theory, this chapter highlights that without proper administrative, technical or enforcement capacities, institutional arrangements can be more symbolic than functional. This sharpens current theoretical assumptions by connecting institutional effectiveness directly to capacity constraints.

Second, the chapter bridges institutional theory with socio-technical transition frameworks through the co-evolution of governance structures and technological systems. Existing transition literature frequently emphasises innovation systems and niche–regime interactions (Geels et al. 2017), while arguably under-theorising the role of public institutions in either enabling or constraining such dynamics. The chapter reveals a more nuanced explanation of why similar technology solutions achieve disparate outcomes throughout governance contexts by considering the role of state capacity in this conversation.

Third, the analysis advances theories of policy implementation by showing how institutional fragmentation and weak coordination and governance risks mediate policy design to outcome linkages. These questions linear approaches to policy making and implementation, while favouring more elaborate views of multi-level governance perspectives that frame comparative approaches to implementation, where web engagements and powerful coalitions drive processes.

Finally, the chapter promotes the notion of institutional reform as an adaptive and iterative process enhanced by policy learning and capacity-building. It underscores the theoretical importance of dynamic governance approaches, emphasizing that institutional resilience and flexibility are important components for managing uncertainties of simple and green transition. Combined, these contributions place state capacity as a central variable when theorising the effectiveness of sustainable development and environmental policy.

18.2 Managerial/Policy Implications

This chapter presents new findings that will be of great significance to policymakers, regulators and public sector managers involved in promoting sustainable transport systems. Strategic investments in administrative capacity, technical know-how, and institutional coordination should be prioritised to enhance state capacity. Governments need to invest in integrated regulatory approaches that reduce fragmentation and are better aligned across transport, environment and energy sectors. Additionally, strengthening enforcement mechanisms and accountability systems is crucial in ensuring compliance with environmental standards and in reducing corruption risks.

Public institutions must have the right skillset to implement complex green transition policies, and they are not there yet; training initiatives (e.g. professional training, knowledge exchange programmes) that support the capacity-building of public institutions are critical. Policymakers can also draw on international cooperation to help them gain technical support and financing as well as learn from best practices in sustainable transport governance. Additionally, either public–private partnerships leverage innovation and resource mobilisation, or participatory governance approaches increase transparency and participation from stakeholders. For these reasons, long-term sustainability outcome institutional reform approaches need to be both well-contextualized and adaptable as well as clearly aligned to wider development goals.

19 Conclusion

This chapter has highlighted state capacity and institutional reform as two additional key conditions for a green transition in the transport system. It has been demonstrated that technological innovation and policy ambition are necessary but not sufficient without the presence of strong governance structures through which this can be executed and enforced successfully. Institutional constraints in the form of bureaucratic fragmentation, weak coordination, limited technical capacity and temptation to corruption still undermine progress in translating policy goals into real outcomes.

Through the lens of institutional theory, the chapter has shed light on how regulatory, normative and cognitive dimensions of governance affect the effectiveness of environmental policy. It has also shown that with regulatory performance, institutional capacity can be extremely successful by using policy learning, capacity-building, and collaboration across borders. It requires not only incremental policy amendments to achieve a sustainable and low-carbon transport system, but also a complete institutional restructuring. The former, which is the consummate governance reform that is a long-term investment in institutional development, thus becomes an imperative for realising the green transition.

References

Abram S, Atkins E, Dietzel A, Jenkins K, Kiamba L, Kirshner J, … Santos Ayllón LM (2022) Just Transition: a whole-systems approach to decarbonisation. Climate Policy 22(8):1033–1049

Andrews M, Pritchett L, Woolcock M (2017) Building state capability: Evidence, analysis, action. Oxford University Press. https://doi.org/10.1093/acprof:oso/9780198747482.001.0001
Andrinos M, Greco L, Menelaou A, Metaxas T, Nikolaidis E, Psatha E, Sirakoulis K (2026) Port city regions in transition: skills, inclusion, and innovative VET pathways for the twin transformation. Sustainability 18(5):2538
Aslam T, Omerbegović NS, Sher F, Bernardo P, Khan MK, Benettayeb A, ... Razmkhah O (2026) Challenges, barriers and solutions in achieving net zero. In Net ZERO: foundations, technologies and strategic innovation, pp 623–706. Elsevier
Barrett BF, Ohta H, Trencher G (2022) Prospects for acceleration of socio-technical transitions for deep decarbonization. In: The routledge handbook of sustainable cities and landscapes in the pacific rim, pp 400–414. Routledge
Bouckaert G, Peters BG, Verhoest K (2022) Policy design for policy coordination. In: Research handbook of policy design, pp 351–370. Edward Elgar Publishing
Chasek PS, Downie DL (2020) Global environmental politics. Routledge
Chausson A, Welden EA, Melanidis MS, Gray E, Hirons M, Seddon N (2023) Going beyond market-based mechanisms to finance nature-based solutions and foster sustainable futures. PLOS Climate 2(4):e0000169
Cheshmehzangi A, Chen H (2021) China's sustainability transitions: low carbon and climate-resilient plan for carbon neutral 2060. Springer Nature
Chikwere D, Chibaro M, Govere E (2024) Leveraging alternative energy sources to enhance sustainable transportation. In: Contemporary solutions for sustainable transportation practices, pp 94–127. IGI Global Scientific Publishing
Chikwere D, Mbudaya J, Chibaro M (2025) exploring green energy alternatives for a sustainable future: opportunities and challenges for developing countries. Adaptive strategies for green economy and sustainability policies, 183–206
Del Vecchio P, Mele G, Siachou E, Schito G (2022) A structured literature review on Big Data for customer relationship management (CRM): toward a future agenda in international marketing. Int Mark Rev 39(5):1069–1092
Di Vaio A, Ali T (2025) Accounting for carbon neutrality: corporate accountability in the hydrogen economy. Routledge
DiMaggio PJ, Powell WW (1983) The iron cage revisited: Institutional isomorphism and collective rationality in organizational fields. Am Sociol Rev 48(2):147–160
Dubash NK (2021) Varieties of climate governance: the emergence and functioning of climate institutions. Environ Politics 30(sup1):1–25
El Hafdaoui H, Khallaayoun A, Al-Majeed S (2025) Renewable energies in Morocco: a comprehensive review and analysis of current status, policy framework, and prospective potential. Energy Conv Manag X 26:100967
Fathoni F, Lovett JC, Rifansha MM (2025) A multi-level perspective on transition to renewable energy in the Indonesian transport sector. Energies 18(21):5723
Fink A (2019) Conducting research literature reviews: from the internet to paper. Sage publications
Fukuyama F (2013) What is governance? Governance 26(3):347–368
Giwah ML, Nwokediegwu ZS, Etukudoh EA, Gbabo EY (2020) Sustainable energy transition framework for emerging economies: Policy pathways and implementation gaps. Int J Multidisciplinary Evolut Res 1(1):1–6
Gordon JA, Balta-Ozkan N, Nabavi SA (2023) Socio-technical barriers to domestic hydrogen futures: Repurposing pipelines, policies, and public perceptions. Appl Energy 336:120850
Hale T (2020) Transnational actors and transnational governance in global environmental politics. Annu Rev Polit Sci 23(1):203–220
Intergovernmental Panel on Climate Change (IPCC) (2022) Climate change 2022: mitigation of climate change. Cambridge University Press
Jepperson RL, Meyer JW (2021) Institutional theory: the cultural construction of organizations, states, and identities. Cambridge University Press

Kamrul Ahsan AHM (2023) Inter-agency Coordination. Global encyclopedia of public administration, public policy, and governance. Springer International Publishing, Cham, pp 6887–6891
Kauppi K, Salmi A, You W (2018) Sourcing from Africa: a systematic review and a research agenda. Int J Manag Rev 20(2):627–650
Khaleel M, Yusupov Z (2025) Advancing sustainable energy transitions: Insights on finance, policy, infrastructure, and demand-side integration. Unconventional Resources, 100274
Khan M, Andreoni A, Roy P (2019) Anti-corruption in adverse contexts: strategies for improving implementation. ACE Working paper series, (013)
Kotilainen K, Aalto P, Valta J, Rautiainen A, Kojo M, Sovacool BK (2019) From path dependence to policy mixes for Nordic electric mobility: lessons for accelerating future transport transitions. Policy Sci 52(4):573–600
Leeuw FL, Rist RC, Sonnichsen RC (Eds) (2020) Can governments learn?: comparative perspectives on evaluation and organizational learning. Routledge
Leong C, Howlett M (2022) Policy learning, policy failure, and the mitigation of policy risks: Re-thinking the lessons of policy success and failure. Administration Soc 54(7):1379–1401
Marzouk OA (2025) Summary of the 2023 report of TCEP (tracking clean energy progress) by the International Energy Agency (IEA), and proposed process for computing a single aggregate rating. In: E3S web of conferences (Vol 601, p 00048). EDP Sciences
Mazzocchi P, Capoani L, Carlucci F, Trincone B (2025) European ports as energy hubs: a sustainability index to assess territorial development. Business strategy and the environment
Meckling J, Nahm J (2019) The politics of technology bans: industrial policy competition and green goals for the auto industry. Energy Policy 126:470–479
Meredith J (1993) Theory building through conceptual methods. Int J Oper Prod Manag 13(5):3–11
Michaelowa K, Michaelowa A (2019) Transnational climate governance initiatives: designed for effective climate change mitigation? In: The comparative politics of transnational climate governance, pp 146–178. Routledge
Munuhwa S (2024) Driving towards emission-free urban centers: exploring Glasgow's sustainable road transport. In: Contemporary solutions for sustainable transportation practices, pp 186–206. IGI Global Scientific Publishing
Newell P (2021) Power shift: the global political economy of energy transitions. Cambridge University Press
Nguyen T, Li ZHOU, Spiegler V, Ieromonachou P, Lin Y (2018) Big data analytics in supply chain management: a state-of-the-art literature review. Comput Oper Res 98:254–264
Nilsson LJ, Åhman M, Bauer F, Ericsson K, Johansson B, van Sluisveld M, … Schiro D (2020) A European industrial development policy for prosperity and zero emissions
North DC (2025) Institutions and the performance of economies over time. In: Handbook of new institutional economics, pp 25–35. Cham: Springer Nature Switzerland
Oyenuga MO, Singh R, Apata SB, Khan S, Kumar V (2025) Smart technologies for climate change and net zero policies: Practical approaches towards sustainability. Springer, Springer Nature Switzerland, Imprint
Pandey N, De Coninck H, Sagar AD (2022) Beyond technology transfer: innovation cooperation to advance sustainable development in developing countries. Wiley Interdisciplinary Rev Energy Environ 11(2):e422
Pode R, Diouf B (2026) Decarbonization of road transportation: the electric vehicle (EV) transition
Radtke J (2025) Understanding the complexity of governing energy transitions: introducing an integrated approach of policy and transition perspectives. Environ Policy Gov 35(4):595–614
Rodrik D (2021) Why does globalization fuel populism? Economics, culture, and the rise of right-wing populism. Ann Rev Econ 13:133–170
Sackitey GM, Oddei JA, Boğa S, Fumey MP, Bekun FV (2025) Green growth in Scandinavia: assessing the role of environmental taxation, green innovation, trade, and urbanization in advancing clean energy consumption. Economics 13(3):27–52
Scott WR (2014) Institutions and organizations: Ideas, interests, and identities (4th ed.). Sage Publications

Siddika A, Sarwar A, Tareq MA, Rahman HM, Haron R (2025) Green innovation and renewable energy sources policies: a systematic and integrative review. Sustain Futures 10:100870
Sovacool BK, Axsen J, Delina LL, Boudet HS, Rai V, Sidortsov R, … Galvin R (2022) Towards codes of practice for navigating the academic peer review process. Energy Res Soc Sci 89:102675
Stern N, Stiglitz J, Taylor C (2022) The economics of immense risk, urgent action and radical change: towards new approaches to the economics of climate change. J Econ Methodol 29(3):181–216
Transparency International (2022) Corruption perceptions index. URL: https://www.transparency.org/cpi2021
United Nations Environment Programme (UNEP) (2023) Emissions Gap Report 2023. UNEP
Wang CL, Chugh H (2014) Entrepreneurial learning: past research and future challenges. Int J Manag Rev 16:24–61
Williams MJ (2021) Beyond state capacity: bureaucratic performance, policy implementation and reform. J Inst Econ 17(2):339–357

Effect of Finance on Sustainable Transport Infrastructure

Wilbert Manyanga, Munyaradzi Chibaro, Tendai Manyanga, James Kanyepe, Maxwell Chufama, and Masimba Elvis Manyangara

Abstract This study focused on the effect of finance on transport infrastructure in Zimbabwe. Transport infrastructure is crucial for sustainability and societal well-being. Transport infrastructure supports trade, consumption and economic growth. Transport infrastructure in most developing countries is deteriorating with minimum investment to replace or maintain such transport infrastructure due to lack of funding. Financing the transport infrastructure is the enabler for sustainability of transport infrastructure. While much of the research on finance on transport infrastructure has been conducted in more developed markets, focusing on Zimbabwe provides new evidence in a different economic environment coupled with financial challenges. The study examines the effect of public funding, private investment, development finance and green bonds on sustainable transport infrastructure. The sustainable development theory anchored the study. Positivist paradigm, cross-sectional survey and structured self-administered questionnaire were implemented. Data were gathered from 307 respondents in Bulawayo, Zimbabwe. The respondents were selected through convenience sampling method. Hypotheses were validated using structural equation modelling. Results reflect that public funding, private investment, development finance and green bonds positively influence sustainable transport infrastructure. The study contributes on theoretical knowledge by bringing in recent financing models such as green bonds in the finance and sustainable transport infrastructure discourse. The study offers new paradigms for understanding how governments may structure different financing models to create sustainable transport infrastructure. Sustainable

W. Manyanga (✉) · M. Chufama
Department of Marketing, Logistics and Sport Management, Faculty of Commerce, Human Sciences and Education, Namibia University of Science and Technology, Windhoek, Namibia
e-mail: wmanyanga@nust.na

M. Chibaro · J. Kanyepe
Department of Logistics and Supply Chain Management, University of Botswana, Gaborone, Botswana

T. Manyanga
Ministry of Primary and Secondary Education, Harare, Zimbabwe

M. E. Manyangara
Department of Marketing, Chinhoyi University of Technology, Chinhoyi, Zimbabwe

J. Kanyepe and I. Chirisa (eds.), *Advancing the Green Agenda in the Transport and Logistics Sector in Global South Countries*, World Sustainability Series,
https://doi.org/10.1007/978-3-032-31374-4_18

transport infrastructure supports trade and economic growth. Hence, governments are encouraged to consider public funding, private investment, development finance and green bonds to enhance sustainable transport infrastructure.

Keywords Development finance · Finance · Green bonds · Private investment · Public funding · Transport infrastructure

1 Introduction

Globally, transport infrastructure has become an enabler in trade and economic growth (Yang et al. 2025). It is also the access for social integration and the link between sustainable production and sustainable consumption in a circular economy (Bakkar et al. 2025). Transport infrastructure supports all sectors of the global economy; however, the traditional transport systems have been criticised for over reliance on fossil fuels and resulting in urban congestion, social inequalities and pollution (Suleiman et al. 2025). Hence, the growth in the phenomenon called sustainable transport infrastructure among transport stakeholders to create a more inclusive, efficient, environmentally responsible, reliable and viable global transport system (Zheng et al. 2025). Therefore, governments are encouraged to invest in public transport infrastructure that are smart and resilient to reduce environmental damage but enhancing efficiency and inclusive (Bakkar et al. 2025). Nevertheless, the crucial role of sustainable transport infrastructure, finance is the main challenge in creating more sustainable and reliable transport infrastructure globally but mainly the impact is much felt in developing economies (Tanko et al. 2025).

Transport infrastructure require a lot of investment in capital and its long term in nature which require stable and reliable funding due to risk involved (Suleiman et al. 2025). Most private sector shun such investment if it does not yield return in the short run due to unforeseen risks associated with such projects such as policy ambiguity (Yang et al. 2025). There is need for government commitment for such high capital projects for them to be successfully implemented (Zheng et al. 2025). There is need to embrace sustainable transport infrastructure that are climate proof, environmentally friendly and linked to emerging technologies (Bakkar et al. 2025). Therefore, structure, accessibility and availability of financing options determine if such sustainable transport developments will start, finished and maintained with success (Yang et al. 2025). Finance is the key success factor in this investment issue on sustainable transport infrastructure as it affects the quality, scale and how sustainable will be the transport infrastructure (Suleiman et al. 2025). Hence this study seeks to examine the effect of finance on sustainable transport infrastructure in the context of a developing country.

2 Literature Review

2.1 Sustainable Development Theory

This theory suggest that development should address the current needs and preserving for the next generations to provide for their needs (Bahman 2025). The theory combines environmentally, social and economic sustainability as key pillars of sustainable development (Tanko et al. 2025). This theory supports this study since it clearly shows that sustainable transport infrastructure enhances environmental protection and social inclusion (Anam et al. 2025). In addition, it shows that sustainable transport infrastructure enhances mobility and economic development (Vorwerg et al. 2025). The theory postulate supports the fact that finance is the enabler of sustainable transport infrastructure (Dindar 2025). This implies that public funding, private investment, development finance and green bonds are instruments that enhance sustainable transport infrastructure that is friendly to the environment, inclusive and resilient (Al-Zubairi et al. 2025).

2.2 Financing

The world over, governments has come to the knowledge that public funding is not enough to finance the huge capital requirements to build sustainable transport infrastructure (Vassallo and Garrido 2023). The gap for financing infrastructure is growing fast in relation to increase in demand for transport especially in underdeveloped countries due to pressure in rural to urban migration, increase in population and industrialisation (Al-Zubairi et al. 2025). This gave rise to the need for different financing options to be considered such as public funding, private investment, development finance and green bonds to enhance sustainable transport infrastructure (Dai et al. 2023). These financing options are regarded as viable options to secure funding for sustainable transport infrastructure (Tanko et al. 2025). Development finance is considered as an alternative global source for funding for low carbon transport systems (Vassallo and Garrido 2023). Whereas green bonds are regarded as new options for financing environmentally sustainable infrastructure (Chupin et al. 2024). Also, private sector investment is considered as a complementary option to public funding in enhancing transport infrastructure (Anam et al. 2025). Hence, sustainable transport infrastructure is considered as an outcome of proper financing since availability of funding can enhance construction of sustainable transport infrastructure (Bahman 2025). Similarly, lack of structured financing options result in unfinished projects, failure to start projects, lack or inadequate infrastructure maintenance and failure to construct environmentally friendly transport infrastructure (Wang et al. 2025). Lack of adequate funding may result in low cost and short-term transport systems compared to more sustainable but costly long-term options (Dindar 2025).

Funding availability influence the rate and pace of transition to modern sustainable transport systems (Vassallo and Garrido 2023).

Despite that finance is considered as one of the key enablers of sustainable transport infrastructure, current literature show contradictions on the effect of various financing options (Chupin et al. 2024; Vassallo and Garrido 2023). Other scholars argue that increase in availability of finance enhance infrastructure and sustainability of the environment are key (Bahman 2025; Dai et al. 2023). While others postulate that there is need for good governance and proper allocation and accountability of funding to get the desired outcomes (Al-Zubairi et al. 2025; Dindar 2025). Most studies were done in developed countries and create a gap in developing countries. Other studies support the need for stakeholder engagement, good policies and innovation, in addition to availability of funding to attain sustainable transport infrastructure (Anam et al. 2025; Tanko et al. 2025). This study test if finance is the major driver or it depends on the availability of other broader socio-economic and institutional capacity and competency.

2.3 Public Funding

Finance options such as public funding are the main sources of finance in many developing countries for transport infrastructure (Dai et al. 2023). This option is wholly funded by government through the fiscal budget and various levies such as tollgates among others (Chupin et al. 2024). Taxpayers' money is used to fund the national projects infrastructure (Zhu et al. 2023). Public funding remains the main source of infrastructure development because of the capital intensive of such projects, hence most private sector shuns such projects if the government is not committed to pay the private in a sustainable and consistent manner (Tanko et al. 2025). Public funding supports infrastructure projects that are capital intensive which no private sector is willing to fund due to low returns in the short run but high capital outlay in the initial stages (Pamidimukkala et al. 2023).

Through public funding government can fund long term projects which have little return but have greater social good and environmental sustainability outcomes (Feldman 2023). This provides inclusive, quality and accessible transport infrastructure services through public funding (Al-Zubairi et al. 2025). Sustainable transport infrastructure funded by public funding is mainly targeted at helping the poor, disabled people, old people, women and those from marginalised communities (Anam et al. 2025). Government may utilise public funding to ensure inclusivity and universal access to transport infrastructure instead of focusing on urban areas only (Giles-Corti et al. 2022). Public funding is therefore supporting provision of transport infrastructure in areas that are shun by the private funders (Wang et al. 2025). This is crucial for developing economies where the majority rely on cheaper and inclusive public transportation for day-to-day movement (Chupin et al. 2024).

Public funding is key in ensuring that infrastructure is maintained for sustainability (Zhu et al. 2023). Public funding may also be used modernise public roads and transport facilities so that they are resilient to climate change and improve road safety and signage for both the public and motorists (Al-Zubairi et al., 2025). Public funding ensure that public transport infrastructure remain in place and reduce the impact on the environment and enhance the sustainability of transport system (Carbajo 2023). Through public funding, other funding options can be mobilised (Jelti et al. 2023). If government is seen as committed to transport infrastructure development, other players such as private sector and global financial institutions may also complement government efforts through structured financing options (Chupin et al. 2024). Also, public funding acts as a guarantee to other financial institutions and this may reduce reluctance by other financiers to complement government efforts in sustainable transport infrastructure (Tanko et al. 2025). The success of public funding as an enabler of sustainable transport infrastructure is hinged on fiscal discipline, government commitment, policy consistency and proper project management capabilities (Pamidimukkala et al. 2023).

Most developing countries are characterised by corruption, small budgets, poor project management and competing financial needs hence limit the impact of public funding on sustainable transport infrastructure (Giles-Corti et al. 2022). For public funding to fully support sustainable transport infrastructure, government should improve on management of public funding and focus on sustainable goals (Khurshid et al. 2023).

Public funding alone is criticised as inadequate because of limited national budget and poor funding allocation (Carbajo 2023; Pamidimukkala et al. 2023). Some scholars are of the opinion that over-reliance on public funding is the source of project delays and reduction in participation by the private sector in infrastructure development (Giles-Corti et al. 2022; Jelti et al. 2023). Others are of the opinion that public funding should be hinged on accountability, good governance and should be complemented with other financing options sustainability (Wang et al. 2025; Zhu et al. 2023).

2.4 Private Investment

One of the financing options that has been regarded as complementary to public funding is private investment (Nhat and Hoang 2025). The demand for sustainable transport infrastructure has increased recently due to the growing awareness on climate change and the need to attain sustainable development goals (Al-Zubairi et al. 2025). Private investment is regarded as funds that are raised by private organisations, investors and construction organisations for constructing or maintaining transport infrastructure (Motowidlak and Motowidlak 2023). Private investment may be used for environmentally friendly infrastructure and smart transport systems to ensure sustainable transport infrastructure (Bahman 2025). Private investment is a vital financing option to complement constrained government funding (Chupin et al.

2024). Private funders may easy pressure on the fiscus through provision of funding for more sustainable and reliable transport infrastructure and systems (Motowidlak and Motowidlak 2023). Transport infrastructure can be successful if sufficient funds are available, hence the private sector is roped in to provide capital and technology (Khurshid et al. 2023). This is vital in developing countries where the demand for public transport infrastructure is higher than the available public funding (Wang et al. 2025). Through private funds the government may complete certain projects and improve the transport systems and infrastructure to be more sustainable (Nhat and Hoang 2025).

Private investment contributes with new technology, innovation and efficient transport infrastructure and transport systems that are climate proof and affordable (Akomea-Frimpong et al. 2023). Through private investment corruption and bureaucracy may be minimised (Motowidlak and Motowidlak 2023). Private investors normally focus on controlling costs and improve on project delivery in the shortest possible time to improve returns on investment (Nhat and Hoang 2025). Also, the private sector is characterised by expertise, technical knowhow and bringing in new technology that may improve quality of transport infrastructure as well as sustainability (Anam et al. 2025). Most private sector investors can provide recent technology such as electric vehicles and smart transport management systems (Tan 2022). These technologies reduce carbon footprint, improve convenience for the public and reduce transportation cost (Ding and Liu 2023). Hence, the private investment improves on technology, operational efficiency and capital provision (Nhat and Hoang 2025). Private investors also enhance the commercial viability and long run management of sustainable transport infrastructure (Chupin et al. 2024). Private investors usually check project risk, returns and feasibility, hence improves on project planning and management (Bahman 2025). Hence, fosters financial sustainability of infrastructure in the long term (Jelti et al. 2023). Most successful sustainable transport projects are normally done by private sector such as toll roads and bult-operate and transfer projects. This reduces pressure on the fiscus and ensure more sustainable transport infrastructure (Tan 2022).

However, private sector funding may not always provide positive results as in some times what we call private investment is simply companies of government officials who may siphon public funds and appear as private investors (Wang et al. 2025). Such private sector players that are conniving with government officials normally causes corruption, favouritism and bureaucracy (Nhat and Hoang 2025). The private sector is normally motivated by profit, hence may charge high for construction of transport infrastructure (Yannis and Chaziris 2022). In addition, projects with low returns may be shunned by the private sector and may result in lack of funding in such projects (Anam et al. 2025). So, projects that are in rural areas and other low-income communities may not receive funding from private sector investors (Tan 2022). Private investment may need to be regulated since their interests are profit over environmental and social equality interests (Motowidlak and Motowidlak 2023). The success of private investment is hinged on provision of a clear road map on how government engage and contract them, good policies, transparency and strong institutional frameworks (Akomea-Frimpong et al. 2023). Hence, the need to jointly plan together between public and private sector in project management (Bahman 2025). The government represent the social and environment interests of the society and provide a conducive working environment (Nhat and Hoang 2025). The government provides policy consistency, risk sharing and incentives to private sector stakeholders so that the projects may be successfully implemented (Tan 2022). The private sector remains a complementary option to public funding and should not replace public funding (Abbasnejad et al. 2025). A balance between private and public funding should be maintained to ensure sustainable transport infrastructure (Jelti et al. 2023).

Private investment is considered a critical source of funding for infrastructure development due to expertise, efficiency and additional funding it brought (Bahman 2025; Ding and Liu 2023). Nevertheless, private investors are criticised for being profit oriented hence may shun less profitable infrastructure projects (Motowidlak and Motowidlak 2023; Nhat and Hoang 2025). Other scholars argue that private investment alone may not provide sustainability but needs to be complemented by public oversight and government regulation (Khurshid et al. 2023; Tan 2022).

2.5 *Development Finance*

Another important financing option for sustainable transport infrastructure development is development finance (Carbajo 2023). This is funding from development financial institution, multilateral banks, international donors and bilateral agencies (Motowidlak and Motowidlak 2023). These funders aim at providing social services to the communities through funding development initiatives such as road constructions that require large amount capital (Chupin et al. 2024). These funds may be used for climate resilient transport infrastructure (Al-Zubairi et al. 2025). Development finance acts as a bridge to cover for short falls in public funding since transport infrastructure require large amounts of capital (Bahman 2025). Development finance ensure that governments achieve their sustainability targets (Nhat and Hoang 2025). Most developing economies are characterised by huge domestic and foreign debt, limited national budgets and many public responsibilities to provide social services, hence development finance plays a critical role in closing the capital gap (Tan 2022). Huge capital outlay projects may be funded by development partners and they may offer loans, technical knowhow and structured financing options for sustainable transport infrastructure (Wang et al. 2025).

Development finance assist governments to modernise their infrastructure and enhance connectivity and mobility as well as implementing climate and environmental initiatives for sustainable transport infrastructure (Carbajo 2023). Hence both development targets and sustainable goals may be achieved for the greater good of society (Motowidlak and Motowidlak 2023). Development finance is more sustainability and inclusive orientation as well as to maintain quality on the project infrastructure (Anam et al. 2025). Development finance is not commercial but strives to offer inclusive, affordable and sustainable transport infrastructure to support the social and economic well-being of the country's development goals (Yannis and

Chaziris 2022). Development finance may support low emission transport systems, infrastructure for the disabled persons and climate proof infrastructure (Tan 2022). This ensures that transport infrastructure and facilities are suitable for current mobility needs and for next generations (Wang et al. 2025).

Development finance capacitate the government through policy consistency and transfer of technical knowhow (Chupin et al. 2024). They also provide with project management skills, advise and other supporting services that ensure that the projects are successfully implemented in a sustainable manner (Jelti et al. 2023). This ensure that investments done in infrastructure are secured and may sustain future generations (Motowidlak and Motowidlak 2023). Through development finance other capital may be secured based on the reputation of development finance partners (Bahman 2025). Similarly, through development finance, project risk is minimised, hence this may attract private investors to join in the projects (Nhat and Hoang 2025).

Nevertheless, the success of development finance depends on how well the government aligns itself in terms of its priorities, implementation and governance systems (Ding and Liu 2023). The government should remove all bureaucratic tendencies, improve on fund disbursements and strengthen institutional capacity for sustainable transportation infrastructure to be a reality through development finance (Anam et al. 2025). Also, the government should manage its debts in a sustainable way to enhance fiscus sustainability (Akomea-Frimpong et al. 2023). Development finance is a good option for infrastructure projects, however, there is need for good governance, good project management and policy consistency (Khurshid et al. 2023).

Development finance is regarded as an enabler of sustainable infrastructure development, especially in developing countries where access to capital is limited (Al-Zubairi et al. 2025; Bahman 2025). Scholars argue that over-dependency on development finance from international institutions create dependency syndrome, balloon debt burdens and may come with conditions that are not in line with the needs of the local country (Anam et al. 2025; Ding and Liu 2023). On the other hand, some scholars argue that development finance work well when it is complemented with good governance, effective utilisation of the development finance and institutional capacity (Khurshid et al. 2023; Yannis and Chaziris 2022).

2.6 Green Bonds

One of the finance options that is innovative and has been regarded as key in sustainability is green bonds (Zheng et al. 2025). These are debt securities that are put in the market by governments or other financial institutions to generate capital for targeted development projects that are perceived as environmentally friendly and sustainable (Ding and Liu 2023). Green bonds particularly fund projects such as low carbon transport systems and infrastructure (Motowidlak and Motowidlak 2023). Due to the increasingly awareness of climate change and sustainable development goals, green bonds are regarded as a financing option that provide funds for sustainable transport infrastructure (Al-Zubairi et al. 2025). Green bonds offer devoted and targeted financing for sustainable transport infrastructure (Chupin et al. 2024). Sustainable transport systems and infrastructure require huge capital to construct modernised and climate proof infrastructure and transport systems (Wang et al. 2025). Other sources of funding may neglect projects that support environmental initiatives, hence green bonds are structured to address that. Green bonds guarantee that resources are structured to transport infrastructure that are environmentally sustainable (Zheng et al. 2025). This will capacitate government and institutions to fund in transport systems that address climate and sustainable development goals (Nhat and Hoang 2025).

Green bonds assist in attracting huge capital for massive projects from investors that are environmentally conscious (Dai et al. 2023). Globally there is a wave by investors to fund climate change and environmentally friendly projects for greener global economy (Mahmood et al. 2024). Green bonds provide an avenue by which such investors may be attracted to invest in infrastructure that provide environmental and financial returns (Abbasnejad et al. 2025). Green bonds will complement government funding and other private investments (Tan 2022). This is crucial especially in developing countries where public funds are limited and there are competing interests for resources to funds many projects (Wang et al. 2025). Green bonds ensure sustainability reporting, transparency and accountability in financing of the sustainable transport infrastructure (Khurshid et al. 2023). This is important since funders of green bonds expect the government to report how the funds were utilised and the impact of funding on outcomes that address environmental issues (Akomea-Frimpong et al. 2023).

Green bonds may be used to fund low carbon transport systems and innovative and sustainable transport infrastructure (Yannis and Chaziris 2022). Hence, green bonds act as a catalyst for transition into more sustainable and greener transport infrastructure and smart transport systems (Motowidlak and Motowidlak 2023). Nevertheless, access to green bonds may be limited, high certification costs and poor governance in developing economies (Bahman 2025). Green washing may be a challenge, where funding is sources purported to represent green initiatives, yet they may be diverted to other projects that does not bring environmental outcomes (Jelti et al. 2023). For green bonds to be successful, there is need for strong policies, credible procedures, clear green and sustainability goals (Carbajo 2023).

Green bonds are regarded as an enabler for innovative financing for sustainable transport infrastructure due to their ability to attract environmentally conscious investors and low-carbon development projects (Ding and Liu 2023; Zheng et al. 2025). Some scholars are of the opinion that green bond market has challenges such as high costs in certification, green washing and lack of standardisation that may reduce its credibility and effectiveness (Dai et al. 2023; Wang et al. 2025). On the other hand, other scholars have the opinion that green bonds provide funding for sustainable transport infrastructure and serve as financial and reputational instruments for issuers (Abbasnejad et al. 2025; Nhat and Hoang 2025).

2.7 Sustainable Transport Infrastructure

This is regarded as provision of transport systems, facilities and networks that enable mobility in a way that is inclusive, environmental conscious and economically viable (Vorwerg et al. 2025). It covers infrastructure such as roads, bicycle lanes, bridges, transit systems, airports, pedestrian walk aways and ports among others that are designed and run in a way that sustain the environment and ensure public safety, affordable, accessible and resilient transport system (Anam et al. 2025). Sustainable transport infrastructure is derived from sustainable development which assume a balance among environment issues, social wellness and economic development (Al-Zubairi et al. 2025). From an economic point of view, sustainable transport infrastructure contributes to access, trade, productivity and tourism (Dindar 2025). From social perspective it provides access to employment, education and healthcare for the public (Wang et al. 2025). From an environmental perspective, it provides reliable, low carbon footprint and climate friendly transport systems (Dadashzada et al. 2025).

All these sustainable transport infrastructure aims to transition for greener transport systems that are environmentally friendly and sustainable for the next generations (Jia et al. 2025). It also addresses issues of accessibility and inclusiveness (Shamsuddoha et al. 2025). Hence, sustainable transport infrastructure should address mobility issues especially for low-income people, disabled people and the elderly (Bahman 2025). Sustainable transport infrastructure requires maintenance, financial resources and institutional support (Dindar 2025). This implies that the infrastructure should be durable, properly managed and should meet the mobility needs for current and next generations while minimising environmental damage and economic costs (Shamsuddoha et al. 2025). Hence, there is need for different funding options such as green bonds, private investment, public funding and development finance (Dadashzada et al. 2025).

3 Research Hypotheses Development

3.1 Effect of Public Funding on Sustainable Transport Infrastructure

This research established that public funding influences sustainable transport infrastructure. Previous researchers also established that public funding positively sustainable transport infrastructure (Carbajo 2023; Chupin et al. 2024; Feldman 2023; Khurshid et al. 2023; Pamidimukkala et al. 2023). As a result, we hypothesise that:

H1: Public funding has a positive effect on sustainable transport infrastructure.

3.2 Effect of Private Investment on Sustainable Transport Infrastructure

This research established that private investment positively influences sustainable transport infrastructure. In support of this finding, past research confirmed a positive relationship between private investment and sustainable transport infrastructure (Akomea-Frimpong et al. 2023; Ding and Liu 2023; Nhat and Hoang 2025; Motowidlak and Motowidlak 2023; Tan 2022). Thus, we proposed that:

H2: Private investment has a positive effect on sustainable transport infrastructure.

3.3 Effect of Development Finance on Sustainable Transport Infrastructure

This research established that development finance positively influences sustainable transport infrastructure. Similarly, past researchers established that development finance positively influences sustainable transport infrastructure (Carbajo 2023; Chupin et al. 2024; Kharlamova et al. 2022; Mahmood et al. 2024; Motowidlak and Motowidlak 2023). Hence, we propose that:

H3: Development finance positively influences sustainable transport infrastructure.

3.4 Effect of Green Bonds on Sustainable Transport Infrastructure

This research established that green bonds influence sustainable transport infrastructure. Previous researchers also established that green bonds positively influence sustainable transport infrastructure (Carbajo 2023; Chupin et al. 2024; Dai et al. 2023; Ding and Liu 2023; Mahmood et al. 2024). As a result, we hypothesise that:

H4: Green bonds have a positive effect on sustainable transport infrastructure.

Figure 1 suggests a research model based on the discussion above.

4 Research Methodology

4.1 Questionnaire

The instrument used for data collection was a structured Likert type questionnaire. Items were adapted from past studies and adopted with minor amendments.

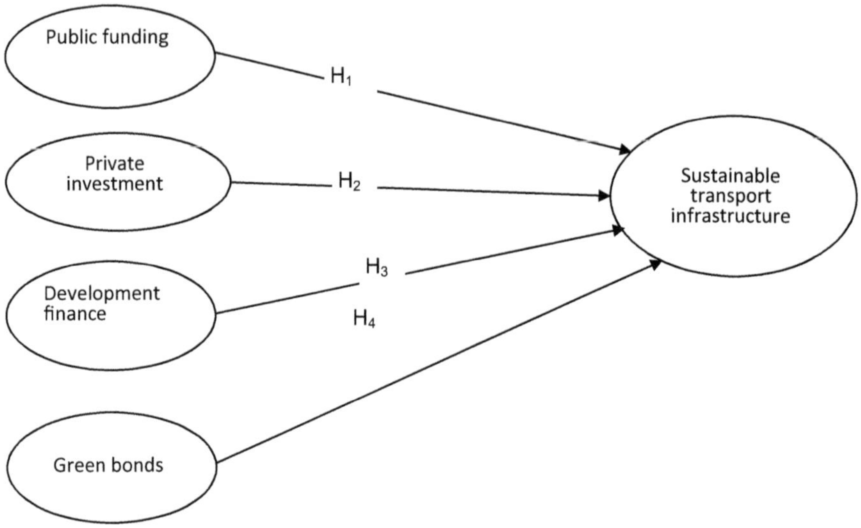

Fig. 1 Research model

4.2 Ethical Considerations

Respondents voluntarily participated and were free to leave the study at any time. Respondents' data was treated with confidentiality and anonymity. Informed consent was sought from respondents.

4.3 Sampling and Data Collection

A positivist philosophy and quantitative method was adopted. Cross-sectional survey was adopted. Respondents were drawn from stakeholders from the infrastructure and built environment who were gathered at the AfriConfex 2025 in Bulawayo, Zimbawe. The size of the population was unknown. Thus, the RAOSOFT sample size calculator assumes that the total population is 20,000 if no actual number is known. Hence, the sample size derived was 377 respondents. Respondents were conveniently visited at their exhibition stands and were asked to complete the questionnaires. The researchers issued 377 questionnaires and received back 307 which were all useable, hence 81% response rate. Descriptive statistics were analysed as shown in Table 1.

Table 1 shows that men (68.1) are the majority respondents. The age group between 31 and 50 was the majority with 65.1%. Those with undergraduate qualifications (50.8) were the majority. Those who earn below USD 1000 accounting for 67.1%.

Table 1 Respondents' profile

Item	Frequency	Percent
Gender		
Female	98	31.9
Male	209	68.1
Age		
18–20	19	6.2
21–30	37	12.1
31–40	104	33.9
41–50	96	31.2
Above 50	51	16.6
Education		
Postgraduate degree	43	14.0
Undergraduate degree	156	50.8
Diploma	108	35.2
Income (USD)		
Below 500	132	43
500–999	74	24.1
1000–1499	46	15.0
1500–1999	39	12.7
At least 2000	16	5.2

5 Results

5.1 *Sampling Adequacy*

KMO was at 0.806 and within the minimum level of 0. Bartlett's Sphericity was significant at $p < 0.001$. Degrees of freedom was at 119 and the Chi-Square was at 9024.103.

5.2 *Convergent Validity*

Convergent validity was measured using confirmatory factor analysis. Table 2 shows that the standardised factor loadings (λ) were above 0.5. Critical ratio (CR) was significant $p < 0.001$. This imply that convergent validity was achieved. Results for reliability are shown in Table 2 and the indices were greater than 0.8. This imply that the measurement items for the study constructs were consistent in measuring what the instrument intent to measure.

Table 2 Convergent validity

Constructs	Items	λ	CR	(α)
Public funding	PUF1	0.614	–	0.903
	PUF2	0.731	7.533***	
	PUF3	0.783	7.721***	
	PUF4	0.742	6.502***	
Private investment	PRI1	0.724	–	0.885
	PRI2	0.721	6.414***	
	PRI3	0.641	6.118***	
	PRI4	0.619	6.057***	
Development finance	DEF1	0.801	–	0.871
	DEF2	0.715	6.309***	
	DEF3	0.702	6.998***	
	DEF4	0.759	6.618***	
Green bonds	GRB1	0.712	–	0.863
	GRB2	0.677	6.203***	
	GRB3	0.718	6.512***	
	GRB4	0.656	6.407***	
Sustainable transport infrastructure	STI1	0.703	–	0.858
	STI2	0.675	6.314***	
	STI3	0.731	6.708***	
	STI4	0.657	6.413***	

Note CR is fixed; *** $p < 0.001$

5.3 Discriminant Validity

Discriminant validity was achieved since the average variances extracted (AVEs) were higher than 0.5 as well as higher than the squared inter-construct correlations (SICCs) as shown in Table 3.

Table 3 (AVEs) and (SICCs)

Construct	PUF	PRI	DEF	GRB	STI
Public funding	**0.682**				
Private investment	0.344	**0.641**			
Development finance	0.353	0.365	**0.593**		
Green bonds	0.367	0.351	0.358	**0.689**	
Sustainable transport infrastructure	0.359	0.337	0.343	0.324	**0.574**

Note Diagonal elements in bold are AVEs

5.4 *Testing Research Hypotheses*

Using AMOS version 30, hypotheses were tested. SEM provides accurate estimations (Field 2013). Table 4 depicts the results. The standardised regression weights are all less than 0.5 confirming positive relationship between the relationship between the measured constructs. Also, the critical ratios were significant were significant and large at $p < 0.001$ signify that the hypotheses were supported.

Table 4 revealed that hypotheses were supported. This implies that public funding, private investment, development finance and green bonds positively influence sustainable transport infrastructure. This implies that transport infrastructure is more sustainable when public funding, private investment, development finance and green bonds are high. These results are supported by the sustainable development theory, which suggest that finance is the enabler of sustainable transport infrastructure. This can be attained by attracting different financing options. This signifies that access to various financing options is key in enhancing sustainable transport infrastructure.

6 Discussion

Despite the transport infrastructure literature's emphasis on finance to enhance sustainable transport infrastructure, there is need to include more financing options that normally targeting greener economies such as development finance and green bonds (Chen et al. 2025; Kaiser and Barstow 2022). Hence, public funding, private investment, development finance and green bonds are crucial in enhancing sustainable transport infrastructure. This is crucial due to the need for going green as suggested by sustainable development goals.

Also, sustainable development theory suggest that development should address the current needs and preserving for the next generations to provide for their needs

Table 4 Hypothesis test

Hypotheses	Description	SRW	CR	Remark
H_1	Public funding → sustainable transport infrastructure	0.457	6.165***	Supported
H_2	Private investment → sustainable transport infrastructure	0.461	7.634***	Supported
H_3	Development finance → sustainable transport infrastructure	0.473	8.728***	Supported
H_4	Green bonds → sustainable transport infrastructure	0.454	6.043***	Supported

NB: standardised regression weight (SRW), critical ratio (CR), *** significant at $p < 0.001$

(Bahman 2025). This theory supports this study since it clearly shows that sustainable transport infrastructure enhances environmental protection and social inclusion (Anam et al. 2025). In addition, it shows that sustainable transport infrastructure enhances mobility and economic development (Vorwerg et al. 2025). The theory postulatc supports the fact that finance is the enabler of sustainable transport infrastructure (Dindar 2025). This implies that public funding, private investment, development finance and green bonds are instruments that enhance sustainable transport infrastructure that is friendly to the environment, inclusive and resilient (Al-Zubairi et al. 2025).

As anticipated, the study confirmed that public funding positively influences sustainable transport infrastructure. The findings are supported by previous (Carbajo 2023; Chupin et al. 2024; Feldman 2023; Khurshid et al. 2023; Pamidimukkala et al. 2023). Literature is in support of this finding, for example, Zhu et al. (2023) postulate that public funding is key in ensuring that infrastructure is maintained for sustainability. Public funding may also be used modernise public roads and transport facilities so that they are resilient to climate change and improve road safety and signage for both the public and motorists (Al-Zubairi et al. 2025). Public funding ensure that public transport infrastructure remain in place and reduce the impact on the environment and enhance the sustainability of transport system (Carbajo 2023). Through public funding, other funding options can be mobilised (Jelti et al. 2023). If government is seen as committed to transport infrastructure development, other players such as private sector and global financial institutions may also complement government efforts through structured financing options (Chupin et al. 2024). Also, public funding acts as a guarantee to other financial institutions and this may reduce reluctance by other financiers to complement government efforts in sustainable transport infrastructure (Tanko et al. 2025). However, the success of public funding as an enabler of sustainable transport infrastructure is hinged on fiscal discipline, government commitment, policy consistency and proper project management capabilities (Pamidimukkala et al. 2023).

Similarly, it was established that private investment positively influences sustainable transport infrastructure. This study validates existing similar previous studies (Akomea-Frimpong et al. 2023; Ding and Liu 2023; Nhat and Hoang 2025; Motowidlak and Motowidlak 2023; Tan 2022). This is also supported by literature, for example, Nhat and Hoang (2025) postulate that through private funds the government may complete certain projects and improve the transport systems and infrastructure to be more sustainable. Private investment contributes with new technology, innovation and efficient transport infrastructure and transport systems that are climate proof and affordable (Akomea-Frimpong et al. 2023). Through private investment corruption and bureaucracy may be minimised (Motowidlak and Motowidlak 2023). Private investors normally focus on controlling costs and improve on project delivery in the shortest possible time to improve returns on investment (Nhat and Hoang 2025). Also, the private sector is characterised by expertise, technical knowhow and bringing in new technology that may improve quality of transport infrastructure as well as sustainability (Anam et al. 2025). Most private sector investors can provide recent technology such as electric vehicles and smart transport management systems

(Tan 2022). These technologies reduce carbon footprint, improve convenience for the public and reduce transportation cost (Ding and Liu 2023).

However, private sector funding may not always provide positive results as in some times what we call private investment is simply companies of government officials who may siphon public funds and appear as private investors (Wang et al. 2025). Such private sector players that are conniving with government officials normally causes corruption, favouritism and bureaucracy (Nhat and Hoang 2025). The private sector is normally motivated by profit, hence may charge high for construction of transport infrastructure (Yannis and Chaziris 2022). In addition, projects with low returns may be shunned by the private sector and may result in lack of funding in such projects (Anam et al. 2025). So, projects that are in rural areas and other low-income communities may not receive funding from private sector investors (Tan 2022). Private investment may need to be regulated since their interests are profit over environmental and social equality interests (Motowidlak and Motowidlak 2023). The success of private investment is hinged on provision of a clear road map on how government engage and contract them, good policies, transparency and strong institutional frameworks (Akomea-Frimpong et al. 2023).

Additionally, it was established that development finance positively influences sustainable transport infrastructure. This supports earlier findings (Carbajo 2023; Chupin et al. 2024; Kharlamova et al. 2022; Mahmood et al. 2024; Motowidlak and Motowidlak 2023). This is in line with the literature, for example, Chupin et al. (2024) postulate that development finance capacitates the government through policy consistency and transfer of technical knowhow. Development funders also provide with project management skills, advise and other supporting services that ensure that the projects are successfully implemented in a sustainable manner (Jelti et al. 2023). This ensure that investments done in infrastructure are secured and may sustain future generations (Motowidlak and Motowidlak 2023). Through development finance other capital may be secured based on the reputation of development finance partners (Bahman 2025). Similarly, through development finance, project risk is minimised, hence this may attract private investors to join in the projects (Nhat and Hoang 2025).

Nevertheless, the success of development finance depends on how well the government aligns itself in terms of its priorities, implementation and governance systems (Ding and Liu 2023). The government should remove all bureaucratic tendencies, improve on fund disbursements and strengthen institutional capacity for sustainable transportation infrastructure to be a reality through development finance (Anam et al. 2025). Also, the government should manage its debts in a sustainable way to enhance fiscus sustainability (Akomea-Frimpong et al. 2023). Development finance is a good option for infrastructure projects, however, there is need for good governance, good project management and policy consistency (Khurshid et al. 2023).

Also, it was established that green bonds positively influence sustainable transport infrastructure. This supports earlier findings (Carbajo 2023; Chupin et al. 2024; Dai et al. 2023; Ding and Liu 2023; Mahmood et al. 2024). This is in line with the literature, for example, Dai et al. (2023), posit that green bonds assist in attracting huge capital for massive projects from investors that are environmentally conscious. Globally there is a wave by investors to fund climate change and environmentally friendly projects for greener global economy (Mahmood et al. 2024). Green bonds provide an avenue by which such investors may be attracted to invest in infrastructure that provide environmental and financial returns (Abbasnejad et al. 2025). Green bonds will complement government funding and other private investments (Tan 2022). This is crucial especially in developing countries where public funds are limited and there are competing interests for resources to funds many projects (Wang et al. 2025). Green bonds ensure sustainability reporting, transparency and accountability in financing of the sustainable transport infrastructure (Khurshid et al. 2023). This is important since funders of green bonds expect the government to report how the funds were utilised and the impact of funding on outcomes that address environmental issues (Akomea-Frimpong et al. 2023).

Green bonds may be used to fund low carbon transport systems and innovative and sustainable transport infrastructure (Yannis and Chaziris 2022). Hence, green bonds act as a catalyst for transition into more sustainable and greener transport infrastructure and smart transport systems (Motowidlak and Motowidlak 2023). Nevertheless, access to green bonds may be limited, high certification costs and poor governance in developing economies (Bahman 2025). Green washing may be a challenge, where funding is sources purported to represent green initiatives, yet they may be diverted to other projects that does not bring environmental outcomes (Jelti et al. 2023). For green bonds to be successful, there is need for strong policies, credible procedures, clear green and sustainability goals (Carbajo 2023).

7 Implications

Below are the study's theoretical, policy and practical implications.

7.1 Theoretical Implications

The study contributes on knowledge by bringing in more sustainable financing options such as development finance and green bonds in the financing and sustainable transport infrastructure discourse. Most studies were done in developed countries, so a study from a developing country bring new ideas to the body of knowledge. Also, previous studies focused mainly on public funding, private funding and the more glorified public private partnerships. Hence, this study expands the finance and sustainable transport infrastructure literature through bringing bonds and development finance as key antecedents of sustainable transport infrastructure.

7.2 Policy Implications

The government must put specific policies and governance systems to attract funding for sustainable transport infrastructure. They must remove bureaucracy, corruption and favouratism in awarding government tenders for sustainable transport infrastructure. These policies may satisfy United Nations Sustainable Development Goal 9, which focus on robust infrastructure for innovation and economic development.

7.3 Practical Implications

The success of public funding as an enabler of sustainable transport infrastructure is hinged on fiscal discipline, government commitment, policy consistency and proper project management capabilities. Government should fund projects with low returns but that provide sustainable transport infrastructure to the public. Projects with low returns may be shunned by the private sector and may result in lack of funding in such projects. The private sector remains a complementary option to public funding and should not replace public funding. A balance between private and public funding should be maintained to ensure sustainable transport infrastructure. To attract development finance, government should ensure that there is good governance, good project management and policy consistency. Green bonds may be used to fund low carbon transport systems and innovative and sustainable transport infrastructure. For green bonds to be successful, there is need for strong policies, credible procedures, clear green and sustainability goals.

8 Limitations and Areas of Future Research

The study faced its challenges; the study focused on Zimbabwe only, hence the research may be extended to other nations. The study was limited to quantitative approach, hence qualitative studies may be done to augment this study. The research was contextualised to the transport sector hence studies on other sectors is recommended for comparisons.

9 Conclusion

The study concluded that financing is key in augmenting sustainable transport infrastructure. The research concluded public funding, private investment, development finance and green bonds positively influence sustainable transport infrastructure. Hence, when government is faced with choices on financing sustainable transport infrastructure it should consider a mix of public funding, private investment, development finance and green bonds based on its needs available, its capacity and accessible. However, all financing options are successful if there are strong policies, credible procedures, clear green and sustainability goals, fiscal discipline, government commitment, policy consistency and proper project management capabilities.

References

Abbasnejad B, Soltani S, Ahankoob A, Kaewunruen S, Vahabi A (2025) Industry 4.0 technologies for sustainable transportation projects: applications, trends, and future research directions in construction. Infrastructures 10(5):104

Akomea-Frimpong I, Jin X, Osei-Kyei R, Kukah AS (2023) Public–private partnerships for sustainable infrastructure development in Ghana: a systematic review and recommendations. Smart Sustain Built Environ 12(2):237–257

Al-Zubairi A, AL-Akheli A, ELfarra B (2025) The impact of financial development, renewable energy and political stability on carbon emissions: sustainable development prospective for Arab economies. Environ Dev Sustain 27(7):15251–15273

Anam MZ, Shakur MS, Bari AM, Debnath B, Bristy FT (2025) Exploring the barriers to decarbonizing the transportation system: a pathway to a cleaner future in emerging economies. Innov Green Dev 4(4):100256

Bahman N, Naser N, Khan E, Mahmood T (2025) Environmental science, policy, and industry nexus: integrating frameworks for better transport sustainability. Glob. Transit. 7:29–40

Bakkar Y, Bark E, Prause G, Ul-Durar S (2025) Green Deal and financing sustainable transport in Europe: a target costing analysis. Transp Policy 163(1):185–198

Carbajo JC (2023) Financing sustainable transport infrastructure in emerging markets and developing economies. In: Handbook on transport pricing and financing. Edward Elgar Publishing, pp 330–347

Chen S, Paramaiah C, Kumar P, Khan S, Haomu Q (2025) Toward sustainable tourism: insights from green financing and renewable energy. Energ Strat Rev 57(1):101618

Chupin A, Sorokin A, Veselko A, Morkovkin D, Pishchik VY, Alekseev PV (2024) Sustainable financing for transport infrastructure: an integral approach for The Russian Federation. Sustainability 16(8):3108

Dadashzada I, Garza-Reyes JA, Roy Ghatak R, Gonzalez-Aleu F (2025) An investigation into the major barriers to the adoption of electric vehicles in last-mile deliveries for sustainable transport. Int J Logist Manag 36(3):906–938

Dai J, Alvarado R, Ali S, Ahmed Z, Meo MS (2023) Transport infrastructure, economic growth, and transport CO2 emissions nexus: does green energy consumption in the transport sector matter? Environ Sci Pollut Res 30(14):40094–40106

Dindar S (2025) A systematic review of urban Regeneration's impact on sustainable transport: traffic dynamics, policy responses, and environmental implications. Sustain Dev 33(1):399–426

Ding X, Liu X (2023) Renewable energy development and transportation infrastructure matters for green economic growth? Empirical evidence from China. Econ Anal Policy 79:634–646

Feldman O (2023) Innovative ways for financing transport infrastructure. In: Capacity building in local authorities for sustainable transport planning. Singapore, Springer Nature Singapore, pp 253–278

Field A (2013) Discovering statistics using IBM SPSS statistics, 4th edn. Sage, London

Giles-Corti B, Moudon AV, Lowe M, Adlakha D, Cerin E, Boeing G, Higgs C, Arundel J, Liu S, Hinckson E, Salvo D, Sallis JF (2022) Creating healthy and sustainable cities: what gets measured, gets done. Lancet Glob Health 10(6):e782–e785

Jelti F, Allouhi A, Tabet Aoul KA (2023) Transition paths towards a sustainable transportation system: a literature review. Sustainability 15(21):15457

Jia Z, Yin J, Cao Z, Wei N, Jiang Z, Zhang Y, Wu L, Zhang Q, Mao H (2025) Sustainable transportation emission reduction through intelligent transportation systems: mitigation drivers, and temporal trends. Environ Impact Assess Rev 112:107767

Kaiser N, Barstow CK (2022) Rural transportation infrastructure in low-and middle-income countries: a review of impacts, implications, and interventions. Sustainability 14(4):2149

Kharlamova T, Desfonteines L, Barykin S, Gavrilova R (2022) Prospects for the development of transport infrastructure to ensure sustainable development. Trans Res Procedia 63:789–797

Khurshid A, Khan K, Cifuentes-Faura J (2023) 2030 agenda of sustainable transport: can current progress lead towards carbon neutrality? Transp Res Part D: Transp Environ 122:103869

Mahmood S, Sun H, Iqbal A, Alhussan AA, El-kenawy ESM (2024) Green finance, sustainable infrastructure, and green technology innovation: pathways to achieving sustainable development goals in the belt and road initiative. Environ Res Commun 6(10):105036

Motowidlak U, Motowidlak T (2023) Development and financing of infrastructure investments to shape sustainable and resilient mobility. Organ Manag Ser 170:349–370

Nhat NM, Hoang NK (2025) Driving sustainable transport infrastructure in urban regions of developing markets: leveraging private sector involvement. Int J Sustain Dev Plan 20(10):42–79

Pamidimukkala A, Kermanshachi S, Rosenberger JM, Hladik G (2023) Evaluation of barriers to electric vehicle adoption: a study of technological, environmental, financial, and infrastructure factors. Transp Res Interdiscip Perspect 22:100962

Shamsuddoha M, Kashem MA, Nasir T (2025) A review of transportation 5.0: advancing sustainable mobility through intelligent technology and renewable energy. *Future*. Transportation 5(1):8

Suleiman H, Magaji S, Musa I (2025) Assessing the effect of green loans and carbon finance on sustainable cities and community development in Nigeria: an analysis of urban sustainability indices. Int J Innov Sci Res Technol 10(5):3098–3109

Tan C (2022) Private investments, public goods: regulating markets for sustainable development. Eur Bus Organ Law Rev 23(1):241–271

Tanko Y, Magaji S, Musa I (2025) Effect of green finance on climate change mitigation in Nigeria. Int J Econ Perspect 19(7):1–22

Vassallo JM, Garrido L (2023) Transport funding and financing: a conceptual overview of theory and practice. In: Handbook on transport pricing and financing, pp 273–294

Vorwerg FC, Ebadi Torkayesh A, Venghaus S (2025) From fossil fuels to alternative fuels: strategy development for a sustainable transport sector in Germany. Energy Sustain Soc 15(1):4

Wang W, Wang J, Wu H (2025) Assessing the impact of new digital infrastructure on enterprise green transformation from a triple performance perspective. J Knowl Econ 16(1):5594–5633

Yang T, Yang F, Tan P, Lu C, Shao C, Yang B (2025) Research on the impact of transportation infrastructure construction on urban sustainable development. Sci Rep 15(1):29725

Yannis G, Chaziris A (2022) Transport system and infrastructure. Transp Res Procedia 60:6–11

Zheng H, Li D, Cai J (2025) Driving green innovation: the impact of digital finance on China's transition to clean energy. Energy 318(1):134760

Zhu J, Xie N, Cai Z, Tang W, Chen X (2023) A comprehensive review of shared mobility for sustainable transportation systems. Int J Sustain Transp 17(5):527–551

Comparative Pathways to Green Transport Transitions in the Global South

Innocent Chirisa, Halleluah Chirisa, Tinashe Magande, Africa Muneti, Nyasha N'onzwe, and Roselin Katsande-Ncube

Abstract Transport emissions and congestion pose entangled environmental and social threats in the Global South and call for low carbon mobility solutions tailored to local realities. Nevertheless, research is still skewed towards experiences in the Global North and case studies of individual cities, thus offering insufficient insights into interactions among governance, financing, technologies, informality, and equity in multiple Southern settings. The primary aim of this qualitative research was to shed light on alternative yet complementary approaches to greening transport in selected African, Asian, and Latin American cities. The data set included results of an extensive desk review of relevant policy, scholarly, and grey literature, while textual analysis served as the method of data analysis. The analysis showed that, firstly, good governance is an indispensable condition for introducing and maintaining low carbon mobility policies because it involves political commitment and specialized institutions. Secondly, diverse and adaptable funding sources, which include state funds, climate funds, and private capital—are responsible for the speed and extent of

I. Chirisa (✉)
Office of the Vice Chancellor, Zimbabwe Ezekiel Guti University, Bindura, Zimbabwe
e-mail: innocent.chirisa@gmail.com

I. Chirisa · T. Magande
Department of Urban and Regional Planning, University of Johannesburg, Johannesburg, South Africa

H. Chirisa
Zimbabwe Ezekiel Guti University, Bindura, Zimbabwe

A. Muneti
Firm Central Property Group, Harare, Zimbabwe

IC University of Technology, Phnom Pehn, Cambodia

N. N'onzwe
Department of Social Work and Applied Psychology, Zimbabwe Ezekiel Guti University, Bindura, Mashonaland Central, Zimbabwe

R. Katsande-Ncube
Faculty of Social and Gender Transformative Sciences, Women's University in Africa, Harare, Zimbabwe

J. Kanyepe and I. Chirisa (eds.), *Advancing the Green Agenda in the Transport and Logistics Sector in Global South Countries*, World Sustainability Series,
https://doi.org/10.1007/978-3-032-31374-4_19

technological innovation. Finally, attempts at improving transport infrastructure and using clean energy sources are likely to be futile if they ignore the issue of informality and pay no attention to equity in peripheral and poor.

Keywords Green transport transitions · Global south · Governance · Financing · Informality · Equity

1 Introduction

There is a rapid increase in transport emissions in the Global South, driven by urbanisation, motorisation, and reliance on fossil fuels. The transport industry contributes to almost a quarter of all energy-related carbon dioxide emissions globally, and the demand is increasing at the most rapid pace in the emerging economies (International Energy Agency (IEA) 2023). African, Asian, and Latin American cities are also not exceptions as a large portion of air pollution has been associated with road transport that negatively affects the health and productivity of the population (World Health Organisation (WHO) 2021). The growth of cities also happens in a rapid manner and in most cases lack proper transportation system and consequently more people depend on use of personal vehicles and informal transport (United Nations Environment Programme (UNEP) 2019). Inadequate infrastructure, narrow fiscal space, and governance limitations complicate further the low-carbon transitions in these areas (Emodi et al. 2022; Naufal et al. 2026) Meanwhile, the risk of leapfrog to cleaner technologies, i.e. electric buses and integrated mass transportation systems, is high (Sims et al. 2014). These dynamics pose the challenges of urgency of sustainability that demand context-sensitive and at the same time scalable solutions.

Despite the increased policy attention in green mobility, there remains a greater challenge that policy implementation in low-carbon transport transitions is slow to none existent particularly in the Global South. It is of great concern that some of the cities have integrated public transport reforms and electric mobility, whereas others continue to struggle with institutional coordination. As such, the major research problem in this chapter is on how the success and failure of green transport shape differing governance structures, financial capacities, and policy approaches in the Global south. The transitions between regions should be compared due to high diversification of the Global South in terms of institutional capacity, economic form and urban form. Although other cities have developed electric mobility initiatives, other cities are unable to formalise simple bus solutions (C40 Cities 2022). The comparative analysis can be used to determine the policy instruments that can be transferred and those that are conditioned by particular governance arrangements (Baker et al. 2014). Socio-technical transition studies indicate that political leadership, coalitions between stakeholders, and financial models influence the results of transitions in different settings in varied ways (Geels 2011). Comparisons between regions also identify the impacts of international climate finance and development banks on transport reforms in less-income environments (Steckel et al. 2017). With no systematic comparison, policy transfer will continue to oversimplify complex realities.

Existing scholarship on sustainable transport transitions has continued to put much focus on cities in the Global North with limited attention to cross regional analysis in the Global South. In studies where the global South is included, single country experiences are examined with no cross regional analysis creating a gap on how political, economic, and institutional settings shape green transport transitions across Africa, Asia and Latin America. The chapter addresses that gap and answers three fundamental research questions. To begin with, what is the variation in the results of transport transition by the chosen countries in Africa, Asia, and Latin America? Second, what are the general trends of success or failure in trying to implement green mobility systems? Third, how far are particular transition strategies applicable in various political and economic settings? To respond to these questions, the chapter investigates the case studies such as the emerging bus reforms in Kigali, the Bus Rapid Transit system in Cape Town, the complete electrification of government buses in Shenzhen, the electric vehicle policies and metro development in Delhi, the TransMilenio in Bogota, and the rapid conversion of electric buses in Santiago. These instances demonstrate various forms of governance, income, and technological ways (Hidalgo and Graftieaux 2008). Through the comparison of these experiences, the chapter aims at capturing common drivers, recurring barriers, context-specific areas that influence the process of green transport transitions.

The remainder of the chapter is organised as follows. Following this introduction is a conceptual framework with theoretical foundations that grounds the chapter. Subsequently, a brief literature review on existing studies at global, regional and local level on the topic follows. The section is followed by the methodology describing the research design, data sources, and procedure used in the chapter. The findings sections follow with six comparative case studies from Africa, Latin America and Asia. There are two case studies from each region. The final section discuss the overall implications for policy transfer, governance, and sustainable urban transport transitions in the Global South.

2 Conceptual Framework

The theoretical framework of the chapter lies in the Multi-Level Perspective (MLP) on socio-technical transitions, coupled with the theories of political economy, governance transitions and policy diffusion to examine the issue of green transport pathways in the Global South. The MLP framework can be used to explain the emergence of green transport systems innovations in niches, resistance to such innovations by established regimes, landscape level forces like climate change and urbanization (Geels 2002). The political economy theory brings into focus the role of power relations, institutional forms and economic interests influence policy decisions and adoption of green transport (Cerny 2010). Governance transitions stress making the shift in institutional structures, as well as the participation of stakeholders, at various levels down to transnational, which support sustainable transport initiatives (Lambert et al. 2026). Policy diffusion also helps to understand how green transport policies move, change and transform in various settings and incorporate global standards and local innovations (Costantini et al. 2017).

The issues to consider as key variables to be compared are governance structures, finance mechanisms, technological adoption, integration of informal sector, and equity issues. Governance studies institutional strength, regulatory consistency, and involvement in decisions that are needed to create transition legitimacy (Kern and Rogge 2016). The finance focuses on the supply and composition of publicly existing investments, climate finance, and green bonds and drives infrastructure and innovation (Voege et al. 2018). The concerns that technology is concerned with are penetration of low-carbon modes of transport, infrastructure preparedness, and socio-technical fit. The idea of informal sector integration acknowledges the presence of informal transit providers that feature the majority of Global South cities and the need to integrate them into the sustainable transport policies (Agyemang et al. 2020). Equity refers to distributive effects of transitions, which focuses on accessibility and inclusion of marginalised groups to stop the further increase of socio-economic polarities (Bulkeley et al. 2014).

The assessment of success refers to changes to low-emission transport modes, inclusive and adaptive governance, diversified and sustainable finance, successful use of technologies, and improvement of equitable access. Success includes carbon emission cuts that are quantifiable as well as social amenity like enhanced mobility among the vulnerable communities (Geels and Schot 2007). Failure on the other hand is expressed through policy stasis, insufficient funding, lock-ins of technologies, the marginalization of the informal actors, and lack of governance that weakens transition credibility and success. The criteria also take into account the capacity of governance structures to change and the degree of mobilization of the financial mechanism to gather adequate resources to scale sustainable transport (Ahire et al. 2024).

3 Literature Review

To the entire world, the social-technical frameworks regarding sustainable transport transitions are the most popular and have analyzed the interaction between technologies, institutions, and societal actors. Multi-Level Perspective (MLP) has played a key role in comprehending how niche innovations such as electric mobility come into existence and may upset the established fossil-fuel-based regimes (Yudha et al. 2022). In addition to this, socio-technical transitions theory prefigures the political economy by paying attention to power relations and governance processes that facilitate or inhibit the mobility of the low-carbon (Kanger and Sovacool 2021). The literature on climate governance also refines these arguments by analyzing the structures through multi-scalar policy networks and the role of international climate regimes in influencing urban transport decarbonization (Van Asselt and Green 2022).

The world shows that the electrification, modernization of transport, and active active policy diffusion are the most fundamental approaches to decarbonize transport, but there are still disagreements about whether changes are made on a small scale or they demand radical governance changes (Stalmokaitė and Yliskylä-Peuralahti 2019). In addition, the prevailing scholarship of the world has been biased toward technological advances and governance reforms mostly seen in the Global North, and have occasionally poorly recognized the social-political complexities and resource limitations of the Global South (Watari et al. 2021).

Literature at the continental level shows conflicting but overlapping issues and enablers that are influencing the development of green transport in Africa, Asia, and Latin America. In Africa, the infrastructure shortages in the system, as well as the widespread transport informality, undermine the formal development of mass transit, which spreads fragmented modal integration and partial electrification efforts (Krüger et al. 2021). The transport shifts in Asia are expressed through the rapid motorization pressures and against the background of strategic industrial policy and new leadership in the electrification, especially in China and South East Asia (Zhou et al. 2021). The innovations in Bus Rapid Transit (BRT) systems and integrative public transport networks are often discussed by Latin American scholarship as one of the response measures to urban sprawl and inequality (Grande-Ayala 2024). In these continents, the differences in institutional capacity (governance maturity to financial mechanisms) are decisional, and structural barriers can often be associated with political fragmentation, lack of fiscal space, and urban informality (Ho and Tirachini 2023; Krüger et al. 2021).

The subtle interaction between local politics, financing models, and social equity considerations in influencing transport transitions are supported by empirical studies on the local and city level. Due to the development of the rapid urbanization environment, Shenzhen is an example of a city with sophisticated urban greenway systems that integrate ecological and transport operations (Song et al. 2022). The example of Delhi switching to Compressed Natural Gas buses is the successful outcome of judicial activism and multi-actor coalitions that are able to instigate change despite technological and institutional barriers (Kanger and Sovacool 2021). Latin American urban centers such as Bogota and Santiago unveil the role of BRT and comprehensive public transport redesign in alleviating the congestion and pollution as they struggle with the issue of scaling and fair service delivery (Grande-Ayala 2024). Innovations are challenged by the unchanging history of spatial segregation and informality in African cities, including Cape Town and Kigali, with inclusion in governance and hybrid transportation emerging as key to solving the problem (Krüger et al. 2021). The modalities of financing can be very diverse, including public subsidies, public-private systems, and donor financing, which is determined by the situation in local institutions and the political leadership (Osieczko et al. 2021).

While the literature offers some important lessons regarding transitions to sustainable transport systems, few pieces of research have managed to break free from regional confines or one-city case studies. Comparativist work investigating how specific governance systems, modes of financing, and socio-technical conditions affect the way in which transport systems undergo transitions between different regions in the Global South is scarce. Not enough comparative research has been done in order to connect experiences of transport system transitions in Africa, Asia, and Latin America in one framework of analysis. This chapter seeks to fill that gap through comparative analysis of specific Global South cities in order to highlight key determinants of their transitions to more sustainable transport systems.

4 Methodology

The study used a qualitative comparative desk-review and document analysis methodology. an in-depth examination of green transport transition pathways in selected countries across Africa, Asia, and Latin America. A comparative case study design was adopted to explore how governance structures, policy instruments, and institutional capacities influence sustainable transport transitions across different urban contexts. The six case study cities Kigali, Cape Town, Shenzhen, Delhi, Bogotá, and Santiago were selected through purposive sampling. The cases were chosen based on four criteria: (1) evidence of active green transport transition initiatives; (2) regional representation across Africa, Asia, and Latin America; (3) variation in governance capacity, economic conditions, and transport policy approaches; and (4) availability of sufficient policy and academic documentation. Kigali and Cape Town were selected to represent emerging and relatively advanced African transport reform experiences respectively, while Shenzhen and Delhi illustrate contrasting Asian approaches to electrification and mass transit development. Bogotá and Santiago were selected because of their internationally recognised Bus Rapid Transit and electric bus transition programmes in Latin America. The selection therefore allowed for comparison across diverse socio-economic and institutional settings rather than statistical representation.

Data for the study were sourced from multiple secondary data material, including books, policy documents, peer-reviewed journal articles and some credible online sources. A comprehensive review was conducted by selecting these sources to capture theoretical views and empirical findings from existing articles. To get access to this literature, the study used electronic search engines that include Science Direct, Scopus, and Google Scholar. The inclusion criteria consisted of: (1) peer-reviewed publications and institutional reports published between 2008 and 2025; (2) studies focusing on green transport transitions and low-carbon urban mobility in Africa, Asia, and Latin America; and (3) English-language sources with direct relevance to governance, financing, technological pathways, informality, and equity in urban transport systems. Exclusion criteria included sources lacking empirical grounding, documents with limited methodological transparency, and studies focused solely on Global North contexts without transferable insights for the Global South.

Textual analysis served as the main analytical method. A systematic examination of selected English-language documents was conducted to identify recurring themes, policy patterns, governance dynamics, and transition challenges across the chosen case studies. Themes were coded comparatively around five core dimensions:

governance structures, financing mechanisms, technological adoption trajectories, approaches to informal transport integration, and distributive equity outcomes. This coding scheme enabled structured cross-case comparison and helped surface both convergences and divergences in transition pathways. The approach enhanced analytical consistency and strengthened the replicability and transparency of the study's qualitative insights. This helped the researchers to ground the study in authoritative and well-established academic discourse. The study used textual analysis as a primary analytical tool for data analysis. A systematic examination of selected English documents was done using this method to identify recurring themes and patterns in literature.

5 Findings

Findings in this chapter were presented in the form of case studies. Six case studies were analysed to allow a comparison across three continents in the developing world: Africa, Latin America and Asia. The section begins with two cases from Africa.

6 Africa

6.1 Case Study 1: Kigali, Rwanda—Bus Rapid Transit (BRT) Development

Kigali, Rwanda, can be used as an example of a rapidly developing African city that has undertaken a transformational public transport project, which is the introduction of Bus Rapid Transit (BRT) system, within the dynamic economic, political, and urban context. Kigali has undergone much population increase and urbanization, and Rwanda after the genocide securing the politically-consolidated state that promotes secure urbanization and modernization as the main national development pillars (Goodfellow and Smith 2013). The new middle-income position of the city is motivated by the development ambitions; however, these projects are limited by the lack of infrastructure and financial difficulties typical of Sub-Saharan urban settings (Kalisa et al. 2025). The pressures of transport demand, the environmental factors, as Kigali experiences the problem of air pollution, and the urge to enhance mobility to continued economic growth were a solid force that reasoned the shift to more efficient and inclusive urban transit system (ibid.). Through a socio-technical transitions lens, Kigali's BRT transition shows how socio-technic pressure factors like urbanization and environment can act as a catalyst for transport regime level change in fast growing cities.

Before developing the BRT, the transport system of Kigali was largely dependent on the informal forms of transportation like moto-taxis and minibuses, which were not regulated and did not work efficiently and caused safety and congestion problems (Finn 2017). Such informality curtailed any credible urban mobility and economic opportunities, particularly to low-income and peri-urban residents (Rizzo 2014). To this, the Rwanda government formulated integrated transport policies to formalize and modernize the public transport infrastructures, to be in line with the larger urban mobility objectives as well as environmental objectives. These policies focused on the development of bus rapid transit schemes to offer fast, efficient and affordable transportation and multimodal connectivity schemes like bike-sharing schemes to maximize last-mile access (Ntamwiza et al. 2025). In this respect, the case highlights the application of governance transition theory in that it shows how the use of coordination and formalization techniques by state led actors resulted in the restructuring of fragmented informal transport systems into an urban transport regime.

The major policy interventions took advantage of the high political goodwill and institutional alignment of Kigali and overcame the normal urban transport challenges. The government focused on the top-down leadership and simplification of the institutional coordination of the BRT implementation through urban planning, transport, and environmental agencies to maintain the momentum of the BRT implementation (Goodfellow and Smith 2013). In this way, the case shows how governance mechanisms facilitated the application of the Multi-Level Perspective (MLP) in the development of the BRT project as regime actors and public organizations worked to facilitate the scaling up of the niche transport innovation. Funding challenges were however also formidable; financial resources were raised using a combination of public investments, collaboration with developmental organizations as well as innovative funding structures designed specifically to work with low-income nations. Kigali did not have many resources but focused on incremental infrastructure projects and staged implementation to make sure that they are financially sustainable (Brizon et al. 2025). Social equity was also integrated in the initiative to enhance transport access among the disadvantaged groups, which was achieved but informal stakeholders like moto-taxis were restricted and tension of displacement was experienced (Finn 2017). Here, the political economy of transport innovation in the case is seen in how transport reform involved political negotiations between the actors involved in the informal transport systems and new formal mobility systems.

Early results of the BRT implementation in Kigali show the effectiveness of urban mobility, decrease in traffic congestion, and positive environmental effects such as the decrease in particulate air pollution index during interventions, such as, Car-Free Days (Kalisa et al. 2025). Nevertheless, there remain difficulties related to the integration of the feeder services, affordability, and transitions off informal transport operators (Rizzo 2014; Guno et al. 2021). The Kigali lessons imply the primary role of long-term political will, consistent institutional structures, and flexible financing plans to achieve scalable BRT systems in high-urbanizing African cities (Goodfellow and Smith, 2013). Also, it is essential to balance the modern transport infrastructure development with considerate social-economic inclinations to contribute to the equitable mobility and urban resilience. Importantly, the case also demonstrates that sustainable transport transitions in the Global South are not purely technological processes but socio-political transformations shaped by informality, governance capacity, and equity considerations.

6.2 Case Study 2: Cape Town, South Africa-MyCiTi BRT

The MyCiTi Bus Rapid Transit (BRT) service in Cape Town, South Africa, emerged in a complicated situation characterized by the post-apartheid transformation of the city, urban sprawl, and economic disparities. Spatial segregation is a distinct feature of Cape town, as the policy of apartheid has pushed the low-income population in the peripheries, which have limited access to social services and infrastructures (Wood 2015). The monetary disparity and high urbanization boosted the need to have effective and inclusive public transportation that would bridge the various neighborhoods to economic centres and services. On the political front, the initiative was in line with national and local interests to pay historical debts by enhancing more balanced urban mobility and spatial integration (Minty 2006). Pre-MyCiTi urban environments were characterised by a disjointed informal transport network that was unable to meet demand due to the rising car ownership and congestion. Within this context, MyCiTi was seen as a new product that would offer safe, reliable, and affordable transport and facilitate social inclusion and spur economic growth. The case shows how the process of transport reform reflected the governance transition theory in that there was a need for addressing spatial inequality and injustices within the state in its attempts at transport reform.

Prior to the MyCiTi BRT, the transport system of Cape Town was controlled by informal minibus taxis that served uncontrolled routes with inconsistent services and safety levels (Vanderschuren et al. 2021). These unofficial types of transport are everywhere, but are inefficient due to lack of schedule and duplication of routes. The system also placed a lot of dependence on the use of private operators, which reduced government control and intertwined with other forms of transportation (ibid.; Du Preez 2025). There were also high levels of traffic jams in the city as people started using their own cars with no proper public alternatives, particularly to the low-income earners who reside in the city outskirts. This disjointed, unfair system was not sustainable in the mobility of urban populations and restricted the access to jobs and services among large numbers of people. These issues led to the municipality seeking a BRT system based on the global best practices, in a bid to transform the informal transport modes to a formal and integrated and financially viable public transport network (Wood 2015). In this way, the case highlights the role of policy diffusion and adaptation to the local conditions in a global south context.

Some of the major policies and interventions in MyCiTi were strategic planning that incorporated the integration of transport and land use, affordable fare structures, and inclusive governance strategies that incorporated informal transport operators (ibid.). Its implementation began with priority signaling, upgraded stations, dedicated bus lanes, to support reliability and efficiency, and ITS-enabled fare collection

(Schalekamp 2011). The project was equity-focused by ensuring that the coverage of services to the low-income townships was extended and fares were also subsidized to vulnerable populations. Notably, MyCiTi tried to serve informal minibus taxi operators into the new system as well using contractual and training programs, which helped an informal transport transition (Schalekamp 2011, 2017). The case illustrates the Multi-Level Perspective by highlighting that the shift towards a new regime in the case required 'negotiation' between existing actors of informal transport and new actors of formal transport systems, which did not involve technological replacement. A mix of government contribution, payment of fees by users, and operational economies of scale was adopted in attaining financial sustainability. The project was also coordinated with other wider urban development to ensure that economic activities are boosted along transit corridors and transit should be part of spatial planning activities that will focus on shortening all-time travelling distances and encouraging transit-oriented development (Wood 2015).

Among the benefits that the MyCiTi BRT has provided are enhanced accessibility, shorter travel time, and increased safety and comfort to passengers, particularly those with low income status and little served previously (Gauthier 2012; Hadders 2019). Nevertheless, such obstacles as full financial sustainability in the context of subsidy reliance, the expansion of service reach, and the experience of tensions in the course of informal-to-formal transition of the operators still exist (Nnene et al. 2023). The increase in equity is somewhat mitigated by network coverage holes and fare patterns that continue to be obstacles to the impoverished groups (Schalekamp 2011; Hadders 2019). These tensions highlight the constraints of technocratic transport reforms, when deep inequalities and informality are entrenched in urban mobility systems. Lessons include the value of learning policies step by step, good stakeholder participation (including informal operators), and using integrated land-use and transport plans to ensure a high potential of impacts on social and economic benefits (Wood 2015). The MyCiTi experience in Cape town highlights the issues of informal urban transport transformation in developing cities and also suggests a scalable approach to informal urban transportation with a strong focus on social inclusion, operational effectiveness, and financial responsibility. The case thus adds to the body of Global South transition scholarship by highlighting that transport transitions are not solely driven by infrastructure and technological innovations, but also by historical spatial inequalities and political negotiations.

7 Asia

7.1 Case Study 3: Shenzhen, China–Full Electric Bus Fleet

The city of Shenzhen, China, provides a good example of urban transportation innovation as it has been able to build the first full electric bus fleet in the world in a booming economic, political, and urban environment. Shenzhen is an urbanized city, a hub of economic development in southern China, and a major industrial region with a booming economy and an escalating environmental issue of traffic emissions and air quality (Li et al. 2020; Zhou et al. 2021). On the political front, a robust governmental dedication accompanied by substantial resources in the hands of the populace facilitated the development of Shenzhen as the first to make efforts in electrifying in line with the national green development policies of China. The high population density and air pollution in the urban environment demanded an immediate shift of the transport system that is based on fossil fuel consumption to systems that are environmentally friendly (Yu et al. 2018). The case of Shenzhen exemplifies the acceleration of regime-level change due to landscape pressures, including environmental crises and national climate priorities, as a result of strong state interventions and institutional coordination from the MLP point of view. Such a political-economic orientation made Shenzhen favourable in order to achieve the rapid scale of electric vehicle (EV) technologies to achieve the goals of environmental sustainability and efficiency in urban mobility.

Before the transition, the transport system of Shenzhen was primarily based on the old-fashioned diesel and hybrid buses, which experienced the usual problem of air pollution, noise, and inefficiency in their work due to fossil-fueled urban transportation (Jiang et al. 2021). There was an increasing straining of infrastructure, commuter experience, and environmental targets due to massive urban demand and increasing fleet size. The progressive policies of the city focused on the total electrification of the city bus park, which involved the inclusion of battery-electric bus purchases, large-scale layouts of charging facilities, and the use of modern management systems to achieve efficiency in work (Li et al. 2020; Wang et al. 2018). Original scheduling structures reduced operational expenses and maintained the reliability of the service despite the specifics of the electric fleet, such as battery capacity controlled and charging conditions (Jiang et al. 2021). The case illustrates the interdependent development of technological innovation, infrastructure development and operational governance that enabled a system-wide process of mobility transition.

The main policy interventions entailed significant state funding and focusing subsidies as well as innovative financing and business models. An inverted U-shaped subsidy curve was employed by Shenzhen as it was starting to stimulate early uptake of electric buses, then wholesittle assistance as the industry matured, which is also indicative of fiscal prudence and market trust (Li et al. 2020). The city enabled multi-stakeholder collaboration among manufacturers, financial institutions, and charging infrastructure operators to lessen initial risk and capital limitations and presented leasing model and battery leasing independent of vehicle ownership (Li et al. 2020). The reliability of operations and the cost-efficiency of operations were improved by investments in the comprehensive coverage of the charging networks and the real-time scheduling technology (Wang et al. 2018). This is the political economy aspects of transition governance, which actively coordinated markets, lowered investment risks, and enabled large-scale uptake of technology by the state. These broad policies were aimed at incentives of the whole value chain and system coordination to aid fast scaling of technology.

Results showed Shenzhen as a global innovation leader, where 100% electric buses make significant improvements in air quality and massive emission cuts in the transport sector (Li et al. 2020). Such obstacles as high entry capital, maintenance complexity, and the need to balance the energy grid requirements against charging infrastructure were also present (Jiang et al. 2021). Although the experience of Shenzhen is not completely transferable to the conditions of lower income due to the specific political will and funding, several concepts can be introduced to be learned, such as integrated funding sources, the development of the technology ecosystem, and sequential subsidies to ensure the sustainability of technology adoption (Yu et al. 2018). It is thus clear from the case that a successful transition in transport relies not only on technological innovations, but also on the institutional and financial ability to coordinate complex socio-technological systems on a large scale. Moreover, Shenzhen highlights that functional operationalization models between service quality and battery and energy limitations are required to scale up electric fleet successfully (Wang et al. 2018).

7.2 *Case Study 4: Delhi, India–EV Policy & Metro Expansion*

Delhi, India, is experiencing serious transport problems associated with the fast urbanisation, the economic development, and political issues. Population growth, motorisation, and sprawling urbanisation have led to a high level of congestion, pollution, and poor mobility in the city (Begam et al. 2024). In terms of politics, the fact that the city of Delhi has a multi-level system of government consisting of municipal, state, and central authorities with overlapping responsibilities and different priorities makes transport policy alignment difficult (Toering et al. 2025). The issue of economic inequalities and space inequalities also contributes to the problem of urban mobility and an equitable access is at the center of development issues. Another essential role is also fulfilled by the informal transport sector that highlights the complexity of the environment in which the transport transformation in Delhi takes place.

The transport system in Delhi used to be dependent on the overcrowded roads occupied by personal vehicles, unofficial buses, and auto-rickshaws before the latest interventions, which increased pollution and inefficiency. The Delhi Metro was opened in the early 2000s as a mass rapid form of transport, but initially it was not sufficiently covered, and the road infrastructure was insufficient, which resulted in constant traffic jams and unsafe travel conditions (Begam et al. 2024). Public transportation was disconnected, and the use of cleaner mobility solutions such as electric vehicles (EVs) was in its infancy because of a gap in infrastructure and regulatory burdens. This is an interpretation of the Multi-Level Perspective, showing how niche innovations like metro systems and EV technologies fail to challenge existing fossil fuel-based mobility regimes in the absence of good institutional coordination.

The major policy solutions encompassed the exponential growth of the Delhi Metro network, combined with an extensive EV policy that facilitated the electrification of the state and personal vehicles. The EV policy has presented incentives in the form of subsidies, lower taxes, and plans to develop charging infrastructure, whereas the metro expansion addressed the expansion of the network coverage and the combination of transport modes (Toering et al. 2025). Nevertheless, regardless of the obvious policy intentions and certain improvements, there were loopholes on the implementation level caused by the difficulties in coordination between government tiers, infrastructure implementation lag, and financial limitations (Toering et al. 2025). Issues on multi-level governance, such as overlapping jurisdictions and irregular regulatory frameworks, became the spoiler in facilitating the smooth provision of integrated transport solutions. The case thus provides an operationalisation of the political economy theory and demonstrates that while strong policy intention can be translated to good implementation of transition, the presence of competing institutional interests and fragmentation of the bureaucratic system can pose a challenge to this process.

Results display half-achievements, such as the metro system in Delhi has grown, making access to the city much easier and minimizing the congestion on some routes, whereas EV-use is slow and only partially successful due to the lack of a power grid to charge them (Begam et al. 2024; Toering et al. 2025). The key issues are still inadequate infrastructure in last-mile connectivity and EV charging, divided governance, and financial unsustainability of expansion plans. The lessons focus on the need to enhance institutional cooperation, comprehensive planning that includes informal sectors, and specifically focus on investments in supporting infrastructure. Importantly, the case illustrates that technological transitions have to be coupled with governance reforms that are able to coordinate the infrastructure, regulation and socio-economic inclusion at different institutional scales. Solving policy-implementation gaps via multi-level governance coordination and long-term allocations of resources is important in realising the vision of sustainable and equitable urban mobility of Delhi.

8 Latin America

8.1 Case Study 5: Santiago, Chile–E-Buses & Integrated Transport Reform

The modern transport system in Santiago had been led by the old-fashioned diesel buses that not only added to air pollution and greenhouse gases but also meant a decline in the state of the buses. The main problem that the city faced was the heavy traffic, ineffective working methods, and social inequality in the use of transport, which still affects low-income citizens depending on informal services. Such circumstances were indicative of an overall absence of overall urban mobility planning, where infrastructure was mostly tailored to fossil fuel-powered vehicles and where modernization was hampered due to the fragmentation of governance (Vecchio et al. 2022; Coq-Huelva and Asian-Chaves 2019; Sagaris et al. 2017). Such a situation highlighted the necessity to have integrated reforms that included cleaner technology and fair service growth. The institutional limits and financial limitations restricted the previous effort, and long-term strategic planning and novel policy guidelines became critically needed in order to deal with environmental and social issues. The urban environment of fast growth and socio-spatial inequalities in Santiago also made the implementation process more difficult and required multi-agency efforts that would integrate climate objectives with mobility justice concerns (Sheehan 2022; Romero-Lankao et al. 2013). Socio-technical transition perspective: Santiago is in a partial restructuring regime, characterized by sustainability innovations taking shape through incremental changes between infrastructure, governance change and environmental policy pressures.

The Chilean government, in turn, has implemented a multi-faceted policy framework that facilitates the integration of electric buses, with regulatory changes, funding infrastructure investment, and financial support being the key priorities. Fleet renewal was supported by subsidies and grants and operational requirements by increased charging infrastructure. One of the prominent innovations was the institutional cooperation of the municipal, regional, and national governments in order to plan and finance, which was not the case previously (Hölzl and Nuissl 2014; Ubilla-Bravo 2018). The pilot tests in certain routes were empirical in nature in terms of the costs of operations, technology, and the maintenance issues, which informed the scaling plans. The key role was played by public-private partnerships, which integrated incentives of private operators with state climate objectives and led to risk-sharing mechanisms. These policies were integrated into a larger urban sustainability strategy and e-public transportation reform was to be incorporated into overall climate adaptation and social justice strategies (Thorne et al. 2021; Duan et al. 2021). This shows how a model of electric mobility crosses the barrier to global and is adapted and embedded into local institutional settings, rather than being taken as a whole.

Initial results showed that it had a strong advantage of lower emissions, noise, and low cost of operation in comparison to other diesel options. Nonetheless, there were some issues related to the large amount of capital needed to start, weaknesses in battery technology due to the climatic conditions in Santiago, and the necessity of specialized maintenance staff (Szürke et al. 2024; Thorne et al. 2021). Also, there were complexities in terms of institutional coordination especially in terms of fair route coverage to poorly served populations. The political economy aspect of transitions is emphasized in these outcomes, where infrastructure investment choices and outcomes impact the tempo and legitimacy of decarbonisation. The lessons of these experiences pointed to the shortcomings of initial success, that technology adoption is not enough without the presence of a strong adaptive infrastructure, capacity building, and institutional learning. It turned out that it is necessary to include socio-spatial equity and climatic resilience in long-term transport planning to maintain and magnify the transition (Romero-Lankao et al. 2013; Vecchio et al.

2022). The case in turn asserts the governance-transition theory, by highlighting the need for maintaining transition momentum in complex urban systems through sustained coordination across multiple actors.

The electric bus shift in Santiago highlights that sustainable urban mobility should be characterized by the long-term strategic consideration and institutional innovation. According to effective reform, there are governance reforms that provide cross-sector cooperation, partnerships among the government and business organizations, and consistent regulatory frameworks. In MLP terms, Santiago also represents the extent to which niche innovations (electric buses) can slowly work their way into, and reconfigure, an already established transport system, including the gradual policy alignment and infrastructure development. It is important to overcome the constraints of the original improvement by adaptive governance that can react to the arising technological and operational needs (Abdul et al. 2024). This kind of innovation promotes mobilization of resources and building infrastructure and uniting climate ambitions and social equity priorities. The future developments demand an investment in data-driven decision-making, infrastructure resilience, and the ability of institutions to keep it moving and to reap maximum benefits. The case of Santiago helps see the intricacies involved in the implementation of electric buses into urban mobility networks but helps confirm that long-term innovation and systematic planning are the ways to cleaner, more equitable, and climate-resilient cities (Romero-Lankao et al. 2013; Sagaris et al. 2017).

8.2 *Case Study 6: Curitiba, Brazil–Early BRT Model*

Curitiba, Brazil, is the place of the Bus Rapid Transit (BRT) concept origin, which is based on the peculiarities of the city development due to the high population growth and socio-economic shifts in the 1960s and 1970s. The economic growth of the city was accompanied with the growing urban sprawl, which forced the policymakers to seek new alternatives to deal with the urban mobility at low costs, unlike the use of rail transport (Lindau et al. 2010a; Hidalgo and Graftieaux 2008). In the political arena, Curitiba had had a stable political leadership with pragmatic long-term sustainable urban planning that incorporated transport and land use policies, which strengthened the social equity and conservation of the environment (Lindau et al. 2010a). Before the BRT initiation, the transport infrastructure in Curitiba was characterised by irregular and unsystematic buses in the traffic jam that resulted in unreliable timetables and unequal access. There was no integration in the fragmented public transport which meant that the commute was long and there was no control over city development. This atmosphere elicited the design of an integrated transport network (RIT) that cohesively connected bus corridors to development locations and open spaces with an emphasis on institutional innovation and coordinated governance to surmount disjointed and poor transit services delivery (Hidalgo and Graftieaux 2008; Krüger et al. 2021). In terms of governance transition, Curitiba is a first case of institutional innovation in which a network of coordination plans facilitated systemic reorganizations in urban mobility.

The main policies behind the BRT of Curitiba were centred on the availability and high capacity of the public transport, in conjunction with long-term city planning. The innovations were special bus lanes, paying in advance to save time on the bus, and buses made of articulation to have the highest number of passengers (Lindau et al. 2010a). The key element was institutional innovation whereby the Institute of Research and Urban Planning of Curitiba was established (IPPUC) which ensured the coordination of the decision-making in land use and mobility where transport infrastructure investment was aligned with urbanization (ibid.). The socio-economic part of the project focused on cost-efficiency through modifying established bus technologies and not the use of expensive rail systems, and the subsidies provided by the government were implemented in a strategic way to keep the cost of fares down. These inventions enabled the use of a corridor-based network that ensured high-quality service at high frequency with fast boarding and offboard collections of fare, and established new standards in the quality of bus transportation in the world (Hidalgo and Graftieaux 2008; Levinson et al. 2003). This case is an example of the "reverse diffusion" of policy, in which Curitiba became a world reference for the subsequent transport transition in the Global South.

The early BRT outcomes of Curitiba highlighted that efficiency and mobility, urban development, and environmental sustainability improved significantly. The capacity of the peak hour transit showed an impressive improvement, leading to the decrease of car reliance and city congestion and facilitating the pursuit of compact and transit-oriented development along bus routes (Lindau et al. 2010b; Wirasinghe et al. 2013). The benefits associated with the environment were realised by way of lower emissions and endorsing green areas in urban spaces with transit routes. However, there were difficulties connected with the institutional scalability and operational complications as the number of people increased; it was necessary to coordinate the actions of various agencies and adjust the infrastructure to new urban patterns due to the continuous innovation of governance (Bouraima et al. 2023; Wood 2015). Initial achievements were forced to be accompanied with the realization of constraints such as variations in service quality, disparities in accessibility to the marginalized groups and the necessity of constant upgrades in the technical aspects to keep the service level in line with the competition. The dynamics are a manifestation of the evolution of the MLP regime, in which innovations that work must continually evolve to retain their compatibility with the infrastructure, governance capacity and users' needs. These limitations emphasised the thin line between expanding BRT and maintaining the institutional and operational capabilities (Krüger et al. 2021; Bouraima et al. 2023).

Lessons learned in the long term in the experience of the BRT at Curitiba dwell on the critical need of visionary, comprehensive urban planning and long-term institutional innovation. The city proved that transit solutions that were grounded in well-established governance systems and sustained by ongoing adaptation and community involvement could yield extensive social and environmental positive returns (Lindau et al. 2010b; Hidalgo and Graftieaux 2008). The boundaries of initial success, however, highlight the fussiness of service quality and fair access in the face of city expansion and technological transformation. The case of Curitiba highlights the need to have adaptable policy structures that enable continuous evaluation and restructuring, which guarantees scalability without altering the overall aims of the services. Moreover, it emphasizes critical roles of land use policies that are coordinated and inclusive urban development as a complement to transit investments in a sustainable manner.

9 Discussion

The comparative analysis reveals that robust and organized governance is the main factor distinguishing a successful green transition in transport and more vulnerable or stalled ones in the Global South. Planned evolution and coordination of sectors and challenging reforms like incorporation or replacement of informal operators in Kigali, Cape Town, Shenzhen, Santiago and Curitiba were made possible by coherent institutional frameworks and political leadership (Geels 2011; Kern and Rogge 2016). These instances prove that the stability of MLP regimes can be strategically guided when political alliances, planning organisations and regulatory regimes converge on the idea of low-carbon mobility, as occurred in Shenzhen in its decisive push towards electrification and in Curitiba in its comprehensive planning via IPPUC (Lindau et al. 2010a; Li et al. 2020). In comparison, in Delhi, multi-level governance is fragmented and policy-implementation lapses have contributed to the political economy argument that dispersion of power, bureaucratic overlap and poor coordination in most cases dilutes ambitious transition agendas (Cerny 2010; Toering et al. 2025). In general, the results support governance-based literature that views transitions as a process of negotiation, not necessarily a techno-economic change, especially in resource-limited cities of the South (Kanger and Sovacool 2021; Krüger et al. 2021). From the findings there is a demonstration that green transport transition in the Global South is fundamentally governance intensive with political bargaining and institutional capacity equally important as technological advancement.

The results also suggest that context-sensitive and diversified financing architectures are essential to scaffold green transport innovations and this finding is consistent with the literature on the political economy of climate and infrastructure finance. The phased subsidy design, battery-leasing models, and coordinated self-financing between the state and market Shenzhen can be seen as examples of how the targeted financial innovation can de-risk new technologies and create a self-sustaining market, also in connection to study of strategic state-led financing in Asian low-carbon transitions (Li et al. 2020; Watari et al. 2021). The comparison critically show that the effectiveness of financing relies on both the amount of investment and the ability of the institution to align financial instruments with local realities. In turn, Kigali and Cape Town demonstrate how incremental BRT development can be supported by blended finance, which combines national subsidies, municipal resources and development partner assistance, under extreme fiscal constraints, as a result of work on climate finance and development banks as key facilitators in lower-income settings (Steckel et al. 2017; Brizon et al. 2025). The fact that Santiago depends on subsidies, long-term contracts and PPPs is also reminiscent of information that credible and operational financial structures are indispensable in large scale e-bus implementation (Thorne et al. 2021; Abdul et al. 2024). Simultaneously, the time spent in the dependency on the subsidies, problems associated with cost-recovery, and susceptibility to macroeconomic shocks witnessed in various cases support scholars who caution that in the absence of structural changes and predictable sources of revenues, the investments in green transport will remain delicate or non-scalable (Ahire et al. 2024; Osieczko et al. 2021). From these findings, it emerged that transition finance in the third world is also an institutional resilience and state capacity issue rather than only a technical funding challenge.

The third important conclusion is that only technological pathways that are socio-technically aligned with local infrastructures, informal transport ecologies, and equity priorities can be translated into a lasting transition. The Curituba, Cape Town and Dar es Salaem-type BRT experiences endorse the literature on socio-technical transitions, which underlines the co-evolution of technology, institutions and everyday practices: lane segregation, prepayment and feeder services only create desired impacts when integrated in enabling planning and regulatory frameworks (Hidalgo and Graftieaux 2008; Wirasinghe et al. 2013; Krüger et al. 2021). This comparison further articulated that successful transition is not guaranteed by infrastructure alone. The cities should also consider embedding technology in both social and technological arrangements for effective outcomes. Shenzhen electrification of its buses can also confirm the thesis that landscape pressures (air quality crises, national climate objectives) can promote niche technologies in the presence of infrastructural preparedness and operational innovations like real-time scheduling and dense charge networks (Geels 2002; Wang et al. 2018; Jiang et al. 2021). On the other hand, the partial EV adoption and final kilometre gaps in Delhi, and the failure to incorporate informal operators in Kigali and Cape Town, are not new, and underscore that reforms that are technology-oriented fail when they do not consider practices in the incumbent industry, as well as informal markets and the daily mobility demands of low-income residents (Agyemang et al. 2020; Ho and Tirachini 2023). The findings supports the socio-technical transition theory, which highlight the need for a context specific adaptive approaches rather than universal blueprint technological solutions.

Lastly, the cross-regional comparison indicates that neither equity nor informality are minor aspect of green transport transitions in the Global South, but rather a constitutive dimension, building on and refining dominant MLP-based narratives (Krüger et al. 2021; Manfriani et al. 2021; Sekabaraga et al. 2011; Stalmokaitė and Yliskylä-Peuralahti 2019; Szürke et al. 2024). In all cases, distributional consequences, such as who benefits, who suffers displacement, and how the informal workers are treated, determine the legitimacy of reforms, as well as their long-term maintenance, and resonate with climate justice and urban governance literature anticipating social inclusion in transition processes (Bulkeley et al. 2014; Romero-Lankao et al. 2013). Empirical findings in Cape Town, Santiago, and Kigali and Dar es Salaam, where transitions only worsen or reduce the current inequalities, are based on attempts to extend services to peripheral and older groups, and the difficulties of informal operators in Kigali and Dar es Salaem, respectively (Finn 2017; Vecchio et al. 2022; Krüger et al. 2021). Its findings, thus, resonate with and build upon recent criticism of standard transition frameworks theorising informality too little, too little, too late, and requiring broadened conceptual lenses to incorporate political economy, transitions in equity and governance even more explicitly (Watari et al. 2021; Van Asselt and Green 2022). The chapter puts forward critiques of the MLP framework as it argue that there is need for a strong engagement with informality, political economy, and climate justice within the theories of sustainability transitions if they are to adequately explain dynamics of the Global South.

10 Conclusion

The conclusion sums it up by emphasizing that the movement of green transport transitions in Global South is determined by the combination of the quality of governance, financial capacity, technological decisions, and the potential socio-spatial inequality entrenched. The relative examination of Kigali, Cape Town, Shenzhen, Delhi, Santiago, and Curitiba shows that with the firm political leadership, specialised planning organisations and coherent multi-level government, more decisive moves toward BRT and electric mobility can be made even when resources are limited. Even contextual diversity, however, restrains the easy transfer of policies, transitions come in the hybrid ways of combining incremental reforms and selective technological leapfrogging. One of the most important contributions of the chapter is that it focuses on informality and equity and reveals that efforts aimed at formalisation of the transport systems and implementation of low-carbon technologies tend to reproduce or reorganise old exclusions unless specific social justice strategies are included. On the whole, the results help to see green transport changes as controversial, path-related processes, in which socio-technical innovation, political economy, and day-to-day mobility practices come into contact.

Recommendations from the study highlight the need to adopt conscious and sensitive strategies to reconcile climate and development agendas in cities in Africa, Asia, and Latin America. The first recommendation is to strengthen integrated and multilevel governance systems to ensure coordination in urban planning, transportation, environmental management, and financial management to fill gaps in implementation and reduce the fragmentation of policies. The second recommendation is to adopt long-term and diversified funding strategies to ensure the integration of domestic resources, climate finance, and innovative instruments to meet capital and operating expenses. The third recommendation is to adopt an inclusive approach to managing informal transportation by negotiating contracts and building capacity and gradually integrating them into the system. The fourth recommendation is to adopt equity-oriented approaches to ensure the needs of the periphery and the poor are met by prioritizing these areas in the design and operation of the transport network. The fifth recommendation is to adopt adaptive and data-driven approaches to the operation of the transport network to ensure the management of uncertainties in technology and travel patterns. The sixth recommendation is to adopt a diffusion approach to policymaking by prioritizing contextual adaptation and using comparative learning to ensure flexible and contextualized transition pathways.

11 Limitations

However, despite the merits mentioned above, there are various limitations associated with this study. Firstly, the use of the qualitative desk review methodology with a secondary dataset hinders the possibility of using primary sources to empirically validate some of the assertions made. Secondly, while the case studies have been selected to ensure coverage across the whole region, they do not comprehensively capture all possible transition trajectories in the Global South. Thirdly, discrepancies in data quality across case studies might affect analysis. Finally, the study mainly concentrates on urban transport transitions without considering linkages with energy systems and rural transport dynamics.

References

Abdul S, Adeghe E, Adegoke B, Adegoke A, Udedeh E (2024) Public-private partnerships in health sector innovation: lessons from around the world. Magna Sci Adv Biol Pharm 12(1):045–059. https://doi.org/10.30574/msabp.2024.12.1.0032

Agyemang KK, Amoako-Sakyi RO, Antwi KB, Mensah CA, Abane AM (2020) Transit oriented development: theory and implementation challenges in Ghana. TeMA-J Land Use, Mobil Environ 13(3):409–425

Ahire JP, Bergman R, Runge T, Mousavi-Avval SH, Bhattacharyya D, Brown T, Wang J (2024) Techno-economic and environmental impacts assessments of sustainable aviation fuel production from forest residues. Sustain Energy Fuels 8(19):4602–4616

Baker L, Newell P, Phillips J (2014) The political economy of energy transitions: the case of South Africa. New Polit Econ 19(6):791–818

Begam S, Jha P, Yadav PK, Joy MS, Rawat P, Bansal T (2024) Urban transformation through transit: the case of Delhi metro. Discov Cities 1(1). https://doi.org/10.1007/s44327-024-00035-1

Bouraima MB, Tengecha NA, Stević Ž, Simić V, Qiu Y (2023) An integrated fuzzy MCDM model for prioritizing strategies for successful implementation and operation of the bus rapid transit system. Ann Oper Res 342(1):141–172. https://doi.org/10.1007/s10479-023-05183-y

Brizon LC, Caetano JA, De Oliveira CM, Orrico Filho RD (2025) BRT Systems in Brazil: technical analysis of advances, challenges, and operational gaps. Urban Sci 9(10):414. https://doi.org/10.3390/urbansci9100414

Bulkeley H, Castán Broto V, Maassen A (2014) Low-carbon transitions and the reconfiguration of urban infrastructure. Urban Stud 51(7):1471–1486. C40 Cities. (2022). The future of electric buses in cities. http://www.c40.org/reports/the-future-of-electric-buses

Cerny PG (2010) Rethinking world politics: a theory of transnational neopluralism. Oxford University Press

Coq-Huelva D, Asián-Chaves R (2019) Urban sprawl and sustainable urban policies. A review of the cases of Lima, Mexico City and Santiago de Chile. Sustainability 11(20):5835. https://doi.org/10.3390/su11205835

Costantini V, Crespi F, Marin G, Paglialunga E (2017) Eco-innovation, sustainable supply chains and environmental performance in European industries. J Cleaner Prod 155:141–154

Du Preez D (2025) A framework for assessing the financial sustainability of minibus taxi route typologies by means of onboard trip data cluster analysis. African Transp Stud. https://doi.org/10.1016/j.aftran.2024.100022

Duan M, Qi G, Guan W, Lu C, Gong C (2021) Reforming mixed operation schedule for electric buses and traditional fuel buses by an optimal framework. IET Intell Transp Syst 15(10):1287–1303. https://doi.org/10.1049/itr2.12098

Emodi N, Okereke C, Abam F, Diemuodeke O, Owebor K, Nnamani U (2022) Transport sector decarbonisation in the global south: a systematic literature review. Energ Strat Rev. https://doi.org/10.1016/j.esr.2022.100925

Finn B (2017) Quietly chasing Kigali: young men and the intolerance of informality in Rwanda's Capital City. Urban Forum 29(2):205–218. https://doi.org/10.1007/s12132-017-9327-y

Gauthier A (2012) Finding Atlantis: Cape Town reconnects. Sustain Transp 23. http://worldcat.org/oclc/31056686

Geels FW (2002) Technological transitions as evolutionary reconfiguration processes: a multi-level perspective and a case-study. Res Policy 31(8–9):1257–1274

Geels FW (2011) The multi-level perspective on sustainability transitions: responses to seven criticisms. Environ Innov Soc Trans 1(1):24–40

Geels FW, Schot J (2007) Typology of sociotechnical transition pathways. Res Policy 36(3):399–417

Goodfellow T, Smith A (2013) From urban catastrophe to 'model' City? Politics, security and development in post-conflict Kigali. Urban Stud 50(15):3185–3202. https://doi.org/10.1177/0042098013487776

Grande-Ayala CE (2024) An assessment of accessibility from a socially sustainable urban mobility approach in mass transit projects: contributions from the northern central American triangle. Sustainability 16(9):3766. https://doi.org/10.3390/su16093766

Guno CS, Collera AA, Agaton CB (2021) Barriers and drivers of transition to sustainable public transport in The Philippines. World Electr Veh J 12(1):46. https://doi.org/10.3390/wevj12010046

Hadders, R. (2019). Access to livelihood opportunities through MyCiTi BRT in Cape Town, South Africa (Master's thesis)

Hidalgo D, Graftieaux P (2008) Bus rapid transit Systems in Latin America and Asia. Transp Res Rec: J Transp Res Board 2072(1):77–88. https://doi.org/10.3141/2072-09

Ho CQ, Tirachini A (2023) Mobility-as-a-service and the role of multimodality in the sustainability of urban mobility in developing and developed countries. Transp Policy 145:161–176. https://doi.org/10.1016/j.tranpol.2023.10.013

Hölzl C, Nuissl H (2014) Urban policy and spatial planning in a Globalized City—a stakeholder view of Santiago de Chile. Plan Pract Res 29:21–40. https://doi.org/10.1080/02697459.2013.872895

International Energy Agency (IEA). (2023). Transport sector CO2 emissions. http://www.iea.org/reports/transport

Jiang M, Zhang Y, Zhang Y (2021) Multi-depot electric bus scheduling considering operational constraint and partial charging: a case study in Shenzhen, China. Sustainability 14(1):255. https://doi.org/10.3390/su14010255

Kalisa E, Sudmant A, Ruberambuga R, Bower J (2025) Natural experiments in urban air quality: lessons from car-free days and COVID-19 lockdowns in Kigali, Rwanda. Cities Health:1–12. https://doi.org/10.1080/23748834.2025.2468017

Kanger L, Sovacool BK (2021) Towards a multi-scalar and multi-horizon framework of energy injustice: a whole systems analysis of Estonian energy transition. Polit Geogr 93:102544. https://doi.org/10.1016/j.polgeo.2021.102544

Kern F, Rogge KS (2016) The pace of governed energy transitions: agency, international dynamics and the global Paris agreement accelerating decarbonisation processes?. Energy Res Soc Sci 22:13–17

Krüger F, Titz A, Arndt R, Groß F, Mehrbach F, Pajung V, Suda L, Wadenstorfer M, Wimmer L (2021) The bus rapid transit (BRT) in Dar Es Salaam: a pilot study on critical infrastructure, sustainable urban development and livelihoods. Sustainability 13(3):1058. https://doi.org/10.3390/su13031058

Lambert IM, Poortinga W, Potoglou D, Xenias D (2026) Preferences for urban street space reallocation to encourage cycling: a best-worst scaling profile case approach. Travel Behav Soc 43:101205

Levinson HS, Zimmerman S, Clinger J, Gast J (2003) Bus rapid transit: synthesis of case studies. Transp Res Rec: J Transp Res Board 1841(1):1–11. https://doi.org/10.3141/1841-01

Li M, Ye H, Liao X, Ji J, Ma X (2020) How Shenzhen, China pioneered the widespread adoption of electric vehicles in a major city: implications for global implementation. WIREs Energy Environ 9(4). https://doi.org/10.1002/wene.373

Lindau LA, Hidalgo D, Facchini D (2010a) Curitiba, the cradle of bus rapid transit. Built Environ 36(3):274–282. https://doi.org/10.2148/benv.36.3.274

Lindau LA, Hidalgo D, Facchini D (2010b) Bus rapid transit in Curitiba, Brazil. Transp Res Rec: J Transp Res Board 2193(1):17–27. https://doi.org/10.3141/2193-03

Manfriani C, Gualdani G, Goli G, Carlson B, Certo AR, Mazzanti P, Fioravanti M (2021) The contribution of IoT to the implementation of preventive conservation according to European standards: the case study of the "cannone" violin and its historical copy. Sustainability 13(4):1900

Minty Z (2006) Post-apartheid public art in Cape Town: symbolic reparations and public space. Urban Stud 43(2):421–440. https://doi.org/10.1080/00420980500406728

Naufal D, Putri HTY, Nurmandi A, Fridayani HD (2026) Low carbon transportation Systems for Sustainable Urban Infrastructure: policy implementation and challenges in Indonesia. In: IOP conference series: earth and environmental science, vol 1610. https://doi.org/10.1088/1755-1315/1610/1/012005

Nnene O, Zuidgeest M, Joubert J (2023) BRT network design for transit cost reduction in Cape Town, South Africa. J Public Transp. https://doi.org/10.1016/j.jpubtr.2023.100042

Ntamwiza J, Bwire H, Nkurunziza A (2025) Enhancing connectivity via GIS-based bike-sharing optimization in Kigali City, Rwanda. J Environ Earth Sci 7(8):191–206. https://doi.org/10.30564/jees.v7i8.10565

Osieczko K, Zimon D, Płaczek E, Prokopiuk I (2021) Factors that influence the expansion of electric delivery vehicles and trucks in EU countries. J Environ Manag 296: 113177

Rizzo M (2014) The political economy of an urban megaproject: the bus rapid transit project in Tanzania. Afr Aff 114(455):249–270. https://doi.org/10.1093/afraf/adu084

Romero-Lankao P, Hughes S, Rosas-Huerta A, Borquez R, Gnatz DM (2013) Institutional capacity for climate change responses: an examination of construction and pathways in Mexico City and Santiago. Environ Plan C: GovMent Policy 31(5):785–805. https://doi.org/10.1068/c12173

Sagaris L, Tiznado-Aitken I, Steiniger S (2017) Exploring the social and spatial potential of an intermodal approach to transport planning. Int J Sustain Transp 11(10):721–736. https://doi.org/10.1080/15568318.2017.1312645

Schalekamp H (2011) Reflecting on Progress with Cape Town's MyCITI Bus System:12

Schalekamp H (2017) Lessons from building paratransit operators' capacity to be partners in Cape Town's public transport reform process. Transp Res Part A-Policy Pract 104:58–66. https://doi.org/10.1016/j.tra.2017.08.002

Sekabaraga C, Diop F, Soucat A (2011) Can innovative health financing policies increase access to MDG-related services? Evidence from Rwanda. Health Policy Plan 26(Suppl 2):ii52–ii62. https://doi.org/10.1093/heapol/czr070

Sheehan M (2022) Everyday verticality: migrant experiences of high-rise living in Santiago, Chile. Urban Stud 61(4):726–742. https://doi.org/10.1177/00420980221114206

Sims R, Schaeffer R, Creutzig F et al (2014) Transport. In: Edenhofer O, Pichs-Madruga R et al (eds) Climate change 2014: mitigation of climate change. Contribution of Working Group III to the Fifth Assessment Report of the Intergovernmental Panel on Climate Change. Cambridge University Press, Cambridge, United Kingdom and New York, USA. https://www.ipcc.ch/pdf/assessment-report/ar5/wg3/ipcc_wg3_ar5_chapter8.pdf

Song Y, Ning H, Ye X, Chandana D, Wang S (2022) Analyze the usage of urban greenways through social media images and computer vision. Environ Plan B: Urban Anal City Sci 49(6):1682–1696. https://doi.org/10.1177/23998083211064624

Stalmokaitė I, Yliskylä-Peuralahti J (2019) Sustainability transitions in Baltic Sea shipping: exploring the responses of firms to regulatory changes. Sustainability 11(7):1916. https://doi.org/10.3390/su11071916

Steckel JC, Jakob M, Flachsland C, Kornek U (2017) From climate finance toward sustainable development finance. Wiley Interdiscip Rev Clim Chang 8(1):e437. https://doi.org/10.1002/wcc.437

Szürke SK, Saly G, Lakatos I (2024) Analyzing energy efficiency and battery supervision in electric bus integration for improved urban transport sustainability. Sustainability 16(18):8182. https://doi.org/10.3390/su16188182

Thorne RJ, Hovi IB, Figenbaum E, Pinchasik DR, Amundsen AH, Hagman R (2021) Facilitating adoption of electric buses through policy: learnings from a trial in Norway. Energy Policy 155:112310. https://doi.org/10.1016/j.enpol.2021.112310

Toering A, De Bruijne M, Veeneman W (2025) Exploring governance challenges of sustainable infrastructure development on the nexus between energy and mobility. Npj Sustain Mobil Transp 2(1). https://doi.org/10.1038/s44333-024-00018-0

Ubilla-Bravo, G. (2018). Building territorial governance: experiences of inter-municipal work through a regional and spatial planning system [Construyendo la Gobernanza territorial: Experiencias de Trabajo Intermunicipal Mediante un Sistema regional de Planificación y Ordenamiento territorial] (no. halshs-01995340)

United Nations Environment Programme (UNEP) (2019) Global environment outlook, p 6. http://www.unep.org/resources/global-environment-outlook-6

Van Asselt H, Green F (2022) COP26 and the dynamics of anti-fossil fuel norms. WIREs. Climate Change 14(3). https://doi.org/10.1002/wcc.816

Vanderschuren M, Cameron R, Newlands A, Schalekamp H (2021) Geographical modelling of transit deserts in Cape Town. Sustainability. https://doi.org/10.3390/su13020997

Vecchio G, Tiznado-Aitken I, Castillo B, Steiniger S (2022) Fair transport policies for older people: accessibility and affordability of public transport in Santiago, Chile. Transportation 2(1):689–715. https://doi.org/10.1007/s11116-022-10346-0

Voege T (2018) Data-led governance of self-driving vehicles for urban shared mobility. In The Governance of Smart Transportation Systems: Towards New Organizational Structures for the Development of Shared, Automated, Electric and Integrated Mobility (pp 81–101). Cham: Springer International Publishing

Waisman H, Bataille C, Briand Y, Torres Gunfaus M, Pye S (2026) Deep decarbonization pathways for national development consistent with carbon neutrality. Clim Policy 26(5):701–708

Wang, G., Xie, X., Zhang, F., Liu, Y., & Zhang, D. (2018). bCharge: data-driven real-time charging scheduling for large-scale electric bus fleets. 45–55. https://doi.org/10.1109/rtss.2018.00015

Watari T, Nansai K, Nakajima K, Giurco D (2021) Sustainable energy transitions require enhanced resource governance. J Clean Prod 312:127698. https://doi.org/10.1016/j.jclepro.2021.127698

Wirasinghe SC, Kattan L, Rahman MM, Hubbell J, Thilakaratne R, Anowar S (2013) Bus rapid transit—a review. Int J Urban Sci 17(1):1–31. https://doi.org/10.1080/12265934.2013.777514

Wood A (2015) Multiple temporalities of policy circulation: gradual, repetitive and delayed processes of BRT adoption in south African cities. Int J Urban Reg Res 39(3):568–580. https://doi.org/10.1111/1468-2427.12216

World Health Organization (WHO). (2021). WHO global air quality guidelines. http://www.who.int/publications/i/item/9789240034228

Yu J, Yang P, Zhang K, Wang F, Miao L (2018) Evaluating the effect of policies and the development of charging infrastructure on electric vehicle diffusion in China. Sustainability 10(10):3394. https://doi.org/10.3390/su10103394

Yudha SW, Tjahjono B, Longhurst P (2022) Sustainable transition from fossil fuel to geothermal energy: a multi-level perspective approach. Energies 15(19):7435. https://doi.org/10.3390/en15197435

Zhou Y, Han J, Li J, Zhou Y, Wang K, Huang Y (2021) Building resilient cities with stringent pollution controls: a case study of robust planning of Shenzhen City's urban agriculture system. J Clean Prod 311:127452. https://doi.org/10.1016/j.jclepro.2021.127452

Informal Transport and Urban Green Transitions in the Global South: Rethinking Mobility Systems and Livelihood Regimes

James Kanyepe and Munyaradzi Chibaro

Abstract This chapter examines the role of informal transportation in urban mobility ecosystems in terms of employment, environmental sustainability, and governance. This study employed a desk review methodology to synthesize literature, policy reports, and case studies from the Global South to develop a nuanced understanding of informal transport through comparative and thematic analyses. The findings highlight that informal transport is essential for offering both direct and indirect employment opportunities to substantial segments of the urban workforce, particularly in regions characterized by high unemployment and labor market informality. However, they also pose environmental challenges, such as emissions from aging vehicles and inadequate regulations. Despite these challenges, shared mobility can reduce per-capita emissions compared with private vehicles. This chapter identifies exclusionary governance approaches as ineffective and disruptive, advocating integrated and inclusive policies that align urban greening and decarbonization with livelihood protection. The study concludes that sustainable mobility transitions in the Global South necessitate gradual formalization, participatory governance, and incentivized fleet renewal, recognizing the structural importance of informal transport and its integration into formal systems.

Keywords Environmental Governance · Transport informality · Job creation · Livelihoods · Sustainable mobility transition · Urban accessibility

1 Introduction

In the urban regions of developing nations, informal and semiformal transportation systems frequently predominate, complement, compete with, or even supplant public transit (Velasco 2025). These systems include minibuses and taxi services,

J. Kanyepe (✉) · M. Chibaro
Department of Logistics and Supply Chain Management, University of Botswana, Gaborone, Botswana
e-mail: jameskanyepe@gmail.com

J. Kanyepe and I. Chirisa (eds.), *Advancing the Green Agenda in the Transport and Logistics Sector in Global South Countries*, World Sustainability Series,
https://doi.org/10.1007/978-3-032-31374-4_20

motorcycle taxis (such as "tuk-tuks"), rickshaws and auto-rickshaws, shared taxis, informal freight services, and other small-scale providers that are crucial for mobility in both urban and rural areas. Millions of people rely on informal transportation for employment, education, healthcare, market access and social interaction (Chaudhry et al. 2024). Consequently, these systems are integral to the social, economic, and physical aspects of urban development in developing nations. They have emerged owing to rapid urbanization, uneven economic growth, and insufficient government investment in transportation infrastructure (Logiodice et al. 2025).

Rapidly expanding cities in Africa, Asia, and Latin America have outpaced government transportation planning, resulting in gaps in public transportation accessibility. Informal operators bridge this gap by offering flexible, demand-responsive options at lower costs and by adjusting routes, schedules, and fares to serve marginalized, informal, and low-income communities that are often overlooked by formal systems (Myers 2021). In addition to mobility, semiformal and informal transport play significant socioeconomic roles, employing millions of drivers, conductors, mechanics, dispatchers, ticket sellers and vehicle owners (Salas-Castillo et al. 2025). Additionally, it bolsters local economies through passenger movement, freight hauling, and connecting communities via last-mile movement, sustaining informal value chains in vehicle maintenance, fuel sales, and terminal activities. Amid high unemployment and labor informality in many developing urban centers, semiformal and informal transport is more than an alternative mode of mobility; it is a crucial livelihood mechanism for the poor (Logiodice et al. 2025).

Despite their critical role in urban society, informal transport systems are frequently viewed unfavorably in urban policy discourse (Gumbo et al. 2022a, b, c). They are often characterized as disorderly, inefficient, congested, poorly regulated, and environmentally detrimental. This perception fuels initiatives to modernize urban transportation, typically aiming to eliminate, replace, or formally integrate these systems into centralized models such as light rail, bus rapid transit, and large-scale bus operations (Desmoulière 2023). While certain reforms focus on enhancing safety, reducing emissions, and improving traffic management, they may have negative consequences if not accompanied by social protection programs, institutional integration, or participatory planning (Dabel 2026). Exclusionary measures have led to job displacement among informal operators, diminished transportation access for low-income communities, and increased urban inequality. In addition, environmental concerns related to informal transportation further complicate mobility discussions (Venter et al. 2019). Many fleets utilize older vehicles with poor fuel efficiency, minimal emission control, and inadequate maintenance, thereby contributing to air quality and greenhouse gas (GHG) challenges (Dabel 2026). Nevertheless, informal transportation provides shared rides that can decrease private car use and promote collective mobility at a lower cost. Consequently, the environmental impact of informal systems is complex and context-dependent, necessitating nuanced evaluations beyond simplistic assumptions that equate informality with unsustainability (Cummins and Monaghan 2025).

Despite the increasing body of research on sustainable mobility options, urban transport transitions, and informality in transportation in the Global South, notable research gaps remain. Existing studies predominantly focus on operational efficiency, negative externalities such as congestion, and environmental concerns associated with informal transportation systems, often depicting them as impediments to modernization and urbanization (Brock et al. 2025). Conversely, there is a dearth of research on achieving environmental sustainability objectives while safeguarding livelihoods and promoting equitable mobility. Most modernization frameworks advocate the replacement or elimination of informal transportation strategies, neglecting the adaptability of hybrid transportation arrangements and their institutional significance. This chapter seeks to address these gaps by examining informal and hybrid transport systems as integral components of mobility ecosystems in the Global South rather than as anomalies to be eradicated. This study aims to (1) explore the contribution of informal transport systems to employment generation within sustainable urban development contexts, (2) assess the environmental impacts of informal systems in relation to urban greening and decarbonization agendas, and (3) investigate policy pathways for greening informal transport systems through improved gradual formalization and transition to low-emission mobility.

Informal and hybrid systems should not be viewed as temporary development deficits in urban governance, transport infrastructure, labor markets, or spatial planning, but rather as responses to gaps in development (Buijs 2025). Their persistence illustrates that mobility in the Global South is characterized by hybrid arrangements involving private enterprises, community coordination, informal labor, and formal institutions. This chapter challenges modernization approaches that seek to eliminate informal transport without recognizing its social role. It posits that sustainable transitions toward mobility necessitate integration-oriented approaches that consider environmental objectives within the framework of livelihood protection, equitable access to mobility, and social justice (Eriskin 2024). Achieving such a balance requires policy frameworks that combine regulatory reform, labor inclusion, environmental upgrading, and participatory governance. To develop this argument, this chapter first examines the historical and structural foundations of transport informality in the Global South. The contributions of the informal sector to employment, accessibility, and last-mile connectivity were analyzed, and their environmental impact and governance challenges were assessed. This chapter critiques elimination-based modernization strategies and explores alternative policy pathways centered on inclusive formalization, incentive-based fleet renewal programs, and collaborative regulatory frameworks. Finally, the implications of hybrid mobility systems for sustainable and equitable urban transportation in the future are considered (Desmoulière 2023).

2 Literature Review

2.1 Informality in Transport Systems

In the fields of urban studies, development studies, and political economy, informality has been extensively examined, particularly in relation to labor markets, housing, and service provision in the Global South. Informality within the transport sector is typically characterized by encompassing all forms of mobility services that operate, either wholly or partially, outside formal regulatory, institutional, and/or legal frameworks (Eriskin 2024). Several scholars contend that informality should not be perceived as a lack of regulation or state control. Rather, it represents an organic organizational method that is both intricate and dynamic, emerging from the interplay between formal institutional regimes, market forces, and individualized survival strategies. Early modernization theorists regarded informal transportation systems as temporary transitional mechanisms linked to either poor governance or low levels of economic development (Cheng et al. 2025). These theorists anticipated that informal modes of transportation would become obsolete with government investment in infrastructure and the development of mature formal public transportation systems (Gumbo et al. 2022a, b, c).

However, subsequent studies have challenged these assumptions. Studies have demonstrated that informal transportation systems persist in many regions worldwide, including rapidly developing cities (RDCs). Research has identified reasons for the development of informal transportation systems beyond governmental failure. For example, informal transportation offers flexible solutions to address high levels of unmet mobility needs, geographical inequality, and limited public investment in transportation. Recently, interest in conceptualizing transportation systems as hybrid combinations of formal and informal practices has increased. This perspective recognizes that many transport providers operate within the context of partial regulation, including licensing schemes, transport unions, route associations, and negotiated governance. Consequently, rather than existing outside the state, informal transportation systems function within a complex network of interactions with public authorities, local politicians, and their associates (Onodugo and Ezeadichie 2020).

2.2 Informal Transport and Urban Mobility

Extensive research has highlighted the pivotal role of informal transportation in facilitating mobility within urban environments across regions worldwide (Duri 2025). Investigations into informal transportation systems in Africa, Asia, and Latin America have revealed that a substantial proportion of daily public transportation journeys in numerous cities are conducted via informal modes, including minibuses, motorcycle taxis, automobile haws, and rideshare services (Duri 2025). Formal transportation infrastructure is often unavailable, fragmented, unaffordable, or inaccessible to many economically disadvantaged urban residents; thus, informal transportation systems are crucial in these contexts. Studies on urban accessibility underscore the operational flexibility of informal transportation systems in urban areas. In contrast to conventional fixed-route transit systems, informal operators can readily modify routes, frequencies, and service coverage in response to fluctuations in passenger demand and development patterns. Consequently, they can service low-density residential neighborhoods and informal settlements, which are frequently overlooked by formal transit agencies (Oraiopoulos et al. 2025).

Research has underscored the importance of informal transportation in providing last mile connectivity (LMC). Numerous informal operators function as feeders to formal transportation systems, conveying passengers to bus terminals, train stations, markets, and other employment sites. In certain urban settings, formal mass transit systems indirectly depend on informal networks to increase access and performance (Cummins et al. 2025). However, researchers have identified significant operational challenges associated with these systems. The primary challenges include traffic congestion, unstable fares, overloading, hazardous vehicle operations, and unreliable services (McKay et al. 2024). These issues arise predominantly from weak regulation of informal transportation systems, inadequate infrastructure, and intense competition among operators. Nonetheless, researchers contend that attributing all these issues solely to the informal nature of these systems is inaccurate. They argued that both ineffective urban governance and deficient infrastructure are critical factors contributing to transportation inefficiency.

2.3 Employment, Livelihoods, and Informal Economies

A substantial body of academic research has examined the role of informal transport systems in generating employment in developing countries. These systems serve as a vital source of income for millions, particularly in urban areas characterized by high unemployment rates, labor market disconnection, and limited industrial development (Gumbo et al. 2022a, b, c). Studies on motorcycle taxi and minibus systems reveal that the transport sector fosters an extensive network of economic relationships that extend beyond drivers and car owners. Employment opportunities are available for mechanics, conductors, dispatchers, fuel suppliers, ticket sellers, spare part vendors, and traders at transportation terminals. Consequently, informal mobility systems are often regarded as essential infrastructure for urban economies. In addition to investigating the precarious nature of work environments within informal transport systems, researchers have explored issues such as unstable income, long working hours, hazardous working conditions, and inadequate access to social protection (Wikstrøm and Røe 2022). Leasing schemes for vehicle acquisition and daily revenue requirements create the potential to exploit labor relations. Nonetheless, many scholars argue that informal transport is crucial for survival in contexts where formal employment opportunities are limited. Recently, there has been a growing emphasis on integrating transportation policy with labor and social protection. Many researchers contend that modernizing transportation while overlooking employment transitions may exacerbate poverty and social exclusion. This focus has intensified advocacy for inclusive formalization approaches that enhance workers' quality of life while safeguarding their livelihoods.

2.4 Environmental Sustainability and Informal Mobility

The environment plays a pivotal role in the current discourse on informal transportation. Informal mobility fleets are significant contributors to particulate emissions (air pollution), greenhouse gas (GHG) emissions, and suboptimal fuel efficiency, primarily because of vehicle aging and inadequate maintenance systems. In numerous urban areas, informal freight vehicles, motorcycle taxis, and minibus operations generate substantial amounts of particulate matter, adversely affecting the environment (Onodugo and Ezeadichie 2020). Although opinions vary regarding the extent of environmental harm caused by informal transportation systems, some scholars argue that these systems reduce individual vehicle usage and enhance ridership per vehicle (Onodugo and Ezeadichie 2020). Consequently, the authors suggest that informal transportation systems may foster environmentally sustainable practices compared to increased reliance on private vehicles.

Onodugo and Ezeadichie (2020) posited that regulatory restrictions alone are insufficient to achieve environmental sustainability. Instead of concentrating solely on the stringent regulation of informal transportation systems, researchers advocate the advancement of technology, improved access to cleaner energy sources, a transition from fossil-based fuels to electricity-based fuels, and investment in long-term sustainable transportation infrastructure (Chibaro et al. 2026b). Scholars have also identified several structural barriers faced by informal operators, including limited access to credit, high costs of new technologies, and insufficient institutional support for fleet replacement and retirement. There is growing scholarly interest in integrating electric mobility into informal transportation systems, such as electric motorcycles and low-emission minibuses with battery-swapping capabilities. Researchers acknowledge that while pilot programs in various countries show promise, a successful transition to an environmentally sustainable transportation system will require financial support, technical assistance, and the establishment of inclusive governance structures.

2.5 Governance, Regulation, and Formalization

Transport governance is a significant theme in scholarly discourse on informal mobility systems. Numerous researchers examining informal transport sectors have conceptualized transport governance as a multifaceted institutional framework encompassing state agencies, transit unions, route associations, local authorities, and informal networks (Cheng et al. 2025). Historically, regulatory strategies have predominantly emphasized stringent control over informal transport sectors through criminalization and eradication. However, the prevailing scholarly consensus suggests that such strategies generally fail to produce sustainable outcomes because they do not adequately address the demand for mobility or the realities of the labor market (Wu 2025). Periodic enforcement actions against informal operators, which often result in vehicle confiscation and route restrictions, typically foster coercive interactions between authorities and operators while exacerbating corruption and disorderly regulations (Onodugo and Ezeadichie 2020).

Contemporary scholarship increasingly focuses on participatory governance. Empirical studies indicate that collaborative engagement among governments, transport associations, civil society organizations, and community groups can enhance policy implementation and mitigate resistance to reforms. Partnership-based organizational models are widely recognized as mechanisms for ensuring accountability, safeguarding workers' rights, and coordinating services (Cherednichenko 2024). The literature on formalization underscores the importance of gradual and incentive-based approaches. Scholars contend that highly bureaucratic licensing systems and costly compliance requirements may marginalize small-scale operators and exaggerate inequalities. Inclusive formalization strategies emphasize phased regulatory changes, simplified registration systems, access to finance, and institutional support systems that facilitate operators' transition to safer and more sustainable practices (Aderibigbe and Gumbo 2024).

3 Methodology

This study used a qualitative desk review to assess the roles of informal and hybrid transport systems in the mobility ecosystems of the Global South. This method is particularly appropriate for this study because it facilitates the integration of existing scientific discussions, policy dialogues, and empirical evidence on transport informality, sustainable mobility transition, job creation, environmental governance, and urban accessibility. Given the diversity of informal transport systems across regions and their interdisciplinary nature, this approach provides a valuable framework for evaluating trends, contextual variations, and developing policies. Furthermore, this chapter draws on a comprehensive range of secondary sources, including scientific journal articles, academic books, policy papers, government reports, publications from international organizations, transport planning documents, and case studies from Africa, Asia and Latin America. These secondary sources were selected based on their relevance to informal mobility systems, hybrid governance of transport, urban transport modernization, environmental sustainability, labor informality, and participatory transport regulation. The specialized literature critically examines the connections between transportation systems, social inclusion, and sustainability transitions in rapidly urbanizing areas (Gumbo et al. 2022a, b, c).

Thematic analysis identified key themes, such as employment creation, accessibility, environmental impact, regulatory challenges, modernization policies, and inclusive formalization, which guided the structure of this chapter. This chapter employs a comparative approach to highlight the similarities and differences between regional experiences. Case studies from the Global South reveal that informal transport systems operate under various conditions, indicating the absence of a common operational pattern in these systems. Transport informality is influenced by local governance, infrastructure, the labor market, and political and economic conditions (Onodugo and Ezeadichie 2020). The methodology spans urban studies, transport geography, development studies, political economy, labor studies, and environmental governance, as informal transport requires more than technical planning frameworks. Employment, governance, social inequality, and sustainability must be considered. While the desk review offers analytical advantages, it relies solely on secondary data and lacks primary empirical information from operators, users, regulators and communities. Consequently, the local mobility dynamics may not be fully captured. Variations in data quality, collection methods, and regional coverage may affect comparability. Despite these limitations, the methodology provides a foundation for understanding informal and hybrid transport systems and proposing sustainable and inclusive mobility governance guidelines for them.

4 Results

4.1 The Role of Informal Transport Systems in Generating Employment Within the Framework of Sustainable Urban Development

The desk review underscores that informal transport systems offer significant employment opportunities in urban and peri-urban regions of the Global South. These systems serve as crucial mechanisms for labor absorption in areas marked by rapid urbanization, structural unemployment, and limited capacity to generate formal employment. Bermúdez and Robledo (2023) suggest that substantial segments of the population are excluded from formal employment owing to educational constraints, skill mismatches, and macroeconomic limitations on job creation. Consequently, informal transport systems play a central role in urban labor markets, extending beyond their function as supplementary mobility providers.

A critical insight from the desk review is that employment within informal transport systems encompasses a wide range of roles beyond the most visible category of vehicle operation. While drivers of minibuses, motorcycle taxis, and automobile workers constitute the frontline workforce, the sector includes a diverse network of interdependent occupations, such as vehicle owners, mechanics, spare parts dealers, fuel vendors, cleaners, dispatchers, ticket sellers, informal terminal operators, and route coordinators (Oluwakoya 2024). These roles are often embedded within locally organized systems governed by informal associations and unions. This illustrates that informal transport functions as an integrated employment ecosystem with significant backward and forward linkages, rather than as a single occupational domain. Thus, its overall contribution to employment generation is much greater than that captured through driver-focused analyses.

The accessibility of informal transport systems as a source of employment is another key aspect highlighted in the literature. Compared with formal sector jobs, which require specific qualifications, formal contracts, and regulatory compliance, entry into the informal transport system is relatively flexible and accessible (Oluwakoya 2024). Individuals can participate through various means, including rental or lease arrangements based on savings, incremental savings-based ownership models, and collective or cooperative ownership. This low barrier to entry is particularly significant in urban environments, where economic opportunities are unevenly distributed. Youth, rural-to-urban, and low-income populations disproportionately rely on informal transportation to secure immediate income. Thus, the sector contributes to economic inclusion by providing accessible livelihood options for marginalized groups, thereby mitigating the social impacts of unemployment and poverty.

While the informal transport sector has significant potential for job creation, a desk review reveals the precarious nature of employment within this domain (Toš 2021). Workers in this sector operate under conditions characterized by highly variable daily earnings, influenced by fluctuations in passenger demand and competition among routes, as well as external cost factors such as rising fuel prices. Additionally, drivers are subject to fixed daily remittances imposed by vehicle owners, requiring them to meet predetermined revenue targets before retaining their earnings. This practice effectively shifts operational and financial risks from asset owners to drivers, thereby exacerbating income vulnerability (Kerzhner 2024). Moreover, the absence of protective regulations exposes workers to exploitative practices and arbitrary enforcement actions, with limited avenues for recourse in the event of a dispute (Agbiboa 2018). Beyond economic vulnerability, workers in informal transport systems face substantial occupational and social risks. For instance, extended working hours, often exceeding standard labor limits, are prevalent owing to the necessity of meeting workers' daily income goals. Moreover, the lack of access to social protection mechanisms, such as health insurance, unemployment benefits, and pension schemes, further increases workers and their households' vulnerability (Faith-Ell et al. 2025). These conditions underscore the inherent tension within informal transport systems: they offer essential employment opportunities but fail to adhere to decent work standards.

Informal transport systems are integral to stabilizing urban economies by absorbing surplus labor through flexible income-generating opportunities (Munuhwa et al. 2024). This function serves as a buffer against the socioeconomic pressures associated with rapid urban growth and the inadequate creation of formal employment opportunities. In many cities, the scale of employment within informal transport is such that any disruption could significantly impact urban stability. Therefore, it is essential to recognize the functionality of mobility solutions and the role of informal transport as a component of urban economic infrastructure. From the perspective of sustainable urban development, the findings of this study have important policy implications for the government. Historically, discourse on sustainable transport has predominantly focused on environmental performance, efficiency, and infrastructure modernization, often neglecting the socioeconomic dimensions of transport systems. The evidence supports the inclusion of employment considerations to facilitate meaningful transitions toward sustainable practice. Strategies aimed at formalizing or restructuring informal transport systems, such as fleet modernization, regulatory tightening, and integration into formal public transport networks, must be carefully designed to avoid adverse outcomes such as job displacement and increased socioeconomic vulnerability (Chaudhry et al. 2024). To address the challenges of transport governance, more inclusive and labor-sensitive strategies should be adopted to harmonize goals related to efficiency and environmental sustainability with the safeguarding and enhancement of local communities' livelihoods. Additionally, recognizing informal transport workers within policy frameworks, implementing gradual processes of formalization, and facilitating access to social protection mechanisms are potential steps toward developing labor-sensitive strategies. Ultimately, incorporating employment considerations into transportation planning is essential for achieving a comprehensive and equitable model of sustainable urban development (McKay et al. 2024).

4.2 Environmental Impacts of Informal Transport Systems in Relation to Urban Greening and Decarbonization Agendas

Informal transportation systems exhibit complex and occasionally contradictory relationships with the environmental dimensions of urbanism, climate change, and air pollution, and with the broader objectives of sustainable urban development. Informal transport has emerged as the predominant form of motorized transportation in numerous cities in developing regions worldwide (Tsavachidis et al. 2022). Consequently, these systems have a substantial impact on the environment and are pivotal to the long-term success of urban development. Empirical studies have identified a positive correlation between informal transportation systems and environmental degradation. This correlation is primarily attributed to the composition and operational characteristics of the vehicles used in informal transport. Notably, a considerable proportion of these vehicles, including minibuses, shared taxis, and motorcycle taxis, are often older, second-hand, or locally refurbished. Consequently, they typically lack contemporary emission control technologies and perform suboptimally in terms of fuel efficiency and emissions. Given that informal transportation frequently occurs in densely populated urban areas with high traffic volumes, the resulting emissions, such as PM10/2.5, NOx, and CO, contribute to the deterioration of air quality and pose significant health risks to the residents.

In addition to vehicle operation, two other categories of factors explain why informal transportation contributes significantly to pollution. One factor pertains to route selection in informal public transportation (Chaudhry et al. 2024). Routes are often chosen based on passenger competition, which prompts drivers to traverse congested streets and neighborhoods to attract passengers, resulting in increased stop-and-go driving and fuel consumption. In addition, inadequate regulatory frameworks in urban areas permit the operation of poorly maintained vehicles in urban areas. Collectively, these factors render informal transportation a substantial contributor to global and urban environmental challenges. However, the review also underscores that the environmental impact of informal transportation is not uniformly negative. Informal transportation can provide environmentally advantageous alternatives to traditional transportation modes. For instance, informal transportation can offer high-occupancy services. In many developing countries, it functions as a mass transit system, transporting more passengers per vehicle than conventional automobiles. This form of "carpooling" can reduce total emissions per capita (McKay et al. 2024).

This example demonstrates the potential of informal transportation to mitigate the emissions associated with the use of private automobiles. When evaluated on a per-capita or per-passenger kilometer basis, shared informal transportation may generate fewer emissions than private automobiles. However, this advantage is contingent upon the inadequacy of formal public transportation systems in a given region, often due to financial constraints or system fragmentation. In such contexts, informal transportation emerges as the primary mode of public transit, serving as a deterrent to the shift toward highly polluting and car-dependent urban mobility. Although informal transportation poses environmental challenges owing to vehicle-level inefficiencies, its overall environmental benefits at the system level may surpass those of formal public transportation. Research indicates that the environmental performance of informal transportation is more closely linked to structural and institutional constraints that impede operators' ability to acquire new vehicles with advanced technology than to their informal status (Tucho 2022). Operators often lack the financial means to purchase newer vehicles with enhanced emission control technologies. In addition, credit schemes, subsidy programs, and other incentives to aid operators in acquiring new vehicles are infrequently available (Cervero and Golub 2007).

Inadequate regulatory frameworks and enforcement capabilities significantly hinder efforts to address the environmental challenges associated with informal transportation. For instance, in numerous urban areas, emissions standards are either nonexistent, outdated, or poorly enforced because of institutional capacity limitations and governance problems. Similarly, vehicle inspection and maintenance programs are often underdeveloped or poorly managed (Tsavachidis et al. 2022). Consequently, vehicles that emit excessive pollutants continue to operate without facing any repercussions. Limited access to cleaner fuels and supporting infrastructure, such as electric charging stations and alternative fuel distribution networks, constrains opportunities for operators in the informal sector to adopt low-carbon alternatives to traditional fossil fuel-based transportation (Teoh et al. 2020). Consequently, the analysis has shifted from identifying deficiencies in operators' behavior to understanding the broader societal and institutional forces that shape operations. Overall, research indicates that informal transportation exists at the intersection of being both detrimental and beneficial to urban greening and decarbonization initiatives (Wallius et al. 2025).

Although it contributes to increased emissions and localized environmental degradation through inefficient vehicles and operational practices, it also promotes a high degree of shared mobility, which can replace private car travel and potentially reduce the subsequent environmental impacts. Understanding this dual nature necessitates an integrated approach to address both the environmental challenges posed by informal transportation systems and their latent potential for sustainable development (Chae and Shin 2025). This study advocates for policies that support the integration of informal transportation into comprehensive decarbonization plans, rather than excluding or displacing it entirely (Nuuyandja et al. 2025). Policies aimed at promoting incremental technological improvements, raising regulatory standards for vehicles, and establishing regulatory structures to promote environmental sustainability should be implemented while maintaining the sector's economic viability. Supporting technologies, such as fleet upgrades through targeted funding mechanisms, incentives to adopt low-carbon energy solutions, and investment in complementary infrastructure, are ways to align policy goals with social and economic realities (Oraiopoulos et al. 2025).

4.3 Policy Pathways for Greening Informal Transport Systems

The desk review highlights a discernible and ongoing shift in policy approaches concerning the management and transformation of informal transportation systems in developing countries. This shift appears to move away from traditional methods characterized by exclusion, punishment, or elimination toward more enabling and incentive-driven frameworks. Historically, most policy models have been designed to either eradicate or replace informal transport through restrictive regulations or by expanding formal public transport options (Nuuyandja et al. 2025). These earlier policy models frequently failed to achieve their intended objectives and often resulted in social disruptions. In contrast, current trends provide increasing evidence to support policy pathways that acknowledge the structural integration of informal transport into urban economies and aim to augment rather than abolish it. This shift is particularly pertinent in the context of green city initiatives and decarbonization plans, which present both environmental and socioeconomic challenges (Oraiopoulos et al. 2025).

A major theme emerging from the reviewed literature is the significance of gradual and contextualized formalization in the learning process. Unlike abrupt regulatory compliance demands that may displace informal operators, effective approaches typically implement incremental measures that reflect the sector's operational realities (Salas-Castillo et al., 2025). Gradual formalization involves introducing incremental steps in terms of improved vehicle standards, safety compliance, emission standards, and service reliability while maintaining the sector's capacity to provide affordable services and generate employment (Chibaro et al. 2026a). The mechanisms employed to facilitate this approach generally include simple and inexpensive licensing schemes, flexible regulatory structures, and progressive compliance timetables that allow operators to gradually adapt to evolving regulatory demands (Caros et al. 2026).

Another essential governance mechanism involves recognizing and integrating existing informal institutions such as transportation associations, unions, and cooperative organizations (Gumbo et al. 2022a, b, c). These entities frequently play critical roles in determining routes, setting fares, resolving disputes, and coordinating the daily operations of informal transportation services in developing nations. By incorporating these organizations as intermediaries within formal regulatory frameworks, regulatory coverage can be expanded, compliance improved, and shared responsibility among stakeholders promoted (Amoako et al. 2025). This approach also helps bridge the gap between government agencies and informal operators, ultimately fostering greater trust and legitimacy in institutions. Incentives to encourage the transition to environmentally friendly transportation constitute another area of focus (Nuuyandja et al. 2025).

One of the most significant challenges in transitioning informal transportation services to cleaner energy forms is the substantial capital required to acquire cleaner technology, such as newer fuel-efficient vehicles, hybrid engines, and electric mobility systems (EMSs), owing to income volatility and limited access to conventional financing channels available to informal operators (Ali 2021). Therefore, policymakers must make financial incentives more accessible. Engagement in participatory frameworks is likely to increase policy legitimacy, promote information sharing among stakeholders, and enhance the potential for compliance by encouraging stakeholder alignment with regulatory objectives (Konijnendijk 2024). Participatory frameworks can also lead to more localized solution designs. Given the diversity of informal transportation systems in cities, local solutions are essential (Ali 2021).

Governance reform encompasses more than augmenting governmental capacity; it involves recalibrating interactions between formal institutions and informal actors (Duri 2025). The desk review identified digitalization as a rapidly evolving and potentially transformative tool for enhancing the environment of informal transportation services. Digital technologies, such as mobile payment systems for fares, GPS-based route management, and data-driven fleet monitoring systems, present opportunities for increased efficiency and reduced unnecessary fuel consumption through improved service delivery coordination (Pauleit et al. 2021). For instance, digital fare systems mitigate the inefficiencies associated with cash transactions and improve revenue generation. Route optimization technologies decrease unnecessary travel time, thereby reducing the emissions resulting from traffic congestion (Larbi et al. 2021). Although the extent of adoption varies considerably based on income level, with lower-income communities exhibiting less frequent utilization, digitalization offers a viable option for integrating informal transportation into intelligent and sustainable urban transportation systems (Petrozziello et al. 2025).

The findings collectively suggest that transitioning informal transportation services to more environmentally sustainable modes necessitates multidimensional and comprehensive policy strategies that extend beyond merely mitigating environmental impacts. Effective strategies must simultaneously address economic, social, and institutional dimensions (Gössling 2025). Moreover, environmental performance is closely linked to securing livelihoods, promoting financial inclusion, and enhancing governance. Consequently, an integrated approach aligns with the principles of a 'just transition', which aims to balance environmental sustainability with social justice, equity, and economic resilience (Wu et al. 2026). From a policy perspective, this implies that decarbonization policies in the transport sector must be crafted to avoid disadvantaging individuals or vulnerable populations who rely on informal transportation for their livelihood. Instead, policies should be developed to facilitate industry upgrades, provide safer working environments, and improve environmental performance (Cummins et al. 2025). Therefore, policymakers must engage in cross-disciplinary collaborations involving transportation planning, financial systems, regulatory frameworks, and technological innovations (Caros et al. 2026).

5 Discussion

The data gathered and presented in this study enhance the existing literature on sustainable urban transportation in the Global South. This study demonstrates how informal transportation systems contribute to social and economic empowerment while presenting environmental challenges. Furthermore, the data show that these systems are not merely peripheral or temporary solutions for urban transportation; instead, they are integral components of the structure of the urban economy. In regions with high unemployment and rapid urbanization, where formal sector growth is limited, acknowledging their role as job-creating mechanisms is essential. This study builds upon and extends previous studies by developing theoretical frameworks that view informal transportation systems not only as modes of transport but also as labor systems offering various direct and indirect linkages. Additionally, our findings suggest that informal transportation systems enable some citizens to engage socially and economically by reducing barriers to employment and livelihood opportunities, particularly for marginalized groups, such as youth and rural-to-urban migrants. However, these positive impacts on social and economic participation for previously disenfranchised citizens are counterbalanced by the precarious nature of employment within informal transportation systems. Issues such as income volatility, lack of social protection, increased risks of workplace injury or illness, and the unequal treatment of workers regarding working conditions and operational processes highlight the contradictions in the role of informal transportation systems. They help alleviate unemployment among certain societal segments, while simultaneously generating vulnerabilities related to employment security and fairness. This conflict mirrors broader debates within development studies regarding the quality, not just the quantity, of jobs in informal economies. From an environmental perspective, we observed a context-dependent relationship between informal transportation systems and urban sustainability.

The widespread use of older vehicles and unregulated operational practices within the fleet contribute to increased urban air pollution and greenhouse gas emissions. Conversely, high occupancy rates and shared usage of informal transportation services can potentially reduce reliance on personal automobiles in cities where formal public transportation is underdeveloped. This duality challenges simplistic policies rooted in binary thinking, which categorizes informal transport users as either entirely detrimental or beneficial to achieving environmental goals. More importantly, this study shifted its focus from analyzing individual operator behavior to examining systemic and institutional factors. We argue that the persistence of environmentally unsustainable practices by operators is primarily due to limited access to capital, ineffective regulation enforcement, inadequate infrastructure, and, most significantly, a lack of government commitment and policy support. This conclusion aligns with the growing body of research advocating for a structural rather than punitive approach to reforming the informal sector. The policy alternatives developed in this study emphasize a paradigmatic shift from exclusionary, enforcement-driven approaches to more inclusive, incentive-based, and participatory models. Incremental formalization appears to be a particularly relevant strategy for achieving gradual improvements in both service quality and environmental performance, without causing significant employment disruptions. Finally, we found that recognizing informal institutions and digital technologies enhances the overall governance capacity, efficiency, and environmental outcomes of informal transportation systems. Our conclusions also support the "just transition" concept, emphasizing that environmental sustainability can be achieved while maintaining social equity and economic stability.

6 Policy and Practical Implications

These findings have important practical implications for policymakers, urban planners, and development practitioners seeking to develop sustainable and inclusive transportation systems in the Global South. Policymakers should redefine informal transportation in their discussions. This means that informal transportation should no longer be viewed as something to eradicate but rather recognized as a fundamental part of the urban economy and mobility system. For this new paradigm to occur, policymakers must create realistic and socially equitable interventions. Policies aimed at reducing carbon emissions without considering the livelihoods of informal transportation workers may exacerbate poverty, inequality, and social instability. Therefore, any future changes to thc transport sector, including decarbonization, must incorporate the principles of "decent" work when addressing the labor aspects of these policies.

Moreover, the findings underscore the importance of adopting gradual and context-specific approaches to formalization. For example, the swift or compulsory implementation of formal regulations might have inadvertently led to negative consequences for local economies and society. Incrementally developed regulatory frameworks, encompassing flexible licensing options, phased compliance structures, and targeted capacity-building programs, would offer a more sustainable trajectory for advancement in this field. Access to financial mechanisms, such as credit, subsidies, and leasing programs, is essential for enabling operators to implement environmental improvements. The availability of these financial tools facilitates the acquisition and utilization of clean vehicles, including fuel-efficient hybrids. In the absence of such mechanisms, operators are likely to persistently employ less environmentally sustainable practices.

In addition, establishing formal relationships with existing transportation associations, unions, and cooperatives fosters greater compliance and improves communication. Furthermore, the creation of formalized institutions facilitates coproductive policy development tailored to the specific needs of each community. In addition, technology can help improve efficiency, decrease emissions, and create more transparent revenue streams for operators through GPS-based routing applications, mobile payment systems, and data-driven fleet management. To ensure that the use of technology does not further widen the gap between those who have access to these technologies and those who do not, investments must be made in digital literacy training and support for digital access. Achieving sustainable outcomes necessitates the integration of cross-sectoral policymaking. Decarbonization of transportation cannot be pursued in isolation from urban planning, energy production, financial markets, and social protection systems.

7 Conclusions

This study examined informal transportation systems through the lens of global sustainable development (urban) globally. Informal transport systems are essential to the livelihoods of people living in cities and are detrimental to the environment. Therefore, policies regarding informal transport must move beyond simplistic solutions to more multifaceted approaches. The results of this study demonstrate that informal transportation systems create jobs and contribute to the functioning of a city. Nonetheless, the hazardous nature of employment and the environmentally unsustainable conditions of informal transportation necessitate immediate intervention to transform these systems. These transformations should not aim to destroy informal transportation systems but instead seek to modernize them to improve the social and environmental aspects. Ultimately, this study advocates for a just transition for workers in the Global South in the field of urban transportation by linking environmental sustainability goals to the protection and enhancement of worker livelihoods. A just transition will require balancing conflicting priorities using incremental forms of formalization, specific financial tools, participatory governance structures, and innovative technologies. Therefore, this study adds to the current discussions about sustainable transportation by moving informal transportation from being peripheral to central to all urban transportation policymaking and academic scholarship.

References

Aderibigbe OO Gumbo T (2024) Smart cities and their impact on urban transportation systems and development. Emerging technologies for smart cities, 105–129, Springer Nature Switzerland. https://doi.org/10.1007/978-3-031-66943-9_5

Agbiboa DE (2018) Conflict analysis in 'world class' cities: urban renewal, informal transport workers, and legal disputes in Lagos. In: Urban forum, Vol 29, No 1, pp 1–18. Springer Netherlands

Ali N (2021) Evaluating sustainable urban transport systems: a review study for the identification of smart mobility indicators. Trans Transp Sci 12(2):16–23, ISSN: 1802-9876, Palacky University Olomouc. https://doi.org/10.5507/tots.2021.008

Amoako C, Dinye IN, Amaka-Otchere A, Baah K (2025) Nuances of informal transport operation. Hybrid urbanisms in secondary cities of the global south, 49–62, Routledge. https://doi.org/10.4324/9781003511137-5

Bermúdez GR, Robledo W (2023) Sustainable urban mobility: evaluation of the monorail as a transportation alternative in Córdoba. Transp Mobility Soc 2:52, ISSN: 3072-8398, South American Publishing. https://doi.org/10.56294/tms202352

Brock C, Lallement J, Findeisen L, Soulat U (2025) Bridging the green gap: understanding the intention–behavior gap in mobility and urban mobility. Lecture Notes in Mobility, 398–402, ISSN: 2196-5544, Springer Nature Switzerland. https://doi.org/10.1007/978-3-031-85578-8_51

Buijs P (2025) Business model implications of innovative solutions for sustainable urban freight transport. Lecture Notes in Mobility, 623–628, ISSN: 2196-5544, Springer Nature Switzerland. https://doi.org/10.1007/978-3-031-95284-5_89

Caros N, Trouvé M, Martinez L (2026) Global trajectories for urban passenger transport decarbonization: a policy-based modeling approach. NPJ Sustain Mobility Transp 3(1), ISSN: 3004-8664, Springer Science and Business Media LLC. https://doi.org/10.1038/s44333-026-00092-6

Cervero R, Golub A (2007) Informal transport: a global perspective. Transp Policy 14(6):445–457

Chae S, Shin H (2025) Urban informality in the making. Resilient cities in the global South, 147–161, Routledge. https://doi.org/10.4324/9781003336709-13

Chaudhry AG, Masoumi H, Dienel HL, Aslam AB, Ahmad M, Shahnaz M (2024) Mobility attitudes and urban form: shaping public transport and shared mobility choices in Dubai and Lahore. Urban Plann Transp Res 12(1):ISSN: 2165-0020, Informa UK Limited. https://doi.org/10.1080/21650020.2024.2420735

Cheng J, Li K, Liang Y, Sun L, Yan J, Wu Y (2025) Rethinking urban mobility prediction: a multivariate time series forecasting approach. IEEE Trans Intell Transp Syst 26(2):2543–2557,

ISSN: 1524-9050, Institute of Electrical and Electronics Engineers (IEEE). https://doi.org/10.1109/tits.2024.3498054

Cherednichenko O (2024) Sustainable urban mobility plan as one of the tools of management of urban transport systems. Spatial Dev (7):594–611, ISSN: 2786-7277, Kyiv National University of Construction and Architecture. https://doi.org/10.32347/2786-7269.2024.7.594-611

Chibaro M, Chikwere D, Chisungo C, Manyanga W (2026a) Transport infrastructure development in Zimbabwe: challenges, policies, and investment strategies. The changing terrain of transport and logistics in Zimbabwe, 73–100, Springer Nature Switzerland. https://doi.org/10.1007/978-3-032-11384-9_4

Chibaro M, Chikwere D, Chisungo C, Manyanga W, Munuhwa S (2026b) Transport and logistics in Zimbabwe: policy and future directions. The changing terrain of transport and logistics in Zimbabwe, 239–277, Springer Nature Switzerland. https://doi.org/10.1007/978-3-032-11384-9_10

Cummins T, Monaghan RFD (2025) An online tool for guiding bus fleet decarbonization through green hydrogen and electrification. Lecture Notes in Mobility, 632–638, ISSN: 2196-5544, Springer Nature Switzerland. https://doi.org/10.1007/978-3-031-89444-2_92

Dabel I (2026) Rethinking public transport adoption in the global south: sustainability legitimacy, safety perception and mobility behavior among university students in Ghana. Elsevier BV. https://doi.org/10.2139/ssrn.6751018

Desmoulière R (2023) Non-centralized urban transport. Urban mobility systems in the world, 135–156, Wiley. https://doi.org/10.1002/9781394256204.ch7

Duri B (2025) Informal transport and the quest for sustainable urban mobility: insights from drivers. Wit transactions on ecology and the environment, 1, 127–136, ISSN: 1743–3541, WIT Press. https://doi.org/10.2495/sc250111

Eriskin E (2024) Collaborative game-theoretic optimization of public transport fare policies: a global framework for sustainable urban mobility. Sustainability 16(24):11199, ISSN: 2071-1050, MDPI AG. https://doi.org/10.3390/su162411199

Faith-Ell C, Arts J, Kalle H (2025) Toward sustainable and inclusive corridors: rethinking planning of infrastructure and spatial development. Lecture Notes in Mobility, 599–605, ISSN: 2196-5544, Springer Nature Switzerland. https://doi.org/10.1007/978-3-031-85578-8_80

Gössling S (2025) Reducing resistance to urban transport system change. Urban Trans 1:100001, ISSN: 3050-6972, Elsevier BV. https://doi.org/10.1016/j.ubtr.2025.100001

Gumbo T, Moyo T, Ndwandwe B, Risimati B, Mbatha SG (2022) Functional integration of public transport systems in Tshwane, Johannesburg, Harare, and bulawayo cities. Urban public transport systems innovation in the fourth industrial revolution Era, 99–109, Springer International Publishing. https://doi.org/10.1007/978-3-030-98717-6_6

Gumbo T, Moyo T, Ndwandwe B, Risimati B, Mbatha SG (2022) Governance and structures in urban public transport in South African and Zimbabwean cities. Urban public transport systems innovation in the fourth industrial revolution era, 27–40, Springer International Publishing. https://doi.org/10.1007/978-3-030-98717-6_2

Gumbo T, Moyo T, Ndwandwe B, Risimati B, Mbatha SG (2022) Innovative urban public transport systems: an evolving continuous expedition in Southern African Cities. Urban public transport systems innovation in the fourth industrial revolution era, 171–180, Springer International Publishing. https://doi.org/10.1007/978-3-030-98717-6_10

Kerzhner T (2024) Creating an informal transport route: implications for mobility, gender, and planning processes. J Am Plann Assoc 90(4):656–671. https://doi.org/10.1080/01944363.2024.2307920

Konijnendijk C (2024) Urban green space use in transition. Rethinking Urban Green Spaces, 24–39, Edward Elgar Publishing. https://doi.org/10.4337/9781803925493.00008

Larbi M, Kellett J, Palazzo E (2021) Urban sustainability transitions in the global south: a case study of Curitiba and Accra. Urban Forum 33(2):223–244, ISSN: 1015-3802, Springer Science and Business Media LLC. https://doi.org/10.1007/s12132-021-09438-4

Logiodice P, Cunha R, Giannotti M (2025) Framing urban mobility injustice from the Global South. Urban Studies 62(11):2258–2274, ISSN: 0042-0980, SAGE Publications. https://doi.org/10.1177/00420980241306676
McKay TJM, Duri B, Gunter A (2024) Navigating the challenges of public transport and urban mobility in Thohoyandou, South Africa. J Transport Supply Chain Manag 18, ISSN: 2310-8789, AOSIS. https://doi.org/10.4102/jtscm.v18i0.1054
Munuhwa S, Chikwere D, Chibaro M (2024) Urban car parking puzzle in the UK. Advances in civil and industrial engineering, 284–308, ISSN: 2326-6139, IGI Global. https://doi.org/10.4018/979-8-3693-3755-4.ch011
Myers G (2021) Urbanisation in the global south. In: Shackleton CM, Cilliers SS, Davoren E, du Toit MJ (eds) Urban ecology in the global south. Cities and Nature. Springer, Cham. https://doi.org/10.1007/978-3-030-67650-6_2
Nuuyandja H, Pisa N, Masoumi H, Chakamera C (2025) Correlates of formal and informal public transport usage in Windhoek, Namibia. J Urban Mobility 8:100149, ISSN: 2667-0917, Elsevier BV. https://doi.org/10.1016/j.urbmob.2025.100149
Oluwakoya AO (2024) The role of informal transport systems in the urban transport landscape of Lagos State, Nigeria. AKSU J Soc Sci 4(2):155–161, ISSN: 2504-933X, Akwa Ibom State University. https://doi.org/10.61090/aksujoss.2024.028
Onodugo VA, Ezeadichie NH (2020) Future planning of global south cities with inclusive informal economic growth in perspective. Sustainability in urban planning and design, IntechOpen. https://doi.org/10.5772/intechopen.89145
Oraiopoulos A, Wieser M, Verdiere M, Lambert R, Fennell P, Ruyssevelt P (2025) Exploring the impact of community urban green infrastructure on outdoor heat stress in informal settlements in the global south., Elsevier BV. https://doi.org/10.2139/ssrn.5279082
Pauleit S, Vasquéz A, Maruthaveeran S, Liu L, Cilliers SS (2021) Urban green infrastructure in the global south. Cities Nat, 107–143, ISSN: 2520-8306, Springer International Publishing. https://doi.org/10.1007/978-3-030-67650-6_5
Petrozziello E, Marotta A, Stępniak M, Cheimariotis I, Lodi C (2025) Achievements and future priorities of artificial intelligence in transport systems. Lecture Notes in Mobility, 41–46, ISSN: 2196-5544, Springer Nature Switzerland. https://doi.org/10.1007/978-3-032-06763-0_6
Salas-Castillo A, Klopp JM (2025) Rethinking (In)formalities in urban global south Mobilities., Elsevier BV. https://doi.org/10.2139/ssrn.5929118
Teoh R, Anciaes P, Jones P (2020) Urban mobility transitions through GDP growth: Policy choices facing cities in developing countries. J Transp Geogr 88:102832, ISSN: 0966-6923, Elsevier BV. https://doi.org/10.1016/j.jtrangeo.2020.102832
Toš I (2021) Use of technology in improving urban transport. Proceedings of the1st international conference public transport and smart mobility 2020: innovative solutions for smart urban mobility, 73–79, Faculty of Transport and Traffic Sciences, University of Zagreb. https://doi.org/10.7307/ptsm.2020.9
Tsavachidis M, Petit YL (2022) Reshaping urban mobility—key to Europe's green transition. J Urban Mobility 2:100014, ISSN: 2667-0917, Elsevier BV. https://doi.org/10.1016/j.urbmob.2022.100014
Tucho GT (2022) A review on the socio-economic impacts of informal transportation and its complementarity to address equity and achieve sustainable development goals. J Eng Appl Sci 69(1):28
Velasco A (2025) Evaluating urban transport systems: a systematic review for the identification of sustainable mobility indicators for Latin America. Springer Science and Business Media LLC. https://doi.org/10.21203/rs.3.rs-6135476/v1
Venter C, Mahendra A, Hidalgo D (2019) From mobility to access for all: expanding urban transportation choices in the global south. World Resources Institute, Washington, DC, pp 1–48
Wallius E, Orta C, Lamuela L, Toni PO, Hautala R (2025) Decarbonizing urban mobility through data-driven services: how to achieve modal shift?. Lecture Notes in Mobility, 576–581, ISSN: 2196-5544, Springer Nature Switzerland. https://doi.org/10.1007/978-3-031-85578-8_77

Wikstrøm RD, Røe PG (2022) Sustainable mobility transitions in suburbia—exploring (dis)connections between transport planning and daily mobility. Urban Res Practice 17(1):72–95, ISSN: 1753-5069, Informa UK Limited. https://doi.org/10.1080/17535069.2022.2119430
Wu H (2025) Sustainable transformation of urban transportation systems: global experiences and policy implications from motorization to green mobility. Adv Econ Manag Res 15(1):1070, ISSN: 2790-1661, Madison Academic Press. https://doi.org/10.56028/aemr.15.1.1070.2025
Wu J, Tian X, Liu Q, Li T, Liu C, Xu J, Zhao H (2026) Generative artificial intelligence and urban transport resilience: Rethinking sustainable mobility. Safety Sci 198:107148, ISSN: 0925-7535, Elsevier BV. https://doi.org/10.1016/j.ssci.2026.107148

Fiscal Sustainability and Green Transport Transition: Intertemporal Fiscal Frameworks and Environmental Tax Reform in Developing Economies

Tinashe Magande, Innocent Chirisa, Takudzwa Mharadze, Nyasha Ndemo-Masimbarasi, Halleluah Chirisa, Africa Muneti, Nyasha N'onzwe, and Roselin Katsande-Ncube

Abstract Public finance holds significant importance in achieving sustainable development due to growing challenges associated with infrastructure development and budget constraints in developing countries. The transportation sector makes a considerable contribution towards emitting greenhouse gases and air pollution,

T. Magande (✉) · I. Chirisa
Department of Urban and Regional Planning, University of Johannesburg, Johannesburg, South Africa
e-mail: magandet@gmail.com

I. Chirisa
Office of the Vice Chancellor, Zimbabwe Ezekiel Guti University, Bindura, Zimbabwe

T. Mharadze
Department of Development Studies, Zimbabwe Open University, Harare, Zimbabwe

N. Ndemo-Masimbarasi
Department of Development Planning and Management, Zimbabwe Ezekiel Guti University, Bindura, Zimbabwe

Department of Development Sciences and Policy, Marondera University of Agricultural Sciences and Technology, Marondera, Zimbabwe

H. Chirisa
Directorate of Research and Innovation, Zimbabwe Ezekiel Guti University, Stand No 1901, Barrassie Road, Off Shamva Road, Bindura, Zimbabwe

A. Muneti
Firm Central Property Group, Harare, Zimbabwe

IC University of Technology, Phnom Pehn, Cambodia

N. N'onzwe
Department of Social Work and Applied Psychology, Zimbabwe Ezekiel Guti University, Mashonaland Central, Bindura, Zimbabwe

R. Katsande-Ncube
Faculty of Social and Gender Transformative Sciences, Women's University in Africa, Harare, Zimbabwe

J. Kanyepe and I. Chirisa (eds.), *Advancing the Green Agenda in the Transport and Logistics Sector in Global South Countries*, World Sustainability Series,
https://doi.org/10.1007/978-3-032-31374-4_21

demanding huge funds in developing sustainable means like public transportation and electric trains. However, macroeconomic instability, debts, and inefficient use of money, especially through fossil fuel subsidies, make sustainable development a tough challenge. The chapter draws from two theories namely, based on the Intertemporal Budget Constraint (IBC) and the Environmental Fiscal Reform (EFR) theory. This is qualitative research that investigates how fiscal sustainability supports green transportation reforms without jeopardizing macroeconomic stability. Data generation involved systematic literature review and desk research of peer-reviewed papers, books, and other credible sources via Google Scholar, Scopus, and ScienceDirect. Through textual analysis, the main results are threefold. First, fiscal sustainability, based on the principle of intertemporal budget constraint, focuses on debt sustainability as a way of funding green infrastructure projects. Second, fiscal policies such as environmental taxation and subsidies rationalization create revenues through double dividends as well as promote low-carbon travel through incentives. Third, effective governance can improve policy implementation, as exemplified by Indonesia's experience of reforming subsidies and Colombia's application of carbon taxes to fund green bonds. This research highlights the importance of fiscal policies in supporting green reforms: reallocation of budgets, utilization of green taxation along with revenue recycling, and innovative finance options like green bonds. In this way, developing countries can develop resilient and sustainable transportation systems.

Keywords Fiscal sustainability · Green transport · Environmental fiscal reform · Public finance · Developing economies · Debt management

1 Introduction

Public finance has gained centre-stage in the discourse on sustainable development, particularly in developing nations that are experiencing an increase in infrastructure, as well as, in the financial strains of government. Most of such economies face the challenges of increasing fiscal shortfall, elevated government liabilities, and inadequate domestic revenue mobilization potential (International Monetary Fund (IMF) 2023). Simultaneously, transport systems can have a substantial impact on greenhouse gas emissions and air pollution in cities, so there is a need to pressure mobility pathways toward being greener (Intergovernmental Panel on Climate Change (IPCC) 2022). The process of converting to sustainable transport entails massive initial investments in mass transport, electrification and resilient infrastructure (International Energy Agency (IEA) 2023). Long-term planning horizons are usually limited by macroeconomic volatility, exchange rate fluctuations, and susceptibility to external shocks, however (World Bank 2022). These colliding fiscal and environmental strains cause the control of money held in the public to be a decisive contributor in facilitating or slowing down green transport reforms.

This paper suggests that fiscal sustainability is not just an inhibitor; it is a strategic facilitator of green transport transitions in cases where revenue, expenditure, and debt management systems are focused in policy design. Capital investment can be secured by well-constructed fiscal policy and plausible medium-term expenditure structures which anchor macroeconomics (Schaechter et al. 2012). It has been demonstrated through empirical evidence that the efficiency of the public investment has great influence on the developmental growth and infrastructure outcomes in the developing economies (IMF 2015). Also, carbon price and environmental taxes reform can be one that raises revenue and simultaneously encourages low-carbon transport systems (World Bank 2023). Even with good governance structures, targeted subsidies and earmarked green funds can also be able to crowd in the private investment (OECD 2021). As such, fiscal architecture determines the size as well as the sustainability of investments in transport infrastructure.

The overall purpose of the study is to determine how developing nations should be able to develop fiscal space to support sustainable transport without jeopardizing macro-fiscal stability. It looks at the dynamics of debt sustainability and how it affects infrastructure financing in the long run (World Bank 2022). It further examines expenditure reallocation policies that favour high-multiplier green investments to ineffective subsidies (IMF 2015). Also, it evaluates the possibility of environmental tax instruments to marshal domestic resources and help achieve climate ambitions (World Bank 2023). The article aims to fill in the gap between the environmental ambition and fiscal realism through insights combined with the research of climate mitigation pathways and fiscal governance (IPCC 2022). The research proposal is expected to develop a logical macro-fiscal model that can inform sustainable transport transitions in developing economies that have limited fiscal space.

2 Background to the Study

Sustainable transport is a leading area of climate mitigation and sustainable development strategy at the global level since the transport sector contributes almost a quarter of all carbon dioxide-related energy emissions in the world (IEA 2023). The trend of rapid urbanization and motorization especially in developing economies is mounting energy demands and putting pressure on the environment (United Nations 2019). According to the Intergovernmental Panel on Climate Change, the transport systems require deep decarbonization to contain global warming to 1.5 °C (IPCC 2022). Nonetheless, this transition would be possible through widespread investment in infrastructure such as public transportation, electrification and non-motorized mobility systems. Estimates of the world show that it requires trillions of dollars in extra annual investments to be in line with climate targets (OECD 2017). Meanwhile, due to their tightening fiscal policies and the increasing vulnerability to debts, many developing countries cannot afford to fund long-term green infrastructure (IMF 2023).

At the regional level, there are developing regions which are currently acutely infrastructure-deficient and fiscally weak like Sub-Saharan Africa, South Asia and some parts of Latin America. The infrastructure gaps in Africa have been estimated to cost the region a lot of productivity loss and less competitiveness (AfDB 2018). The inefficiencies in transport also lead to an increase of high logistics costs and decreased integration of regions and this hurts the prospects of economic growth (World Bank 2020). In the meantime, the levels of public debt in most of the low- and middle-income countries have shot up due to the effect of the global economic shocks, constraining the amount of fiscal space available in capital-intensive projects (World Bank 2022). These issues are even aggravated by climate vulnerability, which harms the available infrastructure and necessitates expensive adaptation strategies (IPCC 2022). The frameworks of regional policies are paying growing attention to the necessity of green growth strategies, and the funding mechanisms are still insufficient to facilitate the presence of sustainable transport changes on a large scale (OECD 2018).

Governments in developing countries at the national level have to juggle competing demands like health, education, debt servicing, and infrastructure development on limited budgets. Governments, in most cases, are not able to finance extensive transport reforms because of weak domestic revenue mobilization and limited tax bases (IMF 2023). Fossil fuel subsidies are still consuming large quantities of government resources in most nations, supplanting the development of alternatives to sustainable mobility (IEA 2023). Transport decarbonisation is a common list of priorities in Nationally Determined Contributions to the Paris Agreement, but a lack of implementation remains because of financial deficits (United Nations Framework Convention on Climate Change (UNFCCC) 2021). The infrastructure investment outcomes are therefore determined by the decisive role played by the public financial management systems and the design of fiscal policy (World Bank 2020). Enhancing macro-fiscal stability and a redistribution of spending in favour of green investment in transport is also an important issue of national development.

3 Research Objectives

This chapter seeks to:

1. Examine the effect of fiscal sustainability on government investment in sustainable transport infrastructure in developing countries.
2. Assess the impact of environmental fiscal reforms on the creation of fiscal space for sustainable transport financing.

While existing macro-fiscal reviews emphasize debt dynamics and expenditure efficiency (e.g., IMF 2023; World Bank 2022), and green-finance literature focuses on carbon pricing or bonds (e.g., OECD 2021; Labandeira et al. 2004), this chapter uniquely bridges these silos. It Combines Intertemporal Budget Constraint for fiscal discipline with Environmental Fiscal Reform theory to model public finance approaches that can facilitate fiscal implementation of 'green transport transition' in the context of fiscal constraint within the developing world. This synthesis contrasts with previous reviews that see fiscal sustainability as such a hurdle; it looks at fiscal sustainability as a strategic enabler and brings in novel aspects from subsidy reforms in Indonesia and carbon tax use in Colombia to feed into specific policy design.

4 Conceptual Model

The Theory of Fiscal Sustainability is based on the Intertemporal Budget Constraint (IBC) model which states that governments should make sure that the present fiscal policy will not impair the capacity to pay future liabilities without the need to over-borrow or default. According to this theory, the value of primary surpluses in the future has to be as large as, or larger than, the current amount of government debt, otherwise "Ponzi finance" and fiscal insolvency will be prevented (Baglioni and Cherubini 1993; Spaventa 1987). Empirical research shows that such sustainability is a challenge to many countries because of the economic vagaries and financial strains (Uctum and Wickens 2000; Greiner and Semmler 1999). The IBC framework provides a quantitative basis to measure deferential to long-term debt solvency requirements to determine the alignment of the public finance strategies, such as those targeted by green infrastructure investment, such as sustainable transport (Collignon 2012). Climate-aligned fiscal rules have been added to post-2020 debates that emphasise the need to clarify the difference between productive green investment and current expenditure. This allows short-term fiscal space for climate transition without compromising long-term debt sustainability (Truger 2020; Dafermos and Nikolaidi 2019b, a).

The Environmental Fiscal Reform (EFR) Theory, including Green Tax Reform, aims at redesigning tax systems to change to taxing to the activities that cause environmental harm like carbon emissions in transport and reduce the distortionary taxes on productive inputs like labour or capital. An alternative approach that could be implemented involves applying the conceptual model, which involves determining the CTE course to develop a strategy employing the physiological method (Labandeira et al. 2004; Freire-Gonzalez and Ho 2018). This is meant to internalise negative externalities to encourage environmentally friendly behaviours and innovations within the transport industry. According to the hypothesis of the double dividend, such reforms have an environmental benefit and an economic performance benefit due to the recycling of revenues (Labandeira et al. 2004). The literature on modelling supports the notion that green taxes that are properly designed can decrease emissions in a sustainable fashion without have significant detrimental effects on economic growth (Freire-Gonzalez and Ho 2018; Martin 2000). Contemporary simulation-based studies reveal that double dividend" is not guaranteed and is highly sensitive to tax design in the developing world (Zhou et al. 2025; Dafermos and Nikolaidi 2019b, a).

The chapter bases its discussion on the two interconnected theories by arguing that fiscal sustainability and green transport transition could be realized together by the implementation of policies that adhere to the IBC and the implementation of environmental taxes. The fiscal discipline concept of the IBC would guarantee the long-term stability of debts is not compromised by the investments in green transport infrastructure (Collignon 2012; Spaventa 1987). Meanwhile, the adequate incentives and financial space required to have the green shift in transport are provided by the environmental fiscal reforms, which tax the revenues of pollution and recycling in a very efficient way and therefore do not harm the fiscal balances (Labandeira et al. 2004; Martin 2000). These frameworks are combined in the chapter to examine the issue of public finance strategies that are favourable to sustainable mobility transitions in the developing economies. The integration perspective goes in line with emerging green fiscal frameworks in which environmental taxation and public investment are designed to support transformation.

The theoretical framework demonstrates the need for policy coherence and institutional credibility in order to address the practical issues of distributional effects, rebound phenomena, and short-term fiscal pressures (Greiner and Semmler 1999; Uctum and Wickens 2000). The IBC-inspired policies need to include adaptive fiscal regulation and reasonable promises, whereas EFR-based tax reforms need to be transparent and socially acceptable to achieve as many environmental and fiscal dividends as possible (Freire-Gonzalez and Ho 2018; Labandeira et al. 2004). This synthesis provides a strong conceptual foundation to direct the empirical analysis and policy proposals in the developing nations in pursuit of allowing fiscal stability and the urgency to undertake green transport transitions. Recent climate policy integration literature also highlights the importance of institutional coordination and social acceptance in implementing green fiscal reforms effectively.

5 Research Methodology

The research made use of qualitative desk review methodology to develop an understanding of the linkages between theory and practice of fiscal sustainability and green transition in transportation systems. The data used were obtained from peer-reviewed scholarly articles, policy documents (IMF, World Bank, OECD), books, and reputable institutions' literature published during 2015–2025 and focused on developing economy contexts. Systematic selection of sources was based on three levels of criteria: (1) relevance to issues of fiscal sustainability (inter-temporal budget constraints, debt dynamics), environmental fiscal reform (carbon taxation and subsidy rationalization), and sustainable transportation funding; (2) scientific rigor (quantitative modelling and case studies) or policy authority; and (3) recency and regional relevance.

Databases searched electronically included: Scopus, ScienceDirect, Google Scholar, and JSTOR through keywords: "fiscal sustainability AND sustainable transport," "environmental fiscal reform AND green infrastructure," "carbon tax revenue recycling developing countries," "fossil fuel subsidy reform transport," with "Indonesia," "Colombia," "Sub-Saharan Africa." Open coding (generation of initial categories), axial coding (links theory-policy), and selective coding (integration) methods were used to analyze textual data. Purposive selection of cases (subsidy reform in Indonesia, carbon tax in Colombia) illustrated successful fiscal-green transition strategies in fiscally constrained middle-income countries, Asia and Latin America, respectively.

6 Literature Review

This section presents the literature review related with the topic of this chapter. Four key areas aligned with the objectives of the chapter are reviewed: Fiscal Sustainability and Sustainable Transport Investment, Environmental Fiscal Reforms and Revenue Mobilisation, Environmental Fiscal Reforms and Sustainable Transport Financing and Institutional Quality, Governance, and the Effectiveness of Fiscal Policy in Green Transport Investment.

7 Fiscal Sustainability and Sustainable Transport Investment

The fields of fiscal sustainability and sustainable transport investment are closely related and are both areas that are becoming subjects of research in the modern economic and policy scene. The long-run capability of governments to finance their expenditure obligations without a high level of borrowing is the fiscal sustainability (fiscal sustainability) that is critical in ensuring sustainable infrastructure development, including transport. The ability to maintain investment capacity is at risk due to state and local fiscal pressures caused by cyclical, structural, and intergovernmental factors and this is a threat to the survival of transport infrastructure projects. New advances in managing revenues and expenses are necessary with the aim of ensuring that governments can sustain capital investment programs without affecting their fiscal health. This especially applies to subnational governments, which sometimes assume substantial transport infrastructure roles but have limited budgets and other pressing social needs (Chapman 2008).

Literature has pointed out that sustainable transport infrastructure plays a major role in economic returns but requires consistent fiscal policy frameworks to assist in the attainment of environmental and social objectives. The importance of emergent paradigms in bibliometric analyses in forms of low-carbon, resilient transport systems incorporated in sustainable urban planning is underscored. These goals, however must be attained using sound and sustainable fiscal tools that will cater to the long term nature of transport investments. As an illustration, a combination of green public investment and subsidy with carbon taxation can lead to ecological efficiency at the expense of the macroeconomy, and such a pro-sustainable fiscal policy should be a balance between fiscal policy. Fiscal policy tools can trigger the shift to green economies, especially at the time of growing ecological and economic uncertainty (Badassa et al. 2020; Dafermos and Nikolaidi 2019b, a). In South Asia, however, the available research indicates that despite highly ambitious targets for urban mobility and the clean energy transition, decarbonization solutions in a transport context are constrained by fiscal pressures and lack of green financing solutions. PERC has uncovered these constraints in its research on India and Bangladesh. Likewise, data from Latin America indicates the fact that even though certain nations, such as Brazil, Chile and Colombia, have been investing in low carbon mobility programs, many challenges remain regarding the execution post the restrictions presented in the fiscal environment and the non-integration of transport funding processes that sustain mass transport and electric mobility projects.

The flows of international green investments are transformative in the enhancement of the local fiscal capacity to spend on the environment, such as transport infrastructure. Statistics of large-scale data evidence that these investments mainly increase capital expenditure at the local level, which in effect promotes sustainable infrastructural development. This capacity is further augmented by fiscal decentralization in that the local governments are able to adjust local transport investments to meet the needs of a region without compromising environmental sustainability. Fiscal sustainability, coupled with environmental innovation and policy, lowers carbon footprints and promotes Sustainable Development Goals (SDGs). Policies that promote international green finance, as well as domestic fiscal adjustments, are essential to allow sustainable investments in transport, particularly to emerging-market economies, where infrastructure gaps and fiscal limitations usually coexist (Xu and Zhang 2024; Udeagha and Muchapondwa 2022).

In the recent past, there have been studies on the concept of regional fiscal sustainability within a framework of sustainable transport that the economic crises had a disruptive impact on the fiscal well-being and infrastructure expenditure. To illustrate, a reduced fiscal sustainability was witnessed in the Euro Area since 2008, and this had strong implications for investment cycles such as transport projects. To reduce shocks, sustainable transport investments should rely on fiscal regimes that are stable, debt management that is prudent and observation of fiscal rules. New financial tools like green bonds and blended public–private financing have been gaining momentum as a way of filling any funding gaps to support sustainable infrastructure. As global and local fiscal policies are unified to environmental sustainability, to keep the fiscal frameworks effective towards resilient and sustainable transport systems around the world, continuous research is necessary to innovate and refine the fiscal frameworks (Afonso and Jalles 2017; Prakash and Sethi 2021).

8 Environmental Fiscal Reforms and Revenue Mobilization

Green taxes and carbon pricing are now a strong means of enhancing sustainability and raising government revenues through the use of environmental fiscal reforms. According to recent analysis, the twin aims of such reforms, to minimize the environmental externality and to raise fiscal revenues, are represented in the so-called hypothesis of a double dividend. The research on green tax reforms in Spain, such as, shows that carbon taxes put in place and decreases of distortionary levies such as social security levies are able to produce both the environmental gains as well as the fiscal gains with minimal adverse distributional impacts (Labandeira et al. 2004). Nevertheless, this depends on how these reforms are designed and recycled because welfare gains can be more than the cost in case the marginal fiscal benefits are more than the fiscal costs (Jaeger 2010). Studies from emerging economies indicate that carbon taxation, fuel excise reforms, and emissions trading schemes can generate substantial fiscal revenues while simultaneously encouraging investment in low-carbon transport and energy systems (World Bank 2023; OECD 2021).

Green taxes are an important meeting point between the voices of environmental conservation and resource gathering in the face of development difficulties in the African context. It has been found out that, although the green tax policies have the potential to increase the income of the governments and lead to a sustainable development goal (SDGs), it can also increase inequality and energy poverty, especially in the population that relies on fossil fuels, among other things. Social costs like health hazards and joblessness can be triggered by such issues as higher energy prices and market distortion. Fiscal environmental reforms in Africa should thus be effective and strike a balance between the environment and socio-economic equity as not to jeopardize access to energy and poverty alleviation initiatives (Mpofu 2022).

The effective use of environmental fiscal policies and revenue mobilization is also affected by fiscal decentralization and institutional reforms. An analysis of OECD nations shows that composite fiscal decentralization (tax power and spending power are balanced) and green innovation and good governance considerably reduce emissions of carbon and promote sustainable growth. These institutional aspects increase the ability of governments to plan and institute environmental taxes that deliver consistent finances and facilitate environmental sustainability. These findings underline the importance of the environmental fiscal reforms working in powerful institutional frameworks to achieve revenue and environmental objectives in an effective manner (Sun and Razzaq 2022). In India, fiscal decentralization has enabled some state governments to experiment with green transport levies and congestion pricing, although inconsistent coordination between national and subnational authorities has limited broader policy effectiveness (Poddar and Bhatnagar 2022).

The general challenges and opportunities of tax reforms for revenue mobilization are captured through empirical evidence in the developing countries. The comparisons of the tax reforms of Nigeria and Ghana show that all-inclusive tax policy reforms enhance revenue collection and buoyancy, but external shocks, institutional capacity, and limitations of tax base balance the results. Also, post-war countries have complicated conditions with fluctuating tax incomes and weakening tax collection, which limit fiscal sustainability. Therefore, environmental fiscal reforms within the larger tax policy should naturally take into consideration the politics and administrative aspects to maximize revenue collection and a sustainable result (Owamagbe 2025; Bekoe et al. 2016; Boogaard et al. 2018). All these results together imply that environmental fiscal reform has a strong revenue mobilization effect, but the success of this type of reform depends on the presence of well-formulated policies, equity in implementation, and positive institutional and political environments. Many empirical investigations show that political economy determinants of a policy, informal economy, mistrust of the public sector and enforcement regimes affect the expected fiscal and environmental effects of green taxes (IMF 2023).

9 Environmental Fiscal Reforms and Sustainable Transport Financing

Environmental fiscal reforms (EFRs) have become crucial tools of promoting sustainable transport financing through the combination of the environmental goals with the fiscal policy that will generate revenue and foster ecological sustainability. Recent research highlights the importance of using EFRs such as environmental taxes and subsidy reform in offering the so-called double dividend by not only ensuring economic growth but also cutting emissions. As an example, dynamic CGE modelling in Spain established that the introduction of higher environmental taxes on transport and energy industries, along with the withdrawal of subsidies, led to a decrease in the consumption of hydrocarbons and the emission of pollutants, with a positive GDP growth observed within a few years (Freire-Gonzalez and Ho 2018). This brings to focus the practicability of environmental fiscal policies as effective tools to fund sustainable transport infrastructures without affecting economic results.

Green bonds and other green finance instruments have taken centre stage in raising capital on sustainable transport. The empirical study in the European Union nations demonstrates that the issuance of green bonds has been a major contributor in the minimization of energy use and carbon emissions, which can serve as drivers of attaining environmental policies. This type of financing helps to finance the development of sustainable transport infrastructure, closing the funding gaps that can be limited by the traditional government budget or the debt ceiling (Petrov et al. 2024). Besides, the dynamics between the fiscal instruments and the development of the financial markets determine the success of the policies since the maturity of the financial sector impacts the environmental impacts of taxation and government spending, which is crucial to the subtle investments in the transport sector (Zioło et al. 2020).

The interaction between the fiscal policy and environmental innovation plays a critical role in converting transport systems into sustainability. A combination of post-Keynesian ecological macroeconomic models. Research indicates that carbon taxes, green subsidies and government green investments will be better than each one taken on its own to achieve ecological efficiency with reduced macroeconomic risk. These hybrid fiscal policy models can support sustainable transport policies by creating a stable source of revenue meant to be used in building green infrastructure and developing new technologies (Dafermos and Nikolaidi 2019b, a). Simultaneously, G7 countries have proven that green financing not only leads to the reduction of CO2 emissions but also promotes so-called eco-innovation needed in sustainable transport technologies, which strengthens the growth of the economy in line with environmental objectives (Elhassan 2025).

Fiscal decentralization would also increase the efficacy of the environmental fiscal reforms and sustainable transport financing by allowing the subnational actors to customize and adjust the fiscal policies to local requirements. The literature in fiscally decentralized states shows that a combination of devolved tax and expenditure authority, and green innovation policy, helps to reduce emissions and make more environmentally sustainable governmental investments, including transport (Ji et al. 2020). Moreover, modern econometric studies suggest focusing more on environmental protection spending and tax reforms since protection-oriented fiscal instruments seem to have greater ecological impacts because diversified fiscal strategies are critical in achieving sustainable transport finance systems (Erdogan and Pata 2024). Collectively, these lessons contribute to the holistic approaches to the combination of fiscal reform, green financing, and multilevel governance in order to support the financing of sustainable transport infrastructure in line with environmental requirements.

10 Institutional Quality, Governance, and the Effectiveness of Fiscal Policy in Green Transport Investment

The quality of institutions and governance has a strong bearing on the efficiency of the fiscal policy in encouraging investment in green transport. Recent studies point out that fiscal transparency and effective institutional structures maximize the effectiveness of government expenditures on the environment with respect to green technological developments. As an example, the statistics of 254 cities in China show that fiscal transparency enhances the correlation between environmental protection expenditures and green technology development and corporate confidence and social investment are attracted, which further promotes sustainable solutions to transport (Zhou et al. 2025). This highlights the fact that clear and responsible institutions are the bedrock for mobilizing financial resources effectively in green transportation.

The green credit policies and environmental fiscal tools also indicate that institutional settings moderate the funding results of industries with high levels of pollution that are also under pressure to transition to cleaner investments. In China, there are indications that environmental protection investments are related to the greater availability of debt financing to polluting companies within the green credit schemes, but are abated by the regional environmental pollution states, the financial development, and fiscal policies (Ji et al. 2021). This implies that strong governance and quality of the regulations do not only impact the implementation of the fiscal policies, but also the affordability and accessibility of funds used to finance the green transport infrastructure.

The institutional reforms coupled with fiscal decentralization are critical in enhancing green innovation that is vital in ensuring the long-term green investment in transport. A cross-country investigation of 32 nations in the OECD countries reveals that composite fiscal decentralization, which incorporates both revenue and spending authority, and institutional governance, will promote environmental sustainability through carbon emissions. This impact is particularly high in those areas where the level of emissions is intensified, which is an indication that decentralized fiscal systems supported by well-developed institutions provide a favourable platform of fiscal policies that can be used to fund green transport (Sun and Razzaq 2022). Therefore, effective green transport investments rely on the presence of governance systems that allow subnational governments to distribute resources in order to achieve efficiency.

Furthermore, the governance quality can pose challenges in the fiscal policies to promote the green transport investments to ensure the policies are free of corruption, effective in regulation and the rule of law. The transition country analysis reveals that instability in institutions worsens the accumulation of public debt, which compromises the fiscal sustainability level of the country, and can change the priorities of fiscal resources to be spent on green investments (Nguyen and Luong 2021). On the other hand, better governance enhances the efficiency of governments when it comes to the expenditures and revenues they can manage and therefore green transport can have sustainable fiscal space. To illustrate, supportive institutional and governance environments in China have lowered the financing cost of high-speed rail projects by attracting private capital through targeted fiscal policies, demonstrating the role of supportive institutional and governance environments in enhancing the efficiency of fiscal policies in developing the green transport infrastructure (Wong et al. 2022). Taken together, these works outline that good governance and the quality of institutions are part of the ability to efficiently harness fiscal policy instruments so that green transport investment is achieved.

11 Results/Discussion

Results in this section are presented in the form of two case studies. The first case study is from Indonesia, Southeast Asia and the second case study is from Colombia, Latin America. The section starts with Indonesia.

11.1 *Case Study 1: Indonesia: Fossil Fuel Subsidy Reform and Fiscal Space Creation*

In Indonesia, the reform of fossil fuel subsidies has been of great importance in enhancing the fiscal balance in the country in terms of the government spending on energy subsidies. In the past, Indonesia used to devote large shares of the budget to fossil fuel subsidies to ensure that domestic fuel prices were kept low and could achieve political patronage. These subsidies, however, were costly to the fiscal budget, as they diverted money to important investments (Skovgaard and Asselt 2019). Subsidy reforms, especially in recent times, have tried to restrain subsidy expenditure on the part of the Indonesian government, which not only boosted fiscal performance by decreasing deficits in the hands of the population but also liberated the budget to meet other needs. Such reforms are in line with the trends in the world where the idea of fossil fuel subsidy reduction is regarded as a necessity in the fiscal sustainability and environmental goals (Droste et al. 2024; Lockwood 2014). Irrespective of fiscal benefits, the reforms tended to face opposition due to politically sensitive constituencies that enjoyed cheap fuels, making the interaction of fiscal policy and the political economy in subsidy reforms very fragile (Strand 2013).

The created fiscal space by rationalizing subsidies has been planned to be distributed strategically to the development of infrastructure, particularly transport and energy, which is critical in promoting sustainable growth of Indonesia. The reallocation happens in favour of the Indonesian aspirations of inclusive economic development and green transitions, prioritizing the funds on projects aimed at improving connectivity and decreasing the intensity of carbon (Sumarno et al. 2022). As an illustration, state spending on infrastructure of public transport and investment in renewable energy sources promoted the Indonesian policy of decreasing greenhouse gas emissions and urban pollution. Not only do these fiscal reallocations spur economic output, but they also influence the participation of the psychology of the private sector by reducing the risks of government fiscal risk (Kyle 2018). However, the efficiency of the utilization of these resources is strongly dependent on the quality of governance and the capacity of these institutions, and these aspects are aspects that still need to be reinforced (Harring et al. 2023).

One of the key features of considering the rate and scale of fossil fuel subsidy reform in Indonesia is the political economy. These reforms are not welcomed because of the vested interests and fears over the social effects of subsidy cuts to the vulnerable groups (Strand 2013). Also, issues of governance, such as corruption at the local level, may compromise the credibility and efficacy of mechanisms of compensation that aim to subsidize subsidy eliminations, including specific social assistance initiatives (Kyle 2018). Research studies on reforms in Indonesia reveal that the greatest resistance occurs where the local governance capacity is low and therefore, there is a necessity to enhance the institutional structures to promote popular confidence in fiscal reforms. Effective reforms in Indonesia show that transparency, stakeholder consultations, and incremental introduction can help overcome the role of politics and society (Skovgaard and Asselt 2019; Lockwood 2014).

Governance structures and institutional quality also have an additional impact on the results of fossil fuel subsidy reform through the establishment of policy coherence, fiscal discipline, and public acceptance. The experience of Indonesia demonstrates that reforms can be more sustainable when they are not influenced by political pressures in the short term and have strong institutional checks (Droste et al. 2024; Nguyen and Luong 2021). The central and local governments should coordinate to implement reforms in a manner that addresses the compensation structures for the targeted beneficiaries. By improving governance in overseeing the implementation of fiscal policy, leakages and corruption will be minimized, which is important in preserving fiscal space and redistributing the funds to support sustainable infrastructure investments (Kyle 2018). The case of Indonesia, therefore, shows the complex interdependent relations between the quality of governance, the politics of political economy and the effectiveness of fiscal policy in the realization of green fiscal goals by reforming fossil fuel subsidies.

11.2 Case Study 2: Colombia: Carbon Tax and Green Investment Financing

The introduction of a carbon tax in Colombia is an excellent move to get to the point of achieving the so-called double dividend, which is both to cut carbon emissions and to enhance the fiscal performance. Not only has the carbon tax stimulated reduced emission of greenhouse gases, but also provided other fiscal resources that have been reinvested into the economy, promoting the overall growth and jobs in the economy (Dafermos and Nikolaidi 2019b, a). This is in accordance with the empirical evidence on other nations that proves that properly designed carbon taxation coupled with revenue recycling may produce both environmental and economic positive effects and still not negatively affect competitiveness (Labandeira et al. 2004). The case of Colombia helps to uphold the idea that carbon taxes will help eliminate negative environmental externalities and give governments additional financial assets to invest in sustainable development priorities.

The integration of fiscal and climate policy in Colombia has been achieved by using the avenue of directing the environmental taxation revenues to finance green investments, especially in renewable energy and sustainable infrastructures. Use of carbon tax revenues by the government in financing green bonds and other green financing platforms has played a greater role in mobilizing capital for climate-resilient projects (Duque-Grisales et al. 2024; Petrov et al. 2024). These fiscal tools reveal Colombia in an overall strategic alignment of tax reform with climate goals, making it easier to grow low-carbon by focusing on cleaner technologies, better transport, and energy efficiency projects. This financial union not only enhances fiscal sustainability but also hastens decarbonization efforts.

The practical connection between environmental fiscal reforms and green infrastructure financing is pointed out by empirical data in Colombia. Research explains the benefits of the carbon tax, together with state green investment and subsidies, in enhancing ecological efficiency and investment performance (Dafermos and Nikolaidi 2019b, a). These reforms have been demonstrated to help allay carbon emissions as well as counter flaws in financial markets through informing of risks related to carbon-intensive activities. The Colombian green fiscal policy uses several tools, such as carbon charges and fiscal subsidies, to stem the influx of investments by the private sector and enhance the efficiency of the investment of renewable energy businesses (Meng et al. 2022). Environmental and economic feedback loops generated result in sustainable development goals of Colombia.

The situation in Colombia also confirms the international theoretical models that a mixed policy of green fiscal policy involving carbon taxes and green subsidies and investments is more effective than the results of isolated fiscal policy (Dafermos and Nikolaidi 2019b, a). Through a comprehensive approach to fiscal policy development, Colombia gets the environmental and socioeconomic payoffs, which support the argument in favor of combined policy frameworks. Moreover, the expanding green bond market in Colombia is an illustration of successful fiscal policies to redirect the environmental tax income to climate-compatible projects, which can become a model to be followed by other emerging economies that want to create climate-compatible fiscal space (Petrov et al. 2024; Duque-Grisales et al. 2024). This synthesis explains the forward policy conditions in Colombia that is congruent with fiscal reform and sustainable finance as a way of dealing with climate change in totality.

12 Linking Theory to Empirical Evidence

This synthesis particularly uses the two concepts in respect to the case studies and determines adherence to IBC solvency concepts and EFR principles. The fossil fuel subsidy reform (2013–2024) was conducive to IBC, as it has resulted in primary surpluses of 1.2% GDP per year on average after implementation of the subsidy reform, the debt-to-GDP ratio fell from 28% (2014) to 39% (2023) without the implementation of a solvency measure, and the r-g differential remained below 2%, resulting in the re-allocation of KRW 300 trillion to sustainable infrastructure without triggering solvency risk. A green bond and an MRT system emerged as a proxy for redirected subsidy savings that relates to the second aspect of double-tax revenue recycling, that of "green" yields. Carbon tax in Colombia affected IBC by the 0.8% increase in GDP revenue that was re-directed to green investment funds, the primary balance improvement helped keep the debt-toGDP ratio at 60%, transport emissions were taxed by taxing fossil fuels, which saw a 12% decrease in emissions through EFR.

13 Further Research

The proposed macro-fiscal model should also be empirically tested in future research that should go beyond the qualitative desk analysis and use quantitative and mixed-method research. The causal relationship between fiscal sustainability indicators, including debt-to-GDP ratios, primary balances, and revenue mobilization, and the public investment in sustainable transport infrastructure could be measured by the panel data analysis of the developing economies. Comparative regional analysis, especially in Sub-Saharan Africa, South and Latin America, would also assist in establishing how the quality of institutions would mediate the effectiveness of environmental fiscal reforms. Further research should also focus on the distributional impacts of carbon taxes and fossil fuel subsidy reforms in the long run and their impact on low-income households and small businesses in particular. Micro-level household data would offer more understanding of the equity implications and popularity of green fiscal reforms. The second prospect to follow is the relationship between green bonds, blended finance schemes, and sovereign debt sustainability models. Studies may be conducted on whether green-labelled debt instruments or otherwise are likely to enhance the borrowing conditions or lower the fiscal risks in the long run. Lastly, further research can combine climatic risk models and intertemporal budget constraint models to determine the effects of climate shocks on the debt path and transport infrastructure resiliency in developing economies that are faced with financial constraints.

14 Managerial/Policy Implications

Results of the current study have significant implications for policy makers, public finance administrators and development planners of the developing economies. Fiscal sustainability must be seen as an enabler rather than a limitation to green transport transitions in the first place. To create fiscal space to invest in sustainable infrastructure, governments need to enhance domestic revenue collection, enhance the efficiency of tax collection and rationalize inefficient spending, especially fossil fuel subsidies. Second, the environmental fiscal reforms, like the carbon taxation, must be formulated with clear recycling of revenue mechanisms. Diversion of a portion of environmental tax revenues to public transport, electric mobility and resilient infrastructure can increase the credibility of the policy and public acceptance. Third, effective institutions and financial transparency are essential to the governance of reallocated funds being used effectively and without corruption. Long-term green investments can be anchored based on medium-term expenditure frameworks and credible fiscal rules consistent with the intertemporal budget constraint without compromising the sustainability of debt. Also, policymakers need to avail of innovative financing tools like green bonds and a partnership between the government and the private sector to mobilize the newly attracted capital without sidelining the healthy debt management behaviour. In sum, to have a stable macro economy and at the same time environmental sustainability, there should be coordination of fiscal and climate policies.

15 Conclusion

This chapter has analysed the connection between green transport transition and fiscal sustainability in developing economies in terms of inter-temporal budget constraint theory and the environmental fiscal reform theory. The discussion shows that fiscal sustainability does not always conflict with environmental ambition; on the contrary, it can serve as an empowering design in cases where revenue systems, expenditure priorities and debt managerial strategies are appropriately balanced. The case evidence in Indonesia and Colombia demonstrates how fossil fuel subsidy reforms and even carbon taxation can generate fiscal space and, at the same time, promote the climate goals. The success of these reforms, however, is strongly dependent on institutional quality, transparency of governance and dynamics of political economy. Poor governance mechanisms may compromise fiscal benefits and diminish people's confidence in the reform agenda. Hence, sustainable transport financing must involve consistent macro-fiscal planning, effective environmental taxation and reputable institutions. Developing countries can achieve low-carbon transport systems by combining fiscal discipline with green investment without undermining the long term sustainability of the debt. Finally, fiscal and environmental policies implemented in a consistent way may offer a good way forward in the direction of resilient infrastructure, macroeconomic stability, and sustainable development.

References

Afonso A, Jalles JT (2017) Euro area time-varying fiscal sustainability. Int J Financ Econ 22(3):244–254. https://doi.org/10.1002/ijfe.1582

African Development Bank (AfDB). (2018). *African economic outlook 2018*. https://www.afdb.org/en/documents/document/african-economic-outlook-aoe-2018-99877

Amin S (2025) Decarbonisation of road transportation in india—a round-robin review on low-carbon strategies and financial policies. Fut Transport 5(2):36. MDPI. https://doi.org/10.3390/futuretransp5020036

Badassa BB, Sun B, Qiao L (2020) Sustainable transport infrastructure and economic returns: a bibliometric and visualization analysis. Sustainability 12(5):2033. https://doi.org/10.3390/su12052033

Baglioni A, Cherubini U (1993) Intertemporal budget constraint and public debt sustainability: the case of Italy. Appl Econ 25(2):275–283. https://doi.org/10.1080/00036849300000033

Baute S (2025) The distributive politics of the green transition: a conjoint experiment on EU climate change mitigation policy. J Eur Publ Policy 32(1):52–80

Bekoe W, Danquah M, Senahey SK (2016) Tax reforms and revenue mobilization in Ghana. Journal of Economic Studies 43(4):522–534. https://doi.org/10.1108/jes-01-2015-0007

Chapman JI (2008) State and local fiscal sustainability: the challenges. Public Adm Rev 68(s1):S115–S131. https://doi.org/10.1111/j.1540-6210.2008.00983.x
Collignon S (2012) Fiscal policy rules and the sustainability of public debt in Europe*. Int Econ Rev 53(2):539–567. https://doi.org/10.1111/j.1468-2354.2012.00691.x
Dafermos Y, Nikolaidi M (2019) Fiscal policy and ecological sustainability: a post-Keynesian perspective (pp. 277–322). Springer. https://doi.org/10.1007/978-3-030-23929-9_7
Dafermos Y, Nikolaidi M (2019b) Fiscal policy and ecological sustainability: a post-Keynesian perspective. Frontiers of heterodox macroeconomics. Springer International Publishing, Cham, pp 277–322
Droste N, Chatterton B, Skovgaard J (2024) A political economy theory of fossil fuel subsidy reforms in OECD countries. Nat Commun **15**(1). https://doi.org/10.1038/s41467-024-49835-4
Duque-Grisales EA, Patiño-Murillo JA, Duque-Marín J, Giraldo-Giraldo S, Acosta-Strobel JA (2024) Can green bonds boost the development of energy projects in Colombia? An opportunity for responsible and sustainable investment. Int J Membr Sci Technol 11(1):487–496. https://doi.org/10.15379/ijmst.v11i1.3685
Elhassan T (2025) Green technology innovation, green financing, and economic growth in G7 countries: implications for environmental sustainability. Economics 13(1):69–91. https://doi.org/10.2478/eoik-2025-0023
Erdogan S, Pata UK (2024) Uncovering time-, frequency- and quantile-based impacts of energy, fiscal policy instruments, and economic growth on sustainable development. Int J Sust Dev World 32(1):65–79. https://doi.org/10.1080/13504509.2024.2398708
Fiscal Sustainability in Theory and Practice. (2005). World Bank. https://doi.org/10.1596/978-0-8213-5874-0
Freire-González J, Ho M (2018) Environmental fiscal reform and the double dividend: evidence from a dynamic general equilibrium model. Sustainability 10(2):501. https://doi.org/10.3390/su10020501
Greiner A, Semmler W (1999) An inquiry into the sustainability of german fiscal policy: some time-series tests. Public Finance Review 27(2):220–236. https://doi.org/10.1177/109114219902700205
Harring N, Jönsson E, Matti S, Mundaca G, Jagers SC (2023) Cross-national analysis of attitudes towards fossil fuel subsidy removal. Nat Clim Chang 13(3):244–249. https://doi.org/10.1038/s41558-023-01597-5
Intergovernmental Panel on Climate Change (IPCC) (2022) Climate change 2022: mitigation of climate change. http://www.ipcc.ch/report/ar6/wg3/
International Energy Agency (IEA) (2023) Transport sector CO_2 emissions. http://www.iea.org/reports/transport
International Monetary Fund (IMF). (2015). Making public investment more efficient. http://www.imf.org/external/np/pp/eng/2015/061115.pdf
International Monetary Fund (IMF) (2023) Fiscal Monitor: Climate crossroads. https://www.imf.org/en/publications/fm/issues/2023/10/10/fiscal-monitor-october-2023
Jaeger WK (2010) The welfare effects of environmental taxation. Environ Resource Econ 49(1):101–119. https://doi.org/10.1007/s10640-010-9426-x
Ji X, Umar M, Ali S, Ali W, Tang K, Khan Z (2020) Does fiscal decentralization and eco-innovation promote sustainable environment? A case study of selected fiscally decentralized countries. Sustain Dev 29(1):79–88. https://doi.org/10.1002/sd.2132
Ji L, Jia P, Yan J (2021) Green credit, environmental protection investment and debt financing for heavily polluting enterprises. PLoS ONE 16(12):e0261311. https://doi.org/10.1371/journal.pone.0261311
Kyle J (2018) Local corruption and popular support for fuel subsidy reform in Indonesia. Comp Pol Stud 51(11):1472–1503. https://doi.org/10.1177/0010414018758755
Labandeira X, Labeaga JM, Rodríguez M (2004) Green tax reforms in Spain. Eur Environ 14(5):290–299. https://doi.org/10.1002/eet.361

Lockwood M (2014) Fossil fuel subsidy reform, rent management and political fragmentation in developing countries. New Political Economy 20(4):475–494. https://doi.org/10.1080/13563467.2014.923826
Martin GM (2000) US deficit sustainability: a new approach based on multiple endogenous breaks. J Appl Economet 15(1):83–105. https://doi.org/10.1002/(sici)1099-1255(200001/02)15:1%3c83::aid-jae543%3e3.0.co;2-j
Meng Z, Sun H, Liu X (2022) Impact of green fiscal policy on the investment efficiency of renewable energy enterprises in China. Environ Sci Pollut Res 29(50):76216–76234. https://doi.org/10.1007/s11356-022-20832-8
MF. (2023). Fiscal monitor: climate crossroads—fiscal policies in a warming world. International Monetary Fund. https://www.imf.org/-/media/files/publications/fiscal-monitor/2023/october/english/ch1.pdf
Mpofu FY (2022) Green taxes in Africa: opportunities and challenges for environmental protection, sustainability, and the attainment of sustainable development goals. Sustainability 14(16):10239. https://doi.org/10.3390/su141610239
Nguyen TAN, Luong TTH (2021) Fiscal policy, institutional quality, and public debt: evidence from transition countries. Sustainability 13(19):10706. https://doi.org/10.3390/su131910706
OECD (2018) Green growth in cities. OECD publishing. https://www.oecd.org/en/publications/green-growth-in-cities_9789264195325-en.html
OECD (2021) Taxing energy use for sustainable development: opportunities for energy tax and subsidy reform in selected developing and emerging economies. OECD Publishing. https://www.oecd.org/content/dam/oecd/en/publications/reports/2021/01/taxing-energy-use-for-sustainable-development_7505cdd6/89167cef-en.pdf
Organisation for Economic Co-operation and Development (OECD). (2021). Financing climate action. http://www.oecd.org/climate-change/finance-usd-100-billion-goal/
Organisation for Economic Co-operation and Development (OECD). (2018). Financing climate futures: Rethinking infrastructure. http://www.oecd.org/environment/cc/climate-futures/
Organisation for Economic Co-operation and Development (OECD) (2017) Investing in climate, investing in growth. http://www.oecd.org/environment/investing-in-climate-investing-in-growth-9789264273528-en.htm
Owamagbe OP (2025) Evaluating the effectiveness of tax reforms on revenue generation in Nigeria: a time-series analysis (1999–2023). Development 8(4):131–147. https://doi.org/10.52589/ajesd-yyezpacm
Petrov S, Aleksandrova S, Kirova S (2024) Environmental effects of green bonds and other forms of financing in the European Union. Int J Econom Sci 13(1):81–105. https://doi.org/10.52950/es.2024.13.1.005
Poddar S, Bhatnagar S (2022) Fiscal decentralization and urban environmental governance in India. [Unpublished manuscript/Institutional publication].
Prakash N, Sethi M (2021) A review of innovative bond instruments for sustainable development in Asia. Int J Innov Sci 14(3/4):630–647. https://doi.org/10.1108/ijis-10-2020-0213
Rietig, K. (2020). The importance of compatible beliefs for effective climate policy integration. In The Future of European Union Environmental Politics and policy (pp. 42–61). Routledge
Schaechter A, Kinda T, Budina N, Weber A (2012) Fiscal rules in response to the crisis—Toward the next-generation rules. International Monetary Fund. http://www.imf.org/external/pubs/ft/wp/2012/wp12187.pdf
Skovgaard J, Van Asselt H (2019) The politics of fossil fuel subsidies and their reform: Implications for climate change mitigation. WIREs Climate Change 10(4). https://doi.org/10.1002/wcc.581
Spaventa L (1987) The growth of public debt: sustainability, fiscal rules, and monetary rules. Staff Papers—International Monetary Fund 34(2):374. https://doi.org/10.2307/3867139
Strand J (2013) Political economy aspects of fuel subsidies. World Bank Washington Dc. https://doi.org/10.1596/1813-9450-6392

Sumarno TB, Sihotang P, Prawiraatmadja W (2022) Exploring Indonesia's energy policy failures through the JUST framework. Energy Policy 164:112914. https://doi.org/10.1016/j.enpol.2022.112914

Sun Y, Razzaq A (2022) Composite fiscal decentralisation and green innovation: Imperative strategy for institutional reforms and sustainable development in OECD countries. Sustain Dev 30(5):944–957. https://doi.org/10.1002/sd.2292

Truger A (2020) Reforming EU fiscal rules: more leeway, investment orientation and democratic coordination. Inter Economics 55(5):277

Uctum M, Wickens M (2000) Debt and deficit ceilings, and sustainability of fiscal policies: an intertemporal analysis. Oxford Bull Econ Stat 62(2):197–222. https://doi.org/10.1111/1468-0084.00168

Udeagha MC, Muchapondwa E (2022) Environmental sustainability in South Africa: Understanding the criticality of economic policy uncertainty, fiscal decentralization, and green innovation. Sustain Dev 31(3):1638–1651. https://doi.org/10.1002/sd.2473

United Nations Framework Convention on Climate Change (UNFCCC) (2021) Nationally determined contributions under the Paris Agreement synthesis report. http://www.unfccc.int/documents/306848

United Nations (2019) World urbanization prospects 2018 revision. https://population.un.org/wup/assets/WUP2018-Report.pdf

Van Den Boogaard V, Prichard W, Benson MS, Milicic N (2018) Tax revenue mobilization in conflict-affected developing countries. J Int Dev 30(2):345–364. https://doi.org/10.1002/jid.3352

Wong Z, Chen A, Shen C, Wu D (2022) Fiscal policy and the development of green transportation infrastructure: the case of China's high-speed railways. Econ Chang Restruct 55(4):2179–2213. https://doi.org/10.1007/s10644-021-09381-1

World Bank Group (2021) Transport decarbonization investment series (TDI): financing low carbon transport solutions in developing countries https://www.worldbank.org/en/topic/transport/publication/transport-decarbonization-investment-series

World Bank (2020) Africa's infrastructure: A time for transformation. https://documents.worldbank.org/en/publication/documents-reports/documentdetail/246961468003355256

World Bank (2022) International debt statistics 2022. https://documents.worldbank.org/en/publication/documents-reports/documentdetail/099435412122316848

World Bank (2023) State and trends of carbon pricing 2023. https://documents.worldbank.org/en/publication/documents-reports/documentdetail/099805106052321586

Xu J, Zhang Y (2024) Global-local nexus in sustainable development: the influence of international green investment on local environmental spending. J Environ Planning Manage 69(2):573–596. https://doi.org/10.1080/09640568.2024.2391549

Zhou Z, Zhang W, Pan X, Hu J, Pu G (2020) Environmental tax reform and the "double dividend" hypothesis in a small open economy. Int J Environ Res Public Health 17(1):217

Zhou C, Li M, Zhao Y (2025) Driving effect of government environmental protection expenditure on green technology from the perspective of fiscal transparency. Human Soc Sci Commun 12(1). https://doi.org/10.1057/s41599-025-04791-2

Zioło M, Kluza K, Kozuba J, Kelemen M, Niedzielski P, Zinczak P (2020) Patterns of interdependence between financial development, fiscal instruments, and environmental degradation in developed and converging EU countries. Int J Environ Res Public Health 17(12):4425. https://doi.org/10.3390/ijerph17124425

Economic Instruments and Market Mechanisms for a Sustainable Transition to Green Transport Systems

Innocent Chirisa, Africa Muneti, Nyasha Ndemo-Masimbarasi, and Tinashe Magande

Abstract Transport systems drive one-fifth of global greenhouse gas emissions, intensifying climate vulnerability, urban congestion, and air quality crises worldwide. Conventional fossil fuel dependence perpetuates market failures where pollution costs remain unpriced, distorting investments away from low-carbon alternatives. Economic instruments like carbon taxes and emissions trading schemes promise to internalize these externalities, spurring efficiency and innovation in vehicle fleets and infrastructure. However, persistent technology lock-ins, financing barriers, and distributional inequities slow progress, particularly in emerging economies racing to expand mobility. Real-world pilots reveal mixed outcomes, with emissions cuts often undermined by leakage and weak complementary policies. This study confronts the challenge of designing integrated mechanisms that deliver environmental gains alongside economic viability and social acceptance. The main objective of this qualitative study was to explore how economic instruments and market mechanisms facilitate sustainable transitions to green transport systems. Data were generated through a comprehensive qualitative desk review of peer-reviewed literature, policy reports, and case studies from regions like China, the EU, and emerging markets. For data analysis, the study used textual analysis to identify patterns, synergies, and institutional contingencies shaping policy effectiveness. The analysis revealed

I. Chirisa (✉)
Office of the Vice Chancellor, Zimbabwe Ezekiel Guti University, Stand No 1901, Barrassie Road, Off Shamva Road, Bindura, Zimbabwe
e-mail: innocent.chirisa@gmail.com

I. Chirisa · T. Magande
Department of Urban and Regional Planning, University of Johannesburg, Johannesburg, South Africa

A. Muneti
Firm Central Property Group, Harare, Zimbabwe

IC University of Technology, Phnom Pehn, Cambodia

N. Ndemo-Masimbarasi
Department of Development Programming and Management, Zimbabwe Ezekiel Guti University, Bindura, Zimbabwe

J. Kanyepe and I. Chirisa (eds.), *Advancing the Green Agenda in the Transport and Logistics Sector in Global South Countries*, World Sustainability Series,
https://doi.org/10.1007/978-3-032-31374-4_22

that (1) carbon pricing reduces emissions by 6–16% in well-designed pilots but fails to trigger radical innovation without subsidies and standards; (2) sequenced policy mixes—starting with R&D support before escalating taxes—minimize lock-ins and build political constituencies; (3) green bonds and public–private partnerships unlock private capital for infrastructure, though success hinges on risk mitigation and governance coordination. The findings suggest that policymakers prioritize dynamic, equity-focused policy packages over standalone pricing to achieve resilient decarbonization. Integrated approaches transform market failures into opportunities for green growth, offering a blueprint for just transport transitions worldwide.

Keywords Carbon pricing · Green transport · Emissions trading · Policy mix · Green bonds · Decarbonization

1 Introduction

The transport sector is a significant source of greenhouse gas emissions to the world, with almost a quarter of the overall anthropogenic CO2 emission going through the transport sector. Increased vehicle ownership and urbanization contribute to congestion, air pollution, and climate effects, especially in the growing economies with infrastructures that are not at par with demand (Sims et al. 2014). Carbon price and market-based economic tools provide potential avenues of internalizing these externalities and leading to the low-carbon transitions. Theorists emphasize that Pigouvian taxes and emissions trading systems provide a realignment of incentives, which encourages efficiency without suffocation (Nordhaus 2013). However, there are still the technology lock-ins and gaps in financing the use of electric vehicles and sustainable public transport. This context highlights the need to have combined policy strategies in the transport green systems.

This paper will posit that the economic devices and market processes when integrated in the context of supplementary policy bundles can well facilitate green transport transitions and overcome equity and innovation hurdles. Carbon pricing is not effective to monumental decarbonization as the pilot programs have shown mixed results (Goulder and Parry 2008). The idea of policy sequencing, which proposes the implementation of subsidies and standards before increasing taxes, creates political support and conquers market failures (Meckling et al. 2017). Green bonds and the partnership with the government are additional ways to access capital to support infrastructure and increase the possibility in a resource-starved environment. The interaction of the environmental economics and the institutional theory synergy brings out environmental gains coupled with the socialeconomic reality. Finally, adaptive governance is revealed as critical when it comes to resilient outcomes.

Although there is widespread advocacy of carbon pricing, there still exist empirical gaps in the comprehension of the role of policy combinations in alleviating distributional inequalities and financing risks in transport decarbonization, particularly in emerging markets. The existing literature tends to ignore institutional coordination of curbing leakage and creating radical innovations (Acemoglu et al. 2012). Pieces of work do not combine the tools of economics and governance systems to provide useful information to the policymakers. This research issue requires an in-depth study of actual examples such as EU ETS and Chinese pilots. The key objective of this paper is to review the efficiency of economic tools in policy mixes to transform transport to green. It finds the best designs to use to facilitate equitable and scalable implementation using qualitative synthesis.

2 Conceptual Framework

The theoretical backdrop of this chapter takes off the Environmental Economics Theory, especially the Pigouvian theory that deals with the market failures that are caused by negative externality like carbon emissions. The theory states that the inefficiencies that arise when the external cost of pollution is not incorporated in the market prices justify intervention in the form of carbon pricing and corrective taxation. Internalizing such externalities through the use of carbon taxes or emissions trading systems causes firms to pay the actual social cost of their operations, and thus, they get effective incentives to cut their emissions (Bromley 1978; Mikhno et al. 2020). Empirical studies point to the fact that properly-designed pricing mechanisms will successfully reduce emissions with minimal economic disturbance to encourage cost-efficient abatement (Scotti et al. 2024). However, also pricing in the form of carbon price might not be effective enough to break technology lock-ins or trigger preferred innovation trajectories, which means that it is important to think of complementary policies in addition to pricing (Scotti et al. 2024; Nykamp 2020). In line with this, Environmental Economics offers the normative basis of utilisation of economic tools in green transport transitions by remediating market failures on the way to the socially optimal.

As a supplement of the Pigouvian foundations, the theory of Institutional and Policy Mix explains the importance of institutional structures, governance mechanisms, and coordination of various policy tools in long term sustainability transitions. The green transport policymaking necessitates coordinated combinations of regulations, subsidies, carbon pricing, and innovation support based on the changes in political and economic conditions (Rogge and Reichardt 2016; Shao and Huang 2023). It has been found that sequencing and integrating policies increases political feasibility, reduces lock in risks, and speeds up decarbonization through leveraging the synergies between the instruments (Scotti et al. 2024; Nykamp 2020). The Institutional Analysis and Development paradigm extends this argument by highlighting the rules-in-use and actor position in the institutional environment to provide a dynamic understanding of the interaction of the policy instruments and the resulting incremental and systemic change (Barton et al. 2017). Therefore, the policy mix theory predicts the complexity of governance facilitating resilient transitions, which supports the conceptual frame of how collaborative multi-level governance and adaptable policy design supports the achievement of sustainable green transport systems (Ekdahl et al. 2024; Mamash et al. 2025). Collectively, the theories present a strong conceptual foundation of the economic tools and mechanisms of the market in the wider governance and innovation frameworks that are required to promote sustainable transport futures.

3 Literature Review

This section presents the literature review of this chapter. The following issues are discussed in the section: economic rationale behind the decarbonisation of transport, carbon pricing and market-based instruments; complementary policies and policy mix design and investment decisions, financing and market design.

3.1 Economic Rationale for Decarbonising Transport (Internal and External Costs)

Decarbonisation of transport it is economically justified because of market failure, especially the negative externalities of carbon emission which are not captured in the market. Carbon emission social costs like health-related, environmental damages, and climate change are not reflected in the decisions made by producers and consumers, yet carbon emissions have social costs (Czermański et al. 2020). This discrepancy between socially optimal results is the reason why policy measures are recommended to internalize these external costs frequently measured as the social cost of carbon (SCC). There is systematic review evidence that carbon pricing, taxes, and tradable permits can bring positive environmental benefits on a general basis by harmonizing social welfare and individual incentives (Peñasco et al. 2021). Nevertheless, they can produce temporary competitive pressures and distributional inequalities, especially affecting small businesses and most disadvantaged socioeconomic groups, so the policy should be carefully structured to reduce the negative impacts.

The instruments that can be used to correct the externalities in the transport sector are the price (carbon taxes and market-based measures, MBMs). Carbon pricing systems directly incorporate the price of emissions into the price of fuel, which encourages switching the fuel, efficiency gains, and innovation (Lagouvardou et al. 2023). Complementary policies such as subsidies or subsidial funding of clean technologies are used to address other market failures such as capital market imperfections and knowledge spillovers that prevent low-carbon investments (Kempa and Moslener 2017). Carbon taxes are very effective in internalizing externalities, however, specific financial tools may stimulate the use of technologies and changes in the structure, but with the threat of inefficient state expenditures when miscalibrated (Gillingham and Sweeney 2012). Balancing between economic efficiency and political and social feasibility in the process of decarbonization transitions is essential and achievable only through the sequencing and combination of policy measures (Meckling et al. 2017).

The distributional effects are an important issue in decarbonization policy. Individualistic carbon taxes and policies can mitigate retrogressive household burdens by taking into account income heterogeneity and existing tax systems but equity benefits may be minimal without social policies (Landis et al. 2017). In economic models, aggressive decarbonization may cause trade-offs, including losses in the competitiveness of the sector or a increase in the number of cases of depopulation, especially of energy-intensive industries and small operators (Peñasco et al. 2021). However, with the implementation of policies that support the affected population, such distributional trade-offs can be converted into co-benefits, which will enable the wider acceptance and sustainability of climate actions (Peñasco et al. 2021).

The economic case of transport decarbonization is still cost-effectiveness. Carbon taxes or trading schemes are commonly viewed as the most efficient methods of reducing emissions in that they offer homogeneous incentives to all participants and reduce economic distortions (Putri et al. 2024). Political economy limits, lock-in of regulations, and unequal distribution of technologies can, however, balance the cost-effectiveness (Meckling et al. 2017). Moreover, the marginal abatement cost curves signify that the need to close the cost gap between conventional fuels and alternative clean fuels demands high carbon prices and favorable policy frameworks to drive innovation and the investment of infrastructure (Lagouvardou et al. 2023). Strong policy frameworks should thus be a mix of a price on carbon and other complementary governance policies as well as incentives on innovations to ensure efficient, fair, and politically feasible decarbonization routes of the transport sector..

3.2 *Carbon Pricing and Market-Based Instruments*

Carbon taxes and emissions trading systems (ETS) are market-based instruments that are the keystone of internalizing environmental cost of emissions in transport markets through pricing of carbon. The results of the EU Emissions Trading System at the firm level have shown its effectiveness, as reported 14 to 16 percent of emissions were lowered without a negative impact on the economic activity, proving that environmental and economic goals can co-exist under a properly designed ETS (Zhang et al. 2020). The carbon taxes can work in the same manner and put pressure on the costs of fuel-intensive carbon-based fuels, offering strong incentives to companies to decrease emissions via cleaner technologies and operations (Samuel 2025). However, these tools frequently require to be supplemented with innovation subsidies or regulation, since carbon pricing alone does not demonstrate much empirical evidence in stimulating deep decarbonization-level technological innovation (Zhang et al. 2020; Digitemie and Ekemezie 2024).

The effectiveness of carbon pricing schemes depends greatly on their design and implementation as they can either help to cut emissions and make innovations. The carbon taxes which are not large enough do not provide incentives to have large emission reductions and optimal pricing can encourage the use of radical green technologies because of direct cost indicators (Qu and Sun 2022). In addition, carbon taxes can be used together with innovation subsidies, this effect may have a synergistic effect, motivating companies to invest in groundbreaking technologies, instead of making incremental gains (Zhang et al. 2020). As quantities-based schemes, emissions trading schemes limit total emissions and enable trading of allowances, which encourages cost-effective cutoffs in emission but can give unpredictable price indicators to innovation than explicit carbon taxes (Millard-Ball 2009).

Although this is the case despite their environmental advantages, the introduction of carbon pricing tools has been faced with distributional fairness and economic competitiveness issues. The empirical research indicates that price on carbon might result in the loss of competitiveness in the energy-intensive industries in the short run and lead to the problem of carbon leakage, i.e., the emissions might be transferred to the jurisdictions that have lenient regulations (Digitemie and Ekemezie 2024). To address negative effects on vulnerable groups and smaller businesses, policymakers should address the concept of differentials pricing or free social policies in order to have a fair transition (Qu and Sun 2022). Combined carbon policies with specific encouragement of the low-carbon innovation can be used to overcome the trade-off between cost-effectiveness and social acceptance, as seen in the successes of carbon taxation across the globe (Samuel 2025; Li et al. 2024).

Moreover, this combination of carbon pricing and other policy instruments like green procurement and investment in infrastructures also creates a comprehensive environment of decarbonization where market forces meet the implementation of technologies and consumer choices (Zhang et al. 2020, 2024). The use of carbon pricing will encourage companies to be energy efficient and innovate, but a systemic transformation requires complementary policies to overcome market failures not related to carbon externalities, such as knowledge spillovers and capital market imperfections (Kempa and Moslener 2017). The effectiveness of carbon pricing is also further enhanced through global cooperation, and coordination of policies to eliminate risks of regulatory arbitrage and stability of prices that is essential in maintaining innovation (Digitemie and Ekemezie 2024). Overall, carbon taxes and ETS are efficiency measures in reducing emissions in the transport markets, but the full potential of these measures can be achieved only in integrated policy frameworks that are balanced in terms of the objectives of the environmental goals, economic competitiveness, incentives to innovate and assure social equity.

3.3 Complementary Policies and Policy Mix Design

To complement the effectiveness and uniformity of non-carbon pricing instruments decarbonization in transport, complementary policies are needed. It has been noted that carbon pricing, when used together with subsidies, regulatory norms, public procurement, or other favourable policies, generate synergies that can be used to adopt technologies and behavioural change towards deep decarbonization (Peñasco et al. 2021). Subsidies especially those offered to clean public urban transport are to bring an increase in ridership and reduce externalities in the environment like greenhouse gas emissions and congestion. Nevertheless, evidence is emerging that most subsidy programs are retrogressive or neutral in distributional terms indicating a need to make sure that subsidies are demand-based and incorporated as a component of broader social welfare systems particularly in the developing environment (Serebrisky et al. 2009). Subsidy design, therefore, should be based on equity and effectiveness at the same time to enhance the outcomes of transport decarbonization.

Pricing is used in conjunction with regulatory instruments such as emissions standards, energy efficiency mandates, and so forth to establish technology or performance floors. Regulation eliminates low-efficiency solutions in the market and speeds up the process of diffusion of potential low-carbon solutions, which is a vital counterbalance to the usually price-unpredictable atmosphere of carbon markets (Peñasco et al. 2021). Another growing influence is the public procurement where governments use their purchasing power to influence the markets towards reduced emissions by imposing sustainable procurement requirement and green standards. EU research reports that irrespective of non-binding country policy, political influences and the high amount of a procurement contract determine the adoption of green public procurement, but such adoption is characterized by the larger projects with decentralized management practices in a way that makes sound policy implementation difficult (Vluggen et al. 2019; Yu et al. 2020). Good public procurement therefore necessitates consistent governance and concordant green standards in the sectors.

The concept of policy sequencing becomes one of the essential aspects of policy mixes to prevent lock-in and opposition to the policy, which is highlighted in the case of the most progressive low-carbon jurisdictions such as California and the European Union (Meckling et al. 2017). Some of the early policies that focus on subsidies in research and development, and industrial policies that foster innovation of clean technologies assist in creating constituencies in favor of decarbonization and reducing the costs of technologies. This incremental strategy makes it easy to introduce carbon pricing instruments and regulatory mechanisms gradually when the political acceptance increases and the institutional capacity becomes stronger. However, sequencing also has issues of rent capture and risk of sustaining an incumbent advantage unless well kept, necessitating adaptive governance and continued stakeholder involvement (Meckling et al. 2017).

The effective decarbonization of transport and equity can be achieved significantly through a diversified and well-sequenced policy package which includes subsidies, regulatory standards, public procurement, carbon pricing, which must be well-coordinated and context-sensitive. When subsidies and regulations do not eradicate market failures and facilitate technology adoption, the operationalization of demand-side commitments is done by public procurement. The policy sequencing is an insurance of political viability and prevention of economic lock-in, which makes the transition to the low-carbon transport systems resilient and inclusive. The focus of future research and policy formulation should be on integrative models that can adjust complementary tools to local circumstances, maximize sequencing, and observe distributional effects so that decarbonization plans can yield numerous environmental and social co-benefits (Peñasco et al. 2021; Yu et al. 2020).

3.4 Investment Decisions, Financing, and Market Design

The capital investment, at large scale, will be core in the transition to green transport systems, and the concept of carbon pricing can be seen as one of the reasons to shift economies toward capital-intensive and low-carbon technologies without causing a significant disruption to the growth of GDP (Okeke et al. 2024). Carbon pricing creates economic policies that will shift the focus of the private investments towards clean energy and green infrastructure. Nevertheless, investment risk, which is caused by unpredictable regulatory relationships, the length of project cycles, and technology risks, tends to discourage the involvement of a private capital in the investment, especially in underdeveloped market countries lacking well-developed financial infrastructure (Okeke et al. 2024; Lwesya 2025). In order to counteract these risks, policy incentives or tax break, subsidies, and regulatory frameworks are essential, since they reduce hurdles and make green investments more appealing to financiers. Moreover, the incorporation of effective risk management techniques such as diversification and public–private partnerships can help to make green transport financing more viable in dissimilar economic environments (Okeke et al. 2024).

Green bonds have become well-known as new financial tools that can enhance the feasibility of financing by attracting socially responsible investors and burden sharing between generations (Alonso-Conde and Rojo-Suarez 2020). Empirical data indicate that in most cases, green bonds are associated with higher internal rates of returns and better credit quality than traditional finance, which motivates the project sponsors to seek low-carbon investments (Alonso-Conde and Rojo-Suarez 2020). Also, green bonds serve to reduce the limitation of financing and enhance corporate reputations, which contribute to green innovation and cooperation (Lian et al. 2023). Nevertheless, the greenbond market is struggling with such obstacles as the inability to regulate the market and establish uniform criteria to warrant the investors and guarantee the transparency of impact (Cheong and Choi 2020).

To enhance green transport investments and capitalsize on the joint resources, knowledge base, and risk-sharing of both governments and the private, the idea of public–private partnerships (PPs) is strategic (Weis and Nikolic 2024). PPP facilitates investment of the private capital in the projects that may otherwise be limited by the budget constraint of the government and encourages the innovation and efficiency gain by virtue of working together. Emerging market case studies point to the fact that institutional frameworks and policy incentives are necessary conditions of successful PPPs to support the development of green infrastructure (Lwesya 2025). The incorporation of innovative tools of green finance, such as the use of fintech platforms, can even streamline the availability of financing and risk management, especially in such regions as India and Africa where problems of the public sector capacity and market maturity still remain (Sreenu 2024; Lwesya 2025).

The literature of market design indicates that a hybrid financial market, with features of both debt and equity financing, a carbon market and green bonds might be needed in order to manage risks related to investments and ensure long-term coordination in green transport (Mukherjee et al. 2024). These hybrid markets offer scalability and deeper capital pools, by offering flexibility through the balances of liquidity and risk distribution and by being able to satisfy the needs of the investors. In addition, permissive regulatory incentives and open reporting can increase market confidence and increase investment inflows. The high-level of decarbonization will probably need an ecosystem approach of policy incentives, creative financial instruments, and adaptive market designs in order to provide long-term financing of green infrastructure of transport (Raman et al. 2025).

4 Methodology

The study used a qualitative methodology that is predominantly a desk review. The qualitative nature allowed the researchers to have a deeper investigation into the sectors that are technically and economically difficult to decarbonise. It further allowed for the exploration of how economic instruments and market mechanisms facilitate sustainable transitions to green transport systems. Data for the study were sourced from a broad range of materials, including books, policy documents, peer-reviewed journal articles, and some credible online sources. A comprehensive review was done through a selection of these different sources to capture theoretical views and empirical findings from previously existing articles. To get access to this literature, the study used electronic search engines that include Science Direct, Scopus, and Google Scholar. This helped the researchers to ground the study in authoritative and well-established academic discourse. The study used textual analysis as a primary analytical tool for data analysis. A systematic examination of selected English documents was done using this method to identify recurring themes and patterns in literature.

5 Findings

Findings in this study are presented in the form of three case studies. The first case study is on China's regional emissions trading systems, followed by a case on the European Union, focusing on carbon pricing and transport policy mix. The last case study is on financing green transitions through carbon pricing and green bonds.

5.1 Case Study 1: China's Regional Emissions Trading Systems (ETS)

To minimize carbon emissions and economic growth in China, it introduced regional carbon emission trading schemes as pilot programs. The introduction of the regional Emissions Trading Systems (ETS) in China was in the form of pilot projects to curb the emission of carbon gases and, at the same time, enjoy a healthy economic growth. Beginning in 2013, these pilots targeted some select provinces and cities with different socioeconomic and industrial characteristics. The ETS was expected to internalize the environmental cost of emissions by putting caps and allowing the trading of carbon allowances to motivate firms to innovate and use cleaner methods. Quasi-natural experiment results indicate that these regional ETSs assisted in a 6.1 per cent decrease in CO2 emissions and a 6.6 per cent decrease in emissions intensity in regulated cities, which supports the effectiveness of these in reducing carbon emissions (Jiang et al. 2024; Cui et al. 2021). Cuts were as a result of increased energy efficiency, cleaner energy mix, and outsourcing of energy-intensive production processes. These pilots had mixed economic results, and some of the negative effects in the employment sector were compensated by the productivity factor, which shows a complex yet positive effect on the paradigms of sustainable development (Chai et al. 2022; Cui et al. 2021).

Nevertheless, one of the most prominent obstacles found in the regional ETSs in China is the leakage of carbon, as the emissions are rising in the neighbouring cities that are not regulated by the policies approximately 1.7% because of the production level and polluting activities moving out of regulatory scope (Jiang et al. 2024; Zhuo et al. 2024). This spillage brings out the dangers of the unequal environmental regulations and the necessity of the extensive coordination of the region. It is important to note that stricter compliance rules and aggressive trading platforms in the market have been demonstrated to reduce leakage effects and enhance emissions abatement in pilot areas (Jiang et al. 2024). Moreover, institutional design and socioeconomic contexts showed variations in the efficacy of ETS, where the central and industrial cities could make greater reductions but were also at the same time more susceptible to leakage. These results highlight the paramount role of institutional power, open surveillance, and penal systems as well as price on the carbon to ensure maximum environmental and economic gains (Duan and Zhou 2017; Jiang et al. 2024).

The example of the regional ETS pilots in China provides some important insights into the lessons of economic instruments that can be implemented in order to achieve sustainable transport and carbon mitigation. To begin with, carbon pricing has the potential to be effective to mitigate the emissions when supported by a sound market design and legal requirements and align the incentives of the firms with the aims of environmental protection (Chai et al. 2022; Cui et al. 2021). Second, it is critical to control carbon leakage by harmonizing policy frameworks between regions so that counterproductive spillovers can be avoided, and overall, reduction in emissions is achieved (Jiang et al. 2024). Lastly, the regional disparity in effects highlights the importance of implementing adaptive policy that considers regional economic arrangements and governing abilities. The lessons of the ETS pilots in China therefore highlight the necessity of institutional design and regulatory intensity in implementing market mechanisms that facilitate a sustainable shift to the low-carbon transport systems and overall climate targets (Duan and Zhou 2017; Zhuo et al. 2024).

5.2 Case Study 2: European Union - Carbon Pricing and Transport Policy Mix

Carbon pricing and other complementary policies have increasingly become part of the European Union (EU) as it strives to achieve its ambitious climate goals especially under the umbrella of the European Green Deal, which seeks to have climate neutrality by the middle of the century. The macroeconomic modeling research demonstrates that the adverse effects of carbon pricing schemes on GDP and consumption have been modest with the ambition carbon pricing schemes including the EU Emissions Trading System (ETS) successively increasing investment in low-carbon technologies and capital-intensive sectors (Fragkos and Fragkiadakis 2022). It is demonstrated that the EU ETS made manufacturing firm emissions decrease by 14–16 percent without affecting the economic output, which indicates that the market-based instrument can stimulate the investment towards energy efficiency and decrease carbon intensity in several industries (Colmer et al. 2024). In spite of the mentioned positive trends, the way of decarbonization is not easy and demands a combination of different policy tools other than price.

But scientific reviews suggest that the carbon charges do not have the strength to result in long-term technological changes or zero-carbon investments that would lead to deep decarbonization (Lilliestam et al. 2020). The development of the transport sector, especially, strongly relies on the national policy decisions and the existence of the corresponding measures like the support of innovations, strict emission regulations, and the development of infrastructure (Dolge et al. 2023). These policies facilitate behavioural change and uptake of technologies as they deal with other market failures, breaking bottlenecks in infrastructures and they also help in the reduction of uncertainty to investors. The policy mix approach of the EU emphasizes that the concept of pricing carbon should be included into the set of institutional and regulatory measures that would help to maximize environmental performance and at the same time remain economically competitive and socially just (Samuel 2025; Weitzel et al. 2022).

There are some important lessons to sustainable transport transitions on the market, as underlined in this case study of the EU. To start with, carbon pricing should be used as a policy tool, and it is necessary but not sufficient to ensure that the economy is well-aligned with the environmental goals of cutting emissions but should be complemented by regulation policies and innovation policies to speed up technological change and adoption (Digitemie and Ekemezie 2024; Laing et al.

2014). Second, the pricing policies have economic effects that might be managed but need to be carefully designed, especially to deal with distributional issues and protect the vulnerable population with complementary social policies (Vandyck et al. 2023). Lastly, the EU experience confirms the necessity of a policy mix framework based on the institutional theory, focusing on the multi-level governance, coordination of stakeholders and designing regulation in a way that facilitates the sustainable and resilient transport future in accordance with climate neutrality goals (Weitzel et al. 2022; Fragkos and Fragkiadakis 2022).

5.3 Case Study 3: Financing Green Transitions Through Carbon Pricing and Green Bonds

Making the green transition in transport is a big upfront financial investment as low-carbon infrastructure and uptake of technology is a capital-intensive process. Political feasibility risks and debt sustainability risks may be mitigated successfully with the use of carbon pricing tools together with financial instruments such as green bonds. Evidence modelling shows that a combined approach of carbon pricing and green bonds is more effective than the application of each one of them separately by allowing prompt climate investments by issuing bonds on the basis of the immediate allocation of funds and using revenue on carbon pricing to maintain fiscal balance and ease the political load on pricing instruments (Heine et al. 2019). Green bonds offer long-term financing as well, which intergenerationally distributes the cost of the investment with the temporal benefits of the investments in emissions reduction and climate resilience enhancements (Sartzetakis 2020).

By its nature, carbon pricing mechanisms create income sources that can be recycled to help sectors impacted, invest in clean technologies and alleviate socioeconomic imbalance due to increased energy prices (Haites et al. 2023). Recycling revenues in the form of subsidies, tax rebates, or other forms of social programs, not only increases equity but also promotes the political acceptability of green fiscal policies, which is essential in the light of the opposition usually faced when it comes to direct carbon taxes or market-based solutions (Dugstad et al. 2024). Complementary to this are financial instruments like green bonds that make it possible to attract huge and stable streams of capital into sustainable transport infrastructure by governments and other actors in the private sector. More investors are also encouraged to invest their money in the green bond market because the market is less volatile, with attractive risk-return profiles that facilitate the scale-up of sustainable projects (Lichtenberger et al. 2022; Eyo-Udo et al. 2024).

Theoretically, the conceptual framework of sustainable transition governance is improved by combining the idea of Pigouvian carbon pricing and intertemporal financing models. Green bonds allow the current generations to share the burden of mitigation and enable future beneficiaries to pay bondholders small amounts over time, through the mechanisms of equity and Pareto improvement in welfare economics (Orlov et al. 2017). Nonetheless, modelling indicates that green bonds will not allow to eradicate intergenerational inequities fully and must be complemented with compensation and low interest rates to maximize the fairness and political plausibility (Heine et al. 2019). Thus, the recycling of carbon pricing revenue and innovative financial instruments such as green bonds become the key to long-term economic sustainability, political and social acceptance of green transportation changes (Fatica et al. 2021). The lesson that this case teaches is that policy integration between environmental economics and financial theory is necessary in order to ensure sustainable development goals are met.

6 Discussion

It is pointed out in the discussion that the use of economic tools to achieve green transport confirms arguments in environmental economics that interiorisation of external costs is the head, and not the end of successful decarbonisation policy. The fact that carbon taxes and ETS has the potential to lower emissions with minimal macroeconomic imbalance is related to Czermański et al. (2020), Peñasco et al. (2021), and Putri et al. (2024) who find that pricing can align the incentives of the private with the social cost of carbon and maintain efficiency. Simultaneously, the acknowledgement of competitiveness pressures, carbon leakage, and distributional tensions are reminiscent of Landis et al. (2017) and Peñasco et al. (2021), who prove that incompetently developed pricing can be retrogressive and sector-draining. In line with Meckling et al. (2017), the results thus confirm the idea of carbon pricing in transport as a larger political-economic strategy not only controlling lock-ins but also equity and industrial adaptation in the short-and long-term.

The discussion of carbon pricing and market-based instruments is aligned with Zhang et al. (2020), Millard-Ball (2009), and Digitemie and Ekemezie (2024) in emphasizing the effectiveness of carbon pricing and sensitivity of the design. Findings resonate with EU ETS and Chinese pilot data that optimally large caps and tax rates can provide large reductions in emissions with no output cost, whereas weak prices cannot provide transformative change as noted by Lilliestam et al. (2020) and Qu and Sun (2022). The small input response to pricing per se, and the specified requirement of innovation subsidies and regulation assurance, is consistent with Kempa and Moslener (2017) and Gillingham and Sweeney (2012), who also identify capital market flaws and technology risk as some other impediments. By doing so, the chapter supports the change in perception of carbon pricing as a silver bullet to a main but partial concept in a broader transport decarbonisation regime.

The focus on the complementary policies and policy-mix design is clearly similar to Peñasco et al. (2021), Meckling et al. (2017), Nykamp (2020), and Rogge and Reichardt (2016). Empirical evidence of the functions of subsidies, standards, public procurement, and sequencing can be compared to that of reviews of various instruments by Peñasco et al. (2021) that reveal that combinations of instruments are required to capitalize on efficiency, equity, and innovation synergies and to alleviate trade-offs. The fear that part of the transport subsidies is retrogressive or distributionally neutral resonates with Serebrisky et al. (2009) who insist on designing subsidies in a demand-sensitive and socially oriented way in the developing environment. Similarly, the significance of standards and green public procurement to rapid technology diffusion is comparable to results by Vluggen et al. (2019) and Yu et al. (2020), whereas the emphasis on sequencing early innovation and infrastructure support can be directly linked to the argument by Meckling et al. (2017) about the virtue of dynamic and path-shaping policy packages reducing opposition and lock-ins.

The chapter delves deeply into the literature of green-finance that is presented in the review because the fields of investment, financing and market design are the focus. The importance of carbon pricing in directing the flow of the privately-owned capital to the low-carbon transport is consistent with the existing studies by Okeke et al. (2024) and Lwesya (2025) that emphasize the role of policies aimed at risk reduction in the emerging markets. The coverage of green bonds, PPPs and hybrid financial markets is consistent with Alonso-Conde and Rojo-Suarez (2020), Cheong and Choi (2020), Lian et al. (2023) and Mukherjee et al. (2024) who demonstrate that these instruments have the potential to soften the financing constraints, enhance credit quality and raise long-term capital. The chapter mirrors the work of Heine et al. (2019), Orlov et al. (2017) and Sartzetakis (2020) in discussing carbon revenues and green bonds as complementary instruments of a fiscally viable and intergenerational just transport transition. As a whole, the discussion summarizes the key conclusion of the literature review that the effective policy of green transport must be based on a combination of economic tools, policy mixes, and financial architecture, which should be designed in accordance with the institutional context.

7 Conclusion and Recommendations

This chapter confirms that economic tools and market mechanisms offer an effective platform on which sustainable green transport systems could be developed by applying specific interventions on the market failure. The discussion can identify the potential of carbon pricing to substantially lower the level of emissions as witnessed in regional pilots and well-established trading schemes but also confirm its limitations in driving radical innovation unless there are supportive policies. Complimentary policies like subsidies, regulatory standards and public procurement generate synergies that expedite the adoption of technology and equity issues, making transitions yet to be viable politically. Green financing solutions such as bonds and partnerships open the doors to private capital infrastructural funding, which helps in alleviating risks of capital-intensive change. The overall success depends on institutional aspects such as the governance coordination and revenue management which ensure leakage is avoided and stakeholder buy-in is built. These lessons emphasize the necessity of the active, combined approaches that would compromise between environment benefit and financial security and social justice. Finally, the chapter also places the market-based strategies as the key contributors to decarbonisation, which are able to transform transport sectors into engines of sustainable development when integrated within the adaptive policy frameworks. Policymakers have an excellent road guide to utilize economic incentives to manoeuvre lock-ins, innovate, and to provide resilient results in a variety of settings.

The findings of the chapter are implemented by policymakers in six practical suggestions to develop green transport. The carbon pricing mechanisms are initially initiated at government level where the revenue is used to subsidize clean fuel and cars to internalize the pollution charges but relieve the burden on the low income earners. Authorities second sequence Policies are coordinated to be developed strategically, starting with the research grants and pilot infrastructure to establish a supportive environment, before bringing in stricter taxes and caps, therefore reducing the opposition and entrenchment of technology. Regulators third-party pricing imposing compulsory efficiency and low-emission area to implement performance floors and accelerate market transition to zero-carbon alternatives. Fourth, financial regulators encourage green bonds and public–private cooperation on the basis of guarantees and direct investments to electrified public transportation, charging systems to surpass capital barriers. Coordinators fifth establish cross-jurisdiction monitoring systems and data-sharing platforms to track and combat emission leakage to achieve equal environmental benefits. Leaders sixth incorporate equity protection, including tiered prices and job retraining, to safeguard vulnerable populations and achieve co-benefits such as cleaner air in cities and green jobs. The steps provide integrated action plans that drive change in the systems.

References

Alonso-Conde A-B, Rojo-Suárez J (2020) On the effect of green bonds on the profitability and credit quality of project financing. Sustainability 12(16):6695. https://doi.org/10.3390/su12166695

Barton DN, Benavides K, Chacon-Cascante A, Le Coq J, Quiros MM, Porras I, Primmer E, Ring I (2017) Payments for ecosystem services as a policy mix: demonstrating the institutional analysis and development framework on conservation policy instruments. Environ Policy Gov 27(5):404–421. https://doi.org/10.1002/eet.1769

Braungardt S, Bürger V, Köhler B (2021) Carbon pricing and complementary policies—consistency of the policy mix for decarbonizing buildings in Germany. Energies 14(21):7143. https://doi.org/10.3390/en14217143

Bromley DW (1978) Property rules, liability rules, and environmental economics. J Econ Issues 12(1):43–60. https://doi.org/10.1080/00213624.1978.11503504

Chai S, Sun R, Zhang K, Ding Y, Wei W (2022) Is emissions trading scheme (ETS) an effective market-incentivized environmental regulation policy? Evidence from China's eight ETS pilots. Int J Environ Res Public Health 19(6):3177. https://doi.org/10.3390/ijerph19063177

Cheong C, Choi J (2020) Green bonds: a survey. J Derivative Quant Stud: 선물연구 28(4):175–189. https://doi.org/10.1108/jdqs-09-2020-0024

Colmer J, Martin R, Muûls M, Wagner UJ (2024) Does pricing carbon mitigate climate change? Firm-level evidence from the european union emissions trading system. Rev Econ Stud 92(3):1625–1660. https://doi.org/10.1093/restud/rdae055

Cui J, Wang C, Zhang J, Zheng Y (2021) The effectiveness of China's regional carbon market pilots in reducing firm emissions. In: Proceedings of the National Academy of Sciences of the United States of America, 118(52). https://doi.org/10.1073/pnas.2109912118

Czermański E, Pawłowska B, Oniszczuk-Jastrząbek A, Cirella GT (2020) Decarbonization of maritime transport: analysis of external costs. Front Energy Res 8. https://doi.org/10.3389/fenrg.2020.00028

Digitemie W, Ekemezie I (2024) Assessing the role of carbon pricing in global climate change mitigation strategies. Mag Sci Adv Res Rev 10(2):022–031. https://doi.org/10.30574/msarr.2024.10.2.0040

Dolge K, Barisa A, Kirsanovs V, Blumberga D (2023) The status quo of the EU transport sector: Cross-country indicator-based comparison and policy evaluation. Appl Energy 334:120700. https://doi.org/10.1016/j.apenergy.2023.120700

Duan M, Zhou L (2017) Key issues in designing China's national carbon emissions trading system. Econom Energy Environ Policy 6(2). https://doi.org/10.5547/2160-5890.6.2.mdua

Dugstad A, Grimsrud KM, Lindhjem H (2024) Carbon pricing acceptance—the role of revenue recycling among households and companies in Norway. Climate Policy 24(10):1365–1380. https://doi.org/10.1080/14693062.2024.2401853

Ekdahl M, Milios L, Dalhammar C (2024) Industrial policy for a circular industrial transition in Sweden: An exploratory analysis. Sustain Prod Consump 47:190–207. https://doi.org/10.1016/j.spc.2024.03.031

Eyo-Udo N, Agho M, Onukwulu E, Sule A, Azubuike C (2024) Advances in green finance solutions for combating climate change and ensuring sustainability. Gulf J Adv Busin Res 2(6):338–375. https://doi.org/10.51594/gjabr.v2i6.53

Fatica S, Panzica R, Rancan M (2021) The pricing of green bonds: Are financial institutions special? J Financ Stab 54:100873. https://doi.org/10.1016/j.jfs.2021.100873

Fragkos P, Fragkiadakis K (2022) Analyzing the macro-economic and employment implications of ambitious mitigation pathways and carbon pricing. Front Climate 4. https://doi.org/10.3389/fclim.2022.785136

Gillingham K, Sweeney J (2012) Barriers to implementing low-carbon technologies. Climate Change Econom 03(04):1250019. https://doi.org/10.1142/s2010007812500194

Heine D, Semmler W, Mazzucato M, Braga JP, Flaherty M, Gevorkyan A, Hayde E, Radpour S (2019) Financing low-carbon transitions through carbon pricing and green bonds. Vierteljahrshefte Zur Wirtschaftsforschung 88(2):29–49. https://doi.org/10.3790/vjh.88.2.29

Jiang J, Ye B, Zeng Z, Yang X, Sun Z, Shao S, Feng K, Tan X (2024) Carbon abatement and leakage in china's regional carbon emission trading. Environ Sci Technol 58(40):17661–17673. https://doi.org/10.1021/acs.est.4c04738

Kempa K, Moslener U (2017) Climate policy with the chequebook: an economic analysis of climate investment support. Econom Energy Environ Policy, 6(1). https://doi.org/10.5547/2160-5890.6.1.kkem

Lagouvardou S, Lagemann B, Psaraftis HN, Lindstad E, Erikstad SO (2023) Marginal abatement cost of alternative marine fuels and the role of market-based measures. Nat Energy 8(11):1209–1220. https://doi.org/10.1038/s41560-023-01334-4

Laing T, Sato M, Grubb M, Comberti C (2014) The effects and side-effects of the EU emissions trading scheme. Wires Clim Change 5(4):509–519. https://doi.org/10.1002/wcc.283

Landis F, Rausch S, Kosch M (2017) Differentiated carbon prices and the economic cost of decarbonization. Environ Resource Econ 70(2):483–516. https://doi.org/10.1007/s10640-017-0130-y

Li C, Lei T, Wang L (2024) Examining the emission reduction effect of carbon emission trading and carbon tax synergism and their impact mechanisms to reduce the carbon emission of company: based on 247 listed companies of China. Technol Anal Strat Manage 37(11):2205–2220. https://doi.org/10.1080/09537325.2024.2346807

Lian J, Huang X, Wu X (2023) How green bonds promote firms' green collaborative innovation? Corp Soc Responsib Environ Manag 31(3):2109–2126. https://doi.org/10.1002/csr.2688

Lichtenberger A, Braga JP, Semmler W (2022) Green bonds for the transition to a low-carbon economy. Econometrics 10(1):11. https://doi.org/10.3390/econometrics10010011

Lilliestam J, Patt A, Bersalli G (2020) The effect of carbon pricing on technological change for full energy decarbonization: a review of empirical ex-post evidence. WIREs Climate Change 12(1). https://doi.org/10.1002/wcc.681

Lwesya F (2025) Green finance in Africa: mapping progress, challenges and prospects. Future Business J 11(1). https://doi.org/10.1186/s43093-025-00596-6

Mamash A, Iyiola K, Aljuhmani HY (2025) The role of circular economy entrepreneurship, cleaner production, and green government subsidy for achieving sustainability goals in business performance. Sustainability 17(9):3990. https://doi.org/10.3390/su17093990

Meckling J, Sterner T, Wagner G (2017) Policy sequencing toward decarbonization. Nat Energy 2(12):918–922. https://doi.org/10.1038/s41560-017-0025-8

Mikhno I, Koval V, Shvets G, Garmatiuk O, Tamošiūnienė R (2020) Green economy in sustainable development and improvement of resource efficiency. Central Eur Business Rev 10(1):99–113. https://doi.org/10.18267/j.cebr.252

Millard-Ball A (2009) Cap-and-Trade. Transportation research record: journal of the transportation research board 2119(1):20–26. https://doi.org/10.3141/2119-03

Mukherjee A, Owen R, Scott JM, Lyon F (2024) Financing green innovation startups: a systematic literature review on early-stage SME funding. Venture Capital, ahead-of-print(ahead-of-print), 1–27. https://doi.org/10.1080/13691066.2024.2410730

Nykamp H (2020) Policy mix for a transition to sustainability: green buildings in Norway. Sustainability 12(2):446. https://doi.org/10.3390/su12020446

Okeke N, Bakare O, Achumie G (2024) Integrating policy incentives and risk management for effective green finance in emerging markets. Int J Front Sci Technol Res 7(1):076–088. https://doi.org/10.53294/ijfstr.2024.7.1.0050

Orlov S, Rovenskaya E, Puaschunder JM, Semmler W (2017) Green bonds, transition to a low-carbon economy, and intergenerational fairness: evidence from an extended DICE model. SSRN Electron J. https://doi.org/10.2139/ssrn.3086483

Peñasco C, Anadón LD, Verdolini E (2021) Systematic review of the outcomes and trade-offs of ten types of decarbonization policy instruments. Nat Clim Chang 11(3):257–265. https://doi.org/10.1038/s41558-020-00971-x

Putri RL, Werastuti DNS, Rustiarini NW, Sutoto A, Wahyono E, Budi Wardono B, Zulham A, Lukas A, Goenawan RD, Anita S, Ardana IK, Normal IN, Daniel M, Yuniarti L, Supriadi K, Vayed DA (2024) Assessing the role of carbon taxes in driving low-carbon transformations: a comparative analysis of implementation policies. Int J Sustain Develop Plan 19(3):1171–1180. https://doi.org/10.18280/ijsdp.190335

Qu X, Sun X (2022) How to improve the function of government carbon tax in promoting enterprise carbon emission reduction: from the perspective of three-stage dynamic game. Environ Sci Pollut Res 29(21):31348–31362. https://doi.org/10.1007/s11356-021-18236-1

Raman R, Ray S, Das D, Nedungadi P (2025). Innovations and barriers in sustainable and green finance for advancing sustainable development goals. Front Environ Sci 12. https://doi.org/10.3389/fenvs.2024.1513204

Rogge KS, Reichardt K (2016) Policy mixes for sustainability transitions: an extended concept and framework for analysis. Res Policy 45(8):1620–1635. https://doi.org/10.1016/j.respol.2016.04.004

Samuel S (2025) Carbon pricing mechanisms for reducing greenhouse gas emissions and encouraging sustainable industrial practices. World J Adv Res Rev 25(2):001–024. https://doi.org/10.30574/wjarr.2025.25.2.0350

Sartzetakis ES (2020) Green bonds as an instrument to finance low carbon transition. Econ Chang Restruct 54(3):755–779. https://doi.org/10.1007/s10644-020-09266-9

Scotti F, Flori A, Crescenzi R, Pammolli F (2024) Demand-pull and technology-push environmental innovation: a policy mix analysis on EU ETS and EU cohesion policy. Climate Policy 25(2):153–170. https://doi.org/10.1080/14693062.2024.2366894

Serebrisky T, Gómez-Lobo A, Estupiñán N, Muñoz-Raskin R (2009) Affordability and subsidies in public urban transport: what do we mean, what can be done? Transp Rev 29(6):715–739. https://doi.org/10.1080/01441640902786415
Shao J, Huang P (2023) The policy mix of green finance in China: an evolutionary and multilevel perspective. Climate Policy 23(6):689–703. https://doi.org/10.1080/14693062.2023.2202181
Sreenu N (2024) Examining the dynamics of risk associated with green investment in India: a study on fintech and green bonds for clean energy production. Journal of Economic Studies 52(3):588–604. https://doi.org/10.1108/jes-12-2023-0680
Vandyck T, Della Valle N, Temursho U, Weitzel M (2023) EU climate action through an energy poverty lens. Scientif Report 13(1). https://doi.org/10.1038/s41598-023-32705-2
Vluggen R, Gelderman CJ, Semeijn J, Van Pelt M (2019) Sustainable public procurement—external forces and accountability. Sustainability 11(20):5696. https://doi.org/10.3390/su11205696
Weis L, Nikolić G (2024) Discussing the role of innovation in green entrepreneurship and development (pp. 1–21). Emerald. https://doi.org/10.1108/978-1-83797-088-920241001
Weitzel M, Vandyck T, Rey Los Santos L, Tamba M, Temursho U, Wojtowicz K (2022) A comprehensive socio-economic assessment of EU climate policy pathways. Ecol Econom 204:107660. https://doi.org/10.1016/j.ecolecon.2022.107660
Yu C, Morotomi T, Yu H (2020) What influences adoption of green award criteria in a public contract? An empirical analysis of 2018 European public procurement contract award notices. Sustainability 12(3):1261. https://doi.org/10.3390/su12031261
Zhang S, Yu Y, Zhu Q, Qiu CM, Tian A (2020) Green innovation mode under carbon tax and innovation subsidy: an evolutionary game analysis for portfolio policies. Sustainability 12(4):1385. https://doi.org/10.3390/su12041385
Zhang R, Hanaoka T, Liu J, Li Z, Sun L (2024) Air pollution reduction co-benefits associated with low-carbon transport initiatives for carbon neutrality in China by 2060. Energy 313:134090. https://doi.org/10.1016/j.energy.2024.134090
Zhuo C, Liu Y, Dai L, Deng Y (2024) Cultural, economic, or transport link: does carbon emissions trading promote "good neighbor" carbon emission reduction? Land 13(11):1762. https://doi.org/10.3390/land13111762

Private Sector Leadership and Corporate Climate Commitments in Logistics

David Chikwere, Justine Mbudaya, Munyaradzi Chibaro, and Shakerod Munuhwa

Abstract The chapter highlights the increasing role of private sector leadership in addressing climate action within the logistics space, including corporate climate commitments, environmental, social and governance (ESG) disclosure frameworks and supply chain decarbonisation approaches. Given that logistics activities represent a large and growing portion of global greenhouse gas emissions, the involvement of private companies, notably multinational logistics providers, has become critical in efforts to achieve global climate goals. The chapter contextualizes these developments within a growing array of hybrid climate governance, in which voluntary corporate actions are increasingly supplementing, and in some cases replacing, traditional regulatory measures. The analysis explores the evolution and scope of corporate climate commitments, highlighting the shift from corporate social responsibility to science-based and net-zero targets. It critiques the difficulties in tackling Scope 3 emissions, which constitute a large majority of the logistics sector's carbon footprint and questions the credibility and accountability of corporate commitments with increasing concerns over greenwash and non-standardized reporting practices. The chapter covers how ESG frameworks help increase transparency and guide corporate behaviour, as well as the challenges of ESG frameworks in examining concrete business performance outcomes. It also explores other operational strategies like electrification, alternative fuels and digital optimisation in supply chain decarbonisation. The chapter assesses the success of corporate voluntary action, contending that though it raises awareness and leads to gradual progress, its overall effectiveness is frustrated

D. Chikwere (✉)
Leeds Business School, Leeds Beckett University, Leeds, UK
e-mail: D.chikwere@leedsbeckett.ac.uk

J. Mbudaya
BA ISAGO University, Gaborone, Botswana

M. Chibaro
University of Botswana, Gaborone, Botswana

S. Munuhwa
Glasgow School for Business and Society, Glasgow Caledonian University, Glasgow, UK
e-mail: shakerod.munuhwa@gcu.ac.uk

J. Kanyepe and I. Chirisa (eds.), *Advancing the Green Agenda in the Transport and Logistics Sector in Global South Countries*, World Sustainability Series,
https://doi.org/10.1007/978-3-032-31374-4_23

by weak enforcement and irregular adoption. Developing economies were also a key focus, with institutional and structural challenges that slow implementation, coupled with opportunities to advance sustainable solutions. The chapter concludes that, in contrast to a piecemeal and corporate-centred approach of the private sector innovation role in climate action, more integrated approaches, coupled with effective frameworks for public policies and accountability systems, are essential if substantial emissions cuts in logistics are to be achieved.

Keywords Sustainable logistics · Corporate climate commitments · Supply chain decarbonisation · ESG disclosure · Private sector climate governance

1 Introduction

The logistics industry is a backbone of the global economy, facilitating the movement of goods across complex and geopolitically dispersed supply chains (Rehana and Qureshi 2024). The growth of global trade, e-commerce and just-in-time production systems has inevitably scaled the activities of logistics like freight transport, warehousing and distribution in both scale and complexity. But this progress has been at a high environmental cost. Freight transport accounts for almost 8–10% of global carbon dioxide emissions (McKinnon 2018; Marzouk 2025), and the logistics sector is responsible for a significant share of global greenhouse gas (GHG) emissions. Without decisive measures, emissions from logistics shall grow even more as demand for goods movement rises, particularly in emerging economies.

Climate governance has traditionally relied upon state-based regulatory regimes, for example, emissions standards and fuel taxes, as well as multilateral treaties such as the Paris Agreement (Marquardt et al. 2022). However, the limitations of public governance, most significantly around enforcement, jurisdictional fragmentation and political constraints, have opened room for private sector actors to play an increasingly key role in climate action. In response to various regulatory pressures, investor expectations and reputational concerns, corporations, especially those in the multinational logistics sector, have begun establishing climate commitments, sustainability strategies, and environmental, social and governance (ESG) disclosure practices (Gündoğdu et al. 2023).

A growing involvement in climate governance by private sector actors is part of wider trends towards hybrid forms of governance where public and private actors work together to design the processes through which policies are implemented and innovated (Boasson et al. 2023). In the logistics industry, this has taken the form of numerous voluntary initiatives such as industrial coalitions or corporate climate commitments to cut emissions and improve sustainability. Despite these efforts, the key questions that remain are whether governments in those countries can sufficiently leverage and allow for participation of all, including private sector actors. Although these corporate commitments are intended to encourage and guide incremental actions, their track record in delivering measurable decarbonisation outcomes in terms of both scale and verifiable impact remains questionable. This is also true in developing economies, where structural constraints (such as poor infrastructure, lack of financial access and weak regulatory institutions) pose challenges for implementation (Chikwere et al. 2025).

Therefore, the purpose of this chapter is to review and discuss critically the contribution workplace leaders play in sustainable logistics systems. These three areas are: (1) Corporate climate commitments; (2) Environmental and sustainability-related reporting frameworks; (3) Decarbonisation of supply chains. It also examines the effectiveness of voluntary initiatives and how corporate actions align with regulatory approaches. The core argument is that some private sector initiatives have increased awareness, encouraged innovation and produced some emissions reduction, but their overall effectiveness is limited by accountability problems, variation in practice and systemic limits. To achieve real decarbonisation in logistics, there is a need for an approach that integrates both voluntary corporate social responsibility, matched with strong and predictable public policy interventions.

2 Corporate Climate Commitments in the Logistics Sector

2.1 The History of Corporate Climate Pledges

Over the decades, corporate engagement with environmental issues has evolved. On the other hand, corporate environmental responsibility is formed based on its broad paradigms of prevailing standards that dictate corporate social responsibility (CSR)'s voluntary practices to create constructive transformations in society and environment (Le et al. 2024). The CSRs were often sidelined in the business operation without much of a detailed focus or a mechanism to hold anyone accountable.

In contrast, contemporary business climate commitments are increasingly integrated into their business models and strategic plans. The emergence of net-zero targets and science-based targets (SBTs) reflects an increasing awareness to customise enterprise action in line with global climate goals identified in the Paris Agreement (Marquardt et al., 2022). The SBTi has been the key driver for this transition through its approach to set emissions reduction targets consistent with holding global warming at 1.5 °C (SBTi, 2022). These include carbon neutrality, net-zero emissions, and reductions in emissions intensity targets across the logistics sector by major companies. These commitments are commonly multi-faceted, from fleet electrification, energy efficiency upgrades, and supply chain engagement.

However, the increase in corporate pledges has also resulted in wide variations in ambition and scope. Other companies set targets that include all scopes, while others still focus on operational emissions (Scope 1 and 2) (Hettler & Graf-Vlachy 2024). This variance makes it difficult to assess the overall contribution of corporate commitments towards global decarbonisation.

2.2 Emission Scope and Coverage

A key feature of the logistics sector is the dominance of indirect (Scope 3) emissions. These emissions originate from the activities along the value chain, such as subcontracted transport, upstream production processes, and downstream distribution activities (Hettler & Graf-Vlachy 2024). For most logistics' companies, Scope 3 emissions constitute the bulk of their carbon footprint. However, the scope 3 emissions are also difficult to manage. Whereas direct emissions can be controlled with operational changes, indirect emissions are dependent on the actions of many actors in the supply chain (Fatorachian 2025). As a result, there are complications around coordination, data availability and accountability.

Moreover, logistics networks are highly fragmented, which poses a challenge. The sector is dominated by small and medium-sized enterprises (SMEs) in developing economies where informal, unregulated transport operators are widespread (Zunarelli 2023). These SME organisations usually have limited resources and technical capability to measure and disclose emissions, complicating the path for larger companies to deliver comprehensive reductions in scope 3 emissions. Consequently, many corporate climate pledges have been restricted in scope, concentrating on aspects where companies have direct influence. It should be highlighted that although these climate commitments may result in noticeable reductions, they fall far short of tackling all logistics-related activities' emissions.

2.3 Credibility and Accountability

Whether corporate promises on climate are credible or not remains a central issue in voluntary climate governance. Although ambitious targets may show leadership and commitment, the lack of robust accountability calls into question their usefulness. In the most extreme cases, it may be categorised as greenwashing, a considerable risk that arises when environmental performance is exaggerated or misrepresented (Gallas et al. 2025). Since they are simply popularising their brand, companies can score all kinds of eye-catching climate pledges without anything materially changing, if reductions are blended across emissions. To address these challenges, several mechanisms have been developed to support and enable accountability, such as third-party verification, standardised reporting frameworks, and independent assessments. However, not all organisations have implemented such systems, and many remain reliant on self-reported data.

The other issue with corporate commitments is their temporality. Goals such as net zero emissions by 2050 are generally ambitious, often lacking measurable interim targets and a pragmatic route to achieve them. This can lead to increasing constraints all the way until the last minute as corporations invest excessively, clinging to the notion that future technological advancements would enable them to reach their targets. Therefore, if corporate climate commitments are an important element going forward in a transition to sustainability, their legitimacy and effectiveness will increasingly depend on a process that would allow for understanding whether those commitments were transparent, verified or enforceable.

3 ESG Frameworks and Disclosure Practices

3.1 Overview of the Key ESG Frameworks

The need for increased transparency and accountability around corporate sustainability regimes has resulted in an explosion of ESG frameworks (Kotsantonis and Serafeim 2019). Until now, frameworks like the Global Reporting Initiative (GRI), the Task Force on Climate-related Financial Disclosures (TCFD), and the International Sustainability Standards Board (ISSB) have provided a global minimum standard for sustainability reporting.

The focus and the scope of these frameworks vary. GRI is more holistic as it emphasizes the broader sustainability implications, covering both environmental and social dimensions, while TCFD looks to the financial impacts of climate-related risks and opportunities (Yusof and Widyasamratri 2025). ISSB seeks to align existing standards with complementary and useful guidelines for sustainability disclosure globally. ESG reporting in logistics usually covers data such as fuel consumption, emissions intensity against travel length, energy used and operational efficiency (Gündoğdu et al. 2023). Such disclosures also provide stakeholders with information needed to evaluate corporate performance and compare firms across an industry.

3.2 Disclosure as a Catalyst for Change

ESG disclosure has become a major driver of corporate conduct by promoting transparency and facilitating stakeholder scrutiny (Yusof and Widyasamratri 2025). What has really changed are investors who have become increasingly powerful drivers of corporate climate action. Institutional investors apply ESG criteria to their investment decisions, which gives companies an incentive to make improvements in environmental performance (Lopez-de-Silanes et al. 2024).

Disclosure also allows for benchmarking and competition between firms (Mba 2024). When companies compare performance indicators against those of other entities, meaningful best practices emerge, and more effective strategies can be adopted. Thus, benchmarking initiatives help in spreading the use of innovations like fuel-efficient technologies and digital optimization tools in the logistics sector (Shee Weng 2025). Furthermore, implementing ESG reporting can help improve corporate reputation and foster trust with stakeholders such as consumers, regulators and the public. This aspect of reputation adds an extra motivation for businesses to implement sustainability measures.

3.3 Limitations of ESG Reporting

Although it is favourable for all parties, ESG disclosure has some drawbacks. The generation of non-comparable and inconsistent data is one of the main challenges regarding a lack of standardization across frameworks (Kotsantonis and Serafeim 2019). Thus, companies often pick the methodology that makes them look best when it comes to how much they pollute.

Another limitation is a focus on disclosure over performance. Companies might therefore choose to disclose activities that are easier for them to measure, or where the disclosure looks positive, but ignore much higher complexity emissions sources (Gallas et al. 2025). This can lead to a disconnect between the abstract metrics and how well the actual environmental performance. This adds to the challenge in developing economies where there are less data availability and a weaker regulatory environment with lower levels of institutional capacity. Consequently, the environmental impact of logistics operations may not be entirely reflected in ESG reporting due to this contextual gap.

4 Supply Chain Decarbonisation Strategies

4.1 Technological Innovations

The logistics sector will be largely dependent on technological innovation for its decarbonisation efforts. Electrification of vehicle fleets is one of the most economically viable routes to cut emissions in urban environments, especially regarding the last mile. Battery and charging technologies improvements and the market preparedness have been fostered during the last decades, thereby establishing EVs as a more feasible solution. However, these days, the EVs' costs, availability and infrastructure remain the key barriers for the technology to be a feasible substitute (Marzouk 2025).

Emerging fuels (e.g. hydrogen, biofuels) are also very promising solutions for addressing those sectors where electrification is less viable, like long-distance trucking, aviation and maritime transport (Fatorachian 2025). However, the visible generation and employment of these technologies demands an infrastructure basis as well as intensive policy guidance. These alternatives are essential in developing countries where electricity has not been fully established. Logistics can function more efficiently owing to digital technologies like artificial intelligence (AI), big data analytics, and the Internet of Things (IoT). These technologies not only help reduce costs through optimized routing but also lower emissions by minimizing empty runs and improving load factors.

4.2 Operational Strategies

Operational strategies constitute a key and often overlooked realm of logistics decarbonisation through supplementing technological solutions by increasing the efficiency and utilisation of current systems. While technological solutions like electrification or alternative fuels require considerable capital outlay and the evolution of long-maturity infrastructures, operational measures can often be adopted more quickly as well as less costly (in both developed and developing economy contexts) (McKinnon 2018).

One of the fundamental approaches is modal shift, where some freight movement (in this case, from high-emission modes such as road freight) can be moved onto more energy-efficient transport modes like sea and rail (Munuhwa 2024). For instance, rail freight may emit as much as 75% less CO_2 per tonne-kilometre than road transport, depending on the energy mix (Marzouk 2025). Similarly, while maritime shipping has its own environmental issues, it remains one of the most carbon-efficient modes for long-distance bulk transport. The potential for modal shift is dependent on the availability and quality of infrastructure at each stage, including inter-modal connectivity as well as appropriate regulatory support. Many of the developing economies do not have adequate rail infrastructure and port facilities, thus hindering large-scale modal shifts (Abu-Aisha et al. 2024).

Route optimization is another critical approach as it uses advanced digital technologies and data analytics to determine the most effective delivery routes. Advanced logistics management systems leverage real-time data on traffic conditions, weather patterns and delivery constraints to help minimize travel distance, reduce fuel consumption and enhance delivery times (Fatorachian 2025). Studies show that routing optimization can reduce fuel consumption and emissions by 10–30% in urban logistics operations (Zunarelli 2023). These efficiencies are more pronounced in last-mile delivery, where inefficiencies tend to be most acute due to fragmented delivery patterns and congestion.

The use of vehicles also includes maximizing utilization in terms of load factor and minimizing empty or near-empty trips. This is due to inefficiency in freight transport associated with empty backhauls, with an estimated 20–30% of truck journeys made without any cargo in some regions (McKinnon 2018). Therefore, optimizing load consolidation by using more effective planning and coordination, sharing logistics platforms, and implementing digital freight matching systems can enhance efficiency as well as reduce emissions.

Operational strategies also involve network optimization, firms redesigning distribution networks to minimize total transport distance and improve supply chain efficiency (Munuhwa 2024). This can mean moving warehouses, merging distribution centres, or implementing hub-and-spoke models. Such changes can achieve significant emissions reductions but often involve detailed trade-offs between cost, service levels and environmental outcomes. In resource-constrained environments like developing economies, where access to capital for technological upgrades is limited, operational strategies become relevant (Chikwere et al. 2025). In such cases, incremental efficiency improvements can achieve significant reductions in emissions at no great financial cost. However, the success of these approaches relies on organizational capacity, data availability, and collaboration between supply chain actors.

However, despite their potential operational strategies have their limitations. Efficiency gains may be offset by the rebound effect, which occurs when cost savings result in increased demand for logistics services that partly negate emissions reduction (Jaehn and Meissner 2022). Moreover, without structural change and technological transition, operational improvements on their own will not be enough to deliver deep decarbonisation. Consequently, they ought to be considered complementarily rather than substitutive measures in wider decarbonisation targets.

4.3 Supplier and Stakeholder Engagement

As the majority of emissions from the logistics sector are Scope 3, it is essential for decarbonisation to actively involve suppliers, contractors and other stakeholders across the value chain. Whereas Scope 1 and Scope 2 emissions are (to a large extent) under direct firm control, Scope 3 emissions result from diverse upstream and downstream processes and therefore present significantly more complicated management (Hettler and Graf-Vlachy 2024). High-volume logistics companies, especially multinationals, drive supply chain conduct with their purchasing power and contractual relationships. By embedding environmental criteria in procurement processes, this would create incentives for suppliers to shift to low-carbon behaviours. For instance, companies might require suppliers to meet certain sustainability standards by reporting emissions figures or setting targets for third-party reductions.

Butt et al. (2025) posits that by creating a ripple effect through supplier networks, such green purchasing policies can result in the broader diffusion of sustainable practices. In addition to procurement-based strategies, companies can invest in capacity-building programmes to help suppliers decarbonise. They could include, but are not limited to, the provision of technical assistance, experience sharing and making financing for low- carbon dioxide technologies affordable. These initiatives are particularly important in developing economies in which suppliers frequently work under heavy limitations of both resource and understanding.

Collaborative initiatives and industry alliances are other significant mechanisms for engaging stakeholders. Efforts around logistical collaboration, carbon disclosure and sector-specific partnerships help firms share information, align methodologies and coordinate action on platforms (Jaehn and Meissner 2022). These projects address collective action problems by providing aligned incentives and reducing uncertainty. However, for collaborative approaches to be effective, stakeholders must be engaged and committed. There are well-known issues with voluntary initiatives, such as selective participation, whereby industry leaders are active while smaller or under-resourced actors get locked out. Consequently, there is unequal advancement and a limited effect of collective action (Walker 2026). Additionally, they face challenges to equitable and effective engagement due to power imbalances within supply chains. Although larger firms can influence suppliers, smaller ones may not always have the leverage needed to attain favourable conditions or implement sustainability efforts (Butt et al. 2025). The issue is more pronounced in developing economies, where informal and fragmented supply chains are common.

The second thing is data transparency and standardization. For organizations to engage effectively with their supply chains on emissions reduction, they need reliable and comparable data about emissions across suppliers, but many suppliers often lack the systems and capabilities needed to gather and disclose this information (Munuhwa 2024). The lack of standards also makes it difficult to combine and calculate value chain emissions without the process losing credibility. The process of stakeholder engagement is a crucial step towards meaningful decarbonisation, despite the odds. Interestingly, companies are also taking multi-stakeholder approaches that include engagement with governments as well as non-governmental organizations like financial institutions. The system approach needs to emphasize the need for multiple and cross-sectoral interventions for effectively and sustainably addressing complex sustainability challenges (Chikwere et al. 2024). In this respect, digital technologies are also gaining ground. Technology tools that share data, track emissions, and gain visibility of supply chains can support transparency and enable more constructive engagement. However, equitable access to technologies such as these remains a fundamental challenge.

4.4 Effectiveness of Voluntary Corporate Initiatives

The rise of voluntary corporate action has been identified as a key feature of modern climate governance, especially in sectors like logistics, where regulatory oversight is dispersed across multiple jurisdictions and supply chains span the globe (Boasson et al. 2023). These range from industry coalitions and emissions accounts platforms to sustainability certification programs and corporate initiatives. These initiatives are often framed as flexible, innovation-friendly ways to supplement traditional regulation when it comes to helping solve climate challenges as a way of tapping into the highly developed resources that private firms have at their disposal (Boasson et al. 2023; Walker 2026).

Rules and regulations may well perform some role in generating positive change, but the extent of their ability to deliver genuine reductions has been an issue of debate (Walker 2026). While such initiatives have undoubtedly been part of raising awareness, facilitating transparency and supporting new research, their effectiveness in driving system-wide decarbonisation is particularly contentious within hard-to-accommodate sectors like logistics. This section critically evaluates the voluntary corporate initiatives and examines their design, effectiveness and limitations.

4.5 Industry-Led Initiatives and Alliances

Over the past two decades, numerous industry-led initiatives and alliances have emerged in the logistics sector as increasing awareness about collective action to confront climate change. These initiatives encompass coalitions targeting specific sectors, carbon disclosure schemes like carbon disclosure projects (CDP), and certification programs designed to encourage environmentally friendly practices (Yusof and Widyasamratri 2025). From the governance perspective, these initiatives can be seen as a type of transnational private governance where non-state actors regulate corporate, social and environmental behaviour internationally through norms, rules and standards (Nedziwe and Tella 2023). In the absence of comprehensive global regulatory frameworks, these initiatives are important to help coordinate action across firms operating in different institutional contexts.

One of the strengths of industry-led initiatives is that they can provide an environment for knowledge and capacity-building. These efforts have been able to form platforms of collaboration, which allow firms to share best practices (as industry-specific case studies) and build a repository of common methodologies and reduce the costs of uncertainty surrounding the adoption of new technology and practices (Nedziwe and Tella 2023). It has helped in the logistics sector, for example, to spread innovations in fuel efficiency measures, emissions tracking tools and digital optimization systems.

Furthermore, voluntary initiatives offer a space for policy experimentation whereby firms test creative solutions that can subsequently inform formal regulation (Radosevic et al. 2025). For instance, voluntary programs can establish industry standards that articulate best practices and even inform government regulation or international agreements. While certainly these efforts can be noble, the leverage provided by voluntary measures has not completely paid off when it comes to achieving meaningful reductions in emissions. Participation is often self-selective, with selection bias toward more environmentally active firms (Gallas et al. 2025). Consequently, a given outcome gain may merely attest to pre-existing divergences instead of an effect generated by participation. These initiatives also do not involve binding commitments or compliance enforcement measures. Even in cases where a few programs do include monitoring and reporting requirements, the consequences for not complying with those are weak or non-existent. This raises further questions about the role and impact of voluntary initiatives in delivering change beyond incrementalism.

4.6 Measuring Impact

Operationally, it is hard to measure the effect of voluntarily corporate-driven initiatives. At the centre of this issue is a patchwork of sometimes inconsistent metrics and reporting frameworks that make it difficult to compare firms and their initiatives on a performance basis (Kotsantonis and Serafeim 2019). Although similar ideas might be proposed in reports, programs might differ in definitions, methodologies and indicators that they use to describe them, as these are not harmonized, which results in inconsistent data. Another challenge relates to causality. It is hard to make inferences directly about emissions reductions because companies use many interventions in a sustainability program that are often jointly determined by multiple context factors, such as regulation, market conditions or technological readiness. That complicates the effort to isolate any voluntary initiatives' specific impact.

The scientific case for such efforts is not clear. In other studies, self-compelled environmental programs induce better ecological performance manifested by lower emissions intensity and firms' energy intensity (Chikwere et al. 2024; Walker 2026). The most common explanations seem to be increased transparency, peer pressure and reputational incentives. But other research points to a much narrower and more uneven impact of voluntary initiatives, especially for absolute emissions reduction (Kotsantonis and Serafeim 2019; Lebdioui 2024). This can happen, for instance, when companies realize efficiency gains and expansion of the activities which lead to aggregate emissions growth. This paradigm bears great relevance to the logistics sector because growing demand for freight transport outstrips effective systems (McKinnon 2018).

For voluntary initiatives to be effective, they must induce emissions reductions that would not have occurred in the absence of participation. In practice, firms can adopt economically efficient measures such as enhanced fuel efficiency regardless of participation in voluntary programs (Lebdioui 2024). Developing economies face added challenges in measurement due to a lack of data availability, weak institutional capacity and informalisation of economic activity. Consequently, the monitoring and verification of emissions reduction is not possible in as much detail as mechanisms are there, thereby lowering the credibility of output reports.

4.7 Critiques of Voluntarism

Literature on environmental governance has provided a long-standing critique of the limits of voluntary corporate initiatives. A key issue with collective voluntary initiatives is that of free-riding, whereby firms benefit from the commercial goodwill associated with a given initiative's collective reputation, without either substantive engagement in emissions mitigation (Radosevic et al. 2025). Without enforcement mechanisms, there is not all that much to keep firms from just signing on in name only. Greenwashing is another closely related phenomenon that occurs when companies adopt voluntary initiatives as tools of image building instead of changing their behaviour (Gallas et al. 2025). A rapid pledge and an environmental initiative can create the illusion of a sustainable company, despite little to no change in practice.

An additional major limitation is uneven participation by firms and regions. Voluntary initiatives favour the interests of major multinationals with resources and capabilities which are not currently available to SMEs. Especially in economies at the subsidiary level, this is much more deeply rooted, especially for the logistics sector, where supply chain actors involve a considerable share of small and medium enterprises (Zunarelli 2023). However, SMEs contribute to a large amount of emissions that are not captured at the global and national levels. This dynamic presents a significant gap with SMEs, which can play an important role in crediting emissions reductions but leave a large proportion of carbon emissions unassessed. However, the participation costs, like data collection or reporting and compliance, could become potential torturers for smaller companies and deepen inequalities in supply chains.

In theory, such difficulties show the constraints of voluntarism as a form of governance. However, without some rules that bind such voluntary efforts and mechanisms for enforcing them, they are unlikely to operate at the scale needed (Boasson et al. 2023). This carries extra weight in highly competitive and price-sensitive domains such as logistics. The success of a voluntary action can depend on the broader institutional context, too. In such robustly regulatory settings with a vibrant civil society, voluntary programs can help to complement and reinforce better institutionalized standards and catalyse more incremental change. Alternatively, in contexts where the administrations that rule them are becoming increasingly fragile, they may embody another route toward regulation with little actual result.

The last category includes the new idea of hybrid governance models that combine autonomy with enforced boundaries and consequences, market mechanisms or penalties, as well as accountability to citizens. Such approaches could go a long way to harnessing the feedback of private sector innovation while placing ceilings on the role of voluntarism as self-regulatory mechanisms that are negotiated and enforced through standards and accountability (Boasson et al. 2023; Radosevic et al. 2025). Such efforts can help support public policy, but they cannot replace it. The governance of deep decarbonisation, however, needs to be more integrated and capable of bridging between voluntary corporate leadership and tough regulation policy-making, as well as broadbased stakeholders' participation in decision-making of common goods.

5 Private Sector–Public Policy Interactions

Understanding the dynamics of climate governance in the logistics sector requires an exploration of how private sector initiatives relate, if at all, to public policy frameworks. Corporate climate actions and government interventions do not operate in isolation but interact in complex and often interdependent ways. This interaction is most properly understood in terms of hybrid governance, where authority and responsibility are distributed among public- and private-sector actors (Boasson et al. 2023). Although innovative leadership can push forward momentum and implementation, in many ways, the success or failure of such efforts is deeply conditioned by the policy landscape in which firms operate.

5.1 *Complementarity in Public Policy and Corporate Action*

Complementary to public policy are initiatives in the private sector that can provide flexibility, nurture innovation and resources not available from only governments. In the logistics sector, companies can rely on a big data landscape where technological advancement and operational complexity loom large to experiment with novel solutions in-house and scale successful innovations. One of the main areas of complementarity is the role that public–private partnerships (PPPs) play in filling infrastructure and financing gaps. A poorly developed transport infrastructure, including rail networks, ports and urban logistics systems in developing economies, is a key factor hindering decarbonisation (UNCTAD 2023). PPP's combination of public support and private expertise/capital can unlock investment in low-carbon infrastructure. Partnerships to create areas (for electric vehicle charging or intermodal freight hubs) can decisively contribute to the viability of sustainable solutions in logistics.

Private sector efforts can also help fast-track the implementation of policies by serving as early adopters for developing standards and practices. Voluntary commitments like science-based targets or pledging a zero-emission fleet can happen in many cases faster than regulatory change. In this sense, corporate leadership is also a form of policy entrepreneurship that shapes the ambitions and goals of public policy (Radosevic et al. 2025). In addition, to the extent that most companies operate across state and national borders or in interconnected global supply chains, corporate engagement can help diffuse sound policy across jurisdictions. Multinational logistics companies tend to operate in differing regulatory environments and can move best practices and standards from more highly regulated jurisdictions to less regulated ones (Gündoğdu et al. 2023). Such an approach can encourage the unification of practices and increase baseline performance levels.

5.2 *Policy Gaps and Misalignment*

Despite the strong potential for synergy, a huge gap and misalignment persist between private sector actions and public sector policy making that limit the impact of both. A significant hurdle involves the fracture of regulatory structures over various jurisdictions (Chandra 2025). The logistics sector is inherently transnational, but climate policies are largely formed at the national or regional level. As a result, it causes unintended misalignments in standards, reporting requirements and enforcement mechanisms, thereby creating challenges of compliance for organisations with global footprints. Such divergences would create regulatory uncertainty and greater operational complexity, by creating differences between countries in terms of emissions standards, fuel regulations and carbon pricing mechanisms (Marzouk 2025). For instance, companies can still have mixed and incoherent demands regarding policy signals, or they cannot make enough incentive for change in low-emission technologies due to fragmentation as well as shortsightedness.

There is also no known region which has effectively utilized the policy architecture available to it with strong regulatory enforcement. The low- and middle-income economies have weak institutional capacity with legal regimes often inactive and misaligned with effective enforcement, leading to problems that are likely negatively salient when it comes to environmental legislation monitoring (Zunarelli 2023). Such an outcome could create an uneven playing field where companies in geographic regions with room for increased enforcement are less likely to pursue sustainable practices than those in more stringent jurisdictions. Corporate strategies and policy timelines could also be misaligned. As pathways that governments take may cover policies in the short run, corporate climate commitments often reach much further out into the future, for example, 2050 net-zero targets. This can also be a gap in short-term action that does not lead to long-term goals being achieved.

Moreover, there are often major gaps between policy goals and practical feasibility. A key investment of the pipeline is seen as creating faster change without considering how this can be realised further down the line at a much larger scale due to infrastructural and financial limitations (Chandra 2025). Conversely, weak policies can have little effect, as firms are able to postpone or avoid required investments. The pressures of the moment have exposed a necessity for greater coherence and coordination within policy frameworks that map to where corporate actors desire to make collaborative strides while signalling clearly what they should be investing in, innovating toward an expectation of an essential walk between conversation and action.

5.3 Role of Incentives and Regulation

It is the policies that shape corporate behaviour that will chart the pace and scale of decarbonisation in the logistics sector. Such instruments can be classified generally as market-based mechanisms, regulatory mandates or supporting incentives. The market-based instruments, like carbon pricing, are generally accepted to include users in environmental costs as an efficient way (Samuel 2025). Such systems operate through putting a price on carbon and thereby providing firms with economic incentives to cut their emissions and invest in low-carbon technologies (Marzouk 2025).

Carbon pricing, for example, can affect fuel decisions, transport modes and supply-chain design in logistics. However, another condition for a successful carbon price is not just its level and coverage, but whether jurisdictions can implement it consistently.

Financial support in the form of subsidies and other economic incentives is also a key promoting factor towards the uptake of low-carbon technologies (Marzouk 2025). Subsidies for electric vehicles, alternative fuels and renewable energy can also help to offset high initial costs and speed market uptake. Access to concessional finance and international climate funding is especially critical among developing economies for enabling investment in sustainable logistics infrastructure (Samuel 2025). Regulatory requirements such as emissions limits and fuel economy standards offer clear and enforceable benchmarks for corporate action. While voluntary initiatives may be easy to draft, they rely on the goodwill of firms and do not prevent free-riding, making them less effective to implement than regulation or government-mandated programs, which create guaranteed obligations that provide built-in compliance by implementation across the sector.

Voluntary initiatives can accelerate early action and innovation, but on their own, they are typically insufficient to lead to deep decarbonisation without the imposition of statutory requirements. Real-world evidence indicates that deep emissions reductions are typically linked to robust regulatory systems, especially when used in tandem with financial motivators (Marzouk 2025). However, the design of policy instruments must balance environmental ambition with economic feasibility. There will be no meaningful changes due to weak policies, and on the other hand, a stronger regulation may apply narrow, unreasonable costs on the SME firms too. Robustness policy design must therefore advance sectoral characteristics, technological maturity and institutional capacity.

6 Challenges and Opportunities in Developing Economies

Decarbonisation of logistics systems in developing economies is a complex problem -opportunity framework. Although those areas are likely to lead the growth in freight demand with their fastest pace, they also have considerable structural and institutional barriers against implementing sustainable practices (UNCTAD 2023). At the same time, developing economies present special opportunities for leapfrogging to low-carbon systems by avoiding the carbon-intensive trajectories of industrialized countries.

6.1 Structural and Institutional Barriers

The transition to sustainable logistics systems is hampered by several structural barriers in developing economies. The most significant challenge, however, has been the absence of adequate infrastructure, ranging from rail networks and ports to urban logistics systems. In many regions, there is no effective alternative to road transport, accounting for a higher emissions intensity (UNCTAD 2023). Firms are further constrained in their ability to invest in low-carbon technologies and infrastructure by limited access to financial resources. High capital costs, along with the limited availability of credit and nascent financial markets, present barriers to investment. This is particularly challenging for SMEs, which make up a substantial portion of the logistics sector in developing economies.

One of these factors, which is also quite relevant, especially in developing countries, is the institutional factors such as weak governance structures and regulatory capacity. Corruption, bad law enforcement that enhances environmental quality and incoherent policy diversification can at one time undermine sustainable development. However, sustainability may take a backseat to developmental priorities such as economic growth and poverty reduction.

6.2 Technology and Capacity Constraints

This mismatch in ability and authority reduces the uptake of low-carbon technologies, especially in developing economies. Most companies do not have the expertise to reach a high level of logistics solutions, such as artificial intelligence optimization systems or emissions monitoring tools (Fatorachian 2025). Nonetheless, higher up-front costs, especially for electric and alternative-fuel vehicles, remain a huge barrier. While these technologies may yield cost savings in the long run, high up-front costs often block their deployment without supportive financing mechanisms. Therefore, decarbonisation strategies should stress capacity-building and knowledge transfer. International collaboration and support, including transfer of technology and other technical assistance can help to bridge capacity gaps in poorer countries (Chikwere et al. 2024). They can also share their own experience and seek to assist their suppliers in taking sustainable measures.

6.3 Emerging Opportunities

Notwithstanding the presents of such challenges, there exist great opportunities for sustainable logistics growth in emerging economies. A significant opportunity lies in possible leap-frog developments, meaning that countries could embrace new low-carbon technologies on state-of-the-art standards and radically change their development pathway without following a conventional carbon-intensive development path (Lebdioui 2024). Investment in digital technologies has the potential for significant contribution. Once digitized, logistics management and freight matching and emissions tracking as well as the ability to increase urban delivery can provide huge efficiency benefits without requiring large-scale physical infrastructure. In such situations, digital solutions can at times bridge infrastructure gaps by making better use of existing assets.

The expansion of renewable energy also facilitates the decarbonisation of logistics systems with particular focus on transport electrification (Lebdioui 2024; Munuhwa 2024). As noted by the chapter, the costs of renewable energy are reducing gradually, at some point integrating clean energy into logistics operations becomes inevitable. Besides, the increase in global commitment towards climate action especially in terms of climate-finance and development assistance offer more avenues for capital access to sanitisation infrastructure and developments. Green financing mechanisms and multilateral development bank programs are examples of mechanisms to mobilize investment and reduce financial barriers (Lopez-de-Silanes et al. 2024). Finally, in many developing economies fast-paced urbanisation presents an opportunity to build sustainable urban logistics systems from scratch through the implementation of low-emission modes of transport and efficient distribution networks and smart technologies.

7 Policy Implications

This analysis reveals key policy priorities to advance sustainability in the logistics sector. The first is the need to enhance regulatory frameworks and enforcement. Requiring certain explicit and mandatory emissions targets, penalties for noncompliance, increased reporting thresholds at the policy level as well as broader, enforceable sectoral limits on emissions, which will be espoused by greater institutional capacity. Second, standards harmonization for ESG disclosure is vital to enhancing the comparability and credibility of data, including standardized approaches to measuring Scope 3 emissions.

Third, building out market-based instruments and financial incentives, which include carbon pricing, subsidies, and green finance, can scale up investment into low-carbon technologies in an accelerated fashion in most cases within significant developing economies where financial barriers are most acute. Fourth, it is important to push for inclusive supply chain engagement in scope 3 emissions. However, the need for SME-targeted support in terms of capacity-building, technical assistance and provision or access to digital tools is crucial for large-scale decarbonisation.

Fifth, focused investments in various forms of sustainable infrastructure and innovation, including rail systems, intermodal networks, renewable energy sources and digital logistics platforms, can lock in systemic gains in efficiency. Such investments need to be mobilized through public–private partnerships. Sixth, context-specific and phased policy approaches can help support realistic transitions for developing economies while allowing leapfrogging to low-carbon systems. Lastly, the chapter proposes that enhancing multi-level and multi-stakeholder governance is essential.

Improved collaboration between governments, the private sector, international organizations and civil society can support policy coherence, scale solutions and ensure more equitable outcomes.

8 Conclusion

This chapter has explored the importance of private sector leadership in delivering climate action in the logistics sector through corporate climate commitments, ESG disclosure frameworks, supply chain decarbonisation strategy and voluntary initiatives. It argues that private sector engagement features prominently in contemporary climate governance, especially amid fragmented global supply chains and evolving regulatory landscapes. The corporate climate commitment, and especially the science-based target setting, shows that a paradigm shift is taking place as corporates are starting to adopt sustainability as the core business strategy. Nonetheless, these advances are limited by deep-seated issues, including the scale of Scope 3 emissions, inconsistent ESG disclosure practices and insufficient accountability mechanisms. These are stricter in logistics, where emissions are allocated across long, complex and multi-actor value chains.

Since they have not followed through on the emissions reductions, corporate voluntary measures do more to raise awareness, drive innovation and foster collaboration, but far less in terms of shutting down other forms of capital to drive emission reductions. The free-rider problem, selective take-up rates, and weak enforcement make their ability to initiate systemic change limited, thereby exposing the limits of voluntarism as a stand-alone governance mechanism. It also illustrates that outcomes are the result of a pas de deux between private sector action and public policy. Corporate efforts can contribute to the strengthening of regulatory action, but they must happen in a pre-existing environment that has comprehensive and consistent policy frameworks with predictable enforcement. Such effort is critical for private sector efforts not to be piecemeal. Lack of coordination leads one to wonder if the efforts in place by the private sector are piecemeal or if at least something could be done to prevent duplication. In developing economies, structural constraints in terms of poor or even non-existent infrastructure make limited finance and weak institutional capacity reduce the progress that natural resources can advance but also present large opportunities to leap-frog through technology and digital innovation.

References

Abu-Aisha T, Audy JF, Ouhimmou M (2024) Toward an efficient sea-rail intermodal transportation system: a systematic literature review. J Ship Trade 9(1):23

Boasson EL, Burns C, Pulver S (2023) The politics of domestic climate governance: making sense of complex participation patterns. J Eur Publ Policy 30(3):513–536

Butt AS, Alghababsheh M, Sindhwani R, Gwalani H (2025) Role of supplier engagement to reduce Scope 3 emissions in circular supply chains. Bus Strateg Environ 34(1):598–611

Chandra S (2025) Flexibility vs harmonization: analysing the environmental regulations and global supply chains (No. 58e4b_v1). Center for Open Science

Chikwere D, Chibaro M, Govere E (2024) Leveraging alternative energy sources to enhance sustainable transportation. In Contemporary solutions for sustainable transportation practices (pp. 94–127). IGI Global Scientific Publishing

Chikwere D, Mbudaya J, Chibaro M (2025) Exploring green energy alternatives for a sustainable future: opportunities and challenges for developing countries. Adaptive Strategies for Green Economy and Sustainability Policies, pp 183–206

Development (UNCTAD) (2023) Review of Maritime Transport 2023: towards a green and just transition. Stylus Publishing, LLC

Fatorachian H (2025) Digital transformation for sustainable transportation: leveraging Industry 4.0 technologies to optimise efficiency and reduce emissions. Future Transport 5(2):34. https://doi.org/10.3390/futuretransp5020034

Gallas S, Chouchene MR, Bouzgarrou H (2025) Greenwashing illusion and financial performance: the moderating role of environmental practices. Bus Strateg Develop 8(4):e70234

Gündoğdu HG, Aytekin A, Toptancı Ş, Korucuk S, Karamaşa Ç (2023) Environmental, social, and governance risks and environmentally sensitive competitive strategies: a case study of a multinational logistics company. Bus Strateg Environ 32(7):4874–4906

Hettler M, Graf-Vlachy L (2024) Corporate scope 3 carbon emission reporting as an enabler of supply chain decarbonization: a systematic review and comprehensive research agenda. Bus Strateg Environ 33(2):263–282

Jaehn F, Meissner F (2022) The rebound effect in transportation. Omega 108:102563

Kotsantonis S, Serafeim G (2019) Four things no one will tell you about ESG data. J Appl Corp Financ 31(2):50–58. https://doi.org/10.1111/jacf.12346

Le TT, Tran PQ, Lam NP, Tra MNL, Uyen PHP (2024) Corporate social responsibility, green innovation, environment strategy and corporate sustainable development. Oper Manag Res 17(1):114–134

Lebdioui A (2024) Survival of the greenest: Economic transformation in a climate-conscious world. Cambridge University Press

Lopez-de-Silanes F, McCahery JA, Pudschedl PC (2024) Institutional investors and ESG preferences. Corp Govern Int Rev 32(6):1060–1086

Marquardt J, Fast C, Grimm J (2022) Non-and sub-state climate action after Paris: from a facilitative regime to a contested governance landscape. Wiley Interdiscip Rev: Climate Change 13(5):e791

Marzouk OA (2025) Summary of the 2023 report of TCEP (tracking clean energy progress) by the International Energy Agency (IEA), and proposed process for computing a single aggregate rating. In E3S Web of Conferences (Vol. 601). EDP Sciences, pp 00048

Mba JU (2024) Advancing sustainability and efficiency in maritime operations: integrating green technologies and autonomous systems in global shipping. Int J Sci Res Arch 13(2):2059–2079

McKinnon A (2018) Decarbonizing logistics: distributing goods in a low carbon world. Kogan Page Publishers

Munuhwa S (2024) Driving towards emission-free urban centers: exploring Glasgow's sustainable road transport. In Contemporary solutions for sustainable transportation practices (pp 186–206). IGI Global Scientific Publishing

Nedziwe CL, Tella O (2023) Non-state actors as transnational agents. In Transnational activities of women-focused civil society actors in Southern Africa. Springer Nature Switzerland, Cham, pp 49–69

Radosevic S, Tsekouras G, Wostner P (2025) Institutionalising experimentation in innovation policy: challenges and solutions in upscaling. JRC Publications Repository

Rehana N, Qureshi S (2024) Supply chain geopolitics: navigating international business in a fragmented world

Samuel SA (2025) Carbon pricing mechanisms for reducing greenhouse gas emissions and encouraging sustainable industrial practices. World J Adv Res Rev 25(2):1–24

SBTi (2022) Science based targets initiative. GEZ Group. Found at https://www.cez.cz/webpublic/file/edee/2022/06/cez-group-tvr.pdf

Shee Weng L (2025) Green logistics: innovations in sustainable transportation and distribution. Available at SSRN 5144977

Walker GE (2026) Ethical leadership and funding strategies in nonprofit organizations. Doctoral dissertation, Walden University

Yusof AFM, Widyasamratri H (2025) Environmental, social, and governance: a review of frameworks, metrics, and reporting for sustainable development. Civil Sustain Urban Eng 5(2):102–116

Zunarelli S (2023) The logistics industry in the digital era: problems and opportunities for the SMEs of the transport sector. In SMEs in the Digital Era. Edward Elgar Publishing, pp 208–223

Leadership as a Catalyst for Embedding Circular Economy into Corporate Strategy in Global South: A Systematic Review

James Kanyepe and Nyarai Kasambuwa

Abstract The circular economy (CE) represents a fundamental organizational transformation, and integrating its principles into corporate strategy requires a shift beyond mere technical or procedural change. This article presents a systematic literature review of peer-reviewed articles published between 2010 and 2025, synthesizing the available literature on the influence of leaders and their role in adopting CE in different sectors and geographical areas. A structured review protocol was used to search six major databases, namely, Scopus, Web of Science, DOAJ, and IBSS, using predefined inclusion and exclusion criteria. Only studies that focused on either the empirical or conceptual aspects of the CE were included, whereas those that solely addressed the technical or operational aspects of the CE were excluded from the review. The data extracted from the studies included the methods employed by leaders to promote CE within their organizations, organizational enablers, positive results experienced by organizations, and other contextual factors that promote CE. Despite the existence of these enablers, organizations continue to experience obstacles in successfully implementing the CE. Furthermore, this study identifies gaps in the current research and suggests possible directions for future studies on in-driving sustainability transitions. It also provides additional insights into how organizations can successfully adopt a CE for long-term environmental and economic benefits.

Keywords Circular economy · Sustainability leadership · Strategic management · Organizational transformation · Systematic review · Corporate strategy

J. Kanyepe (✉)
Department of Logistics and Supply Chain Management, University of Botswana, Gaborone, Botswana
e-mail: jameskanyepe@gmail.com

Faculty of Business, Economics and Law, University of Botswana, Gaborone, Botswana

N. Kasambuwa
Academy of Teaching and Learning, Chinhoyi University of Technology, Chinhoyi, Zimbabwe

J. Kanyepe and I. Chirisa (eds.), *Advancing the Green Agenda in the Transport and Logistics Sector in Global South Countries*, World Sustainability Series,
https://doi.org/10.1007/978-3-032-31374-4_24

1 Introduction

The circular economy (CE) has evolved significantly over the last decade as a new paradigm for sustainable development, focusing on breaking the link between economic growth and resource depletion through improved resource use, waste reduction, and the recovery of materials for their maximum reuse potential (Teixeira 2025). The CE is a regenerative and restorative system that maximizes the value of products, materials, and resources for as long as possible (Morseletto 2020). Therefore, the CE has become a focal point for many governments, businesses, and organizations seeking to develop long-term solutions to environmental deterioration and resource shortages. As the CE continues to grow in terms of discourse, a large volume of literature has developed on the technical, operational, and technological aspects of the CE (e.g., eco-design, reverse logistics, recycling technology, etc.) and circular business models (de Jesus et al. 2021). However, beyond operational considerations, the successful adoption of a CE requires organizational transformation, which necessitates a shift in the organization's strategic orientation, decision-making structure, and organizational culture (Bertassini et al. 2021). This has caused significant attention to be given to the organizational and behavioral aspects of CE implementation, with leadership being one of the most important factors affecting CE adoption.

Leadership has been recognized as a key driver of an organization's sustainability performance (Juicharoen et al. 2025) and can influence an organization's strategic direction, influence stakeholders, provide the means for innovation, and enable an organization to adopt new ways of working (Bashynska et al. 2024). Leadership is also seen as a facilitator of organizational learning, enabling the management of uncertainty and facilitating the overcoming of inertia when undergoing systemic changes (Khan and Soomro 2025). In the case of the CE, leaders are expected to create value over the long term, ensure that environmental and economic goals are aligned, and facilitate the cross-functional and interorganizational collaboration required to close material loops (Metcalf and Hinske 2022).

While leadership, recognized as a critical element of an organization's sustainability efforts, is well established, organizations attempting to implement CE strategies continue to experience significant barriers to implementation (Bertassini et al. 2021). These barriers include resistance to change, poor organizational design and coordination, conflicting incentives, inadequate managerial skills, and time constraints related to short-term financial needs (Mubarik et al. 2024). These barriers indicate the need for leadership approaches that can effectively manage the complexities of sustainability and ensure its full integration into an organization's strategic planning processes. Additionally, while leadership approaches have been used to promote sustainability, the literature on CE is largely fragmented in terms of the treatment of leadership (e.g., leadership is rarely treated as a primary focus of CE research but is instead addressed in isolation or as a secondary factor).

Although leadership has been studied extensively in the broader sustainability and environmental management literature (Boeske 2023; Dominguez-Escrig and

Mallen-Broch 2023; Suriyankietkaew et al. 2022), there is a lack of comprehensive and systematic synthesis of leadership and CE adoption. Studies on leadership and CE adoption have focused primarily on a single leadership style, industry, or geographic region, limiting the opportunity for cumulative knowledge building (Cheffi et al. 2023; Oktrivina et al. 2025). Additionally, the relationships among leadership approaches, organizational enablers, challenges, and performance outcomes associated with CE adoption are largely unexplored and undert-theory; therefore, they remain poorly integrated across studies.

Therefore, this study aimed to conduct a systematic literature review to synthesize and analyze the existing research on leadership and CE adoption. This review aims to identify the major leadership approaches and frameworks that have been employed to integrate CE into corporate strategy, identify the organizational strategies and enablers that have supported the implementation of CE by leadership, and evaluate the challenges and success factors associated with the implementation of CE driven by leadership. Additionally, this study evaluated how different leadership approaches influence CE adoption in different types of organizations across various industries and countries/regions. Therefore, the following research questions guided this study.

RQ1. What leadership approaches have been used to integrate circular economy concepts into corporate strategies?

RQ2. What organizational strategies and enablers have supported the leadership-driven adoption of circular economy practices?

RQ3. What barriers prevent leadership from effectively implementing circular economy initiatives?

This study contributes to the CE and sustainability leadership literature by identifying and resolving the fragmentation of the CE literature, identifying research gaps, and providing empirically grounded recommendations for practitioners, policymakers, and researchers seeking to improve the transition to the CE. The remainder of this paper is organized as follows. Section 2 describes the methodology and systematic review process used to complete this study. Section 3 presents the results of thematic analysis. Section 4 presents a discussion of the results. Section 5 concludes the paper and discusses the implications of the study.

2 Research Methodology

This study adopts a systematic literature review (SLR) to synthesize and critically evaluate existing research on leadership strategies for integrating circular economy (CE) principles into corporate strategy. Systematic literature reviews are widely recognized as rigorous approaches for consolidating fragmented knowledge, particularly in multidisciplinary and emerging research domains (Tranfield et al. 2003). This review is positioned as theory-building rather than theory-testing to generate higher-order conceptual insights. This approach is appropriate given the conceptual diversity and methodological heterogeneity of CE leadership research, where studies draw on multiple theoretical perspectives and employ various research designs. Under such conditions, systematic reviews are particularly valuable for theory development and conceptual integration. This review follows the structured approach proposed by Tranfield et al. (2003) framework and adheres to the PRISMA guidelines to ensure transparency and replicability in the identification, screening, and selection of studies.

2.1 Review Protocol and Search Strategy

A review protocol was developed before data collection to enhance transparency and minimize researcher bias, which is consistent with the best practices established in systematic reviews (Popay et al. 2006). The protocol defined the scope of the review, search strategy, inclusion and exclusion criteria, and analytical procedures. A literature search was conducted in major academic databases, including Scopus, Web of Science, DOAJ, and IBSS. Structured search strings were developed via Boolean operators and truncation to capture terminology variations in the CE and leadership literature. The use of multiple databases and systematic search procedures enhances comprehensiveness and reduces potential database bias (Kitchenham and Charters 2007).

2.2 Inclusion and Exclusion Criteria

Explicit inclusion and exclusion criteria were applied to ensure the analytical focus and theoretical relevance of the studies. The review included only peer-reviewed journal articles published in English, ensuring baseline methodological and theoretical quality. Conceptual and empirical studies were retained to support the integration of theoretical and empirical insights, which is important in emerging research domains (Snyder 2019). Studies were excluded if they focused solely on the technical or engineering aspects of circularity without addressing the organizational, leadership, or strategic dimensions, thereby ensuring alignment with the review's core constructs.

2.3 Screening and Article Selection

The study selection process followed a structured multistage screening procedure in accordance with the PRISMA guidelines. After removing duplicates, titles and abstracts were screened for relevance, followed by a full-text assessment of the eligible studies. All inclusion decisions were based on predefined criteria to ensure consistency and reduce selection bias. A staged screening approach improves the transparency and methodological rigor of systematic reviews (Tranfield et al. 2003).

2.4 *Data Extraction and Analysis*

Data were extracted via a structured template that captured key study characteristics, including the research context, theoretical perspective, leadership approach, and CE outcomes. Structured extraction enhances comparability across studies and supports systematic synthesis (Büchter et al. 2020). The analysis employed a hybrid deductive–inductive coding approach that combined theory-driven categories with emergent themes. The synthesized data were analyzed via thematic synthesis, following Thomas and Harden (2008), to identify the recurring patterns, mechanisms, and contextual factors shaping leadership-driven CE integration.

2.5 *Limitations*

This review has several limitations. For example, restricting the analysis to peer-reviewed journal articles published in English may exclude relevant insights and introduce publication bias. Additionally, the findings depend on the quality and scope of the data included in these studies. Despite these limitations, the transparent and systematic approach used in this study provides a robust foundation for future research.

3 Results

The data from the included studies were analyzed via thematic analysis, a systematic and structured method for identifying, examining, and reporting common themes or trends in qualitative data. The use of thematic analysis allows for the examination of recurrent concepts and relationships between studies and provides insight into leadership practices, facilitators and barriers to circular economy implementation, and outcomes associated with adopting circular economy practices. The thematic analysis approach enabled this study to synthesize evidence across various organizational settings and industry sectors, allowing the development of an integrated conceptual framework that relates leadership approaches to circular economy principles. To enhance analytical transparency and demonstrate how themes were derived from the data, this study adopted a Gioia-based data structure. The analysis progressed from first-order concepts grounded in the literature to second-order themes and, ultimately, to aggregate theoretical dimensions. This structured approach strengthens the rigor of the thematic synthesis and provides a clear link between the reviewed studies and the resulting theoretical insights.

4 Leadership Approaches in the Circular Economy Adoption in the Global South

Research has shown that leadership can be a very important factor in how organizations adopt and sustain circular economy (CE) thinking and actions (Bashynska et al. 2024). Several empirical and theoretical studies have shown that transformational leadership, strategic sustainability leadership, adaptive leadership, and collaborative leadership can all be forms of leadership that help facilitate organizational transition to CE (Khan and Soomro 2025; Alqatan et al. 2025; Ali Esmaeili and Sobhanifard 2025). Transformational leadership is one of the many types of leadership commonly associated with (Huang et al. 2021). Transformational leadership is characterized by the leader's ability to provide an inspiring and motivating vision of the future of the organization, motivate employees to participate in creating that future, and establish an organizational culture that encourages innovation and risk-taking (Gupta 2025).

When developing CE strategies, transformational leaders encourage employees to question current linear production and consumption patterns, prioritize the efficient use of resources, and explore new CE-based approaches (Arora et al. 2023). Empirical research has demonstrated that transformational leadership promotes the behavioral and process changes necessary to develop CE strategies in manufacturing and industrial contexts (Birasnav 2014). However, most studies indicate that transformational leadership is most effective when used in conjunction with other forms of leadership because it relies heavily on the organization being prepared culturally, employee motivation and engagement, and the presence of additional forms of leadership (Budur 2020; Al Marshoudi et al. 2023).

In addition, strategic sustainability leadership is another form of leadership that is often cited as a mechanism to enable CE strategies. Strategic sustainability leadership involves integrating environmental and sustainability goals into the core of an organization's strategy (Alkhodary 2023). Therefore, strategic sustainability leaders incorporate CE strategies into an organization's decision-making, financial investment, and evaluation processes (Bashynska et al. 2024). Prior studies have shown that strategic sustainability leadership promotes interdepartmental collaboration by establishing common sustainability objectives across different departments within an organization (Galpin and Lee Whittington 2012; Kurucz et al. 2017). The literature indicates that strategic sustainability leadership works well with transformational and adaptive leadership to promote the development of sustainable CE-based business models. However, the success of strategic sustainability leadership is dependent on the organization's culture and whether there are structural mechanisms in place to support it (Asgari & Asgari 2021; Bag and Pretorius 2022).

Moreover, adaptive leadership is a common form of leadership that is often referred to in the literature as a response to the complex, dynamic, and uncertain conditions related to the transition to the CE (Korhonen 2020). Adaptive leadership is the ability of leaders to respond effectively to complexity, uncertainty, and change by fostering an environment of organizational learning, experimentation, and flexibility (Fatima and Fatima 2022). Research has demonstrated that adaptive leadership provides organizations with the means to address the regulatory, technological, and market-related uncertainties that accompany the implementation of CE strategies (Sott and Bender 2025; Kokhuis et al. 2025). Research further shows that adaptive leadership is particularly useful in industries that involve rapid innovation (e.g., energy and technology), where the need for continuous adjustments and system-wide problem solving to implement the CE is present (Akinwalere et al. 2025; Boikanyo 2025). Like other leadership approaches, adaptive leadership works best when combined with a transformational vision and strategic sustainability orientation, as these three forms of leadership reinforce the complementarity of leadership approaches to facilitate CE adoption (Malek and Almarri 2023; Miao and Nduneseokwu 2025).

Furthermore, collaborative leadership is considered a major form of leadership that is beneficial for implementing CE strategies in organizations. Collaborative leadership involves coordinating and engaging a variety of stakeholders, both internally and externally, to the organization, such as employees, suppliers, customers, government regulators, and community groups, through dialog, cocreation, and knowledge-sharing activities (Modha 2021). As a result, collaborative leadership creates opportunities for organizations to learn, optimize resource use, and innovate CE strategies (Köhler et al. 2022). Examples of CE strategies that benefit from collaborative leadership include industrial symbiosis and urban waste management programs (Alhosani 2025). Collaborative leadership is also a way for leaders to build on their transformational, strategic sustainability, and adaptive leadership skills, leading to hybrid leadership configurations that enable the strategic and operational aspects of CE strategy implementation (Bashynska et al. 2024; Odiase and Agbadudu 2024).

5 Organizational Enablers and Strategies Supporting Leadership-Driven Adoption of Circular Economy Practices

Many articles highlight strategic alignment as a primary organizational enabler for adopting circular economy (CE) practices (Scipioni et al. 2021; Ahmed et al. 2025). Most of these studies emphasize the need to integrate CE principles into corporate strategy, planning processes, and capital allocation decisions to create sustainable changes (Mendoza et al. 2017; Kristensen et al. 2021). When organizations include circular objectives in their decision-making processes, they can treat sustainability concerns as part of their core priorities rather than peripheral activities (Bertassini et al. 2021). Many scholars agree that strategic alignment allows organizational leaders to send signals of long-term commitment to circularity, legitimizes circular investments, and provides a stable framework for internal and external stakeholders regarding the organization's move toward circularity (Osei et al. 2025; Abu-Bakar & Charnley 2024).

Cross-functional collaboration is another common enabler of CE implementation from an operational perspective. Many studies have reported that cross-functional coordination is necessary to break down organizational silos and enable the systemic design of products, processes, and supply chain systems (Hofstetter et al. 2021; Burke et al. 2023; Zils et al. 2025). Formally established knowledge-sharing structures and cross-functional teams help bring together diverse expertise, increase problem-solving capabilities, and facilitate the dissemination of circular practices throughout an organization (Marra et al. 2018). Therefore, organizations can implement complex CE projects and incorporate circularity into their daily operations. In addition to internal coordination among departments, many researchers have highlighted stakeholder engagement as an important external organizational enabler (Fobbe and Hilletofth 2023).

Given that CE strategies typically extend beyond organizational boundaries, effective implementation requires collaboration with suppliers, customers, regulatory bodies, and other community actors (De Mattos & De Albuquerque 2018). Strategic partnerships can optimize resources, foster joint innovation, and scale up circular solutions, especially in situations where infrastructure or institutional barriers limit what can be accomplished individually. Stakeholder engagement also promotes interest alignment, builds trust, and fosters accountability for circular outcomes (Tanveer et al. 2025).

Organizational culture is another important moderator that affects the effectiveness of strategic and operational enablers (Kwarteng et al. 2022). Research indicates that learning-oriented, experimental, and sustainability-focused organizational cultures are more supportive of CE adoption because they promote employee participation and reduce resistance to change (Scipioni et al. 2021). These cultures provide an environment that encourages the continuous iteration and improvement of circular practices over time, promoting lasting transformation rather than temporary compliance. The ability of organizations to use technology and digital tools has been identified as an additional enabling factor for CE in the literature. Many studies have emphasized the role of digital tools, such as lifecycle assessments, material tracking, and performance monitoring, in providing data-driven decision-making and continuous optimization of circular initiatives (Turskis & Šniokienė 2024). Digital technologies enable organizations to make their entire supply chain transparent and trackable, facilitating the identification of inefficiencies, providing ongoing feedback on progress, and demonstrating the benefits of circularity (Sharma et al. 2025).

6 Challenges and Barriers in the Global South

In recent years, increasing institutional and managerial attention has been given to the circular economy (CE); however, despite this attention, research has shown that the development of circular economies is still a developing process and that there is a large disparity in the rate of development among different organizations and industries. Several researchers have argued that this disparity exists because the transition to a circular economy will require a significant transformation of existing production, consumption, and value creation systems rather than simply improving the efficiency of existing linear systems. Several studies have identified organizational culture as a key barrier to adopting circular economy principles (Rehman and Alqahtani 2025).

The dominant logic of the linear economy is often deeply ingrained in the everyday routines, cognitive frames, and decision-making rules of organizations, reducing the likelihood of acceptance of alternative circular approaches (Nygaard 2022). Additionally, risk-averse organizational cultures discourage experimentation, cross-functional cooperation, and learning from failures, all of which are important for circular innovation (Eisenreich et al. 2021). These cultural barriers are frequently supported by hierarchical organizational structures and centralized decision-making processes that prevent bottom-up efforts to implement circular practices and slow their dissemination throughout organizations (Bourdin and Jacquet 2025).

In addition to cultural barriers, several studies have noted that capability and knowledge deficits in organizations are significant obstacles to implementing circular economy principles (Palombi et al. 2024). Many organizations lack the absorptive capacity needed to learn, integrate, and use circular economy-related knowledge in complex production and supply systems (Ćwiklicki et al. 2024). Hence, circular initiatives are often developed in an uncoordinated and superficial manner, that is, without the systemic transformation required to achieve meaningful circularity (Schulz et al. 2019). This problem is particularly evident in knowledge-intensive areas such as remanufacturing, advanced recycling and digital material flow management. Moreover, organizational structures can contribute to capability and knowledge deficits by limiting cross-functional coordination and life cycle integration, which are necessary for creating circular value (Hopkinson et al. 2020). Consequently, the structural rigidity of organizations decreases their flexibility in conducting iterative experiments and limits the long-term institutionalization of circular practices (de Oliveira Vieira and Guarnieri 2025).

Additionally, costs are generally recognized as the most decisive barrier to the implementation of circular economy principles (Grafström and Aasma 2021). Circular initiatives normally require substantial initial investment, and the time until the investment pays off is usually long and uncertain (Dewick et al. 2020). Therefore, existing financial control systems based on short-term performance indicators and annual budget planning cycles are likely incompatible with the long term. This incompatibility tends to increase the perceived financial risk of circularity, leading to underinvestment in circular solutions (De La Cuesta-González and Morales-García 2022). Technological preparedness is another prerequisite for the feasibility of implementing circular economy principles in the construction industry in developing countries. Most circular strategies require specialized recycling and remanufacturing technologies and digital systems to trace materials throughout their life cycles (Han et al. 2023). Owing to high capital costs, a lack of infrastructure, or technological immaturity, limited access to these technologies can greatly restrict the scope and scale of circular initiatives (Trevisan et al. 2023).

Many researchers emphasize that the successful implementation of circular economy principles requires coordinated action by multiple actors in the value chain, namely, suppliers, customers, and logistics service providers (Eisenreich et al. 2022). However, differences in the capabilities, incentives, and strategic priorities of these actors lead to a lack of alignment and subsequently hinder collaboration and system-wide circularity (Iacovidou et al. 2021). These differences are especially pronounced in highly decentralized, globally organized supply networks characterized by fragmentation, information asymmetry and unequal power relationships. In addition, the institutional context of organizations is shaped by the external regulatory and policy environment, which, in turn, determines the extent to which organizations adopt circular economy principles (Castro-Lopez et al. 2023). Uncertainty regarding regulatory frameworks, inconsistencies in policies, and the lack of stable market incentives for circular economy activities lead to ambiguity and an increased perception of risks concerning circular investments (Tan et al. 2022). Furthermore, fragmented and weakly enforced regulations create unclear expectations regarding compliance, which can impede decision-making and deter long-term strategic commitments to circular economy practices (Kandpal et al. 2024).

Furthermore, limitations related to measuring and evaluating circular economy development hinder the implementation of circular economy principles and their evaluation (Pakuła et al. 2025). Organizations lack standardized and reliable methods and tools for tracking material flows, assessing circular performance, and demonstrating value creation to relevant stakeholders (Brändström and Eriksson 2022). The lack of harmonized metrics and reliable monitoring systems impairs the quality of managerial decision-making and hinders transparency, accountability, and organizational learning in the context of the circular economy (Abu-Bakar and Charnley 2024).

7 Discussion

Transformational leadership and strategic sustainability leadership are two types of leadership that help shape long-term organizational vision, daily operations, and the implementation of circular economy (CE) principles. Geissdoerfer et al. (2017) agree that the implementation of a circular economy (CE) is driven primarily by the actions of organization leaders rather than a technical or operational shift. Sustainable leadership scholars have argued that organizations must embed their environmental objectives within their overall strategy and decision-making processes rather than treating them as peripheral or symbolic initiatives. Kirchherr et al. (2017) provide examples of the necessity of integrating environmental objectives into an organization's overall strategy.

In addition to developing a long-term vision, effective circular economy (CE) leadership is associated with the establishment of organizational enablers that facilitate learning, cross-functional integration, and the ongoing engagement of stakeholders. The review findings support adaptive leadership theory, which defines leadership as the ability to lead an organization through complex times, uncertainties, and systemic changes. Heifetz and Linsky (2002) and Korhonen (2020) define adaptive leadership as the ability of leaders to create the environment and conditions necessary to allow organizations to experiment, make decisions collaboratively, and develop flexible implementation processes to overcome common barriers to adopting circular economy (CE) practices. Examples of these common barriers include organizational inertia, siloed structures, and short-term financial pressures that often counteract sustainability-oriented initiatives. Kirchherr et al. (2017) and Scipioni et al. (2021) provide examples of common barriers to adopting circular economy (CE) practices.

The review revealed that the application of leadership approaches to adopting circular economy (CE) practices is highly dependent on the organizational context in which they occur. Different organizational contexts require different leadership approaches owing to differences in operational logic, resource dependencies, and institutional environments. In the industrial and manufacturing sectors, leadership approaches to implementing circular economy (CE) practices are focused primarily on optimizing processes, reducing waste, and increasing resource efficiency. These approaches reflect the material-intensive and technologically complex nature of these sectors. Lieder and Rashid (2016) provide examples of leadership approaches in the industrial and manufacturing sectors. In contrast, leadership approaches to implementing circular economy (CE) practices in the consumer goods and technology sectors place greater emphasis on aligning strategic objectives, designing new products and business models, and collaborating with external stakeholders. These approaches indicate that the implementation of circular economy (CE) practices in these sectors is heavily reliant on design-oriented and network-based strategies. Geissdoerfer et al. (2017) provide examples of leadership approaches in the consumer goods and technology sectors. In the public sector, leadership approaches to implementing circular economy (CE) practices tend to focus on coordinating stakeholders, aligning policies, and establishing legitimacy. These approaches indicate that the implementation of circular economy (CE) practices in the public sector is heavily reliant on the institutional context. Klein et al. (2020) provide examples of leadership approaches in the public sector.

Theoretically, the findings of this review advance an integrated perspective by defining the relationship between leadership approaches to CE and CE outcomes, using organizational enablers as mediators. Transformational and strategic sustainability leadership are consistently related to CE enablers, such as incorporating CE objectives into strategic planning, creating cross-functional knowledge-sharing mechanisms and formalizing stakeholder engagement processes. Bashynska et al. (2024) and Kurucz et al. (2017) provide examples of the relationships between transformational and strategic sustainability leadership and CE enablers. Overall, the findings suggest that the effectiveness of leaders in facilitating the implementation of CE cannot be determined on the basis of their individual leadership style but rather through the alignment of their strategic intent, the structure of their organization, and the routines of their operations. The synthesis of leadership approaches with specific CE enablers provided in this review offers a coherent analytical framework for determining how leaders facilitate the systemic and sustainable transformation of organizations toward circularity.

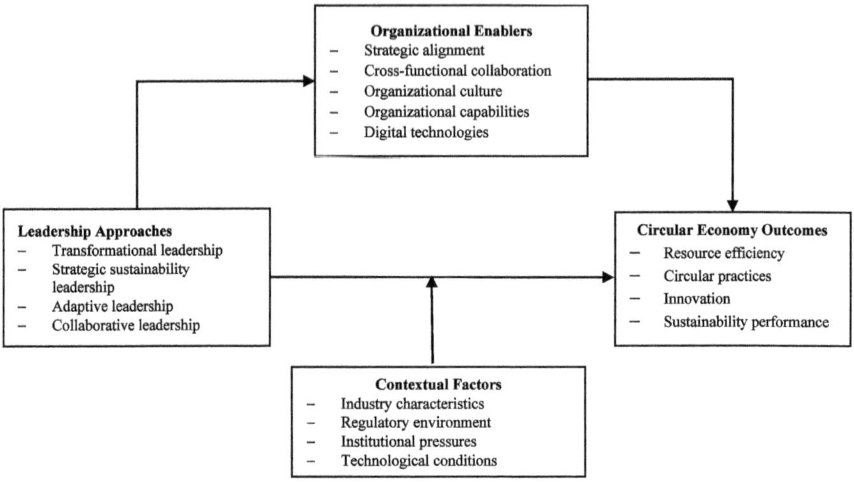

Fig. 1 Integrated conceptual framework for leadership-driven circular economy adoption

To integrate the findings of this review into a coherent theoretical explanation, Fig. 1 presents an integrated conceptual framework for leadership-driven circular economy adoption. The framework illustrates how different leadership approaches influence circular economy outcomes through key organizational enablers while accounting for the moderating role of contextual factors. This synthesis provides a structured representation of the mechanisms through which leadership facilitates circular transformation in organizational settings.

The findings of this review also have significant practical implications for organizational leaders and policymakers. For executive leaders, the findings suggest the need to integrate the objective of implementing CE into core business functions while simultaneously creating opportunities for organizational learning, cross-functional collaboration, and stakeholder engagement. Galpin and Lee Whittington (2012) and Bag and Pretorius (2022) provide examples of the need to integrate CE objectives into core business functions to achieve sustainable development. The findings also suggest that effective CE leadership requires the ability of leaders to collaborate and engage in participatory leadership practices rather than relying on hierarchical authority alone. Köhler et al. (2022) provide examples of the importance of participative leadership practices in achieving the objectives of CE. For policymakers, the findings suggest the need to develop and implement regulatory frameworks, establish incentives for organizations to invest in CE practices, and provide institutional support mechanisms to reduce the uncertainty and risk faced by organizational leaders when they invest in long-term CE practices. Kirchherr et al. (2017) and Kandpal et al. (2024) provide examples of the need for institutional support mechanisms to encourage organizational leaders to invest in CE practices.

Furthermore, this review identifies several areas for future research. Although there has been a considerable increase in scholarly research on the role of leadership in achieving the objectives of CE, the evidence base remains unevenly distributed across sectors and geographic regions, and longitudinal studies investigating the sustained impact of leadership practices on CE outcomes remain underdeveloped. Future research should utilize longitudinal and multilevel designs to investigate the dynamic relationships between leadership, organizational culture, digital technologies, and institutional environments. This type of research will enhance our understanding of the mechanisms through which leadership influences long-term circular transitions within and across organizations and ultimately contribute to the advancement of theory and practice in both the CE and sustainability leadership literatures.

8 Conclusion and Implications

This systematic review examined the role of leadership in enabling the integration of circular economy (CE) principles into corporate strategies and organizational practices. The findings demonstrate that leadership functions as a catalyst and structural enabler of organizational transition toward sustainable development. Across sectors and organizational types, four leadership approaches—transformational, strategic sustainability, adaptive, and collaborative leadership—consistently emerged as the most influential in advancing CE adoption. Collectively, these approaches shape strategic directions, guide decision-making processes, foster stakeholder engagement, and enable operational and business innovations. Thus, the review confirms that effective sustainability leadership is inherently multifaceted, context sensitive, and strategically oriented rather than reliant on a single leadership style.

This review identifies a set of organizational enablers and barriers that interact closely with leadership practices during CE implementation. Strategic planning integration, cross-functional knowledge exchange, and structured stakeholder communication consistently function as critical enablers of CE adoption. The findings further indicate that leadership effectiveness depends on the alignment between strategic vision and everyday organizational operations. Conversely, barriers such as resistance to change, short-term financial pressures, and deficits in CE-related capabilities underscore the need for leaders to complement visionary leadership with robust change management, capacity building, and learning-oriented practices. These insights strengthen the conceptual linkage between leadership approaches and tangible organizational outcomes in the context of circular transformation.

8.1 Theoretical Implications

This study adds to the body of knowledge on sustainability, leadership, and circular economy studies by developing a comprehensive theoretical framework that connects specific leadership approaches to circular economy (CE) principles through organizational enablers (Bashynska et al. 2024; Geissdoerfer et al. 2017). By synthesizing the literature on different types of leadership and the ways in which they support or hinder the implementation of CE mechanisms, this study extends previous work that has provided either style-specific or fragmented explanations of how leaders influence the transition to circular economies (Kirchherr et al. 2017; Kurucz et al. 2017).

The results of this study also add to the literature on sustainability leadership by demonstrating that leadership in CE transitions involves much more than simply articulating and communicating a vision of sustainability. Rather, consistent with both transformational and strategic sustainability leadership, effective leadership requires deliberate and intentional restructuring of organizational processes and decision-making systems and the establishment of new stakeholder relationships to enable organizations to implement circular strategies (Bashynska et al. 2024). These findings are consistent with recent views that describe leadership as an embedded, relational, and processual construct rather than an individual-level construct, especially when complex and systemic sustainability transitions are examined (Korhonen 2020; Heifetz & Linsky 2002).

This study demonstrates that the effectiveness of leadership in achieving CE adoption varies depending on the organizational context. Therefore, the findings provide an important critique of the universalistic assumption made by many researchers in the sustainability leadership literature, suggesting that a single leadership style can be used to achieve circular transitions in all settings (Lieder & Rashid 2016). Instead, this study demonstrates that successful CE adoption requires flexible and hybrid leadership configurations that draw on a combination of transformational, strategic sustainability, adaptive, and collaborative leadership attributes that are responsive to sector-specific resource dependencies, institutional pressures, and technological conditions (Bashynska et al. 2024). Furthermore, this study conceptualizes leadership as a dynamic and integrative organizational capability that enables the alignment of circular economy principles with organizational enablers, thereby creating a foundation for systemic and sustainable long-term sustainability transformations.

8.2 Practical Implications

From a practical standpoint, the findings provide actionable guidance for managers, executives, and sustainability leaders. Organizations seeking to embed CE principles into their corporate strategy should combine participatory decision-making with a clear strategic direction, invest in employee training and capability development, promote cross-functional collaboration, and actively engage with external stakeholders such as suppliers, regulators, and local communities. These practices enhance organizational readiness for circular transitions, strengthen internal commitment, and support the long-term environmental, social, and economic performance of the organization. For policymakers and regulators, the results highlight the value of coherent policy frameworks, stable incentives, and institutional support mechanisms that reduce uncertainty and enable organizational leaders to commit to circular investments.

9 Limitations

Despite providing value-added information on this topic, this study has several limitations. The first limitation is that the study was limited to selected databases and peer-reviewed journals; thus, it could have missed important gray literature and practitioner-oriented studies, in addition to other sources of information. Another limitation of this study was that there were different ways to define leadership constructs and CE outcomes, which made it difficult to directly compare and synthesize the studies included. In addition, many of the studies reviewed in this literature review were from specific geographic regions and industries. This may create an unintended bias toward studies conducted in areas with more developed sustainability infrastructure. Therefore, these limitations should be considered when evaluating the findings of this study and generalizing the results to other populations.

10 Directions for Future Studies

Future research on CE and leadership will benefit from longitudinal and multilevel research designs to investigate how leadership impacts CE adoption over time and across organizational levels. Future research should also focus on geographic regions, SMEs, and industry-specific contexts where institutional conditions for circularity differ from those in more developed countries. Future research can also build upon existing theories related to CE and leadership, including but not limited to dynamic capabilities, institutional theory, and organizational learning, to provide a greater understanding of how leadership can facilitate long-term circular transformation. Additionally, empirical studies investigating the interactions among leadership practices, policy environments, and technological innovations in the CE will provide additional cumulative knowledge in the fields of CE and leadership.

References

Abu-Bakar H, Charnley F (2024) Developing a strategic methodology for circular economy roadmapping: a theoretical framework. Sustainability 16(15):6682

Ahmed S, Ahsan N, Alam SS, Kokash HA (2025) Advancing circular economy implementation in malaysian SMEs: the role of financial resources, operational alignment, and absorptive capacity. Circular Econom Sustain 1–29.

Akinwalere S, Chang K, Barbhuiya S (2025) Sustainability-embedded leadership for successful change management. Sustainability 17(17):7973
Al Marshoudi FB, Jamaluddin Z, Al Balushi FI (2023) Transformational leadership mediates the relationship between employee engagement and performance in production media. Int J Manage 1(2)
Alhosani K (2025) Leveraging public private partnerships for a circular industry economy: advancing economic sustainability in industrial waste management in the Emirate of Ajman, UAE. Challenges 16(3):31
Ali Esmaeili Z, Sobhanifard Y (2025) Leadership style model for commercial supply chain transformation based on capabilities 5.0: sustainability and circular economy. Circular Economy Sustain 1–41
Alkhodary D (2023) Integrating sustainability into strategic management: a path toward long-term business success. Int J Prof Bus Rev 8(4):39
Alqatan A, Simmou W, Shehadeh M, AlReshaid F, Elmarzouky M, Shohaieb D (2025) Strategic pathways to corporate sustainability: the roles of transformational leadership, knowledge sharing, and innovation. Sustainability 17(12):5547
Arora R, Mutz D, Mohanraj P (Eds.). (2023) Innovating for the circular economy: Driving sustainable transformation. CRC Press
Asgari A, Asgari R (2021) How circular economy transforms business models in a transition toward circular ecosystem: the barriers and incentives. Sustain Prod Consump 28:566–579
Bag S, Pretorius JHC (2022) Relationships between Industry 4.0, sustainable manufacturing, and circular economy: proposal of a research framework. Int J Organizat Anal 30(4):864–898
Bashynska I, Malynovska Y, Kolinko N, Bielialov T, Järvis M, Kovalska K, Saiensus M (2024) Performance assessment of sustainable leadership of enterprise's circular economy-driven innovative activities. Sustainability 16(2):558
Bertassini AC, Ometto AR, Severengiz S, Gerolamo MC (2021) Circular economy and sustainability: the role of organizational behavior in the transition journey. Bus Strateg Environ 30(7):3160–3193
Birasnav M (2014) Relationship between transformational leadership behaviors and manufacturing strategy. Int J Organ Anal 22(2):205–223
Boeske J (2023) Leadership toward sustainability: a review of sustainable, sustainability, and environmental leadership. Sustainability 15(16):12626
Boikanyo DH (2025) Adaptive leadership in times of organizational change driven by. Contemp Perspect Organiz Behav 37
Bourdin S, Jacquet N (2025) Closing the loop at the local scale: Investigating the drivers of and barriers to the implementation of the circular economy in cities and regions. Ecol Econ 231:108542
Brändström J, Eriksson O (2022) How circular is a value chain? Proposing a material efficiency metric to evaluate business models. J Clean Prod 342:130973
Büchter RB, Weise A, Pieper D (2020) Development, testing and use of data extraction forms in systematic reviews: a review of methodological guidance. BMC Med Res Methodol 20(1):259
Budur T (2020) Effectiveness of transformational leadership among different cultures. Int J Soc Sci Educ Stud 7(3):119–129
Burke H, Zhang A, Wang JX (2023) Integrating product design and supply chain management for a circular economy. Prod Plan Control 34(11):1097–1113
Castro-Lopez A, Iglesias V, Santos-Vijande ML (2023) Organizational capabilities and institutional pressures in the adoption of circular economy. J Bus Res 161:113823
Cheffi W, Zahir-ul-Hassan MK, Farooq MO, Baqrain A, Mansour MMH (2023) Ethical leadership, management control systems and circular economy in SMEs in an emerging economy, the UAE. J Bus Res 156:113513
Ćwiklicki M, Baryła M, Mirzyńska A, Żabiński M (2024) Circular economy adoption at the regional level: a neo-institutional perspective. Eur Plan Stud 32(10):2258–2278

de Oliveira Vieira B, Guarnieri P (2025) Toward the circular economy from the perspective of new institutional theory: a systematic analysis of the literature with InOrdinatio protocol. Manage Res J Iberoamerican Acad Manage 23(1):65–89
de Jesus A, Lammi M, Domenech T, Vanhuyse F, Mendonça S (2021) Eco-innovation diversity in a circular economy: Toward circular innovation studies. Sustainability 13(19):10974
De La Cuesta-González M, Morales-García M (2022) Does finance as usual work for circular economy transition? A financiers and SMEs qualitative approach. J Environ Planning Manage 65(13):2468–2489
De Mattos CA, De Albuquerque TLM (2018) Enabling factors and strategies for the transition toward a circular economy (CE). Sustainability 10(12):4628
Dewick P, Bengtsson M, Cohen MJ, Sarkis J, Schröder P (2020) Circular economy finance: clear winner or risky proposition? J Ind Ecol 24(6):1192–1200
Dominguez-Escrig E, Mallen-Broch FF (2023) Leadership for sustainability: fostering organizational learning to achieve radical innovations. Eur J Innov Manag 26(2):309–330
Eisenreich A, Füller J, Stuchtey M (2021) Open circular innovation: how companies can develop circular innovations in collaboration with stakeholders. Sustainability 13(23):13456
Eisenreich A, Füller J, Stuchtey M, Gimenez-Jimenez D (2022) Toward a circular value chain: impact of the circular economy on a company's value chain processes. J Clean Prod 378:134375
Fatima R, Fatima H (2022) Adaptive leadership: navigating change and uncertainty in today's organizations. J Leadership Admin Develop 1(1):20–28
Fobbe L, Hilletofth P (2023) Moving toward a circular economy in manufacturing organizations: the role of circular stakeholder engagement practices. Int J Logistics Manage 34(3):674–698
Galpin T, Lee Whittington J (2012) Sustainability leadership: from strategy to results. J Bus Strateg 33(4):40–48
Geissdoerfer M, Savaget P, Bocken NM, Hultink EJ (2017) The circular economy—a new sustainability paradigm? J Clean Prod 143:757–768
Grafström J, Aasma S (2021) Breaking circular economy barriers. J Clean Prod 292:126002
Gupta P (2025) Transformational leadership: inspiring change and innovation. Int J Sci Res (IJSR) 14(2):504–509
Han Y, Shevchenko T, Yannou B, Ranjbari M, Shams Esfandabadi Z, Saidani M, Li G (2023) Exploring how digital technologies enable a circular economy of products. Sustainability 15(3):2067
Heifetz RA, Linsky M (2002) A survival guide for leaders. Harv Bus Rev 80(6):65–74
Hofstetter JS, De Marchi V, Sarkis J, Govindan K, Klassen R, Ometto AR, Vazquez-Brust D (2021) From sustainable global value chains to circular economy—different silos, different perspectives, but many opportunities to build bridges. Circular Econ Sustain 1(1):21–47
Hopkinson P, De Angelis R, Zils M (2020) Systemic building blocks for creating and capturing value from circular economy. Res Conserv Recycl 155:104672
Huang SY, Ting CW, Li MW (2021) The effects of green transformational leadership on adoption of environmentally proactive strategies: the mediating role of green engagement. Sustainability 13(6):3366
Iacovidou E, Hahladakis JN, Purnell P (2021) A systems thinking approach to understanding the challenges of achieving the circular economy. Environ Sci Pollut Res 28(19):24785–24806
Juicharoen N, Na-Nan K, Inmor S, Phanniphong K, Wang X (2025) Driving sustainability in emerging economies: Leadership, culture, and knowledge management in environmental performance. Int J Informat Manage Data Insights 5(2):100381
Kandpal V, Jaswal A, Santibanez Gonzalez ED, Agarwal N (2024). Circular economy principles: shifting toward sustainable prosperity. In Sustainable energy transition: circular economy and sustainable financing for environmental, social and governance (ESG) practices. Springer Nature Switzerland, Cham, pp 125–165
Khan AN, Soomro MA (2025) Empowering change: leadership, flexibility, and the path to circular economy success. J Organizat Change Manage

Kirchherr J, Reike D, Hekkert M (2017) Conceptualizing the circular economy: an analysis of 114 definitions. Res Conserv Recycl 127:221–232
Kitchenham B, Charters S (2007) Guidelines for performing systematic literature reviews in software engineering
Klein N, Ramos TB, Deutz P (2020) Circular economy practices and strategies in public sector organizations: an integrative review. Sustainability 12(10):4181. https://doi.org/10.3390/su12104181
Köhler J, Sönnichsen SD, Beske-Jansen P (2022) Toward a collaboration framework for circular economy: The role of dynamic capabilities and open innovation. Bus Strateg Environ 31(6):2700–2713
Kokhuis K, Van den Berghe K, Straub A (2025) Implementing circular practices in a construction clients' organization. Strategic interventions on intraorganisational barriers for operationalizing TU Delft's circular campus ambitions. J Corporate Real Estate 27(4):281–301
Korhonen J (2020) The circular economy as a complex adaptive system. Handbook of the Circular Economy, pp 28–38
Kristensen HS, Mosgaard MA, Remmen A (2021) Integrating circular principles in environmental management systems. J Clean Prod 286:125485
Kurucz EC, Colbert BA, Lüdeke-Freund F, Upward A, Willard B (2017) Relational leadership for strategic sustainability: practices and capabilities to advance the design and assessment of sustainable business models. J Clean Prod 140:189–204
Kwarteng A, Simpson SNY, Agyenim-Boateng C (2022) The effects of circular economy initiative implementation on business performance: the moderating role of organizational culture. Soc Respons J 18(7):1311–1341
Lieder M, Rashid A (2016) Towards circular economy implementation: a comprehensive review in context of manufacturing industry. J Clean Prod 115:36–51
Malek MA, Almarri K (2023) The role of adaptive and transformational leadership in the successful adoption and implementation of new technologies and innovations in organizations. In European, Mediterranean, and Middle Eastern Conference on Information Systems. Springer Nature Switzerland, Cham, pp 120–132
Marra A, Mazzocchitti M, Sarra A (2018) Knowledge sharing and scientific cooperation in the design of research-based policies: the case of the circular economy. J Clean Prod 194:800–812
Mendoza JMF, Sharmina M, Gallego-Schmid A, Heyes G, Azapagic A (2017) Integrating backcasting and eco-design for the circular economy: the BECE framework. J Ind Ecol 21(3):526–544
Metcalf M, Hinske C (2022) 10 Circular economy leadership. The Impossibilities of the Circular Economy, 103
Miao Q, Nduneseokwu C (2025) Applications of leadership theories and concepts in environmental leadership. In Environmental leadership in a VUCA era: An interdisciplinary handbook. Springer Nature Singapore, Singapore, pp. 237–275
Modha B (2021) Collaborative leadership with a focus on stakeholder identification and engagement and ethical leadership: a dental perspective. Br Dent J 231(6):355–359
Morseletto P (2020) Restorative and regenerative: Exploring the concepts in the circular economy. J Ind Ecol 24(4):763–773
Mubarik MS, Kontoleon A, Shahbaz M (2024) Beyond the hurdles: Exploring policy obstacles in the path to circular economy adoption. J Environ Manage 370:122667
Nygaard A (2022) From linear to circular economy: a transaction cost approach to the ecological transformation of the firm. Circular Econ Sustain 2(3):1127–1142
Odiase B, Agbadudu J (2024) Reengineering leadership for sustainable strategy implementation. J Acad Res Econom 16(3)
Oktrivina A, Sumail LO, Hendryadi H, Effendi S, Budiantoro H, Astuti SB (2025) Green transformational leadership and financial performance in micro and small enterprises: the interplay of circular economy practices and environmental uncertainty. Front Sustain 6:1684185

Osei A, Agyemang AO, Cobbinah J (2025) Strategic sustainability initiatives and the circular economy: insights from firm-level targets, board dynamics, stakeholder pressure, and digital transformation. Bus Ethics Environ Respons

Pakuła N, Łapniewska Z, Dutra CJ (2025) Comparative measurements of circular economy performance among European countries: reviewing approaches and limitations. J Environ Manage 375:124414

Palombi G, Schiaroli V, Fraccascia L, Nonino F (2024) Identifying barriers and good practices for implementing circular economy principles in small and medium enterprises. Bus Strateg Environ 33(7):6773–6794

Popay J, Roberts H, Sowden A, Petticrew M, Arai L, Rodgers M, Duffy S (2006) Guidance on the conduct of narrative synthesis in systematic reviews. ESRC Meth Programm Vers 1(1):b92

Rehman MA, Alqahtani DA (2025) The moderating role of organizational culture on barriers and drivers of sustainable construction practices in Saudi Arabia's construction industry: a circular economy perspective. Buildings 15(20):3663

Schulz C, Hjaltadóttir RE, Hild P (2019) Practising circles: Studying institutional change and circular economy practices. J Clean Prod 237:117749

Scipioni S, Russ M, Niccolini F (2021) From barriers to enablers: the role of organizational learning in transitioning SMEs into the circular economy. Sustainability 13(3):1021

Sharma T, Singh A, Sutrave Y, Azhar M (2025) Circular supply chains: digital tools for enhanced traceability. In: Innovating Sustainability Through Digital Circular Economy. IGI Global Scientific Publishing, pp. 175–202

Snyder H (2019) Literature review as a research methodology: an overview and guidelines. J Bus Res 104:333–339

Sott MK, Bender MS (2025) The role of adaptive leadership in times of crisis: a systematic review and conceptual framework. Merits 5(1):2

Suriyankietkaew S, Krittayaruangroj K, Iamsawan N (2022) Sustainable Leadership practices and competencies of SMEs for sustainability and resilience: a community-based social enterprise study. Sustainability 14(10):5762

Tan J, Tan FJ, Ramakrishna S (2022) Transitioning to a circular economy: a systematic review of its drivers and barriers. Sustainability 14(3):1757

Tanveer U, Hoang TG, Ishaq S, Khalid RU (2025) Public-private partnerships as catalysts for digital transformation and circular economy: Insights from developing countries. Technol Forecast Soc Chang 219:124270

Teixeira N (2025) Circular economy perspectives: challenges, innovations, and sustainable futures. Discover Sustain 6(1):738

Thomas J, Harden A (2008) Methods for the thematic synthesis of qualitative research in systematic reviews. BMC Med Res Methodol 8(1):45

Tranfield D, Denyer D, Smart P (2003) Towards a methodology for developing evidence-informed management knowledge by means of systematic review. Br J Manag 14(3):207–222

Trevisan AH, Lobo A, Guzzo D, de Vasconcelos Gomes LA, Mascarenhas J (2023) Barriers to employing digital technologies for a circular economy: a multilevel perspective. J Environ Manage 332:117437

Turskis Z, Šniokienė V (2024) IoT-driven transformation of circular economy efficiency: an overview. Math Comput Appl 29(4):49

Zils M, Howard M, Hopkinson P (2025) Circular economy implementation in operations & supply chain management: building a pathway to business transformation. Prod Plan Control 36(4):501–520

Blockchain-Driven Blended Finance for Climate Investment in Developing Economies

Justine Mbudaya, Kundayi Chinyongo, David Chikwere, and Munyaradzi Chibaro

Abstract Emerging market and developing economies (EMDEs) face a persistent financing gap in green infrastructure, with annual climate investment needs exceeding $2.4 trillion far beyond what current capital flows can support. Two structural obstacles currency risk and governance deficit always makes viable climate projects unbankable in traditional blended finance approaches. This chapter discusses how Distributed Ledger Technology (DLT) and smart contracts can remove the two barriers at the same time: automation of foreign exchange (FX) hedging and immutable, real-time Monitoring, Reporting and Verification (MRV). Based on a systematic literature review, comparative case study analysis and a proposed conceptual model of Blockchain-Enabled Blended Finance (BEBF), the chapter shows that pilots in Southeast Asia, Sub-Saharan Africa and Latin America have realized a 30 percent decrease in the cost of issuing and a 50 percent decrease in the settlement time relative to traditional instruments. Three-layer BEBF architecture, consisting of concessional capital, blockchain infrastructure and blockchain-based institutional investors is suggested along with specific policy proposals to the governments of EMDEs. The indications are that blockchain-based financial models can significantly reduce the cost of capital and speed up the process of moving private finance into climate infrastructure across the developing world.

J. Mbudaya (✉)
BA ISAGO University, Gaborone, Botswana
e-mail: justine.mbudaya@baisago.ac.bw

K. Chinyongo
Midlands State University, Gweru, Zimbabwe
e-mail: chinyongok@staff.msu.ac.zw

D. Chikwere
Leeds Business School, Leeds Beckett University, Leeds, UK
e-mail: D.chikwere@leedsbeckett.ac.uk

M. Chibaro
University of Botswana, Gaborone, Botswana
e-mail: chibarom@ub.ac.bw

J. Kanyepe and I. Chirisa (eds.), *Advancing the Green Agenda in the Transport and Logistics Sector in Global South Countries*, World Sustainability Series,
https://doi.org/10.1007/978-3-032-31374-4_25

Keywords Blockchain · Climate finance · Blended finance · Currency risk · Distributed ledger technology · Emerging markets · Green bonds · Smart contracts · Governance · Tokenisation

1 Introduction

The global shift towards a low-carbon economy requires mobilisation of capital, as never before, especially in emerging marketplaces and in developing economies (EMDEs). Even though scientific consensus on whether climate action is urgent is beyond doubt, a persistent financing gap persists in blocking the implementation of green infrastructure projects at scale. According to the most recent estimates, to comply with the goals of the Paris Agreement in developing countries, it would be necessary to invest more than 2.4 trillion a year by 2030 (Climate Policy Initiative, 2024). Current inflows of the private capital into these areas are well below that mark (Fig. 1).

The difference between need and reality indicates the perceived and real risk profile of climate projects in EMDEs. These projects are often considered as high-risk by institutional investors such as pension funds and insurance companies and are cited as such due to political instability and regulatory uncertainty and, above all, financial volatility. The most common solution to these issues has become blended

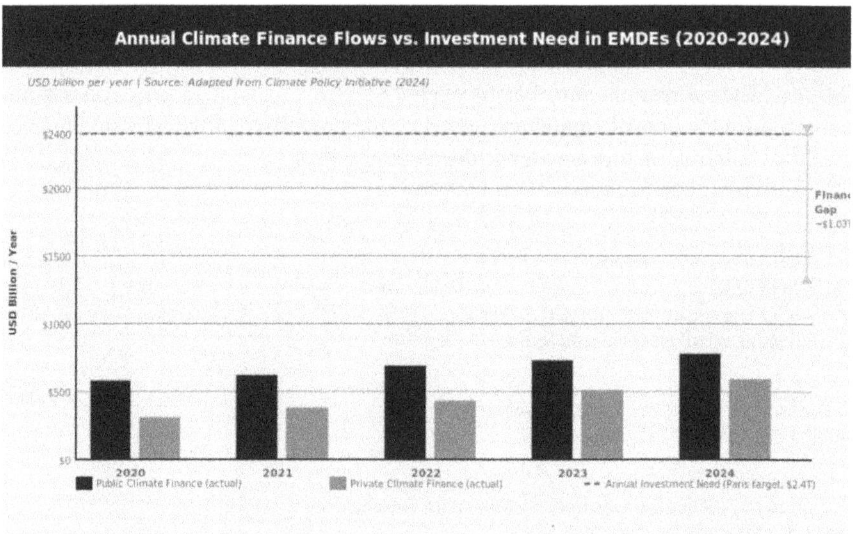

Fig. 1 Annual climate finance flows versus investment need in EMDEs (2020–2024). *Source* Adapted from Climate Policy Initiative (2024). Actual public and private flows are authors' estimates based on CPI annual tracking data. The $2.4 T annual need represents the Paris Agreement target for EMDEs

finance, which uses the capital of the public or philanthropic to strategically provide capital to de-risk private investment (OECD 2023). But even in blended structures, two systemic impediments continue to choke capital flows: currency risk and lack of governance.

Currency risk occurs due to the fact that most climate projects receive revenues in local currencies and their capital and debt are denominated in hard currencies such as the US Dollar or Euro (Climate Policy Initiative 2024). In the case of depreciation of local currencies, the expenses of delivering debt services may increase significantly, which can lead to the default of projects. Conventional hedging instruments like cross currency swaps are either prohibitively expensive or simply unavailable in many EMDEs (Convergence Blended Finance 2024).

The lack of governance gives rise to just as severe a lack of trust. To ensure the financial resources are deployed in the name they purport and to guarantee the environmental outcomes are properly documented, institutional investors must be subjected to rigorous Monitoring, Reporting and Verification (MRV) (World Bank 2023). Transaction costs are inflated by administrative overhead and the risk of funds diversion in many developing contexts, which also exposes the investors to greenwashing liability (Mzoughi 2024).

A combination of these two barriers creates a risk premium that makes many otherwise bankable climate projects unbankable. Blended finance shares some of this risk with concessional capital and guarantees, but the mechanisms used to manage these structures are normally slow, manual and error-prone (IFC 2025). In this chapter, the author argues that the technological infrastructure to automate not only trust but also risk management in climate finance is provided by blockchain technology specifically Distributed Ledger Technology (DLT) and smart contracts.

2 Background to the Study

Blended finance has been a major international development topic since the 2015 Addis Ababa Action Agenda, which called for $100 billion in public finance to be leveraged by private financing to address climate change. It provides a framework in the climate space to connect the financing requirements of green infrastructure projects in developing countries with the risk appetite of private investors (OECD 2023). But although it has been in use for over a decade, blended finance has not yet reached the size that its advocates envisioned.

The primary obstacle is the fact that there are two structural hurdles which have proven themselves to be very difficult to overcome with traditional solutions. The first is currency risk. Infrastructure projects in EMDEs are long-term, multi-million dollar and local currency. If capital is brought in to a project by international lenders in US Dollars or Euros, the depreciation of the local currency immediately raises the real cost of debt service and sometimes leads to project insolvency (Convergence Blended Finance 2024). Cross currency swaps that have traditionally been used to manage this risk come at a high cost, are short-dated and are not widely available in frontier markets.

The second barrier is the governance deficit. The institutional apparatus for regulating intricate financial arrangements is weak in many developing countries. The lack of transparency on fund use, poor impact reporting and a broader reputation issue with a lack of trust in how climate finance is being used have caused credibility issues to stop the private sector from investing even where financial returns might seem sufficient (Schulz and Feist 2021). Standard monitoring and verification systems are post hoc and periodic, thus leaving extended gaps of opportunity for errors, manipulation and diversion to remain unidentified.

It is within this context that the DLT and smart contracts have gained increased academic and practical interest. The promise of blockchain is simple enough: It can solve both structural barriers by automating certain financial transactions and by writing all transactions down on a non-alterable and transparent ledger. Currency hedging can be automated using smart contracts with triggers and governance can be enforced by code rather than just by relying on institutional oversight (Malamas et al. 2024). This chapter looks at the evidence for and the extent of that promise.

3 Research Objectives/Hypothesis

This chapter is guided by three research objectives and a central hypothesis.

4 Research Objectives

The first goal is to explore how DLT-based stablecoins and automated FX trigger using smart contracts can help lower the cost and complexity of foreign exchange (FX) management in emerging market climate projects. The second goal is to evaluate the role and effectiveness of Proof of Green protocols on blockchain and the immutable ledger in delivering real-time transparency and minimizing MRV administrative burden. The third goal is to create the Blockchain-Enabled Blended Finance (BEBF) conceptual model, which visualises the interplay between public capital, private finance and decentralised infrastructure within a coherent digital-first model.

5 Research Hypothesis

The central hypothesis tested is: blockchain-integrated financial models can significantly outperform traditional blended finance structures in terms of transaction speed, cost-efficiency and investor confidence, thereby unlocking greater flows of private capital into green infrastructure in developing economies (Zaher and Roy 2025).

6 Literature Review

The following review synthesises recent scholarship (2021–2026) across four pillars: currency risk in EMDEs, blockchain applications in finance, governance challenges in climate finance and smart contract-based compliance mechanisms.

7 Currency Risk in Emerging Markets

Currency volatility is always cited as the major obstacle to global climate investment in EMDEs. It is estimated by the Climate Policy Initiative (2024) that the FX risk premium can increase the cost of capital in some African and Latin American markets by up to 10%. The standard hedging tool, Cross-Currency Swaps (CCS), are often illiquid with a maturity longer than five years precisely the tenure of the green infrastructure debt (Convergence Blended Finance, 2024). The OECD (2023) states that blended finance structures need to develop beyond being a simple guarantee in order to survive in a very volatile currency environment.

8 Blockchain and DLT Applications in Finance

The years 2024 to 2026 have been characterised by a shift of a speculative cryptocurrency to Regulated DeFi and tokenisation of Real-World Assets (RWA). The IMF (2025) shows that tokenisation has the potential to enhance liquidity in green bond markets and to lower the transaction costs associated with such markets. Li and Chen (2025) also demonstrate that blockchain payment systems have the potential to compress cross-border settlement times that would have taken several days to just a few seconds a feature that would directly apply to the FX risk management process as it would remove the time interval over which currency exposure can be realised.

9 Governance Challenges in Climate Finance

Weaknesses in governance, especially in the institutions, such as the Green Climate Fund, have seen repeated demands to be more transparent (Schulz and Feist 2021). The structural issue of greenwashing where money is misallocated or effects exaggerated is a structural problem within the sector. According to Mzoughi (2024), blockchain can provide a verifiable source of truth accessible to all project stakeholders because of a single, immutable ledger. This decentralised governance structure minimises reliance on manual audits, which are often slow and prone to manipulation.

9.1 Greenwashing Risk and Immutable MRV

The traditional MRV systems are retrospective in nature: audits are usually conducted months or even years after the disbursement. This delay creates a trust gap which poses actual financial and reputational risk to institutional investors. Schulz and Feist (2021) suggest that the immutable ledger of blockchain allows a continuous and real-time MRV model to be used, where all the transactions and impact metrics are recorded at the point when they occur. This change to perpetual auditing is a major governance enhancement of attracting a large-scale institutional capital to emerging markets (World Bank 2023).

9.2 Decentralised Autonomous Organisations (DAOs) and Climate Fund Management

A more structural use of blockchain is in the form of Decentralised Autonomous Organisations (DAOs) to govern climate funds (Zaher and Roy 2025). A DAO is a code-based organization that has no central administration. A Climate Fund DAO would facilitate decision-making by a group of donors and investors in a climate finance context, with each participant holding voting rights proportionate to their contribution. This distributed governance model has the potential to significantly enhance the level of transparency and accountability of international climate funds (Schulz and Feist 2021).

9.3 The 'Proof of Green' Protocol

One of the most technically challenging tasks in climate finance is to verify environmental impact reliably (Malamas et al. 2024). A Proof of Green protocol applies the cryptographic methods to ensure that a project has met the stated environmental goals. An example is a solar energy project, in which the inverter at the site cryptographically signs and stores data on-chain. This automated validation will remove the cost of hiring third-party auditors, which are quite expensive and will provide investors with instant, tamper-proof verification of their capital being used to generate real environmental impacts (Naderi 2022).

9.4 Double-Counting in Carbon Markets

Double-counting the practice of assigning the same carbon reduction to multiple claimants undermines the integrity of international carbon markets (Allahgholi

2026). Blockchain's tokenisation capability offers a structural remedy. By issuing a unique Non-Fungible Token (NFT) for each verified tonne of carbon removed, the ledger ensures that no credit can be used more than once (Mzoughi 2024). This traceability is fundamental to both the credibility of the global carbon market and the viability of blended finance models that rely on carbon credits as a revenue stream (UNEP 2025).

10 Smart Contracts for Automated Compliance

Smart contracts serve as the operational core of the proposed framework. Malamas et al. (2024) describe smart contracts as self-executing code that automates if–then financial logic. Applied to climate finance, this enables automated compliance fund disbursement is triggered directly by verified environmental outcomes, such as carbon sequestration data fed from IoT sensors. Schulz and Feist (2021) contend that this automation materially reduces administrative overhead and increases the pace at which capital reaches projects.

10.1 If–Then Logic Applied to Fund Disbursement

In practice, a smart contract can be programmed to release the next tranche of project funding automatically once a defined milestone is verified. If a solar farm's output exceeds a specified threshold, for example, the contract executes the payment without any manual intervention. This mechanism reduces the administrative burden on fund managers and gives project developers reliable, timely access to capital (Schulz and Feist 2021).

10.2 Oracles and Data Integrity in MRV

Smart contracts depend on external data feeds known as Oracles to trigger execution (Malamas et al. 2024). In climate finance, an Oracle supplies real-time environmental data, such as energy generation figures or verified carbon measurements. Malamas et al. (2024) advocate for decentralised Oracle networks in which multiple independent data providers cross-validate inputs before they are fed to the contract, strengthening the integrity of the underlying MRV system (Schulz and Feist 2021).

10.3 Automated FX Hedging Triggers

One of the more innovative applications of smart contracts is the automated FX hedging trigger. A contract can be configured to execute a pre-funded currency hedge automatically if a local currency depreciates beyond a defined threshold. If the Brazilian Real falls more than 5% against the US Dollar, for instance, the contract activates the hedge without any human instruction, protecting the investor's return in real time (Climate Policy Initiative 2024).

11 Programmable Green Bonds

Tokenised green bonds are programmable instruments that can automate both coupon payments and environmental impact reporting (Malamas et al. 2024). A tokenised bond can be structured to pay investors a performance-based 'green coupon' when the underlying project exceeds its energy generation target (Belkhiria and Abid 2026). This embedded incentive structure makes green infrastructure more attractive to institutional capital, as it directly aligns investor returns with verified environmental performance (Zaher and Roy 2025).

12 Decentralisation of the Climate Governance

The move towards a decentralised, rather than a centralised governance of climate finance is underpinned by Zaher and Roy (2025) theoretical frameworks that state the potential of Decentralised Finance (DeFi) as a more resilient and sovereign-neutral form than traditional banking. Specifically for LDCs, the institutional capacity to provide oversight in complex financial monitoring is limited. The model frees up fragile local institutions by embedding rules of fund management directly into the blockchain (Schulz and Feist 2021). Algorithmic trust replaces human intermediaries who frequently extract a lot out of international development transactions (IMF 2025).

13 Tokenised Carbon Credits for Blended Finance Collateral

Tokenisation moves beyond financial instruments into the environmental assets themselves. Allahgholi (2026) investigates the role of tokenised carbon credits as collateral within blended finance structures. A verified carbon reduction, which is represented by a liquid, tradable digital token, provides project developers the ability to access working capital that they otherwise would not have. The cross-asset synergy is becoming a defining characteristic of next-gen climate investment systems (Mzoughi 2024).

14 ESG Integration and MRV in the Real-Time Process

Environmental, Social and Governance (ESG) criteria have become a data-driven discipline rather than simply a compliance exercise. Malamas et al. (2024) add that blockchain-based MRV brings the level of granularity of real-time data to which institutional investors now demand as the default. Instead of looking at annual sustainability reporting, investors have access to an existing ESG dashboard, on which every unit of renewable energy generated is recorded on the immutable ledger. This degree of verification is crucial to the governance protections that draw large capital across emerging markets (World Bank 2023).

15 Regulatory Sandboxes in EMDEs

Regulatory uncertainty remains a major hurdle for blockchain uptake. However, IMF (2025) report notes the mounting trend of Regulatory Sandboxes in Africa and Southeast Asia. In these controlled conditions, fintech companies have the opportunity to experiment with the blockchain model for hedging and disbursement under central bank supervision. Yilmaz (2026) maintains that sandboxes are important in trying to balance innovation and financial stability, especially in unstable exchange rate regimes in economies.

16 Digital Inclusion and the Threat of Technology Disparity

The socio-economic effects of blockchain-based climate finance warrant careful consideration (Li and Chen 2025). Although the technology promises real equity to capital, it can also lead at the same time to a digital divide which essentially leaves infrastructure projects in LDCs with less advanced digital infrastructure out of world markets. Research on low-bandwidth blockchain solutions and digital inclusion strategies must be targeted to ensure that the benefits of this transition are broadly shared (IMF 2025).

17 Systematic Literature Review

A systematic literature review was performed using the prominent academic databases (Scopus, Web of Science and Springer Link). The review was limited to peer-reviewed journal articles, institutional reports from the World Bank, IMF, OECD and equivalent bodies and working papers published between 2021 and 2026. Inclusion criteria aimed to include empirical and theoretical research at the crossroads between blockchain technology, climate finance, currency risk and governance in EMDEs. Excluding older material helps to ensure that the analysis reflects the most current state of evidence in a rapidly evolving field.

18 Comparative Case Study Analysis

To embed the theoretical model within real-world evidence, a comparative case study was compiled in four settings: Project Genesis 2.0 (HSBC/HKMA) in Hong Kong; the BIS Innovation Hub green bond pilot in Southeast Asia; tokenised carbon credit platforms in Sub-Saharan Africa; and Central Bank Digital Currency (CBDC) pilots in Latin America. For each case, the applicability to research aims was ensured and the geographical diversity was reflected the decision of studying the case (BIS 2023; HSBC 2024). Combined, these examples reveal both the practical advantages and practical difficulties of applying DLT to climate finance in various organisational and regulatory contexts.

19 Qualitative Content Analysis

Data drawn from the literature review and case studies were subjected to qualitative content analysis to identify recurring themes, enabling factors and barriers to blockchain adoption in blended finance. Thematic coding was applied iteratively across the material, with particular attention to the interface between technological capability and institutional readiness. The resulting insights inform both the conceptual model and the policy recommendations.

20 Conceptual Framework: The BEBF Model

The chapter also presents a conceptual model of the Blockchain-Enabled Blended Finance (BEBF) that organizes the proposed digital-first architecture into three integrated layers. The first is the Concessional Layer in which public and philanthropic capital is held in smart-contract-based escrow accounts and released only upon verified milestone achievement. The second layer is the Blockchain Infrastructure Layer the Trust Engine which stores all transactions using a distributed ledger and sets smart contracts to automate FX hedging triggers as well as impact verification using IoT and satellite data (Malamas et al. 2024). The third layer, the Investment Layer, has institutional investors provide project financing, often in the form of tokenised green bonds that offer superior liquidity and fractional entry points compared to traditional infrastructure debt (Belkhiria and Abid 2026). The DLT architecture is shown in Fig. 2 and the full BEBF framework is presented in Fig. 3.

Table 1 sets out the key differences between the traditional blended finance model and the blockchain-enabled approach.

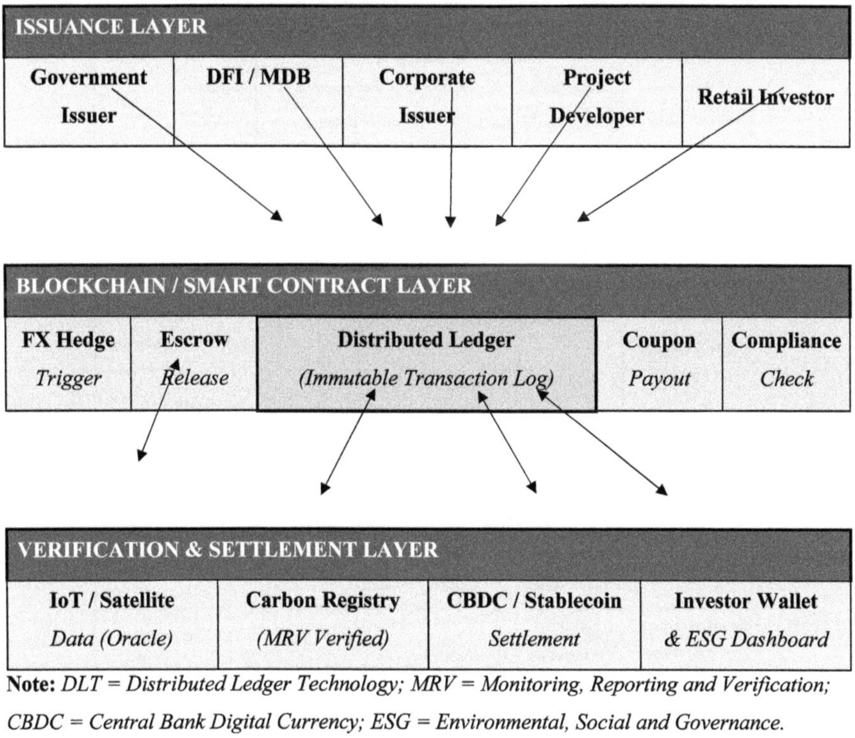

Fig. 2 Blockchain DLT Architecture for Green Bond Issuance and MRV. *Source* Adapted from Malamas et al. (2024) and Schulz and Feist (2021). DLT, Distributed Ledger Technology; MRV, Monitoring, Reporting and Verification; CBDC, Central Bank Digital Currency; Oracle, External data feed provider

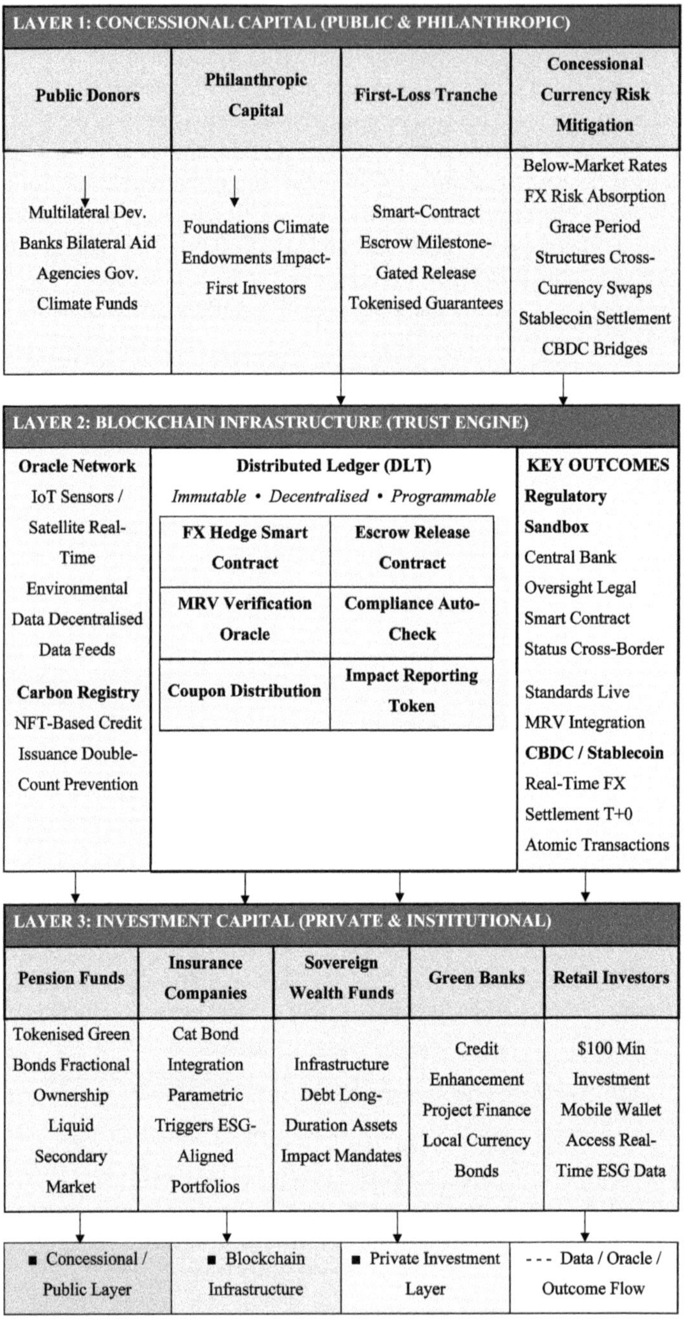

Fig. 3 Blockchain-Enabled Blended Finance (BEBF) Conceptual Model. *Source* Authors' own construction. The BEBF model integrates three layers Concessional Capital, Blockchain Infrastructure and Private Investment connected through smart contracts oracle data feeds and tokenised financial instruments. Arrows indicate capital and data flows; bidirectional arrows indicate feedback loops

Table 1 Traditional versus blockchain-enabled blended finance

Component	Traditional model	Blockchain-enabled model
Risk mitigation	Manual, expensive FX swaps	Automated, smart-contract-based hedging
Governance	Periodic, manual audits (MRV)	Real-time, immutable, automated MRV
Fund disbursement	Slow, multi-step bank transfers	Instant, milestone-based smart contracts
Liquidity	Illiquid, long-term assets	Tokenised assets with secondary market potential

21 Findings and Discussion

These findings are summarised in three thematic outcome areas FX settlement and hedging, fund disbursement accountability and asset tokenisation, four empirical case studies and the closing discussion.

22 Real Time FX Settlement and Automated Hedging

One of the foundational findings of this research is that DLT enables atomic settlement, where assets are exchanged simultaneously and all-or-nothing, thus eliminating counterparty risk. This eliminates the traditional T + 2 settlement window, where FX exposure accumulates over time, in cross-border climate finance (BIS 2023). Stablecoins pegged to the US Dollar that allow investors to lock in exchange rates at the moment of transaction eliminate downside risk that comes with the depreciation of currency at the time of trade execution versus settlement (Climate Policy Initiative 2024). Aside from settlement, smart contracts offer pre-emptive hedging capabilities. If a local currency depreciates past a given threshold, the contract takes action on a pre-funded hedge automatically without manual management. The Brazilian Real case study mentioned in Sect. 2.4.3 demonstrates a means by which this kind of mechanism can be tailored to manage investor yields in unstable environments in the case of volatile financial markets. Decentralised Exchanges (DEXs) also contribute to this ecosystem by providing a pool and market mechanism for a liquid secondary exchange market for tokenised climate assets and they help to reduce the liquidity risk that institutional investors face, which has been a traditional disincentive for long-term green infrastructure-backed institutional debt (Belkhiria and Abid 2026; Zaher and Roy 2025).

23 Smart Contracts and Accountability on Fund Disbursement

Integrating smart-contract-based escrow to fund disbursement helps eliminate one of the strongest governance gaps in the blended finance industry. Fund release under traditional schemes is a manual multi-step process that can take months and is not immune from leakage at multiple points. In smart contracts, concessional capital is tied in a decentralised escrow structure and automatically disbursed upon the attainment of specific, verifiable milestones (Schulz and Feist 2021). Every transaction is posted on an immutable ledger accessed by all authorised stakeholders rather than undergoing a retrospective audit and instead recorded for accountability in near-real-time (Mzoughi 2024). Performance-based interest payments represent a last aspect of governance improvement. A tokenised green bond system could also issue coupon payments linked to determined project output such that if a wind farm exceeds its target amount of generation, bondholders receive the final payment instantly without waiting for administration (Belkhiria and Abid 2026). Such programmable accountability directly addresses the trust deficit, which has been a barrier to accessing institutional capital in emerging market climate projects for decades (World Bank 2023).

24 Tokenise Green Bonds and Climate Assets

Tokenisation radically changes the profile of access to climate investment: It moves ownership into an ownership on a fractional basis. The Project Genesis 2.0 pilot (HSBC 2024) had tokenised green bonds in denominations as low as $100 to allow retail and smaller institutional participants access to large-scale infrastructure finance for the first time (Belkhiria and Abid 2026). Tokenised bonds are programmable, meaning that the instrument itself will come with environmental reporting obligations: each token changes the metrics of its user's impact in real-time as project performance data is examined on-chain (Malamas et al. 2024). Market integrity is reinforced by the automation of compliance. A tokenised bond can be set itself up to automatically suspend coupon payments when the project falls below performance thresholds, inserting investor protection in part into the bond not just through a compliance mechanism elsewhere (World Bank 2023). For the carbon markets, ongoing NFT-based tokenisation of verified carbon credits removes double counting by creating cryptographically unique certificates, a structural change which may help rebuild some confidence in the markets which have been facing problems with credibility in the last few years (Allahgholi 2026; Mzoughi 2024).

25 Case Study Evidence

Four subsequent cases give empirical support to the theories elucidated above.

26 Hong Kong Project Genesis 2.0

The Hong Kong Monetary Authority (HKMA), through the BIS Innovation Hub, piloted retail-accessible tokenised green bonds with real-time carbon credit tracking, through smart contracts by means of Project Genesis 2.0 (HSBC 2024). Deployment time for DLT reduced issuance time from weeks to five days and atomic settlement eradicated T + 2 risk altogether. The project showed that blockchain can democratise access towards climate infrastructure investment, providing entry points to participate at entry points, which conventional bond structures cannot access.

27 BIS Green Bond Pilot Southeast Asia

In another 2023 Southeast Asia BIS Innovation Hub pilot, a private Ethereum-based ledger was implemented in a solar farm green bond that allowed to generate and dispatch coupon payments automatically to bondholders from a solar farm green bond built on the data mined from daily energy output data (BIS 2023). The automated MRV system was able to minimise bond administration costs by nearly 30 per cent and facilitated financing of projects that would, on the other hand, have seemed too small to receive financing on international capital markets.

28 Tokenised Platforms for Carbon Credit Tokens in Sub-Saharan Africa

In Kenya and Rwanda, fintech startups have introduced blockchain-based carbon credit systems that enable reforestation initiatives and smallholder farmers to issue tokens showing verified carbon sequestration (Allahgholi 2026). By attaching a unique on-chain token to each credit, the platforms overcome the double-counting risk that had previously put off international buyers. According to UNEP (2025), platform-based direct-to-buyer sales have enhanced project economics for local developers by up to 20%.

29 CBDC Settlement Pilots in Latin America

Central bank pilots in Brazil and Chile have employed a common DLT platform to settle green infrastructure debt in real time that entirely avoids the SWIFT network (IMF 2025). BIS (2023) finds that this type of direct settlement minimises currency conversion costs and eliminates the liquidity risk associated with carrying local currency for days on the settlement process. This model will set a useful standard for EMDEs to use sovereign digital currency to lower the FX risk premium that now adds several percentage points to the price of international climate capital.

30 Synthesis: Structure and Blended Finance on Blockchain

Collectively, the evidence from the case study and the theoretical literature support a shift in the way climate finance is mobilised. The classical model of manual hedging, periodic auditing and illiquid instruments is being replaced by a digital-first architecture that is automated, auditable 24/7 and structurally liquid. Table 2 presents the risk factors, the blockchain mechanism for each and estimated cost of capital implications.

The most important contribution of blockchain to blended finance is the move from institutional trust towards algorithmic trust. This has direct implications when project performance is verified by code on an immutable ledger, as opposed to through human intermediaries that could be slow, biased or susceptible to capture the governance deficit that has traditionally limited investors from emerging markets (World Bank 2023). Certainly for the concessional nature of blended structures, this is especially true since the legitimacy of the model is built on the certainty that public donor capital is indeed catalysing private investment rather than being absorbed by administrative inefficiency (IFC 2025).

Table 2 Blockchain risk mitigation strategies and impact on cost of capital

Risk category	Blockchain mitigation strategy	Impact on cost of capital
Currency risk	Atomic settlement and smart-contract-based FX triggers	Reduction of 1–2% in FX risk premium
Governance risk	Immutable ledger and real-time, automated MRV	Reduction of 0.5–1% in audit costs
Liquidity risk	Tokenisation and fractional ownership of green assets	Increased secondary market liquidity
Operational risk	Smart-contract-based milestone disbursement	Faster capital recycling, lower overhead

31 Recommendations

The results reveal unique implications for three groups of stakeholders. The successful implementation of blockchain-based climate finance requires coordinated action by the government, the development finance institutions as well as the private investment community.

32 EMDE Governments

Development finance institutions (DFIs) and multilateral development banks (MDBs) will have to shift their role from traditional guarantors to providers of digital risk-sharing infrastructure rather than just traditional guarantors. Future blended finance models oblige DFIs to integrate their de-risking roles directly with programmable rather than manual instruments (OECD 2023). Invest in Digital Risk-Sharing Platforms that employ DLT to dynamically structure guarantees and first-loss capital, with the DFI's exposure adjusted automatically with project performance dynamically reflecting on the fly (IFC 2025). Initiate Digital Finance Capacity Building programmes in local banks, regulators and project developers on Digital Finance to educate regulators, industry insiders on how financial institutions and developers can use and administer blockchain financial instruments such as local banks, regulators or project developers, to make finance (WEF 2025). Facilitate the establishment of cross-chain interoperability standards so that tokenized assets issued under different national frameworks, which work in cross-chain interoperability cross-border trading and settlement mechanisms such that tokenised assets issued by different governments in different territories can be exchanged, traded and settled around and across borders without friction (Malamas et al. 2024).

33 Institutional Investors

The long-term capital that green infrastructure requires comes from pension funds, insurance companies and other institutional investors. The new financial tools that will unlock the benefits of blockchain-enabled instruments require up to date due diligence frameworks and an absolute need to adopt the integration of real-time environmental information in investment and reporting processes. Create targeted due diligence protocols on tokenised assets including validation of smart contract code quality oracle network security and blockchain network governance and third-party audits (Malamas et al. 2024). Integrate live ESG data streams related to blockchain-based MRV systems into impact reports to beneficiaries and regulators, establishing a level of transparency not achievable with traditional reporting (Li and Chen 2025). Include Digital Asset Due Diligence in an integrated ESG Investment framework, one in which a measure of the verifiability and the auditability of blockchain facilitated instruments can be considered separately as a quality element, not just a traditional risk parameter.

34 Priorities for Future Research

Several research needs to be conducted regularly before blockchain-enabled climate finance can truly attain the scale of its potential. The scalability of existing distributed networks to transactions of a real-time global green bond market is unproven (Zaher and Roy 2025). The energy requirements of consensus processes even after Ethereum's evolution to Proof-of-Stake should remain to investigate on the specific scenario of climate finance (Naderi 2022). Cross-chain interoperability and the creation of consistent tokenisation standards are essential conditions for development (IMF 2025; Belkhiria and Abid 2026). Legally enforcing smart contracts across various jurisdictions through comparative legal research needs targeted attention (World Bank 2023). Finally, robust longitudinal studies on socio-economic outcomes (especially for LDC communities) are needed to ensure that the transition to digital climate finance is not replicated or exacerbates existing inequalities (Li and Chen 2025).

35 Conclusion

The climate finance gap in emerging markets is really about trust and risk management and not just about insufficient capital; it will take more than that. The climate finance gap is not simply some issue of money. Institutional investors own the funds necessary for green infrastructure; they lack confidence that those funds will be safeguarded against currency fluctuations and properly redirected toward achieving their stated objective. It has been pointed out in this chapter that blockchain technology through smart contracts, immutable ledgers and tokenised assets responds to both of these shortfalls. As an example, the BEBF model presented here provides a practical conceptual framework for the evolution of this paradigm. It is integrated three levels of concessional capital, blockchain infrastructure and private investment combined in a series of actions to de-risk, verify and scale climate investing in ways that traditional blended finance instruments simply can't replicate. The empirical record from pilots in Hong Kong, Southeast Asia, sub-Saharan Africa and Latin America indicates that the efficiency gains are felt: issuance costs decrease, settlement lags shorten, administrative overhead reduces and transparency increases. Challenges remain. Most EMDEs lack a regulatory framework that can accommodate tokenised assets or smart contract enforcement. National blockchain networks' interoperability is scarce. The idea of a digital divide in which the least connected among them stand to gain the least is real and should be carefully stewarded. That said these are not core challenges to the approach, but engineering and governance issues, with known solutions. With the 2030 targets for climate targets looming in sight and the financing gap failing to close by traditional means, the inclusion of blockchain into the financial architecture of the Paris Agreement has emerged no longer as a speculative proposition. At the speed and scale demanded by the climate emergency, it is now a practical requirement for any meaningful policy to be able to access private capital.

References

Allahgholi A (2026). Tokenization of voluntary carbon markets and its potential for financial inclusion. In: Hoven IG, In S, Puschmann T (Eds.) Sustainable digital finance. financial innovation and technology. Springer, Cham. https://doi.org/10.1007/978-3-032-02983-6_6

Bank for International Settlements [BIS] (2023) Financial stability risks from cryptoassets in emerging market economies. BIS Papers No 138. BIS, Basel

Belkhiria S, Abid E (2026) The impact of tokenization on the trading process costs and carbon emission: Empirical study on the ODDO BHF Bond. Carbon Balance Manage 21:42. https://doi.org/10.1186/s13021-025-00390-5

Climate Policy Initiative [CPI] (2024) Managing currency risk to catalyze climate finance in EMDEs. CPI Global Report. CPI, Washington, DC

Convergence Blended Finance (2024) State of blended finance 2024, Climate. Convergence, Toronto

HSBC (2024) Project genesis 2.0: scaling tokenised green bonds for retail investors. HSBC Sustainable Finance Report. HSBC, Hong Kong

International Finance Corporation [IFC] (2025) Role of blended finance in an evolving global context. IFC Blended Finance Report. World Bank Group, Washington, DC

International Monetary Fund [IMF] (2025) Tokenization and financial market inefficiencies. IMF Fintech Notes 2025/001. IMF, Washington, DC. https://doi.org/10.5089/9798400298905.063

Li T, Chen Y (2025) Blockchain applications in green finance for sustainable development: a banking perspective. Sustainability 17(6):2520. https://doi.org/10.3390/su17062520

Malamas V, Dasaklis TK, Arakelian V, Chondrokoukis G (2024) A blockchain framework for digitizing securities issuance: the case of green bonds. J Sustain Finance Investment 14(1):1–25. https://doi.org/10.1080/20430795.2023.2275212

Mzoughi H (2024) Blockchain markets, green finance investments and environmental impacts. Res Int Bus Financ 69:102142. https://doi.org/10.1016/j.ribaf.2024.102142

Naderi N (2022) Leveraging blockchain technology and tokenizing green assets to fill the green finance gap. Energy Res Lett 3(3). https://doi.org/10.46557/001c.33907

OECD (2023) Blended finance in the poorest countries: the need for a better development finance mix. OECD Development Policy Papers. OECD Publishing, Paris

Schulz K, Feist M (2021) Leveraging blockchain technology for innovative climate finance under the Green Climate Fund. Earth System Govern 7:100084. https://doi.org/10.1016/j.esg.2020.100084

UNEP (2025) Financing the transition: the role of digital assets in climate resilience. UNEP Finance Initiative Report. UNEP, Geneva

World Bank (2023) Blockchain governance and regulation as an enabler of climate action. World Bank Group Climate Change Series. World Bank, Washington, DC

World Economic Forum [WEF] (2025) Blended finance key to addressing Asia's SDG investment gap. WEF Insight Report. WEF, Cologny

Yilmaz A (2026) Cryptocurrency expansion, climate policy uncertainty and global financial stability. Sustainability 18(2):951. https://doi.org/10.3390/su18020951

Zaher HF, Roy JK (2025) Decentralized finance and sustainability: Analysis of global research patterns and emerging themes. Discover Sustain 6(1):42. https://doi.org/10.1007/s43621-025-02311-5

Just and Inclusive Urban Transport Transitions in the Global South: Equity, Policy, and Social Impacts

James Kanyepe, Maxwell Chufama, and Nyarai Kasambuwa

Abstract Decarbonizing transportation is essential for reducing climate change emissions, but its social and economic effects differ across cities. This study explores how low-carbon transportation initiatives impact labor markets in the fossil fuel and transportation sectors, potentially increasing inequality through the distributional and affordability effects of low-cost options from carbon pricing and alternatives to unsustainable modes. The study reviewed peer-reviewed studies, policy documents, and gray literature from 2000 onward, using Scopus, Web of Science, and Google Scholar with terms such as "transport decarbonization," "social equity," "low-carbon mobility," and "Global South." Studies that were published in English, focused on low- and middle-income countries, and were relevant to empirical studies, theory, or policy recommendations were included. The study revealed that transitioning to low-carbon transportation will cause fossil fuel job losses but create jobs in low-carbon industries, although skill gaps hinder transitions. In addition, carbon pricing and infrastructure changes cause unequal cost distributions, creating access barriers for lower-income individuals and informal economy workers. Moreover, gender differences in travel behavior complicate equitable low-carbon options. The results also show that technological fixes alone are insufficient. Furthermore, creating socially fair low-carbon transportation requires coordinated efforts such as workforce training, displaced worker compensation, participatory decision-making for inclusive infrastructure, and gender equity policy measures. This chapter aids policymakers, city planners, and researchers interested in equitable low-carbon transportation for Global South cities.

J. Kanyepe (✉)
Department of Logistics and Supply Chain Management, Faculty of Business, Economics and Law, University of Botswana, Gaborone, Botswana
e-mail: jameskanyepe@gmail.com

M. Chufama
Department of Marketing, Logistics and Sport Management, Faculty of Commerce, Human Sciences and Education, Namibia University of Science and Technology, Windhoek, Namibia

N. Kasambuwa
Academy of Teaching and Learning, Chinhoyi University of Technology, Chinhoyi, Zimbabwe

J. Kanyepe and I. Chirisa (eds.), *Advancing the Green Agenda in the Transport and Logistics Sector in Global South Countries*, World Sustainability Series,
https://doi.org/10.1007/978-3-032-31374-4_26

Keywords Carbon pricing · Gendered mobility · Transport decarbonization · Social equity · Just transition · Low-carbon mobility · Labor dynamics · Urban transport policy

1 Introduction

Transport decarbonization has emerged as one of the most important components of global climate mitigation strategies because of its considerable role in reducing greenhouse gas emissions and air pollution. Owing to the significant proportion of global final energy consumption and carbon emissions accounted for by the transport sector, it is a prime area for climate policy interventions (Solaymani and Botero 2025). Rapid urbanization, increased car ownership, and the spread of metropolitan areas have contributed to an increase in transport-related emissions, especially in cities located in the Global South, which typically possess limited public transport infrastructure and regulatory capability (Thondoo et al. 2020). Consequently, policymakers and planners have begun to promote technological solutions to reduce the carbon footprint of urban mobility, such as electric vehicles, low-emission buses, and renewable fuel transit systems (Acheampong et al. 2022).

Technological solutions are essential for achieving climate mitigation goals, but they are inadequate for ensuring socially sustainable mobility transitions. Transport decarbonization policies may alter the distribution of the costs and benefits of urban mobility among different social groups (Briand et al. 2023). However, if social equity considerations are not explicitly included in such policies, they can reinforce existing social inequalities and undermine public support for climate action (Li et al. 2024). Social equity issues related to transport decarbonization are therefore likely to be particularly pertinent to urban contexts in the Global South, where large sections of the population rely on affordable, albeit carbon-intensive, transport options because of a lack of access to alternative forms of reliable low-carbon mobility (Tammaru et al. 2023).

Socially equitable transport decarbonization occurs via multiple connected pathways. One relevant pathway refers to labor market changes resulting from the transition from fossil fuel-related modes of transportation (Jomaa 2026). Policies aimed at decarbonizing transport have led to a reduced demand for fossil fuels and internal combustion engines. In doing so, they create the possibility of employment disruption in sectors dependent on fossil fuels, such as fuel extraction, refinement, distribution, and conventional automotive manufacturing (Petavratzi and Gunn 2023). Employees in these sectors may experience job losses and declining economic well-being. Furthermore, when few opportunities are available for employees to move into new occupations, their livelihoods are affected. The wider literature on transitions toward sustainability indicates that environmentally oriented policies usually produce disproportionate economic impacts. Therefore, climate policy frameworks should include measures to protect workers during the transition, including training programs to allow workers to acquire new skills and financial assistance packages to compensate for losses experienced during the transition (Wieczorek 2018).

The second dimension refers to the distributive and affordability implications of transport decarbonization policies. Market-based instruments, including fuel taxes, carbon pricing mechanisms, congestion charges, and low-emission zones (LEZs), are widely recognized as being highly effective in encouraging behavioral changes that reduce carbon emissions (Pan et al. 2024). However, if these instruments are used without complementary social protection policies intended to alleviate their social impacts, they may produce regressive outcomes. For example, in numerous cities in the Global South, vulnerable populations, including low-income families, informal sector workers, and inhabitants living in peripheral neighborhoods, rely extensively on fossil fuel-based transport options, as they do not have affordable access to alternatives (Westphal et al. 2017). In such cases, decarbonization measures may increase the cost of mobility and limit access to work, education, and other necessities. Thus, decarbonization measures may exacerbate the current patterns of urban inequality.

Transporting also intersects with gendered mobility patterns. Research in transport planning and mobility studies has demonstrated that women commonly display more complex travel behavior than men do and undertake multipurpose trip chains to facilitate caregiving responsibilities, household management, and informal economic activity (Cresswell and Uteng 2008). Women's access to low-carbon mobility options, that is, public transport, walking, and cycling, may also be restricted by structural constraints, including safety issues, affordability restrictions, and limitations in service coverage. If the varied mobility needs of women are not considered in designing transport planning and infrastructure, decarbonization policies may unintentionally reproduce existing gender inequalities while simultaneously limiting the efficacy of sustainable mobility strategies over time.

To address these challenges, analytical frameworks that can incorporate environmental objectives from a social justice perspective are needed. The concept of a "just transition" provides a useful lens for examining the equity implications of climate policy interventions. Stemming from debates concerning the transformation of the energy sector, the just transition framework highlights the importance of considering distributive justice (the allocation of costs and benefits associated with policy interventions), procedural justice (fairness and inclusion in decision-making processes), and recognitional justice (acknowledging the unique experiences and needs of historically marginalized communities) in environmental governance (Heffron and McCauley 2018). By applying these principles to urban mobility systems, a basis exists for evaluating whether decarbonization policies contribute to environmental sustainability and social equity.

Although scholarly and policy interest in equitable climate transitions is increasing, research investigating the social dimensions of transport decarbonization is currently somewhat unbalanced. Most research has focused primarily on technological innovation, routes to electrification, and the potential for emission reductions provided by transport policies (Creutzig et al. 2015). Although these studies provide valuable insights into mitigation strategies, relatively little research has focused on the broader distributional effects of transport decarbonization policies, specifically in the rapidly urbanizing regions of the Global South (Martens 2017; Lucas et al. 2016). Moreover, research has tended to consider the individual dimensions of equity, such as affordability, labor transitions, and gendered mobility patterns, separately. Therefore, researchers must develop integrated analytical approaches that assess how these interrelated social dimensions influence the equitable outcomes of transport decarbonization policies. This study aims to meet this requirement by investigating the social equity implications of transport decarbonization in the Global South. More specifically, this study pursues three objectives: (1) examine how transport decarbonization influences labor markets within fossil-fuel-dependent and transport-related supply chains; (2) analyze the distributive and affordability implications of carbon pricing and reforming transport infrastructure for vulnerable populations; and (3) investigate how transport infrastructure design shapes gendered mobility patterns and access to low-carbon transport options. This study develops a multidimensional framework for assessing socially just transport decarbonization by combining insights from urban planning, transport economics, and social justice scholarship. The analysis outlines policy mechanisms, including social protection measures, redistributive policy instruments, and inclusive governance practices, that can assist equitable and politically sustainable mobility transitions in rapidly developing regions of the Global South.

This chapter presents a structured approach to ensure coherence throughout the paper. This chapter commences with a review of the theoretical framework, providing the conceptual underpinnings for this research. This is followed by a description of the methodology, detailing the research design, data sources, and methods utilized to maintain rigour and reliability. The chapter then describes the results along with a critique that places these findings within the broader literature and objectives of this research. Finally, this chapter draws together the major conclusions, outlines their implications for policy and practice, and identifies avenues for future research development to improve the knowledge of socially equitable transportation decarbonization.

2 Theoretical Framework

This study is based on a theoretical model that connects urban transport decarbonization policies with their social equity consequences. The theoretical model focuses on achieving an environmentally fair, inclusive, and just transition in developing countries. It includes three complementary theories: transition theory, equity in transport, and sociotechnical transitions. These three complementary theories provide a suitable lens for analyzing the processes and results of transitions toward low-carbon urban mobility. With this theoretical model, we analyze technological developments in the field of transportation and the development of governance systems and social justice simultaneously. This study provides insights into the connection between environmental goals and social equity.

According to Heffron and McCauley (2018), transition theory states that all measures taken to reduce greenhouse gas emissions must consider the socioeconomic rights and livelihoods of people living in vulnerable areas. Therefore, the main goal should be not only to reduce CO2 emissions but also to protect the livelihoods of people living in poverty or on the periphery of society. For example, when new low-carbon urban mobility options are implemented in cities in the Global South, it is important to implement inclusive workforce plans and social protection schemes for workers who lose their jobs due to these developments and to ensure that all people have equal opportunities to use new low-carbon mobility options. For these reasons, both procedural and distributive fairness must be included to ensure that the transition to low-carbon mobility is socially acceptable and sustainable while considering the specific local socioeconomic situation. Equity in transport examines how transport policies influence various social groups on the basis of income level, sex, age, disability status, and geographic location (Cresswell and Uteng 2008). This theory emphasizes the significance of affordability, accessibility, security, and service quality in transportation. Unless deliberate actions are taken in these respects, low-carbon transport policies may exacerbate existing inequalities in the social space.

In the case of cities located in the Global South, these issues are particularly relevant because many residents live in informal housing developments that lack public transportation options, and they often face significant obstacles related to their gender when accessing public transportation. Therefore, including equity considerations in low-carbon urban mobility transitions in the Global South is crucial for preventing unintentional social exclusion. Sociotechnical transitions examine transport decarbonization from a broader system transformation perspective, highlighting the connections between technology, policy, infrastructure, and social practices (Schwanen et al. 2011). From this viewpoint, low-carbon urban mobility represents a multifaceted process in which technological innovations require supportive governance approaches, institutional coordination mechanisms, and behavioral adaptations to produce socially acceptable outcomes. According to this view, successful decarbonization cannot merely depend on technological solutions but must be aligned with societal structures and local contexts, especially in regions characterized by limited resources, typical of those found in the Global South.

3 Methodology

3.1 Research Approach

This study used a desk-based review. A desk-based review is a literature review method that systematically evaluates all available information on a particular topic. The purpose of this literature review is to enable the synthesis of empirical evidence, theoretical frameworks, and institutional analysis regarding the social equity implications of transport decarbonization in the Global South. Literature reviews provide an overarching perspective on complex multidimensional issues (e.g., how transport decarbonization strategies impact social equity) by examining different low-carbon transport measures and their social equity impacts. As this type of literature review does not require the collection of new data (primary data), it can be useful if there are limited resources for collecting data or if there is a need to access data from geographically dispersed areas. Additionally, this methodology enables researchers to identify commonalities across studies and differences and variations in terms of context and location, as well as to develop policy-relevant findings.

3.2 Systematic Literature Search

This study developed a systematic methodology for conducting searches to ensure that all applicable literature was included. The authors searched databases including Scopus, Web of Science, and Google Scholar by utilizing the terms "low-carbon mobility," "gendered mobility," "labor displacement," and "just transition" in their search parameters; these terms allow the authors to capture the technological aspects of reducing carbon emissions in transportation and the social implications. The authors narrowed their search to peer-reviewed articles written in English published from 2000 to the present. This provides the authors with a high level of methodological clarity but introduces the possibility of linguistic bias, as literature that has been researched in languages other than English may have been missed.

3.3 Selection Criteria and Source Justification

Both explicit inclusion and exclusion criteria were used to provide a clear and rigorous analysis. The inclusion criteria emphasized empirical research, conceptual or theoretical studies, and formal institutional reports from low–middle-income countries (LMICs) as the primary focus of interest for this study. Only studies performed in high-income settings were reviewed if they offered insights that could be directly translated into LMIC contexts. The exclusion criteria for this review were opinion articles and all other types of publications with poor methodological transparency. Of the 520 potential resources initially identified for this study, 92 studies used in this study were selected on the basis of their relevance to the topic of interest, overall methodological quality, and comprehensiveness of their data. A comprehensive evaluation of each study's title(s), abstract(s), and full-text document was performed to ensure that each selected study had the best possible evidence available and was contextual to the region.

3.4 Data Extraction and Thematic Analysis

The studies selected for this review were examined through thematic analysis to perform a narrative synthesis. Thematic analysis is an appropriate method for identifying recurring patterns and extracting common themes from diverse sets of research. The identified data were grouped into four categories of analytical themes: labor dynamics, affordability and distributive impacts, gendered mobility inequalities, and infrastructure-related equity challenges. As the coding process progressed, it became increasingly iterative and reflexive, as each coder continually cross-referenced the coding to verify consistency and accuracy. Through this process, both empirical findings and theoretical perspectives were synthesized and contextualized.

3.5 Cross-Case Comparison and Contextual Analysis

To improve the depth of the analysis, cross-case comparisons were made across three categories (regions, types of transportation, and policies). These cases were selected on the basis of their geographic and socioeconomic differences to illustrate the degree of variability associated with outcomes and contextual factors that include, but are not limited to, urban design/structure, government capacity/skill level, and socioeconomic status. By comparing these case studies and identifying similarities and differences within them, a more nuanced understanding of how low-carbon transportation policies impact social equity is developed and can be used to create equitable policy recommendations for various contexts that remain relevant on a broader scale.

3.6 Methodological Rigor and Limitations

The research design utilized an integrative systematic review; both primary and secondary studies with the same inclusion/exclusion criteria across all papers were used, along with thematic synthesis and cross-case analysis, to assure the dependability (reliability), validity, and transparency of the results. Secondary literature was used as the sole source of empirical evidence because a variety of methodological biases can exist when relying solely on secondary literature (e.g., publication bias, language bias) and a lack of available contextual information. The structured process utilized in this study represents a reliable and evidence-based platform for developing an understanding of the challenges and opportunities associated with achieving transport decarbonization policies in equitable and sustainable ways in the Global South.

4 Findings

The thematic analysis of the literature identified three key themes: (1) labor dynamics and workforce transformation, capturing the displacement and creation of employment across fossil fuel-dependent and low-carbon transport sectors; (2) distributive and affordable impacts of carbon pricing and infrastructure reforms, highlighting the socioeconomic consequences of decarbonization policies for vulnerable populations; and (3) gendered mobility patterns and access to low-carbon transport, illustrating how infrastructure and service design influence women's travel behavior and opportunities.

5 Labor Dynamics and Transport Decarbonization

The global shift to sustainable transportation in the developing world is impacting the labor force in various ways within both fossil fuel-related and transportation industries. As a result, workforce impacts are heavily influenced by the relationships among the industrial structure at the regional level, sectoral composition, and socioeconomic factors. Three major dynamics exist within the literature: direct job loss due to the decline in the use of traditional fuels and internal combustion engines, the development of "green" jobs created as part of the process of transitioning away from fossil fuels, and the increasing vulnerability of informal and disadvantaged workers.

First, through policy measures such as fuel efficiency standards, fleet conversion to electric vehicles, mode shifting toward nonmotorized and public modes of transportation, and carbon pricing, there is an immediate reduction in the demand for fossil fuels and internal combustion engine (ICE) vehicles (Briand et al. 2023), which results in declining employment opportunities in oil extraction/refining, fuel distribution, and the manufacture of ICE vehicles. In West Africa, evidence shows a clear correlation between regions dependent on oil and those with extensive employment declines, whereas automotive-focused centers reliant on producing ICEs are facing structural decline. The impacts are disparate geographically and temporally, and fuel networks/assembly plants located in urban/peri-urban areas experience the greatest negative impacts (Hutchings et al. 2022). Employment outcomes can also be differentiated by skill level, where skilled, semiskilled, and unskilled workers each face unique economic and social challenges (Das 2025; Hutchings et al. 2022).

Second, decarbonization creates employment opportunities in low-carbon sectors. The increasing usage of public transport powered by electricity, deployment of charging infrastructure for electric vehicles (EVSs), expansion of renewable energy-based transit systems, and promotion of cycling/walking infrastructure all create demands for technicians/engineers/operators (Bagdia et al. 2022). Southern African cities have shown an increase in employment in EV maintenance, battery installation, urban logistics planning, and low-emission public transit operations (Collett et al. 2020). However, because many displaced workers do not possess the technical/digital/engineering qualifications to meet the requirements for these emerging job types, especially in countries with limited formal education/vocational training options, anticipatory reskilling/upgrading programs are needed to facilitate labor flow from carbon-intense to green employment (Gicha et al. 2024).

Third, informal and disadvantaged workers are highly vulnerable throughout the transition. Actors operating in the informal economy, including taxi drivers, fuel vendors, and motor vehicle mechanics, typically operate outside regulated environments and social welfare protections, thereby exposing them to financial instability caused by changes in fuel prices, enforcement of regulations, and mode shifts (Miti et al. 2021). Gendered dimensions of inequality also compound this risk; women engaged in informal transportation work are significantly represented in entry-level, lower-paid, formerly stable occupations and are more likely to suffer from job insecurity than men are (Wright 2019). Social class, gender, and geographic class/gender/geographic area intersectionality affect exposure; therefore, it is essential to develop and implement policies that protect/increase access to reskilling and ensure equitable inclusion in low-carbon transport transformation (Priya Uteng and Turner 2019).

6 Distributive and Affordability Impacts of Carbon Pricing and Infrastructure Reforms

The socioeconomic impact of transport decarbonization measures varies widely depending on how the environmental benefits of such policies are shared among society. The distribution of the economic burden of transport-carbonizing policies disproportionately falls on those most economically vulnerable in many parts of the world, particularly low-income people and informal workers. Low-income families spend a greater percentage of their total family income on transportation expenses than middle- and upper-class individuals do. Furthermore, affordable and reliable public transportation options are frequently unavailable in many regions. Therefore, the instruments used by governments to internalize the environmental externalities associated with vehicle emissions, such as fuel taxes, congestion fees, and emissions trading programs, are likely to place an additional economic burden on lower-income households. Additionally, spatial inequities within metropolitan areas are likely to amplify this effect. For example, residents living in suburban neighborhoods or informal settlements may be required to make longer and more expensive trips to reach employment opportunities and other necessary goods and services. Consequently, the carbon pricing mechanism could exacerbate preexisting socioeconomic inequalities.

Therefore, it has become common practice for researchers to emphasize the need for compensation mechanisms to mitigate the regressive impacts of carbon pricing. Revenue recycling is one of the most popular methods by which scholars advocate compensating those who would bear the greatest burden from carbon pricing.

Revenue recycling entails redistributing funds raised through carbon-related charges to disadvantaged populations. Governments may implement direct cash transfers, targeted transportation subsidies, reduced fares, and/or reinvestment of carbon revenues into affordable, reliable, and efficient public transportation systems (Venter et al. 2019). When governments reinvest revenues from carbon pricing into low-cost, high-capacity transit infrastructure, they can simultaneously improve accessibility for lower-income populations while achieving urban decarbonization goals (Blanco et al. 2022). Thus, well-designed redistribution mechanisms can contribute to both the enhanced environmental performance of carbon pricing policies and increased social acceptance of such policies.

Equity considerations also arise regarding equitable access to sustainable transportation modes in relation to infrastructure reforms related to low-carbon transport transitions. While there are numerous examples of environmental and urban liveability benefits resulting from low-emission zone policies, transit-oriented development, and pedestrianization measures, these types of measures may also increase housing prices and travel costs in the impacted areas. This may create an opportunity for gentrification or exclusionary displacement, which limits access to lower-income populations (Dindar 2025). To avoid creating barriers to entry for lower-income populations through equitable planning approaches, planners should consider adopting equity-sensitive planning frameworks. These equity-sensitive planning frameworks may involve policies that support affordable housing near transit corridors, progressive fare structures, and improved transportation connections to peripheral communities (Vecchio et al. 2020). Complementary infrastructure investment, such as building safe walking and cycling networks, may also provide affordable alternative forms of mobility while promoting emission reduction and inclusive urban mobility (Cervero 2014).

7 Gendered Mobility and Access to Low-Carbon Transport

Transport systems and their design are important factors in how people travel on the basis of gender and how much access they must have different economic, social, and educational opportunities. As an increasing amount of literature shows that women's travel behaviors differ from those of men and generally show this disparity due to the inequality in the division of labor and childcare responsibility, women are more likely than men to make multiple purpose trips, including work, school drop-offs, grocery shopping, etc., which is known as trip chaining (Borker 2024). However, most city transportation systems are built primarily to support one-way traffic during peak hours, with emphasis placed on direct travel from home to work; therefore, they do not consider the flexibility and connection that women require for their daily travels.

To better serve women who wish to utilize low-carbon methods of transportation, such as buses, trains, bicycles, and walking, transportation planners should develop plans that offer bus feeders, flexible routes, and reliable transfer points to allow for better use of lower-carbon forms of transportation (Benevenuto and Caulfield 2019). In addition to access to transportation, safety and accessibility issues affect the types of transportation women choose to use. Fear of sexual harassment, theft, or poor design of roads and public transportation systems results in a lack of confidence in women using public transportation and other active modes of transportation (Priya Uteng and Turner 2019). Research indicates that having safe, accessible infrastructure for women, including but not limited to street lighting, security at bus stops, elevated walkways, separate bike paths, and increased staffing by females or video monitoring, can greatly reduce the fear of crime and increase feelings of safety while using public transportation (Priya Uteng and Turner 2019). Lutz et al. (2024) stated that institutional measures (i.e., having female employees work for public transportation agencies or installing cameras to monitor activities) can contribute to increased confidence among women in riding public transportation. These two areas combined can lead to an increase in low-carbon mode usage and make public transportation available to all residents of the city.

Finally, the affordability and design of services are directly related to women's ability to use a specific transportation system. The frequency of services offered, number of routes covered, availability of last-mile transportation options, and fares all impact the feasibility of making daily trips for women residing in suburbs, peri-urban, or slum communities whose transportation options are fragmented or unreliable (Baye 2025; Roy et al. 2024). In general, gaps in service delivery hinder women's ability to engage in social and economic activities. Women-friendly transportation policy options, such as changing operating times for some services, offering discounted fares for certain groups, providing subsidies for transportation for specific segments of the population, and integrating microtransit/demand-responsive services into existing public transportation systems, can help fill these gaps. Improving affordability and access to transportation will assist in improving women's mobility while contributing to achieving both social justice and decarbonizing transportation systems (Priya Uteng and Turner 2019).

8 Discussion

The results of this study demonstrate the many interrelated social, economic, and equity factors that affect transport decarbonization in the Global South. Labor issues, as well as those related to distributional and affordability implications and gendered mobility, are shown to be a range of interconnected socioeconomic issues that transport decarbonization is fundamentally about. This is not merely a technological or environmental issue that has traditionally been considered. Therefore, a coordinated policy response is needed. An examination of labor issues reveals two opposing trends in job displacement due to decarbonization policies. On the one hand, it creates new jobs in the low-carbon transport industry while displacing jobs in other fossil fuel-dependent sectors. As such, the impact of workforce displacement varies greatly depending on the sector, region, and skill level. These differences have resulted in a call for proactive approaches to assist workers in transitioning from their traditional carbon-intensive roles to new green employment opportunities. In addition, informal and marginalized workers—predominantly women who work in precarious transport environments—are most negatively impacted by the lack of adequate social protection and the current lack of targeted interventions in low-carbon transport.

Furthermore, an analysis of the distributive and affordability impacts shows that unless low-carbon transport interventions are specifically designed to mitigate them, they exacerbate existing socioeconomic inequities. While pricing mechanisms and infrastructure reform can result in both reduced greenhouse gas emissions and improved urban quality of life, they often create disproportionate burdens for low-income households, informal workers, and residents of underserved peripheral areas. Therefore, it is necessary to include equity-sensitive measures in the design of low-carbon transport interventions (including) targeted subsidies, revenue recycling, affordable housing close to public transport routes, progressive fare structures, and increased connectivity to the periphery. An analysis of gendered mobility revealed that women face unique barriers when using low-carbon transport. However, these barriers are shaped by women's caregiving responsibilities, perceived safety risks, and insufficient transport services. Specifically, conventional transport systems that prioritize direct peak-hour commutes limit women's ability to engage in multistop/multipurpose trips. Improvements in route flexibility, service frequency, safety, and affordability would increase women's accessibility to low-carbon modes, promote greater social equity/inclusivity within urban transport, and improve the overall sustainability of urban transport systems in developing nations. The results suggest that successful transport decarbonization in the Global South requires a coordinated and socially focused approach. To gain acceptance of transport decarbonization policies among all sections of society and prevent unequal social outcomes arising from transport decarbonization, the labor, distributive, and gendered aspects of transport decarbonization should be addressed together. Integrating social protection into environmental goals, developing worker skills, and incorporating inclusive planning into transport policy will enable the development of sustainable and resilient urban mobility systems in developing nations.

9 Theoretical Implications

This study adds new knowledge to our theoretical understanding of how the intersections of transport decarbonization and social equity play out in the Global South. This study demonstrates that creating sustainable transportation systems is more than an environmental issue; it is a social, technical, and economic process. Interventions create different effects in terms of employment opportunities (labor market), income levels, and gendered ways people move around. Therefore, the development of theoretical frameworks that link environmental policies with social justice, urban governance, and spatial equity is needed. In addition, this study provides new knowledge for the theoretical discussion of "just transition" through examples of how the workforce changes depending on specific contexts. Job loss and job gain differ depending on the sector (formal/informal), type of labor (skilled/unskilled), and geographic location. Overall, this study suggests that theories of sustainable urban transportation must recognize both structural and local vulnerability issues. Additionally, the research extends gender and mobility theory by identifying how different aspects of transportation system design, including services and safety, influence women's decision-making processes regarding their travel choices; therefore, it identifies previously missed barriers to sustainability in conventional planning paradigms.

10 Practical Implications

This research provides practitioners working in policy, urban planning, and transportation with guidance on how to develop, implement, and fund low-carbon transport solutions that meet both environmental goals and social justice criteria in the future. Additionally, this study supports the need to establish inclusive workforce strategies to assist in the transition of labor from fossil fuel-dependent industries to new low-carbon jobs. These workforce development strategies should include reskilling/upskilling opportunities and be supported by broad social protections, especially for informal/marginalized workers, who are frequently excluded from formal regulations and social security net provisions. To ensure fairness, developing equitable carbon-pricing mechanisms and infrastructure is essential. Although low-cost, market-based tools such as fuel taxes, congestion charges, and emissions trading can incentivize lower-carbon transportation choices, they risk worsening existing socioeconomic disparities if not paired with complementary measures such as revenue recycling, subsidies, or reinvestment in public transit. The accessibility of infrastructure must be addressed in ways that favor outer/peripheral communities so that geographic inequities do not hinder the fair distribution of environmental/mobility benefits.

Gender-responsive transportation planning is needed to address the deep-seated barriers to mobility experienced by women. Therefore, gender considerations must be incorporated into all aspects of policymaking (e.g., multiple stop trips, caregiving obligations, safety concerns, and limited services). Providing safe walking/biking paths/cycling lanes/flexible route options/staffing female employees in public transit agencies will help increase access and safety for women, ultimately supporting wider social equity. Finally, there is a clear need to coordinate efforts among urban planners, labor regulators, and transportation agency personnel when implementing low-carbon transportation strategies to link environmental goals to social and economic development. In addition to improving the environmental sustainability of low-carbon transportation systems, holistic strategies help build the legitimacy of decarbonization policies, resulting in policies that promote inclusive, sustainable, and resilient urban development. The use of empirical/thematic data to ground these suggestions demonstrates that low-carbon transportation system equity will depend on both technological innovations and social inclusiveness, in addition to recognizing equity as central to sustainable mobility transitions.

11 Limitations of the Study

Despite its comprehensive synthesis, this study is limited by methodological constraints and the specific context of the research. The desk-based review relies almost exclusively on secondary sources, such as academic literature, policy documents, and gray literature. This allows an extremely wide range of evidence to be synthesized. However, there may be some limitations in capturing emerging or localized labor market, mobility, and equity dynamics in informal transport sectors, especially since primary data within these sectors are generally difficult to find. Therefore, while the study captures many of the labor, mobility, and equity impacts that occur at the local level, it does not capture all these impacts.

The study is also constrained by its reliance on material published in English between 2000 and 2025. This approach ensures that all material used within the study is consistent methodologically and relevant; however, it may exclude other studies written in other languages and therefore limit the potential for the study to be generalizable across all contexts in the Global South. In addition to being subject to limitations, the study's findings are highly dependent on the contexts in which they were generated (i.e., the characteristics of the region(s), transport system(s), and/or policy environment(s) described within the reviewed literature). Therefore, while some conclusions made within the study can be generalized across cities/countries with similar socioeconomic, cultural, and governance conditions, others cannot.

To assist in identifying patterns within the themes identified via thematic narrative synthesis and cross-case comparisons, the interpretive process will depend on the overall quality, completeness, and comparability of the primary source(s). Therefore, long-term workforce transition, infrastructure adaptation, and behavioral change processes may be less likely to be represented than shorter-term changes. Furthermore, although this study establishes a strong conceptual framework for examining routes to social justice through transport decarbonization in urban areas with resource constraints, the available data on gendered mobility, informal sector labor, and localized equity outcomes represent significant constraints in fully evaluating the intersecting forms of vulnerability.

12 Directions for Future Research

Future studies should assess the patterns, timing, and continuation of movement among employees within carbon-emitting industries to low-carbon industries via longitudinal workforce transition data. These studies provide information on skills shortages, obstacles to employee migration, and the success of reskilling or upskilling programs. Studies assessing the impact of equity-focused initiatives, such as progressive pricing, gender-responsive infrastructure and revenue recycling, subsidized fares, and other forms of subsidies, may assist in identifying which policy options would have the greatest effect on reducing inequitable impacts while promoting accessible and sustainable transportation. Studies comparing cities and regions in developing countries may be helpful in recognizing common issues by implementing equitable decarbonizing approaches to address them.

In addition, research examining intersectionality (i.e., how gender, socioeconomic status, and geography intersect) will help in understanding the differences in access to transportation. Data derived from these types of studies will help planners develop plans that address a wide range of population groups, including women, informal sector workers, and people living in peripheral communities. Additional studies examining the response of formal sector businesses operating in the informal sector (e.g., taxi operators, fuel vendors, etc.) to decarbonization policies can provide important information on how to protect vulnerable members of society through social safety nets, how to include labor in low-carbon economy development, and ways to mitigate vulnerabilities in the future. Increasing local data collection on gendered mobility, informal labor, and sociospatial inequality related to transportation will enhance the body of knowledge supporting an equitable and sustainable transition to low-carbon transportation systems.

13 Conclusions

This chapter is concerned with three other aspects of transport decarbonization—social, economic, and equity—in the Global South. It examines labor dynamics, distributive justice, and affordability implications for different population groups and gendered travel behavior patterns. Low-carbon transport interventions can create new job opportunities while simultaneously placing additional socioeconomic burdens on vulnerable populations and influencing equitable access to sustainable modes of travel (especially for women). These social aspects of transport decarbonization are interconnected. Therefore, it is essential that all three are treated together if an inclusive and sustainable transformation toward a low-carbon future occurs. The inclusion of workforce adaptation, equity-sensitive infrastructure design, infrastructure design, and gender-responsive interventions in policymaking can improve the effectiveness of transport decarbonization measures and their social acceptability. Transport decarbonization represents not only an environmental or technological challenge but also a major socioeconomic challenge. Furthermore, combining the integration of social equity with environmental objectives will lead to the development of resilient, inclusive, and sustainable urban mobility systems that advance climate change mitigation while providing broader developmental benefits across the Global South.

References

Acheampong RA, Lucas K, Poku-Boansi M, Uzondu C (2022) Introduction: transport and mobility situations of african cities. In: Acheampong RA, Lucas K, Poku-Boansi M, Uzondu C (eds) Transport and mobility futures in urban Africa. The urban book series. Springer, Cham. https://doi.org/10.1007/978-3-031-17327-1_1

Bagdia R, Keche Y, Bagdia K, Agrawal A (2022) Road transport electrification and renewables: opportunities for integration. Electromobil Renew Electr 11

Baye F (2025) Exploring urban infrastructure challenges in informal peri-urban Woldia: barriers, implications, and informal strategies. Front Sustain Res Manage 4:1555564

Blanco JP, Windisch E, Perkins S, Ito A, Leape J (2022) Decarbonizing transport in Latin American cities: a review of policies and key challenges.

Borker G (2024) Understanding the constraints to women's use of urban public transport in developing countries. World Dev 180:106589

Briand Y, Pye S, D'Agosto MDA, Goes GV, Neves Schmitz-Gonçalves D, Garg A, Trollip H (2023). Passenger transport decarbonization in emerging economies: policy lessons from modeling long-term deep decarbonization pathways. Climate Policy 1–21

Cervero R (2014) Transport infrastructure and the environment in the Global South: sustainable mobility and urbanism. J Region City Plan 25(3):174–191

Collett KA, Byamukama M, Crozier C, McCulloch M (2020) Energy and transport in Africa and South Asia. Energy Econ Growth

Cresswell T, Uteng TP (Ed.) (2008) Gendered Mobilities. Ashgate

Creutzig F, Roy J, Lamb WF, Azevedo IL, Bruine de Bruin W, Dalkmann H, Weber EU (2015) Urban infrastructure choices structure climate solutions. Nat Clim Chang 5(12):1055–1057. https://doi.org/10.1038/nclimate2760

Das DK (2025) Resilient and sustainable urban infrastructure in the global south: strategies to address climate change-linked vulnerabilities. Urban Plan 10

Dindar S (2025) A systematic review of urban regeneration's impact on sustainable transport: traffic dynamics, policy responses, and environmental implications. Sustain Dev 33:399–426

Gicha BB, Tufa LT, Lee J (2024) The electric vehicle revolution in Sub-Saharan Africa: trends, challenges, and opportunities. Energ Strat Rev 53:101384

Heffron RJ, McCauley D (2018) What is the "just transition"? Energy Policy 119:1–5. https://doi.org/10.1016/j.enpol.2018.04.014

Hutchings P, Willcock S, Lynch K, Bundhoo D, Brewer T, Cooper S, Welivita I (2022) Understanding rural–urban transitions in the Global South through peri-urban turbulence. Nature Sustainability 5(11):924–930

Jomaa H (2026) Labor market and ecological implications of the shift from fossil fuels to renewable energy: a systematic literature review. Ecocycles 12(1):1–13

Li S, Wang B, Zhou H (2024) Decarbonizing passenger transportation in developing countries: lessons and perspectives1. Reg Sci Urban Econ 107:103977

Lucas K, van Wee B, Maat K (2016) A method to evaluate equitable accessibility: combining ethical theories and accessibility-based approaches. Transportation 43(3):473–495. https://doi.org/10.1007/s11116-015-9616-x

Lutz C, Mwangi M, Namakula H, Tusubira J, Eckart J, Bejarano M (2024) Enhancing pedestrian and cyclist safety in african cities through living labs: an integrative analysis. In: Conference on sustainable urban mobility (pp. 127–141). Springer Nature Switzerland, Cham

Martens K (2017) Transport justice: Designing fair transportation systems. Routledge

Miti JJ, Perkiö M, Metteri A, Atkins S (2021) The informal sector and social protection. In: Handbook on social protection systems (pp. 389–402). Edward Elgar

Pan J, Cross JL, Zou X, Zhang B (2024) To tax or to trade? A global review of carbon emissions reduction strategies. Energ Strat Rev 55:101508

Petavratzi E, Gunn G (2023) Decarbonizing the automotive sector: a primary raw material perspective on targets and timescales: decarbonizing the automotive sector: a primary raw material perspective on targets and timescales. Miner Econ 36(4):545–561
Priya Uteng T, Turner J (2019) Addressing the linkages between gender and transport in low-and middle-income countries. Sustainability 11(17):4555
Roy S, Bailey A, Van Noorloos F (2024) Understanding the barriers affecting women's mobility in the first-and last-mile stretches in low-and middle-income countries: A systematic review. J Transp Geogr 121:104036
Schwanen T, Banister D, Anable J (2011) Scientific research about climate change mitigation in transport: a critical review. Transp Res Part A: Policy Pract 45(10):993–1006
Solaymani S, Botero J (2025) Reducing carbon emissions from transport sector: experience and policy design considerations. Sustainability 17(9):3762
Tammaru T, Sevtsuk A, Witlox F (2023) Towards an equity-centred model of sustainable mobility: integrating inequality and segregation challenges in the green mobility transition. J Transp Geogr 112:103686. https://doi.org/10.1016/j.jtrangeo.2023.103686
Thondoo M, Marquet O, Marquez S, Nieuwenhuijsen MJ (2020) Small cities, big needs: Urban transport planning in cities of developing countries. J Transp Health 19:100944
Vecchio G, Tiznado-Aitken I, Hurtubia R (2020) Transport and equity in Latin America: a critical review of socially oriented accessibility assessments. Transp Rev 40(3):354–381
Venter C, Mahendra A, Hidalgo D (2019) From mobility to access for all: expanding urban transportation choices in the global south. World Resources Institute, Washington, DC, pp 1–48
Westphal MI, Martin S, Zhou L, Satterthwaite D (2017) Powering cities in the global south: how energy access for all benefits. World Resources Institute, Wahsington DC, Estados Unidos
Wieczorek AJ (2018) Sustainability transitions in developing countries: major insights and their implications for research and policy. Environ Sci Policy 84:204–216
Wright T (2019) The gendered impacts of technological change on public transport workers in the Global South. Res Transp Bus Manag 31:100384

Pathways for Advancing the Green Agenda in Transport and Logistics in the Global South: Policy Implications

Innocent Chirisa, Africa Muneti, Tinashe Magande, Halleluah Chirisa, Kingsford Kobina Annan, and James Kanyepe

Abstract Sustainable transport and logistics systems remain critical for economic growth and environmental resilience in Global South countries, yet persistent barriers hinder progress toward decarbonization and efficiency. Infrastructure deficits, financing shortages, fragmented regulations, and restricted access to green technologies undermine efforts to build resilient networks aligned with global sustainability goals like the United Nations Sustainable Development Goals. This qualitative study explores strategic pathways to overcome these challenges and accelerate the green transformation of transport and logistics in developing economies. Drawing on insights synthesized from the book; data were generated through a comprehensive qualitative desk review of key barriers and policy discussions. Textual analysis was employed to interpret patterns, themes, and opportunities within the reviewed materials. The analysis revealed three pivotal insights: first, regulatory fragmentation

I. Chirisa (✉)
Office of the Vice Chancellor, Zimbabwe Ezekiel Guti University, Bindura, Zimbabwe
e-mail: innocent.chirisa@gmail.com

I. Chirisa · T. Magande
Department of Urban and Regional Planning, University of Johannesburg, Johannesburg, South Africa

A. Muneti
Firm Central Property Group, Harare, Zimbabwe

IC University of Technology, Phnom Pehn, Cambodia

H. Chirisa
Directorate of Research and Innovation, Zimbabwe Ezekiel Guti University, Bindura, Zimbabwe

K. K. Annan
Department of Urban Forestry, Environment, and Natural Resources, Southern University and A&M College, Baton Rouge, LA, USA

J. Kanyepe
Department of Logistics and Supply Chain Management, University of Botswana, Gaborone, Botswana

J. Kanyepe and I. Chirisa (eds.), *Advancing the Green Agenda in the Transport and Logistics Sector in Global South Countries*, World Sustainability Series,
https://doi.org/10.1007/978-3-032-31374-4_27

demands unified frameworks to harmonize national policies with international standards; second, targeted investments in green infrastructure and technological innovation can bridge financing gaps and enhance system resilience; third, stronger institutional coordination fosters collaborative governance, enabling scalable adoption of low-carbon solutions. These findings suggest policymakers prioritize integrated strategies that leverage public–private partnerships, mobilize climate finance, and promote knowledge transfer from Global North experiences tailored to local contexts. By strengthening regulatory alignment, scaling green investments, and enhancing coordination, Global South nations can unlock transformative potential in transport and logistics. The study concludes with directions for future research, urging empirical evaluations of proposed pathways to inform adaptive, context-specific green agendas that support equitable development and climate action.

Keywords Sustainable transport · Green logistics · Global South · Policy pathways · Infrastructure barriers · Decarbonization

1 Introduction

The transport and logistics sector is under increasing pressure because of the Global South due to the high rates of urbanisation, economic growth, and climate vulnerabilities, which increase the emission of greenhouse gases and overburden of infrastructure. The emerging economies are major contributors to transport-related emissions in the world, and road freight prevails because of insufficient multimodal facilities (Dharmapriya et al. 2025). Sustainable transport policies are also economically resilient and ecologically safe, but there are still challenges such as gaps in policies that make them difficult to implement (Fourie 2018). Green logistics incorporates low carbon components like optimal route planning and alternative fuels to reduce footprints in the supply chain (Muchenje 2024). The fast motorisation in other parts of the world such as Africa and Latin America contributes to air pollution and congestion and requires immediate changes to effective systems (Ali et al. 2020). International agreements such as the Paris Agreement encourage adjustment to Nationally Determined Contributions in order to reduce sector emissions (Chandran et al. 2025). These dynamics underscore the application of context-related pathways that would integrate growth and sustainability imperatives.

The main argument that this article develops is that the institutional coordination, investments in green infrastructure, and the alignment of regulations are the strategies that can help to eliminate the barriers to continuing the development of the green agenda towards the transport and logistics in the Global South. Divided policies and funding gaps negatively affect decarbonisation yet coherent systems can integrate activities with SDGs (Nagy and Szentesi 2024). Successful examples of electrification, BRT, and biofuels, which involve the use of synergies between the state and businesses, are presented in case studies of China, Kenya, and Brazil (Musau 2024). Institutional silos create inefficiencies, but multi-stakeholder relationships create scalable low-carbon innovations (Tetteh et al. 2025). The Avoid-Shift-Improve framework will be used to transition to the modal and upgrade technology required to reduce emissions (Nikseresht et al. 2023). The academic community underlines the necessity of localized approaches that would transform the Global North experiences to national limitations and introduce transport as an equal developing agent (Lavitt and Sargeant 2024).

The main purpose of the given qualitative desk review study is to synthesise literature and case findings to outline the strategic ways of green transformation in transport and logistics of the Global South. Regulatory, infrastructural, and governance challenges are identified by textual analysis and suggested to produce actionable recommendations (Ma and Chang 2023). The policymakers receive evidence-based advice on how to mobilise climate finance and PPPs to resilient networks. The search is based on the priority of mostly empirical peer-reviewed materials, without any limitation on the year, with English materials being selected because of the reliability. Results guide adaptive policies, which promote SDGs 9, 11, and 13 via innovation and coordination. This publication encourages empirical follow-ups in order to prove ways to make mobilization transitions to sustainable and inclusive mobility.

2 Conceptual Framework

The conceptual framework of developing the green agenda in transport and logistics in Global South is based on the six constructs that are intertwined and supportive to sustainable change. Sustainable transport involves policies related to mitigating environmental influence through the optimisation of mobility systems, increased energy efficiency and low-emission technologies (Nagy and Szentesi 2024). At its core is the concept of green logistics that incorporates the ecological principles of supply chain management into minimising carbon footprints by optimising routes, using alternative fuels, and minimising energy consumption in warehousing (Muchenje 2024; Vienazindiene et al., 2021). The regulatory frameworks are essential as they define policies and standards that encourage environmental compliance and innovation, but their level of effectiveness is different across the regions, especially in emerging economies where the institutional gaps dominate (Chandran et al. 2025). Moreover, green infrastructure can also be viewed as a physical and technological infrastructure that helps to sustain a sustainable transport practice, such as renewable energy plants, environmentally friendly logistical centers, and intelligent transport systems (Musau 2024). Government, private sector and civil society coordination is required to ensure that efforts are synchronized to be sustainable and cross sectoral silos are broken (Tetteh et al. 2025). Finally, decarbonisation pathways characterise the strategic choices and the process of transitioning to minimise the emissions of greenhouse gases in the transport and logistics operations, which conform to the overall climate objectives (Nikseresht et al. 2023).

The interrelated aspects form a whole lens to explore the possibilities of establishing green pathways in transport and logistics in the Global South when the rapid urbanisation and economic development make the environmental issues increasingly intense. Sustainable transport recognises the inherent limits of developing contexts, including the lack of resources and infrastructural underdevelopment, which require dynamic responses to technology adoption and behaviour change (Ali et al. 2020). Green logistics requires the operationalization to incorporate sustainability, use the new technologies, such as IoT and AI, to maximize efficiency and decrease emissions (Musau 2024; Muchenje 2024). The regulatory frameworks need to be modified to facilitate innovation through a clear, enforceable policy and financial incentives, with consideration to the challenges they must encounter in their respective context (Chandran et al. 2025). Green infrastructure can be implemented at an investment level, which enables shifts in the logistics modalities and resilient networks, which is essential in climate-prone areas logistics hubs (Nagy and Szentesi 2024). Institutional coordination reduces fragmentation through collaborative governance models and multi-stakeholder partnerships as consensus in the literature adopts such as precursors of effective green transitions (Tetteh et al. 2025). Decarbonisation pathways represent the strategic frameworks that define the way technologies are adopted, policy changes, and market procedures in order to make the transport and logistic systems less carbon-intensive (Nikseresht et al. 2023). The framework therefore gives a holistic framework through which it can explore and model effective pathways of the green agenda which can fit within the context of the Global South.

3 Literature Review

3.1 Regulatory Frameworks and Policy Alignment with Global Sustainability Goals

Effective regulatory systems are core towards the development of green transport and green logistics since policy is central in devising green mobility paths. Proper regulations can impact the behaviour of transportation and logistics industries by establishing guidelines on how much should be done to lower the carbon emissions, use less energy, and switch modalities to low-carbon options (Muchenje 2024; Nagy and Szentesi 2024). These frameworks allow the governments and stakeholders to align the efforts and have clear guidelines that incorporate the concept of sustainability in operational and strategic decision-making (Musau 2024). An urban mobility policy, carbon pricing, and fuel efficiency standards are policy tools that can be used to encourage a greater uptake of cleaner vehicles, better optimisation of logistics, and increased adoption of more sustainable means of transportation, like rail and water, to minimise the carbon footprint of the sector (Ali et al. 2020). In addition, laws can stimulate the technological change in green logistics, such as implementing electric cars, integration of renewable energy, and automation to increase efficiency (Nicoletti and Appolloni 2024). However, in the absence of coherent regulatory forms, such gains are probably bound to be restricted.

Policy fragmentation is a major issue in the Global South, in which the fragmentation of national and regional policies, as well as municipal transport policies, has compromised consistent response to green transport and logistics (Fourie 2018). Such fragmentation leads to duplication of mandates, discrepant regulations and gaps in governance that complicate policy enforcement and undermine regulatory standards (Mba 2024). The mechanisms of enforcement are normally non-rigorous due to low institutional capacity and resource limitations, which hinder productive compliance monitoring (Palippui 2024). Moreover, policy inconsistency brings about uncertainty among investors and operators and disencourages the practice and use of sustainable technologies and practices despite aspirational policies (Muchenje 2024). Urban inequality is also associated with unequal distribution of transport policies in urban centers and rural territories, which contributes to environmental and social inequality (Ali et al. 2020). To overcome this institutional fragmentation, there is a need to have coordinated policy development and better governance structures that can align the multi-level stakeholders.

A business case Resonance of national policies with international sustainability models, in particular, the UN Sustainable Development Goals (SDGs) and the Paris Agreement can offer a strategic channel of speeding up sustainable transport transitions (Fourie 2018). The objectives of sustainable transport solutions are specifically emphasized in SDGs 9 (Industry, Innovation and Infrastructure), 11 (Sustainable Cities and Communities), 12 (Responsible Consumption and Production), and 13 (Climate Action) as the critical connection between sustainable transport solutions and the overall development and climate goals (Mba 2024). The promising promises of the Paris Accord, the Nationally Determined Contributions (NDCs) identify the targets of the decarbonization of the transport sector and countries are required to include them in their policy agendas (Palippui 2024). The models like the Avoid-Shift-Improve (ASI) model can assist policymakers in implementing policies that remove the necessity of transport, promote modal shifts towards greener options, and enhance energy efficiency by promoting cleaner technologies and fuels (Nagy and Szentesi 2024). Policy alignment at the strategic level facilitates availability of international funds, technical assistance and it also strengthens accountability to global climatic objectives.

In spite of these systems, most of the Global South countries show a gap in policy-implementation where there are on-paper commitments to green transport, but the institutional capacity and lack of resources cause these commitments to be weakened (Ali et al. 2020; Fourie 2018). Although carbon pricing, fuel efficiency and green logistics regulations have become part of national strategies, their implementation and enforcement are weak (Muchenje 2024). The lack of adequate data infrastructure, untrained staff, conflicting developmental priorities, among other constraints, make sustainable transport policies less effective (Musau 2024). All these implementation difficulties demand capacity-building, improved multilevel governance, and development of a public–private cooperation in order to mobilise investment and technological innovations (Nicoletti and Appolloni 2024; Mba 2024). Sealing the policy-implementation divide is therefore of great importance to converting regulatory aspirations into concrete environmental and social welfare in transport and logistics industries in Global South.

3.2 Infrastructure Gaps, Financing Constraints, and Investments in Green Technologies

Inadequacy of transport infrastructure is one of the key impediments to sustainable transport transformation in the developing economies. Inefficient, carbon-intensive transport networks, including road networks, are aggravated by the poor condition of roads and insufficient rail and multimodal networks, which inhibit efficient, low-carbon patterns of mobility (Lavitt and Sargeant 2024; Nagy and Szentesi 2024). Lack of green logistics centres and inefficient charging or refuelling station of electric cars (EVs) and alternative fuels encourage the shift to greener forms of transport further (Musau 2024). The gap of green infrastructure inhibits the ability to scale electric mobility and hydrogen fuel, which are crucial elements of the sustainable transport strategy (Lavitt and Sargeant 2024). Furthermore, multimodal networks do not integrate well, which narrows down the chances to optimize freight and passenger logistics to enhance the energy efficiency of the sector (Ma and Chang 2023). These infrastructural shortcomings enhance the extent of environmental externalities and limit socioeconomic growth with the pressing necessity of focused investments in accordance with green transport goals.

Funding is a deep crisis to decarbonisation of transport among developing economies. Green infrastructure costs a lot in terms of initial capital, which discourages investments and is aggravated by the lack of access to climate finance and governmental budget constraints (Wasan et al. 2024; Lavitt and Sargeant 2024). Perception of risk among the individual investors is another factor that negatively affects capital inflow on sustainable transport projects, and in this case, the issue is the regulatory uncertainty, the feasibility of the project, and the profitability (Deus et al. 2022; Wasan et al. 2024). There is an increasing use of climate finance (green bonds and the lending of development banks, such as the World Bank, Asian Development Bank, and African Development Banks) to fill financing gaps, although their scale is still too small (Deus et al. 2022; Arshad et al. 2024). Models of blended finance that can draw in private capital using public funding are promising to de-risk green investments and scale sustainable transport infrastructure (Ma and Chang 2023; Wang 2025). However, financing hurdles can only be surmounted with policy consistency, good governance and novel financial derivatives depending on the particular localities.

The green technologies will play a crucial role in developing sustainable transport in the Global South, yet there are serious issues associated with the access of technology and its affordability. The way to decarbonization is through electric mobility, hydrogen, and other alternative fuels, which is aided by new opportunities in smart logistics systems, intelligent transport systems (ITS), and digital freight platforms and technologies that make operations more efficient (Nagy and Szentesi 2024; Nicoletti and Appolloni 2024). However, the cost of imports is high, and the local production capacities are minimal due to which technology transfer and diffusion are constrained (Abid et al. 2021). It is important to create strong local innovation systems that will adapt green technologies to local requirements and local socio-economic realities, promote local innovation, and develop technical capacity (Ma and Chang 2023). Governments, academic and industrial roles should invest in research, capacity-building, and cooperation to bridge the technology gap and allow sustainable mobility transitions in developing economies (Wang 2025).

3.3 Institutional Coordination, Public–private Partnerships, and Stakeholder Collaboration

In developing economies, institutional fragmentation has been a major challenge to sustainable transport projects, as various ministries that prevail in transport, environment, energy, and planning easily work in silos with no integrated governance (Sultana et al. 2025). The absence of centralized transport governance leads to the lack of uniformity in the implementation of policy and the lack of opportunities to implement coherent policies that take cross-sectoral effect into account. The policy implementation in the form of integrated national transport strategies is critical to coordinate goals and resource across sectors (Amović et al. 2020). Institutional fragmentation makes it difficult to respond to complex issues like emission reduction, infrastructural development and technological innovation. A literature review emphasizes the need to have central coordination bodies or knowledge centers in order to have policy coherence, strategic planning and regulatory standardisation to break these barriers (Amović et al. 2020; Sultana et al. 2025).

Public–private partnerships (PPP) is considered one of the key tools to fund infrastructure, implement technology, and enhance the functionality of transport and logistics systems (Budnyk et al. 2024; Pinilla-De La Cruz et al. 2022). PPP builds on the efficiency of the private sector, innovation, and investment potentials in line with the sustainable development mandates of the governments (Berrone et al. 2019; Spielman et al. 2010). Effective PPPs must be supported by enabling legal and institutional environments that help to maintain transparency, compatibility, and accountability (Amović et al. 2020). They facilitate deployment of technology like intelligent transport systems and green logistics solutions with the help of mutual risk and resource mobilisation (Budnyk et al. 2024). Nonetheless, inherent risks and long-term sustainability are still issues and it would be critical to standardise the processes and have a strong governance of PPP to be successful (Soomro and Zhang 2015).

The involvement of all stakeholders is an essential part of expanding the range of sustainable transport projects through the participation of transport operators, logistic companies, technology suppliers, civil society, and local governments (Xing et al.

2018; Spielman and Grebmer 2006). The collaborative form of governance promotes inclusive decision-making, sharing of knowledge and capacity building and narrows institutional disjuncture (Owojori and Erasmus 2025). Knowledge exchange at the Global North is characterized by international collaboration, and multilateral development partnerships based on international cooperation can develop local capacities and institutional resilience (Sultana et al. 2025; Pinilla-De La Cruz et al. 2022). It is observed that multi-stakeholder collaboration plays a crucial role in aligning incentives, mobilising resources and strengthening adaptive governance to make transport transitions sustainable. In this way, the creation of coordinated, cooperative governance mechanisms is the key to the successful overcoming of the coordination failures to promote sustainable transport in developing economies (Owojori and Erasmus 2025; Berrone et al. 2019).

3.4 Decarbonisation Strategies, Technological Innovation, and Low-Carbon Transport Solutions

The transport industry has a great contribution to the emission of CO2 in the globe as it contributes to a considerable number of greenhouse gases that accelerate climate change. The transport emissions are also rising at an alarming rate especially in emerging economies, where mobility and freight services are in high demand due to urbanisation, industrialisation and trade developments (Dharmapriya et al. 2025; Jayabal et al. 2025). The speed at which motorised transport is growing in these areas increases the carbon footprints, further exacerbating environmental issues, and leading to the need to urgently implement decarbonisation measures (Dharmapriya et al. 2025). It is within this context that the so-called Avoid-Shift-Improve (ASI) is becoming a prominent strategic direction: to prevent the unnecessary movement of people, to switch to more sustainable transportation modes, such as a bus or a train, and to optimize the technologies of vehicles to make them more efficient and eco-friendly (Karakosta and Papathanasiou 2025; Usman et al. 2024). These integrative strategies are necessary in ensuring a balanced growth of transport requirements and climate commitments.

Sustainable mobility and logistics are also available in the form of technological solutions to low-carbon transport. EVs play a central role, and the increased focus on the expansion of electric mobility, based on renewable energy, is aimed at reducing reliance on fossil fuels (Jayabal et al. 2025; Nagy and Szentesi 2024). Freight and biofuels run on hydrogen are complementary solutions especially where heavy and long-distance transportation has that electrification can only go so far (Matarru and Shin 2025). Electrified intermodal freight and rail systems also allow the reduction of emissions through modal changes (Karakosta and Papathanasiou 2025). Operations efficiency is improved by smart logistics and digitalisation tools such as AI-based route optimisation, digital freight platforms, and blockchain supply chain visibility, which decreases emissions due to logistics activities (Jayabal et al. 2025; Karakosta and Papathanasiou 2025). Nonetheless, there are obstacles to adoption in terms of cost, infrastructure, technical skill, and regulatory conditions (Samuel 2025).

The support and specific investments, in turn, are crucial to the decarbonisation of transport with the help of technological innovation. The subsidies and carbon pricing are among the policies that encourage cleaner technologies and the mechanisms of financing support the development of infrastructure (Samuel 2025; Huo et al. 2025). The issue is still related to high initial expenditure, lack of infrastructure and capacity, particularly in the developing environment (Dharmapriya et al. 2025). The introduction of renewable energy into transport electrification and the development of digital management tools facilitate the reduction of emissions as well as increase system resilience and efficiency (Jayabal et al. 2025). The transformation of technological routes towards vigorous regulatory frameworks and participation of stakeholders is a holistic approach necessary in sustainable decarbonisation of the global transport and logistics industries (Karakosta and Papathanasiou 2025; Samuel 2025).

4 Methodology

Electronic databases such as Google scholar, Scopus, Web of Science and Science Direct were used to access the information needed to explore strategic pathways to overcome challenges and accelerate the green transformation of transport and logistics in developing economies. Keywords such as Sustainable transport, green logistics, Global South, Policy pathways, Infrastructure barriers, Decarbonisation were used in isolation or in combination to search for data. Peer reviewed journal articles that are based on empirical data were prioritised for reliability of findings in this chapter. The chapter was based on English published sources that were related to this topic. There was no year limit for the searched items that were sourced from the published articles. The published sources that were generated from the search process were further filtered based on their relevance to come up with those that are relevant and most appropriate to the chapter. Relevant academic literature, policy documents, and grey literature on transport and logistics sustainability in the Global South were consulted for this chapter. As such, this this chapter was based on the assessment of existing reports and claims rather than fresh empirical data from the field.

5 Findings

5.1 *Case Study 1: Electrification of Urban Public Transport and Logistics in China*

China is a case study of mass electrification of the urban transport and logistics infrastructure, and demonstrates how effective policy support, urban investment in infrastructure, and industrial policies can influence green transitions. Government initiatives, including the 13th and 14th Five-year Plans, focus on electric mobility, which offers incentives and subsidies to manufacturers and users of EVs to encourage their use. The policies go hand in hand with the climate promises and sustainable urban mobility targets of China, which has provided a favorable atmosphere to encourage the transit electrification. The local governments, such as Shenzhen, have embraced specific policies in support of the national goals, which promote the use of EVs in cities faster. This broad policy framework does not only promote the environmental goals but also drives the economic growth and industrial development in the e-mobility industry, which boosts the competitiveness in international markets (Goritiyal et al. 2024).

Simultaneously, China has invested significant amounts in charging infrastructure that is essential to enable electrified transport networks. Big cities such as Shenzhen have implemented large-scale EV charging systems, which guarantee the availability and stability of electric buses and delivery vehicles. Simultaneous improvements of grid infrastructure and renewable energy integration assist in keeping sustainability and minimizing emissions of the electricity-generating process. Shenzhen city boasts of having switched to the fully electric operation of its entire bus fleet, which consists of more than 16,000 buses, making it the first city in the world to do so. Such electrification not only decreases the air pollution in the area, but also reduces the operating expenses and enhances the air quality in the city, thus becoming a standard in other cities of the country and the world (Goritiyal et al. 2024; Mccollum et al. 2013).

In addition to the public transport, urban logistics in China have also been electrified, as the use of electric delivery vans and freight vans are becoming more and more widespread. The shift helps to reduce the emissions in the highly populated regions and improve the energy system transformation and climate stabilisation. Moreover, digital technologies, including IoT and artificial intelligence, are also applied to smart logistics platforms to maximise route efficiency and freight management, minimising costs and negative effects on the environment. The practice of China shows that the combination of supporting policies, infrastructure implementation, and logistics innovation that were based on digital technologies can increase sustainable changes in transport. To the Global South, this case confirms the significance of government leadership, selective subsidies, and urban electrification policies as powerful tools to achieve a huge reduction in emissions and create resilient and inclusive cities and green transport systems).

6 Case Study 2: Sustainable Urban Mobility and Bus Rapid Transit (BRT) in Kenya

The country of Kenya has major transport issues in urban areas and the main cause is the high urbanisation in Nairobi which has been increasing at an alarming rate in the past decades. This high rate of growth has led to more congestions in the city roads, which is mostly influenced by the informal transport systems, which include buses like matatus, which dominate the transport systems. These unofficial systems also cause a high level of air pollution in the city because people mostly use old and inefficient cars and evade the laws. This has led to congestion and pollution of Nairobi, which is affecting the productivity, health, and quality of life in the city, making it urgent to find sustainable solutions to urban mobility. Moreover, these problems are exacerbated by the sprawling nature of Nairobi because it makes it difficult to provide effective transport infrastructure and increases the commutes. Coherent urban transport planning and infrastructure investment is similarly undermined by funding constraints and institutional fragmentation which limit steps toward cleaner more efficient transport opportunities that can meet the rising population of the city.

In reaction, Kenya has embarked on the sustainability infrastructure projects in transport and this is the development of the Nairobi Bus Rapid Transit (BRT) system. The BRT has been internationally recognised as an economical and flexible mode of public transport that may provide high capacity and dependable transportation services at a fraction of the capital investments that railway systems require. The BRT project of Nairobi is meant to establish focused bus corridors, which have contemporary and convenient stations and combined fare mechanisms to enhance the quality of services and mitigate traffic jams. In line with this Kenya has introduced pilot projects to include electric buses in its public transit fleets to mitigate carbon emissions and operating expenses, and is consistent with the global trends of electrifying transport. These pilots determine the possibility of work and contribute to the creation of the requisite charging and maintenance system. Based on lessons learned around the world, the work of the BRT, and electric buses in Kenya is more focused on sustainability and inclusivity with operations constrained by infrastructure and funding constraints typical of developing settings.

The key themes in the sustainable urban mobility in Kenya have been effective collaboration among institutions which has seen partnership with government agencies, international development banks including the World Bank and African Development Bank and other private transport operators. Such partnerships combine technical know-how, funds and operational capacity necessary to execute BRT and electric bus initiatives with limited domestic budgets. Concessional financing and technical assistance are provided to develop infrastructure and build capacity by international partners and the public–private partnerships are adopted to make better efficiency in investments and service delivery. The case of Kenya underscores the need to have well-coordinated governance structures to harmonize policies and investments in urban mobility reforms. The integrated solutions have proved that the sustainable transport projects in developing nations can be actualised via multilateral funding systems and PPPs which offer a template to other cities within the Global South that are interested in cost-effective and low-carbon transit projects.

7 Case Study 3: Green Logistics and Biofuel-Based Transport in Brazil

Brazil is a global leader in transport based on biofuel with its unique program, ethanol transport, called Proalcool, started in the 1970s. This program also played a great part in reducing reliance on fossil fuels because ethanol that was produced out of sugarcane became a major transport fuel. The widespread use of flex-fuel vehicles in Brazil, which operate on mixtures of any quantity of gasoline and ethanol, has strengthened the use of renewable energy in the transportation sector, and thus the country has become an example to others in the Global South. These massive vehicles have helped to achieve massive reduction in greenhouse gas emissions and also enhanced energy security. This leadership has been affirmed by government policies such as ethanol blending requirements and infrastructure facilitation to produce and distribute ethanol. Besides, the biofuel efforts in Brazil involve powerful institutional framework and consumer incentives, which boost production, technology development, and consumer adoption, which could be a model to other emerging economies pursuing transport decarbonization (Qi et al. 2023; Du et al. 2023).

Brazil has also invested in improvement of logistics using rail and inland waterways in parallel with its bio-fuels leadership, to eliminate road freight dependence that has traditionally contributed to high emissions and congestion. The nation has also established effective freight corridors between agricultural production hubs and export ports by utilising the extensive river networks, so as to facilitate water transport which is more eco-friendly over land. Improvements in the rail infrastructure help in the movement of bulk cargo in a lower carbon footprint than the road mode of transportation, whereas special freight routes run to optimise route efficiency and minimise transit periods. These projects are in line with national biofuel policies, and they incorporate renewable energy into the overall transport sector plans. These investments in infrastructure improve the effectiveness of logistics, lower the cost of transport, and minimize emissions, which is an example of a complex attitude towards sustainable mobility balancing the use of alternative fuels with diversification of transport modes (Du et al. 2023; Musau 2024).

The policy environment of Brazil embraces the decarbonisation of the transport sector with the national biofuel policies, such as the mandatory blending ratios, production incentives and the funding of renewable energy technologies in the research. Such policies create a positive atmosphere of biofuel industries ensuring stability of the market and continuing investments. When the energy of renewable sources is integrated in transport, along with the logistics innovations, the emphasis on the transition to sustainability in Brazil through solutions based on domestic resources is highlighted. The model present in the country teaches the Global

South some important lessons: alternative fuels have the potential to significantly decrease transport emissions, the transport mode diversity through more environmentally friendly freight options can lower the environmental effects, and national energy sources can strengthen the sustainable development. The experience of Brazil supports the idea that a robust policy support along with infrastructure and technological innovation are the key to attaining low-carbon transport in the emerging economies (German et al. 2011; Qi et al. 2023; Musau 2024).

8 Discussion

The results of regulatory fragmentation in the promotion of green transport and logistics in the Global South can be well adapted to the scholars who insist on the necessity of integrated policy frameworks to align national efforts with global standards such as the SDGs and Paris Agreement. Similar to this study, Muchenje (2024) and Fourie (2018) emphasize the role of disjointed regulations that provide gaps in enforcement and uncertainty among investors, which is also one of the main barriers of this study, and that of policy inconsistency. Chandran et al. (2025), also suggest that innovation in low-emission technologies is possible by great regulatory alignment, and according to this recommendation, integrated frameworks are recommended to make use of incentives such as carbon pricing. Such an overlap highlights the urgency of multilevel governance as an intervention to reduce the gaps between policy and implementation as is seen in South African situations where aspirational promises fail because of capacity limitations. This is supported by Ali et al. (2020) who point out that coherent standards help to transition to the option of sustainability, confirming the fact that the research calls for harmonised solutions specific to developing economies. These academic observations support the cruciality of regulatory coherence in opening up green transitions and demand Global South policymakers focus on enforceable, congruent approaches to decarbonisation.

The issues of infrastructure shortage, lack of funds, and barriers to technology access, found in the study, are very close to the literature on green investments in developing settings. Nagy and Szentesi (2024) and Lavitt and Sargeant (2024) explain that the existence of inadequate multimodal networks and the high costs of EV infrastructure are reinforcing the use of carbon-intensive road transport, which reflects the message of the document about the focus on green hub development. Wasan et al. (2024) elaborate on the risks on financing deterrent of financing of the privates, which conforms to the arguments in favor of blended finance such as green bonds to de-risk projects, which have been observed in the World Bank-supported projects. Ma and Chang (2023) promote local innovation ecosystems to customize technologies and contributes to the push of renewable charging and smart logistics investments by the government and corporations in the study. These views spot infrastructure as a special area of investment to resilience, and the Vietnam studies on logistics confirm that simultaneous upgrades diminish emissions and expenses. The combined recommendations of scholars to scale climate finance and PPPs to bridge these gaps support the pathway of the findings of the equitable implementation of green infrastructure in resource-limited Global South contexts.

The institutional coordination suggested in the study based on the use of PPPs and stakeholder involvement is strongly supported by the literature of sustainable transport governance. Tetteh et al. (2025) and Sultana et al. (2025) emphasize multi-stakeholder models in the fight against silos, which is similar to the found fragmentation observed between ministries in Global South countries. In their study, Amović et al. (2020) prove the mobilisation of resources towards efficiency via PPPs, where in Kenya, BRT works with development banks, making the focus on transparent frameworks justified. Berrone et al. (2019) also present some PPPs evaluation models in line with SDGs with an emphasis on such risks as disputes that require strong contracts, as is the case with urban mobility issues in South Africa. Owojori and Erasmus (2025) note policy coherence centers on knowledge, which state affirms the international cooperation in capacity building. This congruency is a positive indicator of collaborative governance as a condition of scaling low-carbon solutions and scholars are calling to inclusive partnerships to drive innovation and responsibility in disjointed institutional environments.

The findings on decarbonisation pathways through ASI frameworks and low-carbon technologies can be smoothly coordinated with the academic approaches to the reduction of transport emissions by the Global South. Asymmetric modal optimisation and electrification of modal improvement is advocated by Nikseresht et al. (2023) and Karakosta and Papathanasiou (2025) as case studies such as Brazilian biofuels and Chinese EV fleets. Jayabal et al. (2025) highlight the importance of integrating renewables in heavy-duty logistics, which helps propel the study into technological innovations in the context of the urbanisation pressure. According to Dharmapriya et al. (2025), the spikes in motorisation have become rapid, supporting policy changes such as subsidies as observed in Kenya in electric pilots. Nagy and Szentesi (2024) associate digital instruments with the improvement of efficiency, which supports the holistic strategies of combining regulation and investment. These convergences underscore adaptive, context-specific pathways-based on China, Brazil, and Kenya-as models of resilient, low-carbon systems, and literature recommends empirical assessments to optimize Global South implementations.

9 Conclusion and Recommendations

The potential of sustainable transport and logistics systems to propel economic and environmental sustainability in Global South countries is enormous, but there remain endemic obstacles in the form of regulatory fragmentation, lack of infrastructure, lack of finance and lack of institutional siloing that can hinder decarbonization processes. This research sheds light on the strategic directions by qualitative review of the literature and case studies conducted in China, Kenya, and Brazil that reveal that integrated policy frameworks, green investments with targeted intentions, and collaborative governance can provide revolutionary solutions. Regulatory complement to international standards including the SDGs and Paris Agreement comes out to be cornerstones to align fragmented mandates to drive the use of low-emission technologies and modal change. The improvements in infrastructure, combined with blended financing schemes, seal some of the major gaps in EV charging networks and multimodal hubs, lessening the dependence on carbon-intensive road transportation. The institutional coordination through the use of public-privacy partnerships and stakeholders interaction breaks up silos and allows scalable deployment of smart logistics and alternative fuels. Case studies prove practicality in the field: in China, the policies of electrifying cities reduced city emissions, in Kenya, the BRT initiatives have addressed the problem of congestion in a sustainable way, and in Brazil, the biofuels diversified the freight choices in an effective way. These lessons highlight the pressing necessity of combined approaches that can adjust Global North experience to the local level, making the transport sectors the drivers of equal development and climate justice. To be fine-tuned in the context-specific implementations, future studies ought to seek empirical verifications of such paths.

Based on the findings of the study, policymakers are supposed to focus on six recommendations that are specific and directed towards increasing the speed at which Global South transport and logistics become green. One, governments must create national frameworks that are unified which incorporate international sustainability goals, imposing clear standards on the emissions and efficiency, which will remove overlaps of policies and ensure more investor confidence. Second, leaders should initiate special green infrastructure funds, with the aid of climate finance provided by organisations such as the World Bank to develop EV stations, railways, and logistics centers to improve the quality of multimodality and resilience. Third, countries should seek to encourage the existence of partnership between the state and the people by having clear legal frameworks, whereby the mobilised capital of the private sector is used to deploy technology and the risks and expertise in different sectors are shared. Fourth, the governments will need to invest in capacity-building initiatives that educate staff about green technologies and develop local chains of innovation to fit the solutions such as AI route optimisation and biofuels to the regions. Fifth, nations need multi-stakeholder forums and coordination of ministries, operators, and communities to ensure inclusive decision-making to deal with urban–rural inequalities in sustainable mobility. Sixth, the decision-makers ought to drive Avoid-Shift-Improve programs, which can expand electric fleets and digital logistics in high-impact cities to show rapid successes in emissions reduction. These measures will guarantee a quantifiable move towards low-carbon systems, which would open socioeconomic opportunities to developing economies.

References

Abid N, Ceci F, Ikram M (2021) Green growth and sustainable development: dynamic linkage between technological innovation, ISO 14001, and environmental challenges. Environ Sci Pollut Res 29(17):25428–25447. https://doi.org/10.1007/s11356-021-17518-y

Ali SS, Kaur R, Ersöz F, Altaf B, Basu A, Weber G-W (2020) Measuring carbon performance for sustainable green supply chain practices: a developing country scenario. CEJOR 28(4):1389–1416. https://doi.org/10.1007/s10100-020-00673-x

Amović G, Maksimović R, Bunčić S (2020) Critical success factors for sustainable public-private partnership (PPP) in transition conditions: an empirical study in Bosnia and Herzegovina. Sustainability 12(17):7121. https://doi.org/10.3390/su12177121

Arshad A, Parveen S, Mir FN (2024) The role of green bonds in reducing CO_2 emissions: a case of developing countries. J Econom Admin Sci. https://doi.org/10.1108/jeas-09-2023-0242

Berrone P, Ricart J, Duch A, Bernardo V, Salvador J, Piedra Peña J, Rodríguez Planas M (2019) EASIER: an evaluation model for public-private partnerships contributing to the sustainable development goals. Sustainability 11(8):2339. https://doi.org/10.3390/su11082339

Budnyk V, Vasylieva N, Artemenko A, Shcherbak V (2024) Public-private partnerships as a catalyst for sustainable development. J Lifestyle SDGs Rev 5(2):e03635. https://doi.org/10.47172/2965-730x.sdgsreview.v5.n02.pe03635

Chandran R, Chandran MCS, Achuthan K, Sahib PR (2025) Theoretical perspectives on green banking adoption in India: regulatory uncertainty, institutional barriers, and policy solutions. Discover Sustain 6(1). https://doi.org/10.1007/s43621-025-01406-3

Čižiūnienė K, Matijošius J, Sokolovskij E, Balevičiūtė J (2024) Assessment of implementing green logistics principles in railway transport: the case of Lithuania. Sustainability 16(7):2716. https://doi.org/10.3390/su16072716

De Deus JL, Crocco M, Silva FF (2022) The green transition in emerging economies: green bond issuance in Brazil and China. Climate Policy 22(9–10):1252–1265. https://doi.org/10.1080/14693062.2022.2116381

Dharmapriya N, Gunawardena V, Methmini D, Jayathilaka R, Rathnayake N (2025) Carbon emissions across income groups: exploring the role of trade, energy use, and economic growth. Discover Sustain 6(1). https://doi.org/10.1007/s43621-025-01261-2

Du J, Cheng J, Ali K (2023) Modelling the green logistics and financial innovation on carbon neutrality goal, a fresh insight for BRICS-T. Geol J 58(7):2742–2756. https://doi.org/10.1002/gj.4732

Fourie W (2018) Aligning South Africa's National Development Plan with the 2030 agenda's sustainable development goals: guidelines from the policy coherence for development movement. Sustain Dev 26(6):765–771. https://doi.org/10.1002/sd.1745

German L, Schoneveld GC, Pacheco P (2011) Local social and environmental impacts of biofuels: global comparative assessment and implications for governance. Ecol Soc 16(4). https://doi.org/10.5751/es-04516-160429

Goritiyal C, Ramana NV, Matcha A, Sukania P, Boopathi S (2024) Economic growth study and inferences from developments in E-mobility. In: A sustainable future with E-mobility: concepts, challenges, and implementations (pp. 178–199). IGI Global Scientific Publishing. https://doi.org/10.4018/979-8-3693-5247-2.ch010

Huo S, Ni L, Shah SAA (2025) Renewable energy dynamics in BRICS: assessing economic growth, carbon emissions, and public health benefits. J Renew Sustain Energy 17(2). https://doi.org/10.1063/5.0239553

Jayabal R, Sivanraju R, Prabhakar P (2025) Low-carbon agricultural strategies: toward environmental protection and energy efficiency. Energy Sci Eng. https://doi.org/10.1002/ese3.70320

Karakosta C, Papathanasiou J (2025) Decarbonizing the construction sector: strategies and pathways for greenhouse gas emissions reduction. Energies 18(5):1285. https://doi.org/10.3390/en18051285

Krüger F, Titz A, Arndt R, Groß F, Mehrbach F, Pajung V, Suda L, Wadenstorfer M, Wimmer L (2021) The bus rapid transit (BRT) in Dar es Salaam: a pilot study on critical infrastructure sustainable urban development and livelihoods. Sustainability 13(3):1058. https://doi.org/10.3390/su13031058

Lavitt T, Sargeant H (2024) Green infrastructure in developing economies: a unique target for risk reduction and financial investment. Climate Policy 25(5):792–805. https://doi.org/10.1080/14693062.2024.2409805

Ma J, Chang C-P (2023) The role of green finance in green innovation: global perspective from 75 developing countries. Emerg Mark Financ Trade 59(10):3109–3128. https://doi.org/10.1080/1540496x.2023.2210720

Matarru AA, Shin D (2025) Biochar production technology as a negative emission strategy: a review. Energies 18(18):4898. https://doi.org/10.3390/en18184898

Mba J (2024) Advancing sustainability and efficiency in maritime operations: integrating green technologies and autonomous systems in global shipping. Int J Sci Res Archive 13(2):2059–2079. https://doi.org/10.30574/ijsra.2024.13.2.2419

Mccollum D, Krey V, Kolp P, Nagai Y, Riahi K (2013) Transport electrification: a key element for energy system transformation and climate stabilization. Clim Change 123(3–4):651–664. https://doi.org/10.1007/s10584-013-0969-z

Muchenje KC (2024) Green logistics and supply chain management (pp. 33–61). Igi Global. https://doi.org/10.4018/979-8-3693-3755-4.ch002

Musau EG (2024) Optimizing transportation and distribution for environmental sustainability, pp. 84–101. Igi Global. https://doi.org/10.4018/979-8-3693-3486-7.ch005

Nagy G, Szentesi S (2024). Green logistics: transforming supply chains for a sustainable future. Adv Logistic Syst—Theory Pract 18(3):29–42. https://doi.org/10.32971/als.2024.026

Nicoletti B, Appolloni A (2024) Green Logistics 5.0: a review of sustainability-oriented innovation with foundation models in logistics. Eur J Innov Manage 27(9):542–561. https://doi.org/10.1108/ejim-07-2024-0787

Nikseresht A, Golmohammadi D, Zandieh M (2023) Sustainable green logistics and remanufacturing: a bibliometric analysis and future research directions. Int J Logistics Manage 35(3):755–803. https://doi.org/10.1108/ijlm-03-2023-0085

Owojori OM, Erasmus LJ (2025) Public–private partnerships as catalysts for green infrastructure: a three-pronged analysis of economic, environmental, and institutional factors. Front Sustain Cities 7. https://doi.org/10.3389/frsc.2025.1591278

Palippui H (2024) Integration of technology and regulations for safe and efficient marine logistics. Collaborate Engineering Daily Book Series 2(1):1–7. https://doi.org/10.62012/collaborate.v2i1.53

Pinilla-De La Cruz GA, Rabetino R, Kantola J (2022) Unveiling the shades of partnerships for the energy transition and sustainable development: connecting public–private partnerships and emerging hybrid schemes. Sustain Dev 30(5):1370–1386. https://doi.org/10.1002/sd.2288

Qi Y, Ibrahim RL, Saleh Al-Faryan MA (2023) Exploring aggregated and disaggregated environmental impacts of biofuels: Do affluence, green technological innovation and green finance matter for top biofuel-abundant economies? Energy Environ 36(1):291–322. https://doi.org/10.1177/0958305x231181673

Samuel S (2025) Carbon pricing mechanisms for reducing greenhouse gas emissions and encouraging sustainable industrial practices. World J Adv Res Rev 25(2):001–024. https://doi.org/10.30574/wjarr.2025.25.2.0350

Sénquiz-Díaz C (2021) Transport infrastructure quality and logistics performance in exports. Economics 9(1):107–124. https://doi.org/10.2478/eoik-2021-0008

Setiadi S, Muharam H (2024) Logistics innovation in developing economies: integrating digital solutions in e-commerce supply chains. Sinergi Int J Logistic 2(4):239–251. https://doi.org/10.61194/sijl.v2i4.733

Soomro MA, Zhang X (2015) Evaluation of the functions of public sector partners in transportation public-private partnerships failures. J Manage Eng 32(1). https://doi.org/10.1061/(asce)me.1943-5479.0000387

Spielman DJ, Von Grebmer K (2006) Public-private partnerships in international agricultural research: an analysis of constraints. J Technol Transf 31(2):291–300. https://doi.org/10.1007/s10961-005-6112-1

Spielman DJ, Hartwich F, Grebmer K (2010) Public–private partnerships and developing-country agriculture: Evidence from the international agricultural research system. Public Admin Develop 30(4):261–276. https://doi.org/10.1002/pad.574

Sultana R, Chowdhury MAH, Chowdhury TA, Tazminur S, Ahmed I, Ahmed N, Baky AA, Shahriar A, Kafy AA (2025) Bridging business strategy and educational development: private sector engagement and value creation framework for sustainable E-learning models in emerging markets. Bus Strat Develop 8(1). https://doi.org/10.1002/bsd2.70098

Tetteh FK, Kwateng KO, Tukue T, Mensah J (2025) Green supply chain management practices: review, framework and future research directions. J Respons Prod Consump 2(1):112–150. https://doi.org/10.1108/jrpc-08-2024-0039

Usman F, Ani E, Ebirim W, Montero D, Olu-Lawal K, Ninduwezuor-Ehiobu N (2024) Integrating renewable energy solutions in the manufacturing industry: challenges and opportunities: a review. Eng Sci Technol J 5(3):674–703. https://doi.org/10.51594/estj.v5i3.865

Vienažindienė M, Tamulienė V, Zaleckienė J (2021) Green logistics practices seeking development of sustainability: evidence from lithuanian transportation and logistics companies. Energies 14(22):7500. https://doi.org/10.3390/en14227500

Wang H (2025) Green finance and foreign investment: catalysts for sustainable prosperity in emerging economies. Front Environ Sci 13. https://doi.org/10.3389/fenvs.2025.1561838

Wasan P, Kumar A, Luthra S (2024) Green finance barriers and solution strategies for emerging economies: the case of India. IEEE Trans Eng Manage 71:414–425. https://doi.org/10.1109/tem.2021.3123185

Xing Y, Liu Y, Cooper SCL (2018) Local government as institutional entrepreneur: public-private collaborative partnerships in fostering regional entrepreneurship. Br J Manag 29(4):670–690. https://doi.org/10.1111/1467-8551.12282

GPSR Compliance
The European Union's (EU) General Product Safety Regulation (GPSR) is a set of rules that requires consumer products to be safe and our obligations to ensure this.

If you have any concerns about our products, you can contact us on

ProductSafety@springernature.com

In case Publisher is established outside the EU, the EU authorized representative is:

Springer Nature Customer Service Center GmbH
Europaplatz 3
69115 Heidelberg, Germany

www.ingramcontent.com/pod-product-compliance
Ingram Content Group UK Ltd.
Pitfield, Milton Keynes, MK11 3LW, UK
UKHW020436270926
475829UK00013B/62
* 9 7 8 3 0 3 2 3 1 3 7 3 7 *